THE VETERINARY
LABORATORY AND
FIELD MANUAL

THE VETERINARY LABORATORY AND FIELD MANUAL

3RD EDITION

Edited by

Susan C. Cork and Roy W. Halliwell

First published 2019

Copyright © 5m Publishing 2019

Published by
5M Publishing Ltd,
Benchmark House,
8 Smithy Wood Drive,
Sheffield, S35 1QN, UK
Tel: +44 (0) 1234 81 81 80
www.5mpublishing.com

A Catalogue record for this book is available from the British Library

ISBN 9781789180459

Disclaimer
Every reasonable effort has been made to ensure that the material in this book is true, correct, complete and appropriate at the time of writing. Nevertheless, the publishers and the authors do not accept responsibility for any omission or error, or for any injury, damage, loss or financial consequences arising from use of the book. Views expressed are those of the authors and not of the editor or publisher.

Book layout by
Keystroke, Neville Lodge, Tettenhall, Wolverhampton

Printed and bound in Wales by Gomer Press Ltd

Photos and illustrations by the authors unless otherwise indicated

Additional and supporting files are available on the book's website:
http://fieldmanual.5mcreative.com/

This book is dedicated to the animal health and veterinary laboratory staff who work hard to provide support for farmers and communities that depend on their livestock.

Contents

Contributors

Dr M. Sarjoon Abdul-Cader, BVSc, MSc. Research Associate. Department of Ecosystem & Public Health, Faculty of Veterinary Medicine, University of Calgary, Alberta, Canada.

Dr M. Faizal Abdul-Careem, BVSc (hons), MVM, PhD, DACPV, DACVM (virology) Associate Professor. Department of Ecosystem & Public Health, Faculty of Veterinary Medicine, University of Calgary, Alberta, Canada. email mfabdulc@ucalgary.ca

Dr Niamh Caffrey, PhD, Research Associate, Department of Ecosystem & Public Health, Faculty of Veterinary Medicine, University of Calgary, Alberta, Canada.

Dr Sylvia Checkley, DVM, PhD (Epidemiology) Associate Professor, Department of Ecosystem & Public Health, Faculty of Veterinary Medicine, University of Calgary, Alberta, Canada.

Dr Susan C. Cork, BVSc, PhD, PG Dip. Public Policy, MRCVS. Professor, Department of Ecosystem & Public Health, Faculty of Veterinary Medicine, University of Calgary, Alberta, Canada. email sccork@ucalgary.ca

Dr Julie Collins Emerson, BSc (Hons) PhD, Senior Research Officer, Infectious Disease Research Centre, Hopkirk Research Institute, IVABS (institute for Veterinary, Animal & Biomedical Sciences), New Zealand.

Dr Patricia Curry, DVM, PhD, independent consultant, British Columbia, Canada.

Roy Halliwell, MIBMS (specialist subject bacteriology). International laboratory consultant, Southport, United Kingdom.

Jennifer Lasley, MPH. World Organisation for Animal Health, Sustainable Laboratory Initiatives, Programmes Department, OIE, Rue de Prony, Paris, France.

Dr Karen Liljebjelke, BSc, MSc, DVM, PhD. Assistant Professor (Bacteriology), Department of Ecosystem & Public Health, Faculty of Veterinary Medicine, University of Calgary, Alberta, Canada. email kliljebj@ucalgary.ca

Dr Manigandan Lejeune, PhD, DipACVM (Parasitology) Director of Clinical Parasitology Population Medicine & Diagnostic Sciences College of Veterinary Medicine – Cornell University. email ml872@cornell.edu

Willy Schauwers, BSc, international laboratory consultant, Gent, Belgium.

Dr Samuel Sharpe, BSc (Hons), BVSc, (Anatomic Pathologist). Dip. ECVP, MRCVS. Diagnostic Services Unit, Faculty of Veterinary Medicine, University of Calgary, Alberta, Canada.

Dr Judit E.G. Smits. BSc, DVM, MVetSc, PhD, Professor (Wildlife & Ecotoxicology). Department of Ecosystem & Public Health, Faculty of Veterinary Medicine, University of Calgary, Alberta, Canada.,

Dr Matilde Tomaselli, DVM, PhD, Department of Ecosystem & Public Health, Faculty of

Veterinary Medicine, University of Calgary, Alberta, Canada.

Dr Karen Tang, MD, MSc, Assistant Professor, Cumming School of Medicine, University of Calgary, Alberta, Canada.

Dr Regula Waeckerlin, DVM, PhD, Research Associate, Comparative Biology & Experimental

Medicine, Faculty of Veterinary Medicine, University of Calgary, Alberta, Canada.

Dr John Woodford, BVM, MSc, MRCVS. international veterinary consultant, France. email J.d.woodford@gmail.com

Foreword

As with the two preceding editions, this third edition of the *Veterinary Laboratory and Field Manual* is a book made to be used. Dr Susan Cork and her contributing colleagues, have drawn on their extensive experience in training, laboratory diagnostics and field development activities to produce a practical, informative and accessible guide to the organization and operation of district and regional level diagnostic laboratories.

Since the original publication of this book, the need for expanded veterinary diagnostic laboratory networks that extend into rural areas, has continued to grow. The ever increasing demand for animal protein and the concomitant increase in livestock populations, the emergence of new diseases and re-emergence of existing diseases, the expanded movement and trade of livestock, the increased contact of livestock and humans with wild animal species due to habitat incursions, and the growing risk of intentional disease introduction through bioterrorism, all underscore the need for improved disease surveillance and rapid diagnosis – efforts that require well-functioning district and regional laboratories as part of a national laboratory network. Early detection and early response are the fundamental elements of effective disease control and the ready availability of reliable basic laboratory services are essential in this regard.

The new edition maintains the well-organized structure of past editions, starting with the most useful Part 1: Laboratory and equip-ment. Chapter 1 focuses on setting up and using a laboratory service. This is followed by the practical and informative Chapter 2 on the selection, use and maintenance of laboratory equipment and supplies. Then, in Part II, in six subsequent chapters, the key diagnostic laboratory disciplines are addressed, including parasitology, microbiology, haematology, serology/immunology, clinical chemistry/toxicology, and pathology/cytology, each including the general underlying scientific principles, discussions of common diseases and procedures for sample collection and conduct of diagnostic tests. Part III then focuses on special topics including epidemiology, surveillance, common clinical presentations, wildlife health monitoring and disease surveillance, antimicrobial resistance, and arthropod vectors and related diseases. The book concludes with a range of appendices, which feature a review of common zoonotic diseases.

At various locations throughout the text, Dr Cork and colleagues have updated existing content and added some new topics. Among the new subjects considered are: training, information management and the value of laboratory networks; hazard assessment and risk management; audits for quality assurance; the benefits and limitations of penside tests; assessing diagnostic sensitivity and specificity for field tests; the applications, implementation and limitations of new technologies; and a chapter on the OIE's core missions in relation to laboratories working in the veterinary domain, contributed by the World Organisation for Animal Health (OIE).

The OIE is pleased to be associated with this new edition of the *Veterinary Laboratory and Field Manual*. The OIE's core strategic objectives, including risk management through transboundary disease control and international standard setting for the safe trade in animals and animal products, maintaining the global reporting system for animal diseases, and strengthening of national veterinary services, are all underpinned by the capacity and performance of national networks of veterinary diagnostic laboratories and the animal health investigation systems that deliver them quality samples. This newly revised manual can make an important contribution to ensuring the capacity and performance of those networks and therefore deserves a wide readership.

Dr Matthew Stone
Deputy Director General,
International Standards and Science
World Organisation for
Animal Health (OIE)
Paris, France
June 2018

Preface

Donor agencies have made significant investments to support the improvement of veterinary extension services in low- to middle-income countries. This has often been accompanied by the establishment of laboratory networks. Despite a growth in both the number and capability of diagnostic laboratories, it remains evident that these services have often had a limited impact on the well-being of farmers in isolated rural communities or on the health and productivity of their livestock. Investment has often been focused on larger-scale agricultural enterprises and infrastructure and reporting systems at the national or regional level. Farmer and community contact remains predominantly with animal health extension services at the district level where resources are generally limited and laboratory support and facilities are poor.

With the origin of many emerging and re-emerging human and livestock diseases often traced back to rural settings, where there is close interaction between humans, domestic animals and wildlife, it is essential that a broad One Health approach to disease detection is adopted and that the sustainable development and delivery of effective animal health extension services is supported. To achieve this, it is important to enhance collaborations and information exchange between and within government agencies. Early detection and rapid response to disease outbreaks in animals also requires access to a reliable and affordable veterinary diagnostic service. The most important role of district and smaller regional veterinary diagnostic laboratories is to provide technical support for veterinary and animal health extension staff so that they can provide reliable animal health advice to the end-user, that is the farmer/community. This support includes the provision of technical advice, as well as diagnostic support, especially with regard to the submission of appropriate diagnostic samples and the interpretation of test results. The quality of the service provided depends on the level of training and experience of the staff, the availability of resources and on fostering good communication between field and laboratory based staff. In many cases the availability of training opportunities and suitable text books is restricted to staff based at well-resourced central and urban facilities.

Despite wider access to the internet and a proliferation of online learning opportunities, there remains a need for a general handbook to cover commonly used procedures for animal health and veterinary laboratory staff working in remote areas. This revised and updated handbook aims to provide an easy to follow summary of the laboratory techniques and sample collection guidelines that will be of practical value for routine diagnostic work in small regional and district veterinary laboratories. The new edition is supplemented by additional online material for those able to readily access the internet. The technology described is selected to emphasize the practical aspects of laboratory diagnosis and veterinary extension work rather than the theory. In addition to the basic laboratory disciplines of parasitology, microbiology, clinical chemistry, haematology and pathology, this updated edition includes some specialist sections on molecular diagnostics, syndromic surveillance in livestock, disease monitoring in wildlife and taking a One Health approach to antimicrobial resistance. There is also a special section on the role of the World Organization for Animal Health (OIE) in supporting veterinary diagnostic laboratories. For additional information, a bibliographic list is provided at the end of each chapter.

Acknowledgements

The first edition of this book was published in 2002 as a result of a collaboration between Roy Halliwell and myself. We began the project in 1997 while working on an EU-funded project to strengthen veterinary services in Bhutan. At that time, we realized that there was a need for a general diagnostic manual suitable for animal health extension staff and veterinary laboratory staff working together at the district level. The second edition was published a decade later and we included new and updated sections with the input of Willy Schauwers and other contributors. In this third edition we have engaged a number of additional subject experts: Dr Mani Lejeune, Chapter 3; Dr M. Faizal Abdul-Careem and M. Sarjoon Abdul-Cader, Chapter 4, Virology and Chapter 6; Dr Julie Collins Emerson, Chapter 4 molecular supplement; Dr Judit Smits, Chapter 7, Toxicology; Dr Sam Sharpe, Appendix 2 (Pathology SOPs) and Web content – necropsy procedures, Pathology; Dr Niamh Caffrey and Dr Karen Tang MD, Chapter 12 (Antimicrobial resistance); Dr John Woodford, Chapter 9; Jennifer Lasley, Chapter 13 (OIE); Dr Matilde Tomaselli and Dr Patricia Curry, Chapter 11. We would like to thank all of the contributors for their engagement in this third edition and also the following for their comments on specific chapters: Barbara Martin (Chapter 1, Quality Management), Dr Karen Liljebjelke (Chapter 4, Bacteriology), Dr Cathy Monteith (Chapter 5, Haematology) and also the following for their assistance in the office, Robert Forsyth, Abir Bachir, Joy Punsalan and Katrine Maurer and the staff of 5M Publishing (Sarah Hulbert, Alessandro Pasini and Jeremy Toynbee) for editorial support.

Dr Susan Catherine Cork
Calgary, Alberta

List of figures, plates and tables

Figures

Plates (positioned between pages 386 and 387)

Tables

PART I

LABORATORY AND EQUIPMENT

chapter 1

Setting up and using a laboratory service

Susan C. Cork, Roy Halliwell and Willy Schauwers

1.1 The role of the veterinary laboratory network within animal health extension services

Most veterinary laboratory networks consist of a central research and/or referral laboratory and a number of regional laboratories, which are less well equipped than the central facility but are able to support most of the routine diagnostic work required by the field staff. Staff based at the central facility are usually responsible for compiling disease status reports, meeting the reporting requirements of regional, national and international authorities, the development of disease surveillance plans and the provision of a range of diagnostic tests and technical expertise. Although the level of development of the central and regional diagnostic facilities will vary from country to country, the general administrative and technical structure is fairly standard. The livestock owner/farmer generally does not have direct access to diagnostic services from a national or regional veterinary laboratory, the service to the farmer is usually provided through the animal health/veterinary or livestock field/extension services.[1] Therefore, a key element in the provision and utilization of veterinary laboratory services is the link between the livestock extension officer and the farmer at the district level. This chapter will focus on the factors that need to be considered when setting up and encouraging the use of a veterinary laboratory service.

Infrastructure and function of the regional or district veterinary laboratory

Veterinary laboratory staff have a varied and important role within the animal health services. The main contributions from laboratory staff include: (1) the development and delivery of the diagnostic service; (2) provision of technical support and training; (3) participation in disease surveillance programmes; and (4) disease reporting.

A range of veterinary professionals, technical experts, laboratory technicians, laboratory assistants and auxiliary personnel will staff veterinary diagnostic laboratories. The number and nature of staff depends on the size and functions of the facility, for example, a central or reference laboratory will have a diverse range of staff including veterinarians and discipline scientists with expertise in specific diseases or diagnostic procedures, whereas small facilities usually require all staff to have a very broad spectrum of expertise. In the larger laboratories, there may also be a research component to the work done with trainees and graduate students engaged in the work. The general requirements for veterinary laboratory facilities, including specialized facilities, are well set out in the standards adopted by the 182 Member Countries of the World Organisation for Animal Health (OIE)[2] and in guidelines prepared by the OIE and the Food and Agriculture Organization of the United Nations (FAO) (see also the bibliography at the end of the chapter). It is the smaller regional and district laboratories that are the focus for this book.

In regional facilities, livestock extension staff should be encouraged to liaise with laboratory staff on a regular basis, this may be via the veterinary officer in charge, or directly with the laboratory technical staff, such as when visiting the facilities to drop off samples, or when working alongside laboratory staff investigating disease outbreaks. Laboratories can provide better diagnostic services when veterinary field and extension staff are well trained so it is important that a team approach is encouraged and developed. Extension staff and field veterinarians need good training in sample selection, preservation and transportation as well as in the submission of supporting information. They will also benefit from ongoing feedback with respect to sample quality.

To maintain support for the laboratory it is important to ensure a timely turnaround of test results and an efficient response to requests for technical assistance. Regular community and professional updates from the head of laboratory facility, for example, with regard to the prevalent disease problems in an area, can also be a good way to ensure that the role of the laboratory and its staff is recognized by the wider local and professional community. Veterinary staff in the laboratory network should be encouraged to facilitate, and participate in, targeted local disease surveillance and animal health awareness programmes, in collaboration with the field and extension staff.

It is important that the laboratory service has an efficient and reliable, reporting system. This is true for reporting results out to the field veterinarians and extension staff as well as providing timely updates to the relevant regulatory authorities and reference laboratories. Regular information bulletins outlining new diagnostic procedures or disease control/treatment recommendations may also be useful to motivate field and extension staff to utilize the diagnostic services available.

Legislation and responsibility

Before planning a new laboratory facility local and national regulatory authorities should be contacted in order to check what, if any, specific restrictions and bylaws may apply to the construction of new buildings. The local regulations and provision for waste disposal, power and water supply, as well as the proximity of residential areas and commercial animal housing, must also be considered, as well as the international standards. This will be discussed in later sections of this chapter.

Education and public relations

Practical training (workshops)

The development and delivery of an effective veterinary service requires good communication and teamwork. Workshops can provide a good forum for different cadres to meet and discuss current animal health issues and through which to gain a better perspective of what each component of the animal health service can provide. Workshops can also be organized for regional and national groups of veterinary field and extension workers to promote new disease investigation campaigns and to provide a background on current diagnostic procedures. It is helpful if sample collection and sample submission guidelines are accompanied by practical demonstrations. Workshops provide the opportunity to motivate veterinary field, extension and laboratory staff to jointly assess the performance of the laboratory and to ensure the effective delivery of services to the end user, that is, the farmer. During any planned joint activities, laboratory staff should make sure that veterinary field and extension staff have an adequate supply of laboratory submission forms and sampling materials, preservatives and transport boxes as well as providing technical advice. Specific training suggestions for laboratory technicians is outlined in section 1.3.

Field visits

Training programmes should incorporate participation in field visits, disease investigations and targeted surveillance programmes for laboratory, livestock extension and veterinary field and support staff. Joint field visits provide the opportunity for training in sample collection, as well as ensuring that good case history notes are taken, and that the quality and type of samples submitted are appropriate.

Laboratory staff should be encouraged to join field teams on a regular basis so that they can experience the practical limitations placed on livestock extension staff and field veterinarians, for example, the lack of facilities for livestock restraint, problems with sample collection and so on. Meeting the end users of the diagnostic service, that is, the farmer, can also help to highlight the importance of handling samples submitted to the laboratory with due diligence and also the need to report results in a timely manner.

The exact requirements for laboratory support and practical assistance during fieldwork will depend on the level of technical competence of the field extension staff and the availability of trained auxiliary staff to facilitate animal handling. The resources available will vary with the country and the region within the country. In urban or central laboratories, there may be little scope for laboratory staff to participate in fieldwork but in the regional and district centres there is often a strong emphasis on team effort with laboratory/extension staff taking an important role in sample collection and advisory services for the farmer.

Planning fieldwork and the use of a mobile laboratory

The emphasis of the laboratory programme will be determined by regional and national requirements. In regional centres the main role of the laboratory may be to act as a diagnostic unit whereas for many of the central facilities, research and the production of biological products will also be an important feature. Veterinary staff in regional laboratories are often responsible for providing technical advice and support as well as organizing and facilitating disease surveillance programmes, monitoring livestock at borders and quarantine units, the assessment of animal health on farms and monitoring slaughterhouse hygiene. Laboratory-based veterinarians may need to be available to provide additional technical backup for livestock extension and veterinary field staff and may be directly involved in their training.

In situations where the laboratory veterinarians, and other technical staff, are expected to provide technical backup for the field extension staff it is important that reliable and appropriate transport is made available along with an adequate budget to facilitate field visits. A robust four-wheel drive vehicle is generally suitable but may need to be upgraded if long distance journeys are required. For individual staff members, a bicycle or a motorbike may be useful for short journeys but a larger vehicle is generally preferable where equipment and sample collection materials need to be transported. Mobile laboratory units may be a good investment in some regions. These allow laboratory and extension staff to go out on tour for several days or weeks at a time and process samples en route (Figure 1.1a).

The mobile unit is especially valuable in remote areas where infrastructure is limited. The basic design for a mobile unit will depend on the work to be carried out. In most cases a refrigeration unit, work bench, suitable lighting and a wash facility for sample preparation and processing are required within the vehicle. A generator and/or a solar panel, small incubator, microscope and disposables can be added from the laboratory store. A field microscope (that is, one with a mirror rather than requiring an electric light source) may be useful where there is no power source, with the caveat that these give a poorer resolution than a standard light microscope.

To develop a functional mobile laboratory the following questions should be considered.

- What type of laboratory equipment will be needed?
- What are the energy requirements to operate the equipment?
- How many staff will be in the team?

Figure 1.1 (a) Mobile laboratory with field team. Regular visits to rural farming communities and district extension units improves visibility and enhances public relations. (b) A district facility with mobile laboratory visiting from the regional veterinary facility. (c) Regional veterinary laboratory with small generator shed at the back of the building. (d) Regional veterinary laboratory parasitology section. (e) Regional veterinary laboratory microbiology staining sink.

- What is the expected duration of the field visits?
- How far will the vehicle need to travel each day?
- What are the road conditions likely to be?

Typical laboratory equipment for field visits might include the following:

- 18 l, 70 W cooling unit(s)
- small-size fridge/incubator: continuous use, 12 V
- LED microscope: 3 W, in use for 2 hours/day, 220 V
- microhaematocrit centrifuge: 600 W, in use for 30 min/day, 220 V.

For the above laboratory equipment, electricity could be supplied by a power inverter of around 1 kW capacity, which is typically run from a rechargeable 12 V lead acid battery or automotive electrical outlet.

A secondary benefit of using the mobile laboratory is that it raises the public profile of the laboratory facility, which encourages farmer participation in animal health initiatives.

1.2 Buildings and maintenance

The site for a veterinary laboratory must be chosen carefully as the location can determine the success or failure of the diagnostic service. The availability of effective transport and communication services, and good access are especially important. In most cases regional laboratories are built near a regional centre and district facilities are located in more rural areas were the demand for additional services justifies the investment (Figures 1.1b and 1.1c – illustrate a small district facility and a larger regional facility).

Location and design

Laboratory buildings should be located on a carefully selected site that has a well-defined compound and, preferably, land available around the perimeter to allow for future development. Owing to the potential biohazard risks the buildings should not be located near commercial animal rearing units or residential housing.

The design of laboratory buildings generally follows a standard plan. An example of a design submitted for a regional veterinary laboratory is illustrated in Figure 1.2. Plans for district laboratories will depend on the specific needs of the area and the budget available. In many cases district laboratories are attached to district livestock extension or veterinary centres. This facilitates sharing of resources and can enhance communication between different cadres of the animal health service. The details of the design for district and regional laboratories tend to depend on local building regulations. Reliable plumbing and power connections are especially important. The style of the out-buildings and the interior fixtures will vary depending on the budget and the building materials available locally but piping and drainage should be of the highest quality. Short-term cost savings often result in significant maintenance problems in the longer term so the quality of construction should be suitable and supervised by a site manager familiar with acceptable laboratory construction standards. For example, the flooring should be smooth and have a continuous join with the walls to facilitate disinfection and cleaning. Ventilation and waste disposal should be well planned. In laboratories that deal with zoonotic diseases and high-risk pathogens there needs to be a designated section with suitable isolation and bio-containment facilities.[3] Ensuring that there is effective waste disposal for biological and chemical wastes from laboratories is essential. This should be efficient and environmentally safe. In most countries, there will be local environmental and public health regulations that will need to be taken into consideration before the laboratory is built. Most laboratories will require a sump tank and septic tank for waste water and sewage to avoid contamination of local water supplies and waterways. Some biohazardous waste may be taken away by municipal authorities but this service is often not available in remote areas. Toxic chemicals should be stored in containment facilities until suitable bulk disposal can be identified. Carcasses and biological wastes that cannot be incinerated will need to be disposed of in a specially built biological pit as illustrated in Figure 1.4.

Services

The laboratory will not function effectively if the relevant services are not available for the routine day-to-day work in the facility, that is, reliable water and power supply. Transport and communication will be considered later.

Power supply: electricity and gas

A reliable power supply is essential for the efficient running of a laboratory.

ELECTRICITY
Electricity is generally provided by the following sources:

- municipal power
- fuel generator
- dry alkaline batteries
- solar energy panels.

In most central veterinary diagnostic laboratories, the power supply is likely to be from a municipal electricity grid system. This source may not be available or reliable in remote rural areas.

When the quality of the power source is irregular (power surges, voltage drops), sensitive equipment such as computers, microscopes and freezers should be linked up to a power

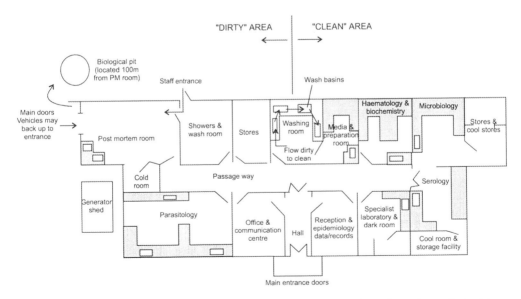

Figure 1.2 Simplified ground plan for a regional laboratory. Note that the flow of specimens in the laboratory should generally be split between the 'clean' and 'dirty' areas. The latter includes the post-mortem and parasitology sections. The size and design for a district laboratory will depend on work requirements, budget and animal health priorities identified in an area.

stabilizer. Apparatus that cannot withstand power cuts (polymerase chain reaction [PCR] cyclers, spectrophotometers, fluorescence microscope and so on) should be connected to an uninterruptible power supply (UPS) or to a 'no-break' battery power supply system. Preferably a central UPS or 'no-break' system, which can automatically switch between the mains power supply and battery power with continuous charging when mains power is available. The capacity of the UPS/'no-break' and power stabilizers should be selected to match the requirements of the apparatus that could be damaged by fluctuating power. Instead of using a stabilizer for dedicated equipment is usually preferable to install a more powerful stabilizer for the whole laboratory, for example, via a generator.

There are a number of suitable generator units available commercially (typically in the range of 8 to 12 kVA), which are generally reliable and economical to run. In most cases, unless the laboratory is very small, the generator will be set up to supply only essential pieces of equipment during a power failure so it is important to identify and mark the outlet sockets that will be supplied with power when the generator is operating.

Where the electricity supply is unreliable, consideration can be given to purchasing equipment that does not require electricity. Some refrigerators and autoclaves, for example, use a gas or kerosene power supply.

For ensuring the provision of power to maintain cold facilities it may also be possible to use solid carbon dioxide or liquid nitrogen, if stocks of these are readily available and are not too costly.

Solar energy supply systems can also be used in remote laboratories and in areas where the electricity supply is erratic. Solar powered refrigerator/freezers are already widely used. Maintenance of solar panels (that is, cleaning and dusting regularly) and the batteries is important to extend the life span of the system.

Some equipment can be run on alkaline batteries or rechargeable internal batteries such as the EKF Diagnostics Tm haemoglobin meter.[4] Microscopes with LED-illumination equipped with small solar panels can be used for several hours without mains supply. Other technologies are being developed to help facilitate work in areas where power supplies are unreliable, because of this, it is good to keep up to date with scientific developments via local equipment suppliers and professional organizations.

GAS

In some areas gas may be supplied via a public mains system or alternatively it may be supplied in liquid form in storage cylinders that can usually be obtained through commercial suppliers. Individual gas cylinders can also be placed under the benches in some laboratory sections (for example, microbiology) so that they can be connected to single outlet points. This is especially useful for small laboratory units where piped gas is not feasible. However, storage of such cylinders is subject to local safety regulations. Spare cylinders should be kept in a well-ventilated storage room in a building separated from the laboratory and suitable fire precautions should be in place.

A laboratory may have a set of large gas storage cylinders maintained outside the building. These should have suitable protection from weather and safety measures in place in case of fire.

Water supply

A reliable supply of clean water is essential for a laboratory to operate. Water is used for all aspects of laboratory work from sample preparation to test procedures, washing and general disinfection. The water quality may vary considerably so the addition of appropriate filtration and purification systems is required to keep the supply suitable for laboratory requirements (see

Chapter 2). If there is no reliable piped public water supply it is advisable to install storage tanks with as large a capacity as possible. These can be fixed to the roof to facilitate water flow, however, a well-fitted pump system is better than utilizing gravity flow alone. The water pump should be linked up with a power stabilizer or a power trip switch-voltage protection unit to prevent damage caused by a fluctuating power supply.

In district laboratories that have an intermittent water and electricity supply, it is essential to have a backup to ensure that water is available each day to allow the laboratory to function. For example, an underground storage tank built outside the laboratory that can be replenished either by rainwater (run off from the roof), or in the dry season, with water brought by bowser from a central supply. Water can then be hand pumped into an overhead tank and supplied by gravity to the lab. If a storage tank is built it is important to add suitable filters to prevent the build-up of silt, which can soon cause a blockage in the pipes.

Hot water can be supplied from electric- or gas-heated geysers; this is especially important for the wash room. To reduce the use of hot water from boilers, try to install the distillation apparatus in the washing room. This way, the cooling water from the distillation apparatus (temperature about 60°C) can be used for cleaning laboratory glassware and other items that need washing.

Distillation and de-ionization of water is outlined in more detail in Chapter 2.

Transport and communication services

The availability of reliable local transport and communication services is very important for the functioning of the laboratory and also for the well-being of laboratory staff. If a laboratory is built in a place remote from normal public services it may be difficult to recruit and retain staff. Lack of good transportation and communication networks can also make it difficult to encourage sample submission.

Managing supplies

The selection, storage and distribution of supplies is an important aspect of laboratory maintenance and service provision (see Chapter 2). When purchasing equipment, appropriate technology should be selected depending on practical considerations as well as the needs of the unit. Careful thought must be given to maintenance costs as well as the longer-term requirements of a laboratory unit. This is especially true with regard to reagents and disposables, for example. Kit tests may be preferable to more laborious methods in the short term but if the cost of replacement reagents or kits is too high then such tests may not be sustainable.

To ensure a regular flow of general supplies and spare parts for equipment it is important to set up an efficient store system. If stores are not well managed the day-to-day work of the diagnostic unit may be impaired especially where deliveries take several days (or weeks) to arrive or where the funds for new purchases are unavailable. Careful stock control and good forecasting of the requirements for each laboratory section at the beginning of each financial year are essential.

Stock control

The functioning and maintenance of a good store room requires a well-organized stock control and ordering system. The amount of stock held in a central or regional store will depend on the availability of funds, and the projected needs of the laboratory and the associated animal health network. The ease of ordering will also determine how often purchases can be made. Planning ahead is important but until a laboratory has been in operation for a few years it may be difficult to accurately assess the requirements ahead of time.

Local distributors and dealers for hardware and equipment can usually provide guidance on, or facilitate the procurement of, associated reagents and disposables. However, in most developing countries it can take some time to procure laboratory supplies. There are several reasons for this.

1 Supplies are often not available in the country and there are no reliable agents.
2 Procurement procedures are cumbersome.
3 Delivery and transport times are long.
4 Foreign currency may be required for online purchases, but is unavailable.

Difficulties in getting adequate supplies of equipment, consumables and reagents can account for major disruptions to work programmes. The time between a request for supplies and receipt of the order in regional and district laboratories can often be several months. This will depend, to some extent, on whether the laboratories have their own budget and stock control system or whether every order needs to be directed through a central laboratory. In most cases the regional and district facilities will have a small budget for consumables but larger expensive items are purchased centrally. However, this varies from country to country due to administrative variation and logistics.

Where there are a lot of remote district laboratory units it may be preferable to place all orders for disposables, and general reagents, through a regional or central facility. Buying in bulk is often cheaper and stock control can be better maintained. If specific disease surveillance projects are planned, the requirements for consumables should be outlined at the

beginning and a specific budget identified to support the planned work. Stock control can be computerized but in the smaller centres it is often still based on manual records. Both systems require frequent checking, updating and forward planning. To ensure consistency a designated staff member should be given the responsibility of maintaining up-to-date stock records.

Store sections

The store room(s) will need to be set up in a tidy and well-organized manner so that the levels of stock can easily be checked by eye as well as by checking the stock books or the computer database. In most cases chemicals will be stored in a separate room from the consumables and equipment. Dangerous or flammable chemicals will need to be stored in a separate, concrete lined, room. All stock rooms should be free of vermin and readily cleaned. Access should usually be limited to one or two store keepers and the laboratory supervisor. Up-to-date records of in-coming and out-going deliveries should be kept and delays in either the arrival or dispatch of orders should be followed up as early as possible.

A store list will usually be organized into sections. The following is a suggestion of the categories to consider:

1 glassware
2 plastic ware
3 laboratory ware
4 media
5 stains and poisonous chemicals
6 acids and other corrosives
7 alcohols and other flammable materials (these should be kept stored in purpose-built fire-proof rooms)
8 instruments and equipment
9 spare parts for equipment
10 perishable reagents (including refrigerator).

A note on isolated district laboratories

District or 'satellite' laboratories are often established in isolated rural areas as a focus point from which to provide a basic diagnostic service and technical advice to the rural livestock extension network. This is especially common where the transportation between regions is slow or services limited. The district facilities are usually supplied with basic equipment, test reagents and consumables through the nearest regional laboratory. District animal health extension staff may also have an office located close by the laboratory, where they will have basic medicines and vaccines in stock. The tests performed in these smaller facilities are usually restricted to basic parasitology procedures and simple microbiological tests, such as screening milk samples for mastitis. The staff responsible for the district facility will often take samples to a regional laboratory if additional laboratory tests are required. District facilities can improve direct services to the farmer through collaborative efforts between laboratory and extension staff. However, it is important to ensure that isolated laboratory units are given sufficient logistical and technical backup from regional and central units. Staff morale can deteriorate rapidly if resources are too restricted and regular contact is not maintained.

1.3   Staff requirements

Fieldwork

Field programmes based at small district laboratories can be run in conjunction with other livestock extension projects (such as artificial insemination schemes, vaccination or educational programmes). This can save time and resources and also encourages teamwork. The number and cadre of staff required for fieldwork will depend on the nature and the size of the project. At the district level, it would be common for all members of the team to share in routine work, such as preparing sample collection equipment, collection of samples and initial preparation of specimens for field based laboratory tests. It should be noted, however, that there may be specific regulatory requirements outlining job descriptions for staff, and government or provincial legislation outlining what technical staff may do in the field, which should be checked for the country or area in which the work will be conducted.

Laboratory

The quality of a laboratory is directly dependent on the training and performance of the laboratory personnel. The laboratory management in charge of the laboratory is generally responsible for the following.

- Deciding the grade (laboratory assistants, technicians, senior technicians, veterinary scientists/specialists, team supervisors and so on) and number of laboratory personnel required to staff the service.
- Preparing job descriptions for each grade of laboratory worker and determining the qualifications required for each grade.
- Employing suitably qualified personnel and provision of training and career development.
- Developing standard operating procedures (SOPs).
- Ensuring that health and safety regulations are complied with.
- Ensuring that quality standards are maintained (for example, for laboratory accreditation).

The requirement for laboratory staff depends on the size of the laboratory unit and the anticipated workload, as well as on the degree of technical competency required. A small regional laboratory may function efficiently with a team of two or three technicians and a senior technical supervisor or a veterinary officer, plus two or three auxiliary staff. Some of the administrative aspects of the laboratory may need to be handled by qualified veterinary officers, especially where diagnostic advice and decisions on disease control and prevention procedures need to be made. The legal requirements will vary from country to country and will, to some extent, dictate the staffing policy.

In larger central and regional laboratories, there may be a senior veterinary officer and several other veterinary professionals working in an aligned disease surveillance unit as well as those placed within each diagnostic discipline in the laboratory. Where laboratory training has been emphasized there may also be senior laboratory technicians and research staff who take a lot of responsibility for the running of specialist laboratory sections and who may supervise teams of discipline specific technicians. The staffing levels will be determined by the needs of the animal health services, the budget available for the service and the availability of competent trained professionals.

Training laboratory technicians

Curricula for training laboratory technicians will often depend on the scope and level of laboratory in which they work, that is, district, regional or central level, and will vary according to the animal health needs prevalent in the region. Core competencies recommended for entry-level veterinary laboratory technicians are outlined in the OIE Competency Guidelines for Veterinary Paraprofessionals published in 2018.[5]

Basic training, in both the theory and practical aspects of laboratory testing is required for newly appointed laboratory technicians. In some cases, laboratory technicians entering the veterinary sector may have already had good foundational training through available biomedical laboratory training programmes. While these

biomedical laboratory technicians may be highly competent to perform some of the core functions in a veterinary laboratory, they will need some additional training to become fully competent in the veterinary diagnostic service.

In some countries, there are already well-developed training programmes established for veterinary laboratory technicians. These generally offer well accepted qualifications at the certificate, diploma or degree level. Formal qualifications may be obtained over one or several years or may be gained later in conjunction with 'on-the-job' training. However, in some cases tailored, short courses may be needed to provide enhanced technical capacity for new animal health programmes requiring more veterinary laboratory support.

For small district laboratories, basic entrance-level training might be achieved in a short intensive course (see below) followed by supervised 'on-the-job' training. In many cases the theoretical aspects of the course material can be delivered in modular form and laboratory staff can build their qualifications from the certificate level through to a more advanced level over time.

Key topics for short intensive courses could include the following:

- laboratory biosafety and biosecurity
- quality management
- laboratory techniques (practical and theory)

  - microbiology
  - parasitology
  - clinical chemistry
  - pathology
  - haematology
  - serology

- professional ethics
- communication skills
- applied anatomy and physiology
- animal diseases common in the region
- field sampling.

Curricula for trainees in larger laboratories, where a larger range of tests might be performed, would need to be more comprehensive and may take up to 1 year or more to complete. The formal requirements for obtaining specific qualifications in laboratory technology will vary from country to country depending on what is supported by 'in country' academic institutions and the relevant regulatory authorities. In most cases, it is preferable to select trainees that already have a good high-school education in biological sciences, applied mathematics, physics and chemistry. Entrance-level qualifications for laboratory assistants and support staff will vary depending on the scope of the job.

It should be noted that in some situations there might also be highly competent laboratory technicians without formal qualifications but who have gained years of valuable experience on the job.

At the end of a formal training programme, laboratory technicians should be assessed on both their theoretical knowledge and practical skills. Newly trained laboratory technicians will need ongoing supervision and support once they reach their province/district. This should be done by a designated laboratory trainer or supervisor. Continuous training in the workplace is one of the most effective ways of maintaining and upgrading the knowledge and technical skills of laboratory staff.

To ensure the quality of the veterinary laboratory service it is recommended that there are regular visits (by experienced staff from the central laboratories to provide ongoing mentorship and technical training support) scheduled to district facilities to assess the performance of personnel. In addition, regular (at least once a year) refresher training opportunities should be available through the larger regional or central veterinary laboratory. These courses provide a good opportunity to introduce and explain the use of new techniques and technologies.

Training in basic laboratory techniques can also be provided to livestock extension and veterinary staff so that they have a better appreciation of tests done in the laboratory. Specific training can also be provided in the use and interpretation of results from common field tests (for example, mastitis screening tests, slide agglutination tests, pen side diagnostic kits and so on).

The importance of ongoing educational support and mentorship for all staff in the animal health services cannot be overemphasized.

1.4 Safety in the laboratory

Access to the core part of the diagnostic laboratory should be restricted to authorized employees. Clear signage should be in place to ensure that unauthorized visitors do not enter beyond the sample submission or registration area. Every member of staff working in a veterinary laboratory, including administrative staff, should be made aware of the possible risks associated with handling potentially hazardous biological material.

Some animal diseases, for example, rabies, hydatids, bovine tuberculosis, brucellosis and salmonellosis are zoonoses, and material from suspected cases of these and other diseases must be handled very carefully. Staff should have knowledge of common zoonotic diseases (see Appendix A1) and if the risk of exposure to specific diseases of concern (for example, rabies, tuberculosis and so on) is high, vaccinations (if available) are recommended. Guidelines for personal protection are provided by the World Health Organization (WHO) but each laboratory should also develop a policy and associated operational guidelines to ensure the health and safety of staff with specific attention to the diseases and hazards common to the region and/or facility. The OIE provides standards for biosafety and biosecurity for veterinary laboratories and animal facilities in Chapter 1.1.4 of the Manual of Diagnostic Tests and Vaccines for Terrestrial Animals. A few general considerations are provided below.

In general, any laboratory sample from a dead or sick animal should be considered potentially hazardous to humans. Precautions should be taken to prevent contamination of the laboratory benches and equipment as well as making sure that strict procedures for personal hygiene are observed. It is generally recommended that disposable gloves are worn when handling specimens. It is recommended practice to undertake risk assessments for handling different types of samples, or for undertaking specific activities, and these are then used to generate SOPs. The microbiology section of the laboratory should always be considered a restricted area and only the staff directly involved in microbiology work permitted to enter. When centrifuging samples of potentially hazardous material, the sample container should be securely sealed before being placed in the centrifuge. There should be protective clothing (laboratory coats, masks, safety glasses, gloves and so on) and biosafety cabinets available for staff handling hazardous specimens, that is, from cases where a zoonotic disease is suspected (for example, tuberculosis, psittacosis and other biological agents). These points are discussed further below.

Handling potentially infectious material

The senior laboratory supervisor will often contact a consulting medical practitioner or public health officer for advice on disease prevention protocols and for the selection of personal protective equipment. Basic guidelines as well as pathogen specific recommendations are provided by the WHO[6] although each laboratory should develop its own guidelines based on a locally relevant risk assessment for both infectious

and non-infectious hazards in the laboratory. Identified risks can be mitigated using SOP and good laboratory practice but staff will also need to be trained. This training should be updated when new risks are identified. Some general principles are outlined below.

1 All specimens received in the laboratory should be regarded as potentially hazardous and handled with due care.

2 Prevention of exposure to potentially infectious agents is very important. This requires training staff in both good laboratory practice as well as making them aware of common zoonotic diseases and how to prevent disease transmission.

3 Provision of appropriate protective gear, including masks and respirators for handling highly infectious pathogens, is essential in laboratories where samples may contain organisms that may be transmitted by aerosol. Such samples can also be handled in a biosafety cabinet.

4 Protective clothing should be worn at work, and removed when leaving the designated work area. This can include coveralls or plain white laboratory coats and disposable gloves, sturdy footwear and so on.

5 Special protective clothing (that is, masks, protective goggles, rubber boots, washable or disposable aprons and so on) should be worn when working with material from cases of suspected rabies or other zoonotic diseases, and at post-mortems. In countries where rabies is common, all staff should be vaccinated and the post vaccination titres checked to ensure protection.

6 Always ensure that staff wash their hands thoroughly with soap and water after handling cultures and/or specimens and before leaving the laboratory.

7 Do not smoke or eat in the laboratory.

8 Spillage of potentially infectious material should be handled as follows:

i   disinfect immediately (wear gloves!)
ii   cover with disposable tissue (to make the spill visible)
iii warn colleagues that there is a spill
iv   keep the spill remains covered with the disinfectant for about 30 min
v   wipe up the spill using absorbent paper and discard in the biological waste bin.

9   Soiled swabs, microbiology samples, cultures and all potentially pathogenic material should be discarded in the biologic waste bin (not in the waste paper basket).
10 All glassware and containers used for potentially pathogenic material must be placed in a disinfectant before sterilization and washing.
11 Benches should be wiped down every day with disinfectant in the morning and before leaving the laboratory.
12 Used, contaminated sharps (needles, Pasteur pipettes and so on) should be discarded in a safe 'sharps container'. In many countries, there are contractors who will provide waste disposal for veterinary and medical facilities, they usually supply sharps containers and other receptacles which are collected for disposal. These contractors must abide by local and national byelaws.

In addition to biological hazards, there are many potentially dangerous chemicals stored and used in the laboratory, such as strong alkalis and acids, which can cause burns and damage eyes. Many reagents, and their vapours, such as alcohol and ether are flammable and only small amounts should be kept in the laboratory in order to reduce the associated fire hazard.

A designated member of staff should be identified and trained to take responsibility for first aid. However, all staff should know which chemicals are hazardous and what measures should be taken to prevent accidents. Every laboratory should also have a first aid box containing emollient creams, an eye bath and an assortment of bandages and plasters. Many laboratory facilities also provide an emergency shower in case of skin contamination. The regulatory requirements for occupational health and safety will vary from country to country and also depend on the designated biosafety level for each division of the laboratory. Most diagnostic laboratories will have sections with level 1 and level 2 biosafety designations. Those with level 3 and higher are highly specialized and will have very clear health and safety guidelines for handling specific pathogens. WHO provides a good summary of what is expected with respect to health and safety training and best practice (see also the bibliography at the end of the chapter). In most countries, the national authorities responsible for public health will also have specific guidelines that must be adhered to.

As outlined earlier, potentially hazardous biological wastes and carcasses are usually disposed of in a specially constructed pit in which biodegradation can occur (Figure 1.4). The pit should have a sealed lockable lid and a lime-sealed concrete surface, which can be washed down after performing post-mortems. No disinfectant or non-biodegradable material should be put in this pit as this will delay decomposition.

Potentially contaminated material from microbiology and parasitology sections should be soaked in disinfectant (for example, phenolics) before being disposed of (not in the biological pit). Contaminated equipment, glassware and consumable materials for re-use (for example, Petri dishes, microscope slides) should be disinfected before being thoroughly washed in detergent and several changes of distilled water. Many laboratory chemicals and biological wastes are hazardous to the environment and should be disposed of carefully to prevent pollution of local water supplies.

Legislation (national and local authorities)

Legislative requirements with regard to health and safety procedures, as well as biosecurity requirements, vary from country to country. It is important to contact local authorities *before* a laboratory is built to determine how any local regulations may affect the day-to-day functioning of the facility. Special rules may apply to the training of staff and the use of safety equipment, provision of containment facilities and so on.

Waste disposal and biosecurity

Laboratory staff have the responsibility to protect themselves, customers, the community and the environment from injury or damage originating from infectious or toxic laboratory waste and to minimize the hazards involved in decontamination, recycling and disposal. These risks can be minimized by laboratory staff knowing and following correct methods for:

- separating infectious materials and laboratory waste
- decontaminating and disposing of non-reusable laboratory consumables and infectious waste
- cleaning and sterilizing reusable consumables and equipment.

It is the responsibility of the laboratory supervisor to avoid environmental pollution and to minimize the potential danger to the staff within the facility and the general public.

The waste fractions in Table 1.1 are gathered separately by the people working in the laboratory. Some of these fractions should be treated before disposal.

Non-biological, non-combustible wastes (that is, broken glass, decontaminated sharps, some plastics and metals) can be stored in a well-constructed waste store (that is, with solid concrete

Table 1.1 Handling of common laboratory wastes.

Waste fractions	Treatment	Disposal
Infectious waste (used bacteriological media, used swabs, etc.)	Steam sterilization	Remainder waste (incinerator or approved dumping ground)
Carcasses	Biological pit or incinerator	If incinerator is used: ash (remainder waste – dumping ground)
Used sharps (infected or not) stored in sharps containers	Steam sterilization	Stored in plastic containers in a waste store until a professional waste treating company can take care of it. Never bury sharps
Dirty reusable not infected ware (glassware, plastic ware)	Washing	The water used to clean the laboratory ware should be removed through a separate sewage system if possible
Dirty non-reusable not infected items		Remainder waste (incinerator or approved dumping ground)
Paper, non-infectious plastic		Remainder waste (incinerator or dumping ground)
Dangerous chemicals		Should be disposed of according to local authority regulations through professional waste disposal services (if available).

walls). Dangerous chemicals and some plastic materials should be disposed of according to local authority regulations through professional waste disposal services (if available). Chemicals should never be buried because they may leach out into the water table and poison the water and soil. Avoid burying or burning sharp objects, such as needles, as these can be picked up by the public or their animals and can cause injury.

Combustible waste (that is, paper, cardboard) can be burned. A suitable container for burning such waste can be made from an old oil drum by making holes in the sides and bottom and removing the top (Figure 1.3). The drum should be set on a foundation of concrete, bricks or stones to allow airflow underneath and it should be held firm to prevent it from falling over. It is important to construct a secure fence around the waste disposal area to reduce the risk of interference from local children or animals.

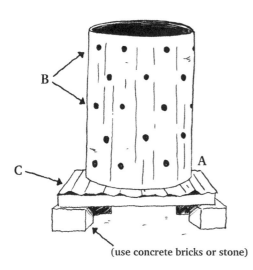

(use concrete bricks or stone)

Figure 1.3 An old oil drum or a metal rubbish bin can be used as an incinerator. (A) Old metal drum or bin with holes made (B) to allow air flow, this improves combustion. (C) Stone or concrete platform to elevate the base of the incinerator from the ground. Note that there must be holes at the bottom of the drum to ensure air flow so do not use a solid platform. Illustration: Louis Wood.

Incinerators

Incinerators can be an appropriate investment for some laboratories as a means of disposing of biological waste material. Large industrial incinerators can be used to dispose of carcasses and other biological material but are very expensive to set up and maintain. In addition, spare parts may be expensive or hard to obtain. A simple low-cost incinerator for combustible waste or non-hazardous material is illustrated in Figure 1.3. This type of incinerator is often used to deal with low-volume waste materials generated by small laboratory units.

Biological pits

A simple, inexpensive and efficient method of dealing with biological waste (including carcasses) is to construct a biological pit. A biological pit may be constructed to back up an incinerator in case of servicing or repair, or as the main means of disposing of biological material, which includes carcasses of large and small animals and fresh, unfixed necropsy specimens.

The pit should be situated and built with due regard to any underground fresh water sources such as wells and springs. The proximity of rivers and other natural water sources should also be considered. To avoid the odour of decomposing materials the pit should be a minimum of 5 m deep. A plan for a pit is given in Figure 1.4 and this can be adjusted to suit local conditions. The dimensions of the pit will depend on the expected amount of material to be disposed of. The surface of the pit will need a well-constructed impervious cover on which post-mortems can be conducted if necessary and to allow regular surface washing and easy disposal of animal waste into the pit. The lid should be strong and preferably have a lock to prevent unauthorized access. Disinfectants, preservatives or any antibacterial chemical should not be used on the surface of the pit cover nor should

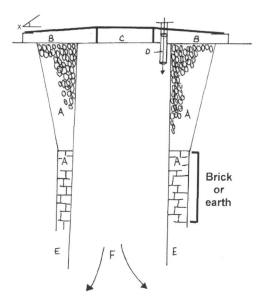

Figure 1.4 Diagram of a biological pit (longitudinal section). The diameter and depth of the pit will depend on the volume of material expected to be put in it. Most are circular in cross section with a diameter of 2–3 m and a depth of 4–6 m. Only biological materials should be added and no chemicals such as disinfectants or antibacterials as these will delay (or even prevent) biological breakdown. (A) Brick, earth or stone lining to a depth of 2 m. (B) Concrete apron (easy to keep clean), surface edge sloped for drainage (X). (C) Metal lid. (D) Fly trap (plastic tube with clear elevated plastic top). (E) Earth. (F) Open end to allow natural drainage.

such materials be washed into the pit as this will interfere with decomposition.

Vermin and flies should be controlled to avoid the risk of spreading disease. For this purpose, most pits are built with a fly trap inserted near the lid. A well-kept biological pit relies on bacterial breakdown of organic matter and can provide an efficient and safe way of disposing of biological wastes. To assist natural decomposition, various commercial mixtures of bacterial culture can be added, that is, those used in some domestic septic tanks. Earthworms may also be

used to provide aeration. The pit must have a strong, preferably metallic and lockable lid.

Procedures and protocols

As outlined earlier, legal requirements may vary from country to country and should be reviewed and obeyed with regard to health and safety regulations. The following are examples of rules that could be posted within the laboratory:

Apparatus

All equipment is potentially dangerous if it is faulty or not operated according to the manual/operating instructions. Any faulty or damaged piece of apparatus should be immediately reported to the chief technician or laboratory manager and not used before it is repaired.

Glassware

Glassware with damaged edges should not be used because it is dangerous and may be inaccurate. Damaged glassware should be discarded into designated receptacles and not in waste paper baskets. Working space should be kept clear of unnecessary glassware.

Pipettes

Pipetting by mouth is not acceptable. Always use a pipette teat or a mechanical unit for pipetting acids, alkali, poisons and samples of potentially infectious material.

Knives and sharp implements

Knives and other sharp or pointed instruments should be cleaned and put away carefully in a designated box immediately after use. Do not leave sharp instruments on the bench or loose in a drawer.

Chemicals and reagents

1 All bottles and containers should be labelled clearly to show their content and date of preparation. Observe warnings on containers and act accordingly.
2 Know the harmful effects and potential danger of chemicals used in the laboratory and how to store them correctly.
3 After using strong acids or alkalis be sure to wipe the neck of the bottle before returning it to the shelf.
4 Neutralize and wipe up immediately any acid or alkali that is spilled.
5 Take extra precautions when working with chemicals which produce a toxic or irritant vapour (that is, only use in a biosafety cabinet, wear protective glasses, masks, and gloves and so on).

Fire prevention and control

When working with highly flammable chemicals the danger of fire should always be kept in mind and adequate precautions taken. Flammable chemicals include ether, benzene, xylene, toluene, acetone and alcohol. It must be remembered that there is a fire hazard from the fumes given off from some chemicals, for example, ether, so all combustible chemicals must be securely stoppered when not in use. All members of staff should be familiar with the location and use of the fire apparatus adjacent to their laboratory. No one should smoke in a laboratory.

1.5 Clinical examination, sample selection, submission and clinical diagnosis

Clinical examination

A clinical examination of any animal should be thorough and systematic. It is important to follow the same procedure every time to ensure that as much information as possible is obtained from each case. In order to determine whether or not there is an abnormality it is necessary to be familiar with what is normal for the species of animal to be examined. The normal range for body temperature, heart rate and respiratory rate for common domestic species are provided in Table 1.2.

A clinical examination should be made with the animal at rest. Temperature (T), heart rate (HR) and respiration rate (RR) should be evaluated early in the course of clinical examination. The HR, RR and T may rise with excitement or fear and will also be elevated after exercise.

Before a clinical examination can take place, it is important that the animal is safely restrained. Some aspects of clinical examination, sample collection and restraint are illustrated in Figures 1.5 to 1.31. Care must be taken that neither the animal nor the handlers are injured. For this purpose, it is important to be familiar with animal handling techniques and always carry extra restraint ropes. If it is known that the animal(s) will be difficult to handle make sure that there is extra trained help and veterinary assistance to

Table 1.2 Normal clinical parameters for various domestic animals.

Species	Dog	Sheep	Horse	Cow	Pig
Temperature °C	38.9 +/– 0.5	39.1 +/– 0.5	37.6 +/– 0.5	38.5 +/– 0.5	39.5 +/– 0.5
Temperature °F	102 +/– 1	102.3 +/– 1	100 +/– 1	101 +/– 1	102.5 +/– 1
Heart rate/min	100–130	75	44	60–70	55–86
Respiratory rate/min	22	19	12	30	20 +/– 5

Figure 1.5 Restraining an indigenous cow for examination and blood sample collection, Khaling, Eastern Bhutan.

Figure 1.6 Collecting samples for the diagnosis of foot and mouth disease using a sterile swab in a foot and mouth disease endemic area.

Figure 1.7 Field team restraining a young cow for sample collection, Khaling, Eastern Bhutan.

Figure 1.8 Restraint of a sheep for examination and trimming of the feet.

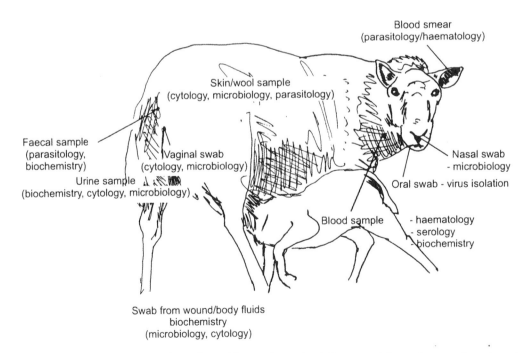

Blood smear
(parasitology/haematology)

Skin/wool sample
(cytology, microbiology, parasitology)

Faecal sample
(parasitology,
biochemistry)

Vaginal swab
(cytology, microbiology)

Urine sample
(biochemistry, cytology, microbiology)

Nasal swab
- microbiology

Oral swab - virus isolation

Blood sample
- haematology
- serology
- biochemistry

Swab from wound/body fluids
biochemistry
(microbiology, cytology)

Figure 1.9 Samples which can be collected from a live animal. For more information see the relevant sections in Chapters 3–8.

provide chemical restraint if required. It is the responsibility of all veterinary and animal health staff to make sure that neither they nor the people assisting are injured. When part of a disease investigation it is important to make sure that a good health history of the farm or village livestock is taken. The following guidelines are general, more detailed information about specific problems is provided in other chapters (see index). The samples which may be selected for specific clinical presentations are outlined in Chapter 9 and in the relevant chapters on specific disciplines.

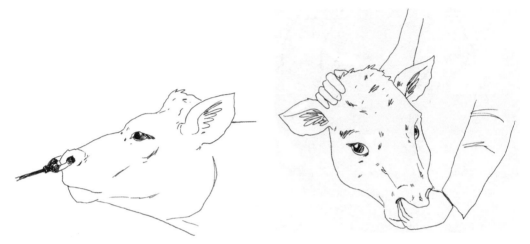

Figure 1.10 Restraining a cow by means of a bull holder. Illustration: Louis Wood.

Figure 1.11 Restraining a cow by grasping the nasal septum and one horn. Illustration: Louis Wood.

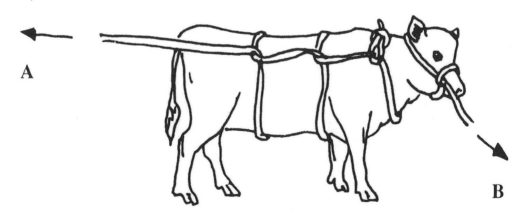

Figure 1.12 Ropes can be used for casting cattle for examination especially where there are no suitable handling facilities. If animal handlers pull slowly on the head ropes (B) and the body ropes (A) the animal will fall to the far side. Attend a training course to learn the correct way of applying the ropes.

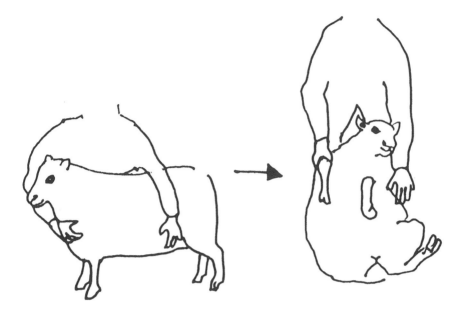

Figure 1.13 Restraining a sheep by turning it on its rump. Blood samples can readily be collected from the jugular vein in the neck (see also Figure 1.18).

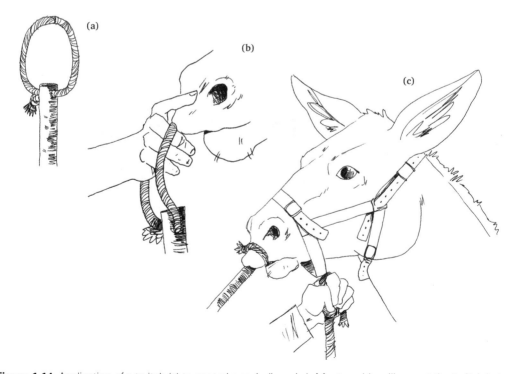

Figure 1.14 Application of a twitch (a) to control a mule (b and c). Most equids will accept the twitch but do not twist the pole too hard as this will damage the upper lip. Illustration: Louis Wood.

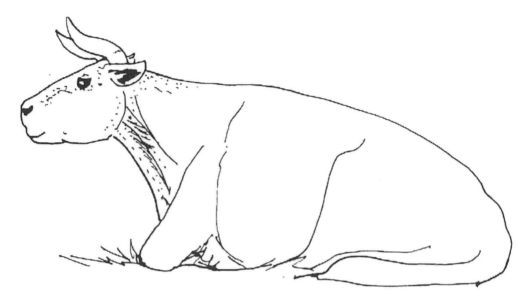

Figure 1.15 The initial examination of a 'downer' cow may be straightforward if the animal is resting comfortably but to determine the cause of the problem it is important to obtain a full clinical history (that is, has the cow calved recently? Was it a difficult calving? Is the cow a high producing animal likely to develop hypocalcaemia and so on). Temperature, heart rate and respiratory rate should be noted as well as general condition, presence or absence of normal abdominal sounds, evidence of trauma, colour of mucous membranes, consistency of faecal material, colour of urine, and so on. Blood samples may need to be collected to assess the severity of the animal's condition.

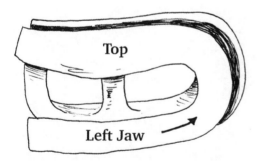

Figure 1.16 Mouth gag for cattle (drinkwater gag), which is placed on one side of the jaw to hold the mouth open. If a bovine animal has difficulty eating, often drops food, or appears to be salivating a lot it is important to look in the mouth. In some cases, it is necessary to use a gag and a torch to allow good visualization of the throat. If rabies is suspected seek advice from the regional veterinary officer *before* examining the animal. Illustration: Louis Wood.

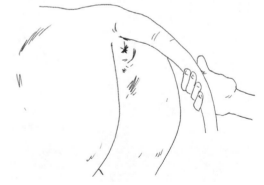

Figure 1.17 Using the ball of the finger (not the thumb) the pulse can be measured at the middle coccygeal (tail) artery in the bovine. Blood samples can be collected from the middle coccygeal vein. Illustration: Louis Wood.

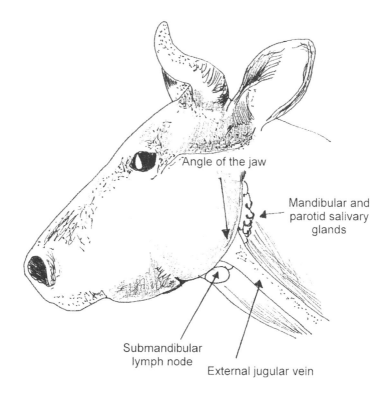

Angle of the jaw

Mandibular and parotid salivary glands

Submandibular lymph node

External jugular vein

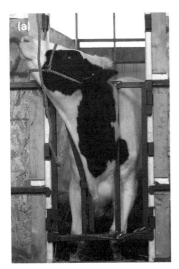 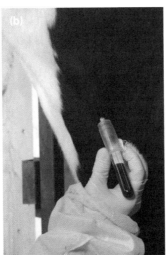 

Figure 1.18a–d (a) The anatomy of the neck of the ruminant (topographical features). This illustrates the location of the external jugular vein which is a common site for the collection of blood samples. The paired jugular veins extend the length of the neck and terminate within the thoracic inlet. The jugular veins return blood from the head to the heart via the cranial vena cavae. (b, c and d) These photographs demonstrate blood collection from the jugular vein of a dairy cow using a wide bore needle and vacutainer. Photo: Dr Regula Waeckerlin, Faculty of Veterinary Medicine, University of Calgary.

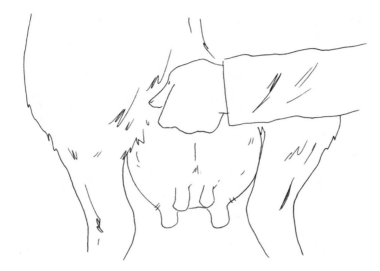

Figure 1.19 Examination of the udder of a cow. Note the presence of any swelling, unusual heat or redness or evidence of pain during palpation. Examine each teat and the milk secreted from each quarter. If the milk is discoloured or very thick it may indicate that the cow has mastitis but it could also mean that the cow has recently calved so make sure that the animal is examined carefully and check the case history. Illustration: Louis Wood.

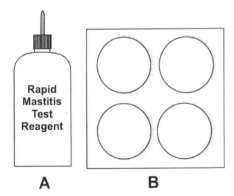

Figure 1.20 A simple test for mastitis is the California Mastitis Test which is available as a kit. Illustrated is a representative plastic squeeze bottle (A) which contains the reagent for the test and a typical plastic paddle (B). The paddle has four shallow cups which can be marked to indicate which quarter of the cow's udder the milk sample was collected from. A few drops of reagent are added to 6–7 drops of milk from each quarter. If the milk sample precipitates it indicates that there are inflammatory proteins and cells present. Often only one quarter of the cow's udder may be affected. If it is necessary to collect a milk sample for microbiological culture collect it aseptically. To do this, wipe the teat(s) with an antiseptic before squeezing the teat to collect the milk. Usually the first few drops are discarded and 3–5 ml collected into a sterile jar. The jar should be labelled to indicate the date, the identity of the animal (age, breed, tag number and so on) the quarter of the udder affected, the name of the farmer and the submitting animal health professional. In most cases the veterinary officer would add a case history to indicate the presence or absence of other clinical signs and the health history of the animal. Illustration: Louis Wood.

Figure 1.21 Examination of breeding and neonatal animals. If a farmer suspects that there are health problems in breeding or neonatal animals it is important to get a good history for the entire group of animals to allow an assessment of the extent of the problem. There are often a number of factors which cause poor neonatal survival and/or abortion and infertility. Epidemiological information (that is, looking at the pattern of the disease) is often more useful than collection of laboratory samples unless specific causes can be investigated. To examine individual cases, follow the normal routine of clinical examination and collection of a case history. There are a range of good text books available which outline the main diseases affecting various age groups of livestock, these are referred to at the end of the chapter.

Figure 1.22 Examination of a mule. The pulse can be taken at the submandibular artery. Illustration: Louis Wood.

Figure 1.23 If it is necessary to look into the mouth of equine species or to rasp the teeth, a gag such as the Haussman–Dunn gag (illustrated) may be required. Illustration: Louis Wood.

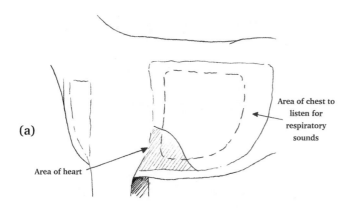

(a)

Area of heart

Area of chest to listen for respiratory sounds

Figure 1.24 Illustration of the area of the chest to listen to in the horse (a) or the cow (b) if it is necessary to assess changes in respiratory sounds. The location of the heart and rib cage in the bovine is indicated in (b). It is important to be familiar with the normal anatomy and physiology of common species examined. Illustration: Louis Wood.

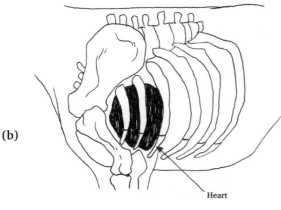

(b)

Heart

Padding

Figure 1.25 Lifting a forelimb of a horse may allow examination of the limb/foot and will restrain the animal if it will not keep still. Ropes should be held firmly. In most cases, it is preferable to get an experienced horse handler to assist. Only try to collect blood samples (usually from the jugular vein) if the animal is appropriately restrained or sedated. Illustration: Louis Wood.

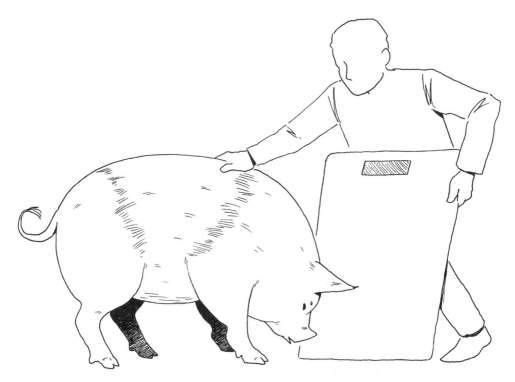

Figure 1.26 Using a small board to assist in moving a pig to a suitable area for sample collection. Illustration: Louis Wood.

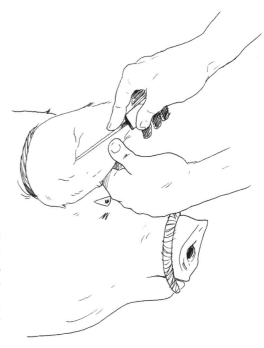

Figure 1.27 Collection of a blood sample from the ear vein of a pig using a 22 gauge needle and a syringe. For some samples, it may be preferable to use a vacutainer (this is discussed further in Chapter 5). A twitch has been applied to the upper jaw of the pig and ear veins have been raised using a rubber band around the ear base. Illustration: Louis Wood.

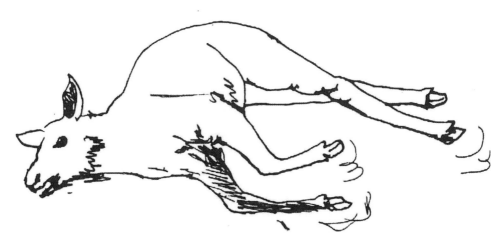

Figure 1.28 The approach to a post-mortem may differ depending on whether one or many animals have died from a disease. In some cases, sick live animals (Figure 1.29) may need to be euthanized to provide fresh material for laboratory tests. This approach may save other animals and, if the animal cannot be treated, may be considered more humane than leaving it to die (it should be noted, however, that in some cultures this may not be permitted for religious reasons). If an animal is found dead and the carcass is still reasonably fresh a full post-mortem (PM) or necropsy can provide a lot of valuable information. If the carcass has already begun to rot then a quick PM may still be worthwhile but it may not be worth trying to collect a lot of samples. The procedure for performing a PM is outlined in Chapter 8. A simple post-mortem kit is outlined in Table 1.4. Practical limitations as well as expense will often dictate the range of samples taken but usually (unless the cause of death is obvious) the following will be collected: (1) Tissues (healthy and diseased) from the main organ system (lungs, heart, liver, kidney etc.) and tissues demonstrating specific lesions (that is, abscesses, vesicles, ulcers and so on) will be collected and preserved for histological examination (see Chapter 8). (2) Swabs of fluids from lesions and/or tissue samples may be collected for microbiological examination (see Chapter 4). (3) Parasitic organisms (internal and external) (see Chapter 3). (4) Blood may be collected for culture. If anthrax is suspected then you should *not* open the carcass (see Chapter 4). (5) Specific samples may be collected for specific diagnoses, that is, rumen contents for suspected poisoning (see Chapter 7). (6) Blood/tissue smears for cytology (see Chapter 8). (7) Urine/eye fluid for microbiology (see Chapter 4). Detailed notes should be kept throughout the procedure to describe any abnormalities found during the post-mortem (see post-mortem submission in Appendix A2).

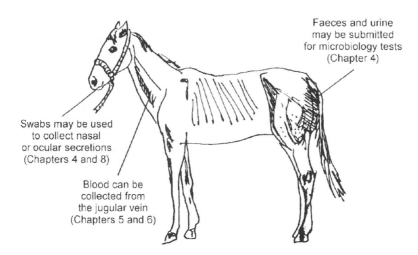

Faeces and urine
may be submitted
for microbiology tests
(Chapter 4)

Swabs may be used
to collect nasal
or ocular secretions
(Chapters 4 and 8)

Blood can be
collected from
the jugular vein
(Chapters 5 and 6)

Figure 1.29 Sick animals may need to be euthanized for humane reasons or for disease control. In many cases ante-mortem samples (especially for haematology and serology) from these animals provide the greatest likelihood of making a diagnosis. Perform a thorough clinical examination *before* samples are collected from the animal.

Figure 1.30 Mule with load. In many rural areas, the mule may be the main method of transport. Practical constrains often limit what can be done to treat disease(s) in animals required for the normal day-to-day activities in rural areas. Treatment may be possible for some conditions but practical factors may dictate if and when appropriate treatment is available for administration (for example, antibiotics/pain relief) and it is not always possible to persuade animal owners to follow advice which may cause significant inconvenience (for example, prolonged rest). To get full compliance from the owner of an animal it is essential that the cause(s) of the problem and the nature of the disease(s) present are fully explained and understood. This takes time and requires a good relationship between the extension staff, veterinary officer(s) and animal owner. Illustration: Louis Wood.

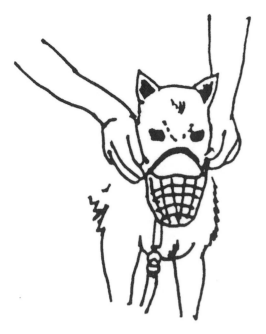

Figure 1.31 Application of a muzzle prior to examination of a dog. Care should be taken before approaching any dog especially in areas where rabies is endemic. In some cases, the animal may appear to be friendly but it is not worth taking the risk of being bitten. Blood samples may be easily collected from a peripheral limb vein for example, the brachial vein (forelimb) but ensure that an experienced handler is available to restrain the animal correctly.

General guidelines for taking a clinical history should include capturing the following (these should be included on most laboratory submission forms or on an attached case history form):

1 date of examination, farmer name, district, village and so on
2 number, species, breed, age groups and sex of animals involved
3 management policies, for example, nutrition, breeding programme, housing, recent transportation and so on
4 case history and veterinary procedures, for example, vaccination, de-worming programme, recent disease problems and treatment provided
5 duration of current problem(s), clinical signs, mortality (number dead) vs. morbidity (number sick)
6 production records (if on a farm), for example, milk production, growth rates, egg output and so on.

General guidelines for a clinical examination should include the following.

1 Evaluate the animal(s) from a distance, observe posture and general appearance.
2 Check hair coat, skin and eyes. Assess the colour of the mucous membranes, for example, if they are pale the animal could be anaemic. Look for any oral (mouth), ocular (eye) or nasal (nose) discharges. Look for the presence of any ectoparasites.
3 Take the temperature, heart rate and respiratory rate and record the findings along with other observations (for example, abnormal abdominal and chest sounds).
4 Check for vaginal (female) or preputial (male) discharges and for faecal staining around the tail and anus. Check the milk (in female ruminants), faeces and urine.
5 Note the animal's behaviour before and during the examination.
6 If a post-mortem is required it is important to get as much information as possible about the clinical signs prior to death, examine other sick animals and check for any sub-clinical cases. Samples collected from in-contact and clinically sick animals can often provide valuable supplementary information to compliment the necropsy findings in cases where a group of animals is involved.

Sample submission, preservation and communication with the laboratory

Examples of sample submission forms are given in Appendices A2. Check also Chapter 8 for further details on their use. It is important to supply as much information as possible on the submission form. The layout of the submission form should allow for addition of any further relevant information. The laboratory service should provide forms that are easy to use. These will usually include epidemiological as well as case specific data. In many cases one style of submission form may not be ideally suited to all situations and for certain disease investigations, or for surveillance programmes, the laboratory staff will need to design a project specific form. The name, work address and telephone number of the submitting animal health professional should always be included to ensure that the results of any laboratory tests are sent back to the correct person.

The collection of blood samples requires experience, and good restraint of animals is essential (see Figures 1.18b, c and d, 1.27). Do not try to collect samples from animals that are likely to become aggressive or difficult to handle unless appropriate facilities and experienced staff are available.

For faecal samples, it may be possible to collect freshly voided material passed by the animal you are interested in sampling, however, for some tests a rectal sample may be preferred. Fresh faecal samples collected from the ground are usually suitable for parasitological and biochemical testing but are not ideal for microbiological testing.

Urine samples, usually mid-stream, can be collected readily from some animals but nervous animals may urinate when approached and therefore such animals may need to be yarded and sampled later.

Collection of skin scrapings, milk samples and wound swabs will usually require restraint of the animal.

In most cases, a full clinical examination requires close inspection and therefore it would be advisable to learn as much as possible about animal behaviour and handling. Good clinical practice and the collection of good quality samples requires training and experience. Figures 1.9–1.32 illustrate some aspects of clinical examination, restraint and sample collection. More information can be found in the bibliography given at the end of the chapter.

General guidelines for submitting samples to the laboratory

1 Always provide as much information as possible on the submission form(s).
2 Take a wide selection of samples unless it is clearly evident what specific disease is present. Specific guidelines for collecting samples for selected clinical cases are given in Chapter 10.
3 Inform the laboratory that samples are to be sent and dispatch them to the laboratory as quickly as possible using appropriate packaging.
4 Sample preservation will be required where there will be a delay in testing. (For details see in specific chapters and the summary box.)

Sample preservation

Microbiology samples and faecal samples for parasitology may be stored in the refrigerator if there is going to be a delay before examination. If it is not possible to send parasitology samples within a day or so, add a fixative (10% formal saline or 70% alcohol) to prevent parasite eggs hatching and subsequent larval development. Tissue samples for microbiology may be stored in the freezer for some virus examination procedures but this may kill bacterial pathogens. Tissues for histological examination may be fixed in 10% buffered formalin for later examination. If samples are submitted for histology ensure that there is a full description of any gross abnormalities along with relevant clinical information. Without good-quality samples, and the necessary background information, the interpretation of histology results can be difficult. Tissue impression smears and smears of material from discharges and faeces can also be prepared. Blood smears can be made directly after blood sampling and can be fixed in methanol and stored for later staining and examination at the laboratory. Whole blood can be collected from representative sick and healthy animals and left to clot overnight at room temperature. Placing the blood tubes in the refrigerator (4°C) for another 4 h often enhances the contraction of the clot. Serum can be separated the following day and placed in labelled containers before dispatch to the laboratory.

5 All samples must be labelled with the animal's identification number, a note of what the sample is, from where it was collected, and the date of collection.

6 Veterinary laboratories can provide a lot of useful information to assist in the selection of suitable samples and can also assist with making a diagnosis. Contact the laboratory directly if specific guidance is required before samples are collected.

Sample collection kits

In the early development of a new veterinary laboratory service it is often difficult to provide sufficient sample collection materials and sample collection services for the evolving field service. Once a system of communication has been set up with submission guidelines for veterinary field and extension staff and transportation and/or postal links for sample delivery the provision of diagnostic services becomes easier. Initially, however, it may be necessary for laboratory staff to pay frequent visits to animal health extension units to deliver sample collection kits (see Figure 1.32) and to collect samples. A list of suggested materials for a sample collection kit is provided in Table 1.3. The requirements for such kits will depend on the resources already available to livestock extension/animal health staff. If the regional laboratory units are also responsible for the storage and distribution of vaccines then the infrastructural network already present may provide a useful distribution source for samples and sample collection equipment to and from the laboratory or extension units. Another benefit of linking the two services is the ready availability of a cold chain. Each diagnostic facility supplied with a post-mortem kit should have an autoclave available for sterilization purposes. If no regular electricity is available, a gas-heated autoclave or a pressure cooker can be used.

Figure 1.32 Sample collection box designed for carriage in a vehicle. The box has several compartments to limit shaking and breakage of the contents during transportation for field work. This example is a wooden box. When available, plastic boxes are more suitable because they are easier to clean.

Table 1.3 Suggested materials for a simple sample collection kit.

Item	Number	Comments
Cool box, 2 cool packs (foam type preferred) or dry ice	1	
Pen, pencil, pencil sharpener, marker, note book and/or data sheets	1	Fill in the disease report form and laboratory submission form!
Scalpel and scalpel blades size 22	10	Always bring a small plastic container to put used needles and scalpel blades in – do not leave them behind as someone may injure themself.
Syringes and needles (20–26 Gauge), 7 ml vacutainers EDTA and plain, vacutainer holder	10	More vacutainers will be required if it is planned to collect serum samples for disease screening.
Cryovials 5 ml	10	
Cryovials with 1 ml brain heart infusion transport medium for avian influenza and Newcastle disease swabs	5	
Universal bottle with 10 ml PBS	5	
Pasteur pipettes, plastic	10	
Transport medium for bacteriology	5	
Cotton wool and/or swabs	1	
Slides (one frosted end), fixative and slide box for blood smears	50 slides 1 slide box	
Ethanol Coplin jar (50 ml)	200 ml 1	Use ethanol for fixing blood smears.
Tooth picks	1 pack	For simple slide agglutination tests.
70% alcohol and 10% buffered formalin in a well-sealed plastic jar	500 ml	Use for small tissue sections for histology/biopsy material and/or preservation of parasite specimen.*
Small plastic containers, sterilin type (20 ml)	10	Use for small specimens.
Sealable plastic bags. Wide mouth bottles with screw cap	30 2	Use for collection of faecal samples. Use for PM sampling.
Small torch, scissors and tweezers and any specific diagnostic reagents required for simple rapid field tests**	1	Make up your own kit of 'useful' items which may be required.
Swabs with and without transport medium	20 of each	Collection of microbiology samples may not be very productive unless there is the facility to keep them cool.
Disposable surgical gloves Plastic aprons. Heavy duty rubber gloves	1 pack 2 2	Make sure that there is protective clothing for sample collection and /or PM work.
Disinfectant	Small bottle or sachets	Make sure that staff is advised about the risk of zoonotic diseases.

Item	Number	Comments
Face mask	5	
Labels	10	
Laminated file with sampling instructions on clinical entities	1	

Notes: The above kit is suitable for a vehicle and can be housed in a wooden or plastic box with handles. It should be easy to clean. The box should have compartments to store standard sample bottles. Ideally there should be a separate cooler box for chilled items. This can operate off a battery or a KVA generator.

This is not meant to be a comprehensive list as there may be specific requirements indicated by the nature of the field work but the list given reflects what we have found practical and useful for small scale field work.

*Not ideal for long-term preservation of museum specimens.

**Performing simple diagnostic tests at the site of sample collection often promotes 'good will' between rural farmers and extension/laboratory staff in the field as it demonstrates that the field program is designed to help the farmer solve his/her animal health problems. Other tests will need to be done at the laboratory.

Table 1.4 Suggested materials for a simple post-mortem kit.

	1 cleanable metal box with tray
Scissors	1 fine scissors, sharp tips, 12 cm
	1 large scissors, 1 sharp tip, 1 blunt tip, 14 cm
	1 curved scissors, blunt tips, 14 cm
	1 bone scissors, 23 cm
	1 pruning shears, autoclavable
Knives	1 autopsy knife, 3 blade sizes, autoclavable plastic handle
	1 amputation knife (Catlin knife)
	1 cartilage knife
	1 brain knife
Forceps	1 fine forceps, serrated, sharp tips, 12 cm
	1 large forceps, serrated, blunt tips, 14 cm
	1 claw forceps
	1 haemostatic forceps
Scalpels	1 one-piece scalpel, 4 cm blade
	1 scalpel handle, 14 cm
	Scalpel blades No. 18
	Disposable scalpel, 1 pack of 10
Other	1 seeker mall probe
	1 bone saw
	1 hammer
	1 tape measure
	5 disposable syringes 5 ml with needles
	Distilled water and larger supplies of preservatives for fixing post-mortem samples

Collection of material for laboratory examination

The quality of diagnostic service offered by the laboratory is determined, to a large extent, by the quality and suitability of the specimens or samples that are submitted and the information on the case(s) that is provided. The laboratory cannot be expected to assist with making an accurate diagnosis without relevant information and properly selected and submitted samples. It is the responsibility of the veterinarian, if present, to select specimens but in his/her absence, especially in rural areas, it is often the livestock or animal health extension worker who has to be aware of the criteria used to select good samples.

Sampling and laboratory diagnosis

To test or not to test? This is the first question to ask. Is a laboratory test desirable, necessary or possible? A number of factors, including feasibility and cost need to be considered. However, once the decision is made to collect samples it is important to ensure that they reach the laboratory in a timely manner and in good condition. Some general guidelines on how to collect, preserve and submit samples are provided below.

GENERAL RULES ON HOW TO BEST PRESERVE SAMPLES FOR SUBMISSION TO THE LABORATORY

A: Some key concepts to keep in mind for tissue samples include:

1 Keep the tissues cool (for example, in a cool box at 4°C). Bacterial contaminants will be present in ANY sample unless they have been collected aseptically. Unfortunately, the contaminants often replicate faster than the disease-causing bacteria of interest. Keeping the samples chilled will slow down bacterial growth and enhance the chances of culturing the pathogen. If marrow from a long bone is to be sampled, contamination can be minimized by swabbing the outside of the bone with disinfectant, splitting the bone open and then aseptic extraction of the marrow by using a sterile swab.

2 Never freeze samples for bacteriology.

3 Samples for bacteriology should preferably be: delivered to the laboratory in a cool box/chiller within 8 h in summer and within 12 h in winter (although this will depend on the local climatic conditions).

4 Try to keep the various tissues separated from one another. If tissue samples are grouped together, bacteria from one sample can contaminate another. This is especially true if intestinal samples or faeces are included.

5 Samples should be kept moist. If the sample dries out, any pathogens present might die, therefore culture will not be possible.

6 For histopathology, 10% buffered formalin preservative should be added to the tissue (add as a 10 : 1 proportion of formalin to tissue volume).

B: Key concepts for handling and preserving other samples

1 For swabs, immerse the swab in transport media and keep at 4°C until it can be sent to the laboratory. Different transport media are required for specific pathogens.

2 For blood samples, keep the EDTA tube at 4°C and mix the tube gently after sampling. If the EDTA blood cannot be examined the same day, a thin blood smear should be prepared and fixed directly after taking the blood. This can then be stained for examination at the laboratory.

3 For serum samples, let the blood sit in a holding rack until it has coagulated. This is best done by collecting the sample in a plain vacutainer (that is, red topped without anticoagulant) and leaving for 12–24 h (in summertime the samples can often be left on

the bench, in wintertime the samples should be kept in an incubator at 37°C). The serum can then be removed and put into a clean, labelled tube. If you have a centrifuge you can get more serum by spinning the tube and then decanting the serum into another vial. If you do not remove the serum in a timely manner the red cells in the clot will slowly break down, releasing their haemoglobin into the serum, causing the serum to take on a pink-red colour which may interfere with subsequent testing. Once the serum is removed into a separate tube, keep it cool until you can get it to the laboratory. If it will be more than one week before the sample can be shipped to the laboratory, freeze it, because antibody levels will slowly fall, even at 4°C. Paired serum samples (collected at the time of, or just before clinical disease and again 2–3 weeks later) can be used to confirm a diagnosis if it is demonstrated that the antibody levels have risen over a period of time. For external parasites, mites, fleas, and ticks can all be collected with forceps and kept in 70% alcohol indefinitely for later identification.

4 For faecal samples, fresh faecal samples should be kept cool until they can be sent to the laboratory. Nematode eggs usually survive well at 4°C but can be destroyed by freezing, so do not freeze samples. The addition of 10% buffered formalin to the faecal sample aids preservation for parasitology examination. For bacteriology examination of faecal samples preservative should not be added to the sample.

SENDING SAMPLES TO THE LABORATORY
Some key principles to consider when sending samples to the laboratory include the following.

1 Speed in getting the samples to the laboratory
The longer the package is in transit, the greater the probability that bacterial contaminants will proliferate or the pathogenic agent will die, so a definitive diagnosis cannot be made. Molecular screening methods may still be possible and are now quite commonly used instead of culture. However, in cases where the cause of the disease is unknown or where we are dealing with a new disease, isolation of the pathogen is still preferable.

Transport the specimens, if possible, within 48 h. Bacteriology samples should ideally be transported to the laboratory within 12 h after sampling.

2 Keep the samples cool on the way to the laboratory
Use cold packs to help ensure that the package does not get warm en route but avoid direct contact between delicate samples and the cooling units

3 Use packaging that will prevent leakage and crushing
Leakage of contents is a biosafety and biosecurity hazard and will cause a package to be refused by the shipper. Also, the package should be sturdy because if it is crushed, it is more likely that samples will be ruined or that leakage will occur. If courier services are used for transporting samples there are specific guidelines for packaging potentially infectious material. Check with your local shipper/courier to find out what the required packaging specifications are.

4 Be sure that all your samples are well-labelled
The laboratory needs to know what kind of samples are being submitted and where they are from. All samples must be labelled with the animal's identification number, a note of what sample it is, from where it is collected and the date of collection.

5 Be sure that appropriate paperwork is included with the samples
Determining the cause of disease is a team effort. By giving the laboratory staff a complete history of the case(s) and noting any gross necropsy findings, the laboratory is much more likely to be successful in helping

to provide an accurate diagnosis. Supply all necessary information on the laboratory submission form or a supplementary disease report form. Always make note of any current or previous treatment used, for example, antibiotic treatment may interfere with the culture of a pathogen even if the antibiotic therapy was not successful in eliminating it.

6   Alert the laboratory

If you let the laboratory know that a submission is coming, testing procedures can be prepared in time for when the sample arrives.

One of the most frequently asked questions is, 'What samples should I collect in order for the laboratory to be able to diagnose a specific disease?' Although some diseases are diagnosed based on gross examination of lesions in specific tissues most are not. Therefore, it is better to collect specimens in a systematic manner. Examples of the samples that should be collected for particular clinical presentations are provided in Chapter 8 and in Appendix A2.

The following tissues are considered 'core' for general disease diagnostic purposes:

- spleen
- liver
- kidney
- lung
- lymph nodes, for example, prescapular or prefemoral lymph node, or any visceral lymph nodes that look abnormal
- tied off piece of small intestine and large intestine, 3 cm each (tied off at both ends).

For transportation of samples to reference laboratories the packaging must comply with domestic transportation requirements and should be appropriately labelled. For international shipments, there are hazardous goods regulations and packaging standards that must be followed and which are well outlined by the International Air Transport Association (IATA)[7].

1.6   Quality assurance and control

Quality control and assurance are the operational techniques used to ensure, demonstrate and provide confidence that laboratory test results are accurate and compare well with those of other laboratories.

A reliable, high-quality laboratory service is achieved and sustained by implementing quality control of laboratory tests, but this is only part of what is needed. A quality management system (QMS) is a more comprehensive and user-orientated approach to quality. A QMS addresses those areas of laboratory practice that most influence how a laboratory service functions and uses its resources to provide a high-quality and relevant service.

A QMS incorporates both the technical aspects of quality assurance (pre-analytical, analytical and post-analytical stages) and those aspects of quality that are important to the users of a laboratory service, such as information provided, its correctness and presentation, the time it takes to get a test result, and the professionalism and helpfulness of laboratory staff.

A QMS includes the following:

- the correct use of the laboratory equipment
- ensuring a timely and reliable service to clients (for example, the livestock extension staff/veterinary staff)
- efficient management of finances, equipment and supplies
- the training and competence of staff
- quality assurance to obtain correct test results (this often requires sending duplicate samples to other laboratories so that results can be compared)
- continuous improvement in quality (this may involve accreditation visits).

Providing a quality service

Understanding and responding to the needs and expectations of clients (for example, veterinarians/animal health professionals/extension staff) requesting laboratory tests are key components of QMS.

The service provided must be:

- **reliable**: with tests competently performed using SOPs, competently under routine and emergency conditions and reports issued 'on time'
- **accurate**: test results free from error
- **accessible and available**: through a network of regional and provincial laboratories and an efficient specimen collection and transport system
- **professional**: laboratory staff knowing their job, presenting clear and informative reports
- **user friendly**: laboratory staff communicating courteously and knowledgably
- **dependable**: laboratory staff arriving at work on time, not being absent unnecessarily, and not allowing tests to be discontinued because reagents have not been ordered correctly or in good time, or equipment has failed because preventive maintenance has not been carried out or replacement parts ordered
- **flexible**: to allow for the introduction of new technologies in response to the needs of users and changing animal health care strategies.

Ensuring the quality of laboratory test results

Within the laboratory

1   Limit the laboratory tests carried out to those which can be done well, that is, keep the work simple and check that routine protocols (SOPs) are developed and followed for specific tests.

2   Ensure that samples are adequately identified and tracked through the system. Some of the larger laboratories rely on barcode tracking but smaller facilities still number samples and track manually.

3   Run quality standards and controls with all tests (this should be stipulated in the SOPs).

4   Implement a regular review of staff performance and provide appropriate training and continuing professional development.

5   Ensure an efficient and reliable recording system.

6   Ensure that the reagents and consumables required are of an adequate quality for the work being done (that is, check formulation and expiry dates).

7   Check the analytical and diagnostic sensitivity and specificity of all routine and specific tests. This, and the need to determine 'normal' values for the population of interest, may require field studies done in collaboration with veterinary field staff and disease surveillance teams.

8   Check the quality of all samples submitted, that is, are the samples in good condition? If not then a note should be placed on the result sheet to indicate that the result might not be reliable.

9   Keep a check on the environmental conditions in the laboratory, that is, temperature and humidity, air flow.

10   Check that laboratory equipment is correctly used and calibrated, for example, glassware/pipettes.

11   Check the water supply daily for the presence of dirt, blockages and so on, and the quality of distilled water produced from it.

12   Check the electricity supply daily for voltage levels and fluctuation.

13   Keep a check on the washing and sterilization of glassware and other items such as equipment used for routine work.

14   Ensure regular maintenance and servicing of all equipment.

15 Information technology (IT) support may also be required – this can be outsourced but in larger laboratories reliant on computing to run equipment and to generate results, invoices, reports and so on there may be a need to employ a full time IT officer.

Outside the laboratory

1 Control sample quality outside as well as inside the laboratory by integrating laboratory and field activities and providing regular training for field and extension personnel.
2 Use reference laboratories to send specimens for confirmatory or duplicate 'blind' testing.
3 Consider building up a laboratory network at the national, and potentially international, level to gain opportunities for knowledge transfer and staff exchange with more established facilities.

1.7 Recording, reporting and interpretation of results

Reporting results effectively and efficiently is important. The interpretation of laboratory results and subsequent guidance on disease control or case management, provided by the laboratory-based veterinary officer, depends on the provision of relevant information by the submitting veterinary or livestock extension staff. If veterinary officers are working in the field the guidance and advice required may be fairly specific and brief. However, if the extension services utilize paraprofessionals then additional clinical explanation and/or technical back up for follow up of cases may be required.

A common complaint about laboratory services is that results often do not reach the end user, the farmer. This is often due to inadequate communication or a lack of follow up between extension, veterinary and laboratory staff. This is an important issue that must be addressed by veterinary laboratory services which are essentially there to support rural development and therefore should meet the needs of the end user, the farmer. In a busy regional centre, it may not be possible for the laboratory staff to directly address the needs of farmers so it is important to ensure that extension staff have sufficient technical support to develop this role.

In most regional laboratories, there will also be a requirement to supply epidemiological data to the central laboratory from district centres. For this purpose, rural centres are often provided with computer assistance and electronic information systems. This will not be considered in detail in this handbook but basic training in the use of computers and in the software packages used in the laboratory is important and regular courses on the use of more specialized programmes may also be needed.

Endnotes

1 The titles given to field service personnel vary from country to country as does the level of training and the degree of involvement of the different cadres, that is, veterinary officers, veterinary field staff, veterinary paraprofessionals, animal health technicians and livestock extension staff. The regulations pertaining to what activities different cadres are permitted to undertake also vary from country to country, and even within a country, so the relevant regional authorities should be consulted to ensure compliance with relevant bylaws. In this book, we will generally use the term veterinary field and extension staff when referring to the key users of regional laboratory services.
2 Founded in 1924 as the Office International des Epizooties, explaining the acronym OIE, it became the World Organisation for Animal Health in 2003. *OIE Manual of Diagnostic Tests and Vaccines for Terrestrial Animals*, http://www.oie.int/standard-setting/terrestrial-manual/access-online/.
3 Standards for biosafety and biosecurity for veterinary laboratories and animal facilities can be found in the OIE Manual of Diagnostic Tests and Vaccines for Terrestrial Animals Chapter 1.1.4, http://www.

oie.int/fileadmin/Home/eng/Health_standards/
tahm/1.01.04_BIOSAFETY_BIOSECURITY.pdf.

General guidelines for Biosafety in the laboratory are also provided by the World Health Organization (WHO) but most countries will have their own set of requirements and regulations with respect to biosafety and biosecurity. Additional information is provided in the bibliography at the end of the chapter.

4 https://www.ekfdiagnostics.com/.
5 http://www.oie.int/fileadmin/Home/eng/
Support_to_OIE_Members/pdf/A_Competence.
pdf.
6 http://www.who.int/csr/resources/publications/
biosafety/Biosafety7.pdf.
7 http://www.iata.org/publications/dgr/Pages/
index.aspx. See also Chapter 1.1.4 of the OIE
Manual of Diagnostic Tests and Vaccines for
Terrestrial Animals, http://www.oie.int/standard-
setting/terrestrial-manual/access-online/.

Bibliography

Cheesbrough, M.C. (2005) Medical Laboratory Manual for Tropical Countries. Volumes I and II. Butterworths, London.

Christienson, D.E. (1996) Veterinary Medical Terminology. W.B. Saunders, Philadelphia, PA.

Constable, P.D., Hinchcliff, K.W., Done, S.H., Gruenberg, W. (2017) Veterinary Medicine: A Textbook of the Diseases of Cattle, Horses, Sheep, Pigs and Goats, 2 vols, 11th edn. Saunders Co, Philadelphia, PA.

Fleming, D.O., Hunt, D.L. (2006) Biological Safety. Principles and Practices, 4th edn. ASM Press, Washington, DC.

Furr, A.K. (2000) CRC Handbook of Laboratory Safety, 5th edn. CRC Press, Boca Raton, FL.

Hendrix, C.M., Sirois, M. (2007) Laboratory Procedures for Veterinary Technicians, 5th edn. Mosby Elsevier, St. Louis, MO.

OIE Manual of Diagnostic Tests and Vaccines for Terrestrial Animals. Chapter 1.1.1 Management of Veterinary Diagnostic Laboratories. http://www.oie.int/fileadmin/Home/eng/Health_standards/tahm/1.01.01_MANAGING_VET_LABS.pdf.

OIE Manual of Diagnostic Tests and Vaccines for Terrestrial Animals. Chapter 1.1.2 Collection, submission and storage of diagnostic specimens. http://www.oie.int/fileadmin/Home/eng/Health_standards/tahm/1.01.02_COLLECTION_DIAG_SPECIMENS.pdf.

OIE Manual of Diagnostic Tests and Vaccines for Terrestrial Animals. Chapter 1.1.3 Transport of biological materials. http://www.oie.int/fileadmin/Home/eng/Health_standards/tahm/1.01.03_TRANSPORT.pdf.

OIE Manual of Diagnostic Tests and Vaccines for Terrestrial Animals. Chapter 1.1.4 Biosafety and biosecurity. http://www.oie.int/fileadmin/Home/eng/Health_standards/tahm/1.01.04_BIOSAFETY_BIOSECURITY.pdf.

OIE Manual of Diagnostic Tests and Vaccines for Terrestrial Animals. Chapter 1.1.5. Quality Management in Veterinary Testing Laboratories. http://www.oie.int/fileadmin/Home/eng/Health_standards/tahm/1.01.05_QUALITY_MANAGEMENT.pdf.

OIE Manual of Diagnostic Tests and Vaccines for Terrestrial Animals. Chapter 1.1.7 Standards for high throughput sequencing, bioinformatics and computational genomics. http://www.oie.int/fileadmin/Home/eng/Health_standards/tahm/1.01.07_HTS_BGC.pdf.

OIE Manual of Diagnostic Tests and Vaccines for Terrestrial Animals. Chapter 3.5 Managing biorisk: examples of aligning risk management strategies with assessed biorisks. http://www.oie.int/fileadmin/Home/eng/Health_standards/tahm/3.5_BIOL_AGENT_SPECIF_RA.pdf.

OIE Manual of Diagnostic Tests and Vaccines for Terrestrial Animals. Chapter 3.6 Recommendations for validation of diagnostic tests. http://www.oie.int/fileadmin/Home/eng/Health_standards/tahm/3.6.00_INTRODUCTION.pdf.

WHO (2004) Laboratory Biosafety Manual, 3rd edn. WHO, Geneva.

chapter 2

The selection, use, maintenance and quality control of laboratory equipment and supplies

Willy Schauwers

2.1 Criteria for the selection of appropriate equipment

The role of the laboratory and the number and range of tests that have to be performed will dictate the type and amount of equipment required.

It is an unfortunate fact that expensive equipment, to the value of tens of millions of euros, lies idle in many laboratories in developing countries. This can occur due to the lack of spare parts or skilled personnel, or because the model supplied was inappropriate for the use required.

Equipment selection has to be performed very carefully.

Providing performance is adequate, it is usually advisable to select simple equipment, because of its robustness, rather than electronically complicated equipment. All equipment should be chosen with due consideration of the current and long-term needs of the laboratory, as well as the ease of use, cost effectiveness and availability of appropriate servicing facilities and spare parts.

When selecting equipment, the following points are worth considering.

- It is important to evaluate the proximity of suppliers, ordering times, availability of local technical expertise and the shelf life of spare parts.

- In all cases equipment should be accompanied by operating manuals in a language understood by the recipient country. A copy of the manuals for each piece of equipment should be made available at the point of use and the original held in a safe central place for reference.

- If technical problems arise the manufacturer's advice should be sought. Warranty options and service agreements will vary from supplier to supplier, but taking these out should be considered for expensive pieces of equipment. For laboratories, located in more remote places, getting equipment serviced can present logistical problems.

- Specifications have to be developed accurately according to the needs of the laboratory, to avoid disappointment and unnecessary discussions with the supplier. The same goes for consumables.

- A demonstration of the installation and use of a piece of equipment is useful.

- Ask the (authorized) supplier for a quotation and details of the power requirements, maintenance schedule, consumables and spare parts needed.

- Ensure that all paperwork is supplied with the delivery and keep the packaging in case of relocation or servicing.

- If equipment cannot be bought nationally, buy it from an overseas supplier (remembering to include transport cost and import charges).

When equipment is purchased through a donor, the receiving laboratory should communicate properly with the donor in order to purchase the right equipment/spare parts/consumables. The donor should be fully informed regarding the specifications of the equipment/spare parts/consumables, maintenance needs, laboratory conditions (humidity, climate hot/cold, dust, ventilation, available power supply and voltage protection) and on-site training requirements for more advanced equipment. Only accept equipment from a donor that is fully functional and comes with the right consumables, spare parts and documentation.

Only when the new equipment has been tested in the laboratory, should the final payment be made. Examples of order lists for reagents/kits/consumables and equipment (prices 2017) can be found on the website. The following companies can help with the purchase of laboratory items:

- VWR: www.vwr.com (choose 'shop VWR globally' and select your country)
- Thermo Fisher Scientific: www.thermofisher.com
- Tactile: https://www.tactilab-rdc.com/, especially for haemoglobinometer and Rayto equipment.

In many parts of the world the conditions in which equipment is expected to operate are not ideal. For example, high ambient temperatures, humidity, dust, the presence of cockroaches (which lay their eggs inside equipment), rats and other vermin (with a liking for plastic wiring and so on) can all take their toll on laboratory equipment. High humidity can encourage fungal growth, which rapidly ruins microscope optics (although most manufacturers are now claiming to have models with components which are resistant to fungal growth). Regular maintenance and cleaning of equipment and premises is essential.

2.2 Ensuring the electricity supply

Fluctuating voltage is a common problem, especially in rural areas, therefore equipment must be protected. The current and voltage supplied to equipment from each power point in a new laboratory should be checked regularly. The local electricity supplier may be able to assist with this. Where voltage stabilizers (Figure 2.1) are used, the input and output for each unit should be checked and faulty stabilizers repaired. Low capacity (0.5 or 1.0 kV) voltage stabilizers may not provide adequate protection for equipment such as computers and spectrophotometers so it is advisable to add a surge protector (Figure 2.2). A surge protector, or surge suppressor or surge diverter, is an appliance or device designed to protect electric devices from voltage spikes. A surge protector attempts to limit the voltage supplied to an electric device by either blocking or shorting to ground any unwanted voltages above a safe threshold.

The voltage stabilizer and surge protector should be switched on *before* the equipment is connected to allow time for the current to stabilize. When there is a break in the electricity supply the equipment should be switched off immediately to avoid damage from power surges, which frequently follow interruptions in power supply.

Equipment sensitive to power breaks (like PCR cyclers) can be connected to an uninterruptible power supply (UPS) or no-break battery power supply system (Figures 2.3 and 2.4), which takes over from the municipal power supply immediately in the event of a power cut.

In places with no or erratic electricity supply it is preferable to install a solar photovoltaic (PV) system (Figures 2.5 and 2.6) and/or to purchase

Figure 2.1 Voltage stabilizers can be used to protect laboratory equipment (refrigerators, freezers) against electricity power fluctuations. Photo: Dr Ziay Ghulam, Dr Wahidullah Bahaer, Central Veterinary Diagnostic and Research Laboratory, Kabul, Afghanistan.

Figure 2.2 Surge protector. Photo: Willy Schauwers.

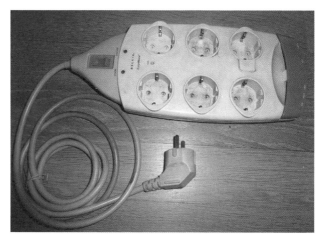

Figure 2.3 No-break system with battery back-up (in Figure 2.4) for PCR-cycler (on left). Photo: Dr Ziay Ghulam, Dr Wahidullah Bahaer, Central Veterinary Diagnostic and Research Laboratory, Kabul, Afghanistan.

Figure 2.4 Equipment sensitive to power breaks (like PCR cyclers) can be connected to a no-break battery power supply system. Photo: Dr Ziay Ghulam, Dr Wahidullah Bahaer, Central Veterinary Diagnostic and Research Laboratory, Kabul, Afghanistan.

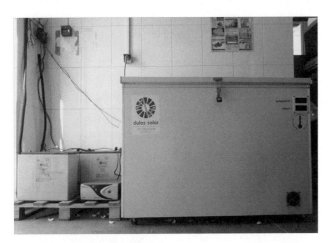

Figure 2.5 Solar PV refrigerator/freezer with battery-system and inverter. Photo: Dr Ziay Ghulam, Dr Wahidullah Bahaer, Central Veterinary Diagnostic and Research Laboratory, Kabul, Afghanistan.

Figure 2.6 Solar PV panels. Photo: Dr Ziay Ghulam, Dr Wahidullah Bahaer, Central Veterinary Diagnostic and Research Laboratory, Kabul, Afghanistan.

Figure 2.7 Generator. Photo: Dr Ziay Ghulam, Dr Wahidullah Bahaer, Central Veterinary Diagnostic and Research Laboratory, Kabul, Afghanistan.

and use a generator (Figure 2.7) to ensure that work can continue. It is important to properly maintain and monitor the solar panels, inverter and batteries, and to store oil and fuel to run the generator and to ensure that the generator is serviced and well maintained.

Solar photovoltaic (PV) systems

Introduction

The sun delivers its energy to us in two main forms: heat and light. There are two main types of solar power systems, namely, solar thermal systems that trap heat to warm up water (mostly for domestic use), and solar PV systems that convert sunlight directly into electricity, as shown in Figure 2.8.

When the PV modules are exposed to sunlight, they generate direct current (DC) electricity. An inverter then converts the DC into alternating current (AC) electricity, so that it can feed into the municipal power grid (grid-tied) or a battery (standalone, off-grid).

Off-grid solar PV systems are applicable for areas without connection to a power grid or with a poor energy supply. An off-grid solar PV system needs deep-cycle rechargeable batteries, such as lead-acid, nickel-cadmium or lithium-ion batteries, to store electricity for use under conditions where there is little or no output from

the solar PV system, such as during the night, as shown in Figure 2.9. Solar PV only produces electricity when sunlight is available. The output of a solar PV system varies with its rated output, temperature, weather conditions and time of the day.

Design and installation

Proper system design and installation are the first steps required for sustainable on-site energy systems. Off-grid energy system design is quite complex and should typically be done by a trained professional. This paragraph does not attempt to provide instructions on how to design and install a system, but rather covers some general principles and lessons learned from previous installations.

General principles

Foremost, it is important to emphasize the importance of doing a design. One-size fits all approaches to health facility energy system procurement are rarely effective. A proper design must be done by a professional (include input from several stakeholders), and should be reviewed after system installation to check whether assumptions made were correct and that it is operating as expected.

Consider using an energy audit template to collect information for the design process

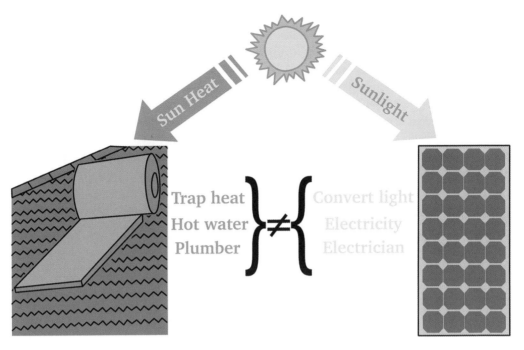

Figure 2.8 The two main types of solar power systems: solar thermal systems, which trap sun heat to warm up water, and solar PV systems, which convert sunlight directly into electricity. Illustration: Louis Wood.

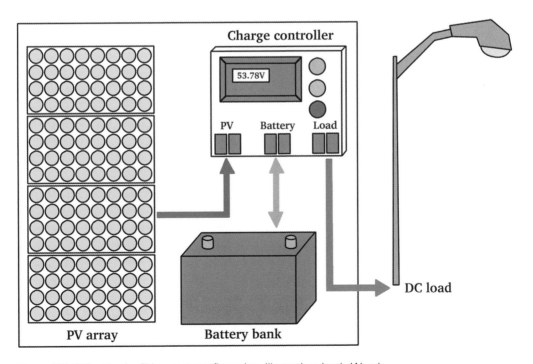

Figure 2.9 Off-grid solar PV system configuration. Illustration: Louis Wood.

and compare different technology choices. Professional installation of energy equipment is critical for proper operation. Several donor programmes have been observed in the field which have provided a health facility with energy equipment, but did not cover installation costs. The results are often that the health facility lets the equipment sit idle, or that they use a local inexperienced technician resulting in incorrect installation of the equipment.

Key lessons learned

- Energy system cost evaluations based only upon initial cost may discourage the choice of renewable energy sources. Life-cycle costs and levelized cost of energy should be used and can be calculated using, for example, the online homer tool (http://tools. poweringhealth.org/). Fuel supply logistics should not be underestimated when comparing PV and fuel-driven generator systems.
- Perceptions are often inaccurate or oversimplified. Common misperceptions are that renewable energy power systems are unaffordable or that they require no maintenance.
- Anti-theft measures should be considered in all design plans.
- 'Maintenance free' sealed batteries are a good option for health facilities.
- Professional installation with proper labelling is critical for long-term system operation.
- Battery-based systems designed for and dedicated to a specific load (for example, computer, vaccine refrigerator) have the highest success rate.
- Oversizing battery banks to provide multiple days of autonomy often results in continuously discharged batteries unless the size of the solar array is also increased or a generator is added to the system.
- In order to recharge batteries and reduce system size and cost, diesel-powered generator/ PV hybrid systems are recommended for all

but the smallest health centres (as opposed to PV-only systems).
- Invertors that easily allow users to override low voltage cut offs (or inverters that can easily override protection against deep discharge of the batteries) are not recommended. Low voltage cut off settings should often be increased from factory set default values to prolong battery life in developing country settings.
- Access to locally available spare parts should be a consideration in initial equipment procurement.

2.3 Equipment maintenance, servicing and repair

Laboratory workshop

The laboratory (network) can rely for repair, maintenance and servicing on the private sector (electricians and laboratory equipment service engineers from the manufacturer) or they can have their own designated laboratory workshop.

The justification for a designated laboratory network workshop facility to repair and maintain laboratory equipment will depend on the amount of equipment involved and availability of local agents and outside expertise. To support one central laboratory and four to six regional laboratories, one small workshop is usually adequate. This can be staffed by one or possibly two service technicians depending on the amount and degree of repairs required. Expertise in the maintenance and repair of equipment, especially electronics and refrigeration is essential and can be a shared resource used for laboratory and cold-chain support. More sophisticated pieces of equipment will usually need to be sent back to the manufacturer or local supplier for maintenance and repair. This, or a visit from the supplier, is a service that can be arranged through the workshop staff.

Table 2.1 Tools required for a one- or two-person laboratory workshop.

	Item	Number
1	Drilling bits	1 set
2	Electric hand-drill	1
3	Electric screwdriver	1
4	Multimeter analogue	1
5	Soldering iron	1
6	Tool kit* for instrument and electronic use (portable)	1
7	Tool kit*, laboratory workshop (metal box)	1
8	Oxygen propane super heating torch (heavy brass shank)	1
9	Socket set (metric), bio-hexagonal socket	1 set
10	Screwdriver set	1 set
11	Screwdriver set, precision (small size)	1 set
12	Screwdriver bits	1 set
13	Screwdriver with marker	2 sets
14	Pliers, combination type	4
15	Electronic pliers (side cutting nose)	2
16	Multi-grip pliers	2
17	Spanners (adjustable)	1 set
18	Tube cutter	3
19	Flaring tools	2
20	Swaging punch kit	2
21	Propane fuel cartridges	2
22	Refrigerant leak detector	1
23	Refrigerant charging manifold	1
24	Cutting torch	1
25	Ratchet wrench	1 set
26	Mains testing screwdrivers	2
27	Wire stripper with spring	1
28	Diamond glass cutter	2

Note: *Specific components depend on work requirements and personal preference.

A list of tools that are required for a small on-site workshop is available in Table 2.1.

Maintenance

Maintenance should be performed regularly (by the laboratory or workshop personnel) as suggested by the manufacturer. This ensures longevity of the equipment and facilities. Therefore, an annual plan for maintenance, servicing and quality assurance for equipment and facilities should be developed (Table 2.2). All documents concerning maintenance, servicing and repair of equipment should be stored in the equipment logbook.

The following remarks are worth considering.

• Technicians using specific equipment should be trained to perform maintenance (as outlined in the relevant standard operating procedure [SOP]).
• Always turn off and disconnect equipment before starting maintenance, repair or servicing.
• Ensure that equipment is inspected regularly, that is, check for corrosion, worn out components, damaged insulation, loose connections, fungal growth and insect/rodent infestation.
• Stocks of essential spare parts should be held in storage with a record of their part numbers and sources of supply.

The following files (available on the website) show an example of a SOP and forms for an annual plan for maintenance and quality assurance:

• SOP-LABORATORY-EQUIP-06 Planning Maintenance and Quality Assurance
• FORM-SOP-LABORATORY-EQUIP-06–01- Equipment Maintenance and Service Annual Plan

Table 2.2 Example of an annual equipment maintenance and service plan.

Year: ....

Equipment	JAN	FEB	MAR	APR	MAY	JUN	JUL	AUG	SEP	OCT	NOV	DEC
Bio safety cabinets												
Microscopes												
Fridge and freezer												
Centrifuges												
Roller-mixers and shakers												
Stirrer-heaters												
Vortex												
Computers												
Ups systems												
Water-baths												
Incubators												
Balances												
Autoclaves												
Dry ovens												
PCR machines												
Micro-pipettes												
Timers												

- FORM-SOP-LABORATORY-EQUIP-06–02-Checklist for Maintenance of Equipment
- FORM-SOP-LABORATORY-EQUIP-06–03-Checklist for Corrective Maintenance
- FORM-SOP-LABORATORY-EQUIP-06–04-Quality assurance annual plan.

2.4 Specific equipment

The following sections outline the equipment/instruments commonly used in a regional, provincial or district veterinary diagnostic laboratory. General principles on the function, care, maintenance and quality assurance of equipment and instruments are provided in each section but the operating manual should always be consulted before using equipment because every model has its own unique properties.

Equipment logbook

Every piece of equipment should have a logbook where the following information is recorded:

- title page
- equipment properties
- authorized personnel

  - responsible person
  - users
  - personnel being trained

- specifications of preventive maintenance

  - users authorized to perform preventive maintenance
  - responsible person
  - company performing preventive maintenance

- control programme.

All information concerning maintenance, defects, calibration, controls are preferably registered and kept in the equipment logbook. For more sophisticated equipment like GC, MS, LC that need a lot of maintenance and repair, a notebook can be used, it is more convenient than a logbook.

The file 'Logbook equipment template', see the website, is a template for an equipment logbook.

How to use the equipment logbook

The equipment logbook is either near the equipment or all equipment logbooks can be stored on a dedicated shelf. All data concerning the equipment (maintenance, defects, calibration, irregularities) are registered in the logbook, if not, it should be mentioned in the logbook where and how this information is stored. Fill in all data immediately and the laboratory person following up/performing maintenance/calibration, reporting defects should include his/her initials.

Check all equipment logbooks on a yearly basis. See 'FORM-SOP-LABORATORY-EQUIP-06–04-Quality assurance annual plan'.

Refrigerators, freezers and cool boxes

There are many types of cooling equipment.
In the laboratory:

- conventional electric refrigerators (+4°C) and freezers (–20°C)
- refrigerators and freezers running on solar PV energy, gas or other fuel
- Sure Chill® refrigerators (Figure 2.10)
- –80°C freezers (Figure 2.11)
- cold rooms.

In the field:

- conventional cool boxes
- digital car refrigerators (like Dometic CFX28, 26 l), can be plugged in to a 12 V (car or motorbike) or a 220 V outlet.

Figure 2.10 Sure Chill open refrigerator.
Photo: Sure Chill® Company, Cardiff, UK.

Figure 2.11 −80°C freezer. Photo: Dr Ziay Ghulam, Dr Wahidullah Bahaer, Central Veterinary Diagnostic and Research Laboratory, Kabul, Afghanistan.

Cooling units used in the laboratory

Cold storage units are an essential component of laboratory equipment. A standard domestic fridge unit (4±3°C) and a freezer unit (−20±3°C) are adequate for most purposes. However, for long-term storage of microbial organisms and some biological material (for example, serum) a −80°C freezer unit is preferable. All refrigerators and freezers (especially the −80°C freezers) should be kept on a stabilizer in locations suffering from regular power cuts and poor-quality electricity supply.

In cases where the power supply is regularly interrupted it may be worth considering the purchase of a solar-powered refrigerator/freezer, gas-fired or paraffin-fuel cold storage units.

Nowadays new technologies are available for cooling. The patented Sure Chill® technology works on the principle that water is at its densest at 4°C. Water, not ice, surrounds a Sure Chill® refrigeration compartment. When power is supplied to the refrigerator, the water cools and ice is formed above the compartment, leaving only the denser water at 4°C cooling the contents. When the power is switched off, the water warms, becomes less dense and rises while the ice begins to melt and the cooler water descends. This ensures water at a perfect 4°C remains cooling the contents of the compartment. The system can operate like this, without any power whatsoever, for many days. Sure Chill® is a cooling technology which is currently being used in medical refrigerators. The refrigerator can stay at 4°C for more than 10 days without power (if not opened – it is important to keep cooling devices closed as much as possible during power breaks), and is used mainly in Africa to store vaccines and other medical supplies. It can be powered by electricity or solar PV power and

Figure 2.12 Cold room 1. Photo: Willy Schauwers, Provincial Institute for Hygiene, Antwerp, Belgium.

Figure 2.13 Cold room 2. Photo: Willy Schauwers, Provincial Institute for Hygiene, Antwerp, Belgium.

uses the physics of water to store energy, thus not relying on batteries.[1]

Cold rooms are mainly used in bigger laboratories (Figures 2.12 and 2.13). If larger amounts of samples need to be stored and/or samples have to be kept for a longer period of time, it might be useful to have a cold room.

Cooling units used in the field

CONVENTIONAL COOL BOXES

Most of the time, conventional cool boxes with cool packs are used in the field to preserve samples at a lower temperature. Cool boxes can be used to keep items between 2–8°C or at –20°C.

- Between 2–8°C: cool packs (wet ice) are stored in a freezer at –20°C. When preparing a cool box that is to keep samples between 2–8°C, the frozen cool packs should be held under running tap water until the cool pack content starts to defreeze (until the content of the frozen ice pack starts to liquefy). Samples should usually be kept cool after sampling. But, make sure there is no direct contact between the cool packs and the samples. Direct contact might freeze the samples and make them unfit for use in the laboratory (for example, haemolysed EDTA blood due to freeze-damage to the red blood cells) (Figure 2.14).
- At –20°C: the cool box is filled with frozen ice packs. Or, instead of frozen cool packs (wet ice), dry ice (frozen, solid carbon dioxide – CO_2) can be used as cooling unit for a cool box. Most cool boxes these days are dry-ice compatible. Dry ice keeps items frozen, not just cold (dry ice has a temperature of –78.5°C). Dry ice does not make things wet (like wet ice). When water goes from ice to liquid it takes up less room. However, when dry ice goes from a solid to a gas it takes up significantly more room, building up pressure

Figure 2.14 Simple cool box, to be used with cool packs. Photo: Dr Ziay Ghulam, Dr Wahidullah Bahaer, Central Veterinary Diagnostic and Research Laboratory, Kabul, Afghanistan.

Figure 2.15 Dry ice maker. Photo: Scilogex, Rocky Hill, Connecticut, USA.

if not allowed to escape. Always use protective gloves when dealing with dry ice. It is so cold that it will either burn your skin, or more likely cause frostbite. By using a pair of gloves, you can safely handle the dry ice without injury.

PORTABLE DRY ICE MAKER

A portable dry ice maker generates approximately 1 kg of solid carbon dioxide in just 1 min. A dry ice maker is safe and simple to use and requires no electrical power. The unit is simply attached to a cylinder of compressed liquid carbon dioxide via a hose with an internal siphon and valve (Figure 2.15).

The dry ice maker creates one solid block of solid carbon dioxide weighing approximately 1 kg, the carbon dioxide sublimates slowly because of the relatively small surface area.[2]

DIGITAL CAR REFRIGERATORS

Digital car refrigerators run on electricity. Either a 12 V source, if plugged in to a motorbike or car, or 220 V if plugged in to a mains supply. Smaller size cool boxes can be used on a motorbike while the larger (Figure 2.16) ones can be used in a car.[3]

The more sophisticated car refrigerators can be adjusted to 37°C. In a cold climate, these cool box/incubators can be used to store whole blood samples (the clotting of blood is a biological process that needs at least room temperature) otherwise the serum-production will be very poor if the whole blood samples are kept at 4°C.

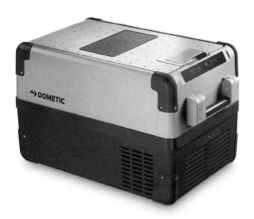

Figure 2.16 Digital cool box, larger size, can be used in cars. Photo: Dometic company.

Care and maintenance

Bear in mind the following when maintaining cooling equipment.

- If parts need to be replaced try to ensure that the correct specifications are met before installation.
- While transporting, keep refrigerators in an upright position if possible.
- Wait 24 h before switching on the refrigerator after transportation.
- Do not place the refrigerator with its backside against the wall. Keep a gap of at least 10 cm.

- Defrost and clean refrigerators regularly.
- Do not use a knife or metal to remove ice formation in the freezing compartment.
- Use a voltage stabilizer of at least 4 amps.
- Do not place hot material inside the refrigerator.
- Never overload a refrigerator, the air inside the refrigerator must be able to move freely, otherwise the cooling capacity is not optimal. Therefore, never put items against the back wall of the refrigerator.
- Make sure that the refrigerator ventilation grill is not blocked or obstructed.
- Check the amount of ice forming around the freezer compartment.

For solar-powered cooling equipment the following are also relevant.

- Check the indicator lights and any other meters.
- Clean the solar panel.
- Check for shadowing of the solar panel.
- If the batteries are of the sealed type, no maintenance is needed, if open (non-sealed) batteries are used, the level of the electrolyte (acid mixture) in any of the cells in the batteries has to be checked and filled up with distilled water if the level has dropped.

Quality assurance

- Check daily the inside minimum and maximum temperature of the refrigerators/freezers (at the same time each day, for example, in the morning when entering the laboratory) (use the form 'FORM-SOP-LABORATORY-EQUIP-05–01 Temperature registration min max thermometer').
- Keep the temperature probe in a small 100 ml plastic bottle filled with water and covered with a lid. This will avoid measuring temperature fluctuations when opening the refrigerator.

Figure 2.17 A maximum–minimum thermometer. The scales are Fahrenheit on the inside of the U and Celsius on the outside. The current temperature is 23 degrees Celsius, the maximum recorded is 25°C, and the minimum is 15°C. Photo: Lumos3 – Wikipedia.

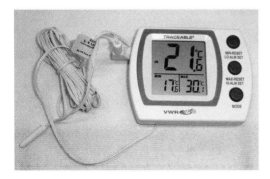

Figure 2.18 Digital thermometer min/max with external probe. Photo: Willy Schauwers, Provincial Institute for Hygiene, Antwerp, Belgium.

For solar-powered cooling equipment you should also record the state of charge level of the batteries (as a percentage).

Temperature registration quality assurance

See 'SOP-LABORATORY-EQUIP-05 Temperature registration'.

THERMOMETERS AND RECORDING EQUIPMENT
Many different types of thermometers exist:

- the common thermometer (always use thermometers without mercury, because the mercury vapours – released when a thermometer is broken – are hazardous)
- certified thermometers
- minimum–maximum thermometers: thermometer with gliders (can be reset by using a magnet) (Figure 2.17)
- digital thermometer with internal and external temperature measuring probe (Figure 2.18)
- digital thermometer with data logger:

  - data logger with internal temperature probe (Figure 2.19) (the device has to be put in the equipment you want to measure the temperature from, for instance a refrigerator)
  - data logger with multiple temperature probes (connected to the data logger), one data logger can monitor the temperature of one or several (nearby) devices (Figure 2.20)

- a wireless temperature monitoring device which registers the temperature continuously and is able to send an SMS or email when the equipment temperature exceeds certain limits
- chemical temperature monitoring devices, which register temperature to detect rising

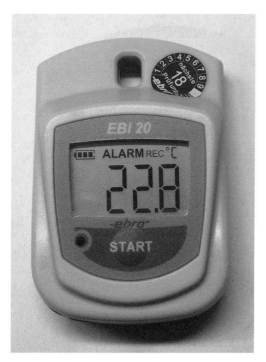

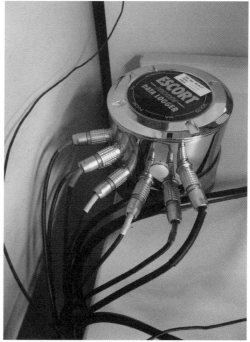

Figure 2.19 Datalogger with internal probe. Photo: Willy Schauwers, Provincial Institute for Hygiene, Antwerp, Belgium.

Figure 2.20 Digital datalogger with multiple probes. Photo: Willy Schauwers, Provincial Institute for Hygiene, Antwerp, Belgium.

and falling temperatures (like 3M Freeze Watch and 3M MonitorMark), are:

- an inexpensive solution for monitoring product exposure
- self-adhesive: stick securely to most exterior packaging materials
- easily interpreted: visual results.

Remark: sometimes the mercury column of (mercury) thermometers is broken (interrupted) after they have been transported by air. This happens quite often with mercury minimum/maximum thermometers. Thermometers with interrupted mercury columns are not fit for use in the laboratory. Another reason not to use non-mercury thermometers.

QUALITY ASSURANCE PROTOCOL

- Daily temperature registration, minimum/maximum temperatures or actual temperatures are registered for each temperature control device (refrigerator, freezer and incubator) and for the laboratory room(s) with HVAC (heating, ventilation and air conditioning). This should be done on a daily basis (the same time in the morning when entering the laboratory). See 'FORM-SOP-LABORATORY-EQUIP-05–01 Temperature registration min max thermometer' or 'FORM-SOP-LABORATORY-EQUIP-05–02 Temperature registration thermometer'.
- If the temperature of a laboratory device is not within the prescribed range, an inspection and corrective measures need to be taken. See 'FORM-SOP-LABORATORY-EQUIP-05–03 Corrective measures'.

- Minimum and maximum temperature limits are defined for each laboratory device with temperature control. See 'FORM-SOP-LABORATORY-EQUIP-05–05 Temperature limits Minimum and Maximum'.
- Each temperature control device being used in the laboratory is checked on a 6 monthly basis against a certified thermometer (these are usually kept at the quality control department). See 'FORM-SOP-LABORATORY-EQUIP-05–04-Thermometer control' and 'FORM-SOP-LABORATORY-EQUIP-05–06-Compare temperature with certified thermometer'. The 'correction temperature' is mentioned on each daily temperature registration form.
- An annual plan for periodic controls (including temperature control) is available in 'FORM-SOP-LABORATORY-EQUIP-06–04-Quality assurance annual plan'.

Autoclaves

Autoclaves use pressurized steam to ensure high temperatures for sterilization. Simple models are metal (aluminium or stainless steel) containers in which water can be boiled. They have:

- a high-pressure vessel with lid and gasket
- a pressure gauge (1) on the lid to record the pressure inside the vessel
- a 'blow off' valve (2) to release steam after the desired pressure is reached
- a safety valve (3) (Figure 2.21).

Autoclaves range from small, simple domestic pressure cookers heated on a stove, to extremely large computerized models. The form of heating used can be a simple flame (for example, gas), electric element or steam supplied from an external source through a jacket around the autoclave chamber.

Figure 2.21 Autoclaves can be used to sterilize equipment for simple surgical procedures as well as laboratory ware, laboratory waste and some instruments. Photo: Willy Schauwers, Provincial Institute for Hygiene, Antwerp, Belgium.

When water boils at sea-level atmospheric pressure (sea-level pressure is 1013.25 mbar, 760.00 mm Hg), the temperature of the boiling water is 100°C. When the pressure is increased inside the autoclave, water will boil at a higher temperature. The pressure used for sterilization is usually 15 psi (1.1 bar = 100 kPa = 15 lb/in^2 = 15 psi) and at this pressure water will boil at 121°C. When this pressure and temperature is applied for 15–20 min most of all known living organisms are destroyed.

Sterilization of articles inside the autoclave takes place when steam comes into contact with the surface. This moist heat has better penetration than dry heat in an oven and is therefore a more efficient sterilizer, particularly for bulky items.

To ensure that an autoclave is working efficiently it is necessary to check that the temperature/pressure readings are correct. The usual method to verify this is to apply special autoclave tape to objects for sterilization. This strip of self-adhesive 'autoclave tape' is impregnated with a chemical that changes colour if the correct temperature/time has been employed. A large range of chemical sterilization products is available, such as Bowie and Dick test sheets, helix test.

The basic steam sterilization process

Air is about twice as heavy as pure steam of the same temperature. Therefore, empty containers should be put upside down, the heavier air will sink down and will be replaced by the lighter steam (downward displacement). In order to let the steam reach all parts, leave sufficient space between the goods.

A full sterilization cycle based on the removal of air by downward displacement will go through the following phases (Figure 2.22).

- Heating up of the water and air removal. The water is heated up to its boiling point of 100°C. During this heating up the air is being removed from the chamber. In order to improve the removal of air, the water is allowed to boil some additional time after the water has reached 100°C. The air in the chamber will be flushed out with the steam. Care should be taken to expel all air from the autoclave before it is put under pressure otherwise pockets of air will hinder penetration of steam into the articles to be sterilized.
- Building up pressure. Temperature increases to sterilization temperature. By closing a valve, the vessel is sealed and pressure and temperature increase to the required level.
- Sterilization time (holding time). During this time, the temperature/pressure is held at the required level for sterilization.
- Reducing the pressure to atmospheric by releasing the steam. By opening a valve the steam can escape and the pressure goes down.
- Cooling down of the load. After sterilization the pressure and the temperature should

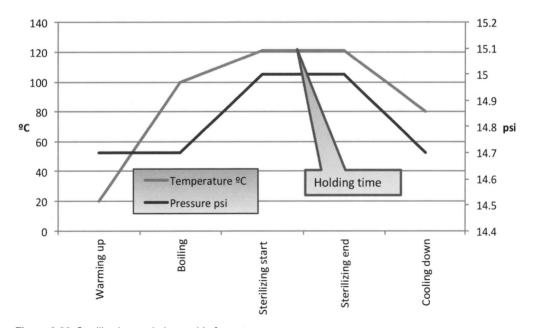

Figure 2.22 Sterilization cycle in graphic format.

be allowed to come down slowly (to atmospheric pressure and temperature around 80°C) to prevent liquids inside the autoclave from boiling and spilling. It is important to make sure that the autoclave is not still under residual pressure before opening it.

Articles for autoclaving should be cleaned (before autoclaving) of any residual organic material. Timing of the autoclaving process should start when the sterilization pressure (15 psi) is reached.

A SOP of a portable Dixons autoclave[4] can be found in 'SOP-LABORATORY-EQUIP-04 Autoclave Dixons ST19T and ST3028'.

Standard care and maintenance of a simple bench top autoclave

- Fill the chamber with distilled water.
- Place the load properly prepared for sterilization in the inner liner, into a wire mesh basket or a bottle rack.

- Close the wing nuts properly (normal manual tightness is adequate).
- Allow the autoclave to cool to 80°C or less and check that the pressure gauge is reading zero before opening the lid.
- The use of alkalis (for example, sodium bicarbonate) for cleaning may cause corrosion of the pot and lid and should therefore be avoided. Use normal household detergent to clean the interior of the autoclave.
- An annual safety check will normally be sufficient unless the autoclave is in very regular use. See 'FORM-SOP-LABORATORY-EQUIP-04–01 Autoclave Dixons Inspection'.
- Electric autoclaves consume easily 2000 W, therefore the electric wiring system should be of sufficient quality to avoid electric failure.

Quality control

- Check the silicone gasket for distortion or wear and tear.

Figure 2.23 Autoclave tape *before* sterilization. Photo: Dr Ziay Ghulam, Dr Wahidullah Bahaer, Central Veterinary Diagnostic and Research Laboratory, Kabul, Afghanistan.

Figure 2.24 Autoclave tape *after* sterilization. Photo: Dr Ziay Ghulam, Dr Wahidullah Bahaer, Central Veterinary Diagnostic and Research Laboratory, Kabul, Afghanistan.

Figure 2.25 Autoclave tape *after* sterilization (prominent black lines). Photo: Dr Ziay Ghulam, Dr Wahidullah Bahaer, Central Veterinary Diagnostic and Research Laboratory, Kabul, Afghanistan.

- Check the rim of the body where the lid gasket seats for damage to the metal casting.
- Check the draw-off cock for signs of leaking around the seals and from the valve.
- Always use a sterilization indicator for each cycle: autoclave tape (Figures 2.23–2.25), Bowie and Dick test sheets, helix test.
- Use a biological indicator at least once per year to check the sterilization cycle. Biological indicators are test systems containing viable microorganisms providing a defined resistance to a specific sterilization process. A biological indicator provides information on whether necessary conditions were met to kill a specified number of microorganisms for a given sterilization process, providing a level of confidence in the process. Bacterial spores (for example, *Geobacillus stearothermophilus* spores) are the microorganisms primarily used.[5]

A laboratory centrifuge consists of an electrically driven motorized spindle that spins a rotor head containing tubes inside a heavy metal case. The unit has a timing control mechanism and modern units have a cut-out switch incorporated in the lid mechanism enabling them to operate only when the lid is closed.

Most small diagnostic laboratories will have one or two standard centrifuges and possibly a portable unit. Field centrifuges can be used with a generator for parasitology work or for determination of haematocrit value.

Centrifuges commonly used in a regional laboratory will operate at maximum speed of approximately 6000 rpm. If the centrifuge is not used properly at this high speed the instrument is potentially dangerous. It is therefore, very important to read the manufacturers operating instructions and carefully follow a few simple operating rules.

Centrifuges

A centrifuge is used to spin tubes that contain substances suspended in liquids, at a high speed. The suspended material will be deposited, in order of molecular weight, with heavier substances at the bottom of the tube. Speed, spinning time and the difference in density of suspended particles and the medium will all affect the settling of particles (Figure 2.26).

The speed of a centrifuge is usually referred to in terms of revolutions per minute (rpm). That is the number of times the centrifuge head completes a cycle each minute. The actual force that the material is subjected to is called the relative centrifugal force (RCF) and this is measured for each centrifuge by using the formula:

$$RCF = 1.118 \times R \times rpm^2 \times 10^{-5}$$

where R = the radius in centimetres measured from the centre of the centrifuge shaft to the tip of the extended centrifuge tube.

Figure 2.26 Centrifuge. Photo: Willy Schauwers, Provincial Institute for Hygiene, Antwerp, Belgium.

Figure 2.27 Different bucket sizes, to accommodate a range of specimen containers/tubes. Photo: Dr Ziay Ghulam, Dr Wahidullah Bahaer, Central Veterinary Diagnostic and Research Laboratory, Kabul, Afghanistan.

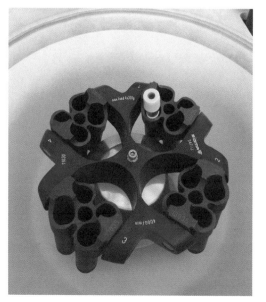

Centrifuges are available in a range of sizes and some have a selection of 'bucket' sizes to accommodate a range of specimen containers/tubes (Figures 2.27 and 2.28). Some models can be expensive and complex, with built-in cooling systems.

There are three types of centrifuges:

- hand-operated centrifuges
- microhaematocrit centrifuges
- general purpose centrifuges.

Figure 2.28 Different bucket sizes. Photo: Dr Ziay Ghulam, Dr Wahidullah Bahaer, Central Veterinary Diagnostic and Research Laboratory, Kabul, Afghanistan.

Hand-operated centrifuge

Great care should be taken when handling hand-operated centrifuges. Keep your distance while operating the centrifuge. They can cause serious injury when it is not attached firmly on a stable support (preferably the edge of the table). This type of centrifuge can be used to examine urinary deposits and to concentrate parasites in stool suspensions. Usually the speed of the centrifuge is too low to separate red blood cells from plasma or serum (Figure 2.29).

Microhaematocrit centrifuge

Haematocrit centrifuges are used to spin haematocrit tubes for haematological tests (see Chapter 5). The microhaematocrit centrifuge (see Figure 2.30) is operated at a standard fixed speed of RCF 12,000 for 5 min.

The microhaematocrit centrifuge helps to diagnose anaemia and is used:

- to determine the PCV value
- to concentrate motile trypanosomes and microfilaria in the buffy coat area.

Figure 2.29 Hand-operated centrifuge. Photo: Dr Ziay Ghulam, Dr Wahidullah Bahaer, Central Veterinary Diagnostic and Research Laboratory, Kabul, Afghanistan.

Figure 2.30 Microhaematocrit centrifuge. Photo: Dr Ziay Ghulam, Dr Wahidullah Bahaer, Central Veterinary Diagnostic and Research Laboratory, Kabul, Afghanistan.

General purpose centrifuge

The general purpose centrifuge is used:

- to centrifuge cells, bacteria and parasites
- to separate the serum/plasma or to wash red blood cells
- to concentrate parasites in a suspension.

A general purpose centrifuge is used with two types of head: the 'fixed angle' head and the 'swing out' head. The 'swing out' head is the one mostly used in the laboratory.

Several types of tube holders are available:

- buckets to hold one round-bottomed tube
- buckets to hold more than one round-bottomed tube
- buckets which can be covered by a lid (screw cap).

Care, maintenance and general operating rules

- Fill the tubes no more than three-quarters full to prevent spillage.
- Check that the tubes are not too large for the centrifuge. Overlong tubes will break during centrifugation.
- Do not use thin-walled glass tubes, they will break easily during centrifugation.
- Close the centrifuge lid before use.
- Nowadays centrifuges with brushless drive induction motors are available (no need to replace carbon brushes).
- If required, keep spare carbon brushes in stock when the centrifuge has a brushed drive motor. If it is difficult to change the brushes, request help from a qualified electrician.
- Check that cushions are in the buckets to reduce the risk of breakages.
- Before spinning tubes, ensure that the samples are balanced. To do this balance them on a two-pan balance (by adding water to the bucket containing the lighter tube) then place

them in the centrifuge diametrically opposite to each other or balance them.

- Check that the centrifuge is on a firm, level base and is steady.
- Ensure that tubes containing potentially pathogenic material are sealed before centrifugation to avoid aerosol forming and spreading.
- Increase the speed of the centrifuge gradually (if mentioned in the instruction manual).
- Wipe the inside of the centrifuge regularly (for example, at least weekly) with disinfectant (70% ethanol).
- If a breakage occurs:

  - switch off the centrifuge
  - wait for minimum 30 min to allow the aerosol to settle
  - remove all debris (wear protective gloves)
  - disinfect the inside of the centrifuge and the buckets (use 70% ethanol).

pH meters

It is critical to know the exact degree of acidity or alkalinity (pH) of reagents in many procedures carried out in a veterinary laboratory, for example, the preparation of media for microbiology and for solutions used in serology and biochemistry. If the pH is not correct the culture or test often will be unsuccessful or invalid.

The principle of pH

pH is a unit of measurement that defines the degree of acidity or alkalinity of a solution. It is measured on a scale of 0 to 14. The pH value is essentially a measure of the hydrogen ion concentration.

The internationally accepted symbol, pH, is derived from 'p', the mathematical symbol of the negative logarithm and 'H', the chemical symbol for hydrogen. The pH value is the negative logarithm of hydrogen ion activity as shown in the mathematical relationship $pH = -\log[H^+]$.

The pH value of a substance is directly related to the ratio of the hydrogen ion $[H^+]$ and the hydroxyl ion $[OH^-]$ concentrations. If the concentration of H^+ is greater than OH^-, the material is acidic and has a pH value of less than 7. Conversely, if the concentration of OH^- is greater than H^+ the material is basic, with a pH value greater than 7. If the concentrations of H^+ and OH^-. are equal the material is neutral with a pH value of 7.

> More H^+ ions less OH^- ions
> Acidic (0 < pH < 7.0)

> Less H^+ ions more OH^- ions
> Alkaline (7.0 < pH < 14)

It can, therefore, be seen that pH is a measurement of both acidity and alkalinity, even though by definition it is a selective measurement of hydrogen ion activity. The logarithmic relationship between hydrogen ion concentration and the pH unit means that a change of one pH unit represents a ten-fold change in hydrogen ion concentration.

pH can be measured by using either pH papers/indicators or a pH meter, dependent on the level of accuracy required. pH papers or indicators change colour as the pH level varies. For greater accuracy, the use of a pH meter is recommended, together with a pH measuring electrode and reference electrode.

PH PAPER STRIP

pH can be measured by using a (paper) strip (see Figure 2.31) impregnated with indicators that show colour variations depending on the pH of the solution. This method is subjective and the colour change is gradual and not very accurate. These can be used as a guide to the pH level, but can be limited in accuracy and dif-

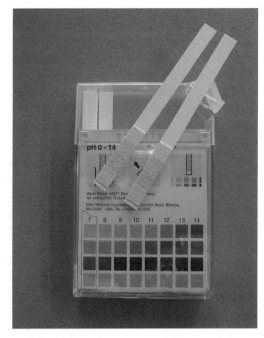

Figure 2.31 pH strip. Photo: Willy Schauwers, Provincial Institute for Hygiene, Antwerp, Belgium.

ficult to interpret correctly in murky or coloured samples.

PH METERS

The use of pH meters (Figure 2.32) is the most accurate method to determine pH but these do require a power supply (battery or mains). A pH meter consists of a glass electrode with a bulb-shaped bottom that is sensitive to hydrogen ions. Inside the glass electrode is a reference electrode immersed in a solution of saturated potassium chloride (KCl). Inside the glass electrode, if the correct conditions are maintained, the H^+ ion concentration remains constant. When the glass electrode is placed in a solution of unknown pH an electrical potential is developed across the glass bulb between the solutions (between reference and unknown solution) and this relates to the difference in pH inside and outside the glass. This electric potential is measured and shown on a scale.

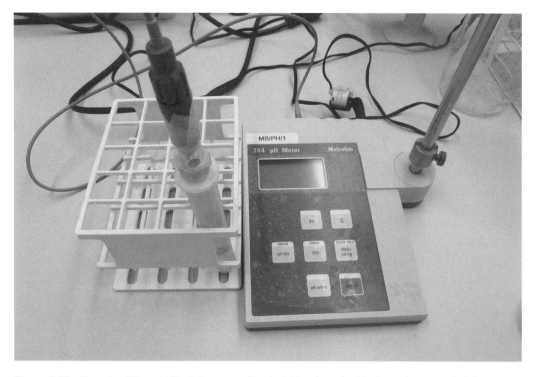

Figure 2.32 pH meter. Photo: Willy Schauwers, Provincial Institute for Hygiene, Antwerp, Belgium.

Each time the pH meter is used (or at least once every day) it should be 'standardized' with a solution of a known pH. Thereafter solutions of unknown pH can be measured.

Two SOPs for pH-meters can be found in 'SOP-LABORATORY-EQUIP-02 pH-mV-Temperature Meter Jenway 3510' and 'SOP-LABORATORY-EQUIP-03 pH-Conductivity-TDS-Salinity-Temperature Meter'.

GUIDELINES FOR USE

Some general rules must be followed to ensure reliable results.

- Take great care of the sensitive and fragile tip (bulb) of the electrode. Rinse it carefully in distilled water before and after it is used and do not handle it roughly.
- Do not wipe the bulb but carefully absorb any hanging drops with tissue paper.
- Take care that the tip does not become covered with a layer of protein or lipids (for example, from media). If this does occur, protein can be removed using 0.1% pepsin or 0.1 N HCl and lipids can be removed using acetone.
- If possible, adjust the pH meter to the temperature of the solution under test.
- Allow the pH meter to warm up for a few minutes and 'settle' before taking a reading.
- When not in use, keep the electrode immersed in liquid.
- Make sure that the solutions used to standardize the pH meter are pure, that is, no turbidity or dirt.
- The date when the standard solution was prepared should be noted so that it can be replaced with fresh solution if necessary (for example, monthly).
- Maintain the correct level of potassium chloride inside the electrode.

- Use similar containers for the standard and the test solutions and standardize the meter at a pH close to that anticipated for the test solution.

Incubators

Incubators (see Figure 2.33) are heated cabinets that are used to maintain a desired constant temperature. The cabinets are heated by electricity passed through heating elements situated around the lagged body of the cabinet and controlled by a thermostat. Most incubators have timing switches. Most have an automatic thermostat and a (digital) thermometer to allow regular checking of temperature (on a daily basis). Some units also have a humidity regulator (which will require the addition of water) or incubators can be designed to create a CO_2 environment (CO_2 gas bottle needed). The incubator should be heated through natural convection. Fan-assisted incubators cause drying out of bacteriology media and are therefore not very suitable for bacteriology (unless the Petri dishes are wrapped up in a plastic bag). The incubator should be fitted with an internal glass door.

In the laboratory, incubators may be used for microbiological culture and for regulating chemical reactions for biochemical tests (ELISA, enzyme-linked immunosorbent assay). The microbiology incubator should be kept for culture use only due to the risk of contamination of other samples.

In countries with a land-climate, summers can be very warm and winters extremely cold. In these countries incubators can be used in wintertime to provide room temperature (for incubating ELISA plates and for clotting of whole blood) and incubators with cooling capacity can be used for bacteriology when the surrounding temperature rises above 37°C.

There are small and inexpensive incubators available for district laboratories with a few specimens requiring culture or where there is a need of portability. The DC-powered Cultura M incubator can be powered from a car cigarette lighter plus the mains for maximum flexibility.

Care and maintenance

- An incubator should be fitted with a digital temperature display or supplied with a thermometer for insertion through the vent hole in the roof of the incubator. Adjust the thermostat until the display or thermometer shows the correct reading, that is, 36°C, for routine incubation of bacteriological cultures.
- Make sure the ventilation holes are not obstructed.
- Do not overload the cabinet. There should be adequate space for air convection around contents.
- Switch off the incubator before disinfection and cleaning.

Figure 2.33 Incubator. Photo: Willy Schauwers, Provincial Institute for Hygiene, Antwerp, Belgium.

- Wipe down the inside of the unit with disinfectant or soapy water as required or between batches of contents.

Quality assurance

Record, on a daily basis, the minimum/maximum temperature of the incubator on the temperature daily control worksheet. For this purpose, use the form 'FORM-SOP-LABORATORY-EQUIP-05–01 Temperature registration min max thermometer'.

Ovens

If the laboratory is fairly large, a small hot-air oven (Figure 2.34) is useful for drying glassware (a drying cabinet is even more suitable for drying glassware) and for dry-heat sterilization of glass or metal laboratory ware. Ovens come in a range of sizes with a variety of shelving options. An oven can be set to a specific temperature and most models have a fan to ensure air circulation. Generally, items should not be added to the oven until the desired temperature has been reached. Let the oven cool down to at least 40°C before removing sterilized items.

Quality assurance

Check the temperature on a regular basis against a certified thermometer. See 'SOP-LABORATORY-EQUIP-05 Temperature registration' and 'FORM-SOP-LABORATORY-EQUIP-05–06-Compare temperature with certified thermometer'.

Balances

The sensitivity of balances commonly used in veterinary diagnostic laboratories ranges from 0.1 g to 0.0001 g with a capacity of 50–2000 g.

A regional laboratory is likely to have two types of balances. In a district laboratory, it might be suitable to have a balance that can operate on battery as well as mains. A double pan balance for counter weighing, for example, for centrifugation, and for more accurate weighing a single pan (Figure 2.35), digital read-out electronic balance. The more sensitive electronic analytical balances (Figure 2.36) are significantly more expensive. Weighing procedures must be carried out with care to obtain the correct weight and to protect the instrument.

Use and maintenance

- The balance should be placed on a solid, level stable surface away from draughts, vibration and direct sunlight.
- Always zero (tare) the balance before commencing weighing.
- Weigh chemicals at room temperature.

Figure 2.34 Oven. Photo: Willy Schauwers, Provincial Institute for Hygiene, Antwerp, Belgium.

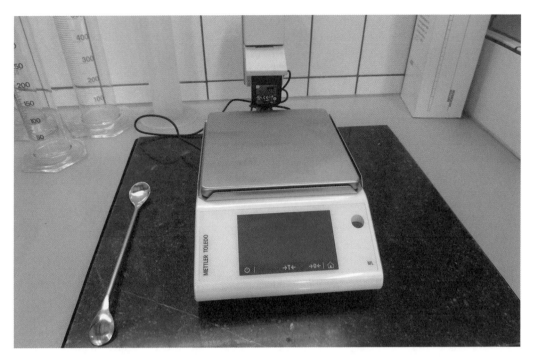

Figure 2.35 Single pan balance. Photo: Willy Schauwers, Provincial Institute for Hygiene, Antwerp, Belgium.

- Never weigh directly onto the pan but use a suitable container (for example, foil caps, weigh boats) or, after taring, weigh directly into the vessel being used to make a solution/preparation.
- When adding or removing chemicals to/from the balance, avoid spilling on the balance pan.
- Clean up any spills immediately.
- Cover the balance when not in use.
- Balances (especially high precision, analytical balances) should stand on a stable (vibration free) table.

Quality assurance

CONTROL WEIGHT

Every laboratory and every balance should have its own control weight(s). The control weight should be measured on a daily basis, preferably in the morning (the control weight(s) should be chosen, based on the weight range of the balance used).

Figure 2.36 Analytical balance. Photo: Willy Schauwers, Provincial Institute for Hygiene, Antwerp, Belgium.

Per control weight a form for daily weight control is available (see 'FORM-SOP-LABORATORY-EQUIP-01–01-Daily control balance').

On this form, the following control weight specifications are mentioned:

- mass of the control weight and mass minimum/maximum limits, as listed by the manufacturer
- make and type.

CERTIFIED CONTROL WEIGHT

The certified control weights should be checked on a 3-yearly basis by a certified laboratory (a certificate should be provided). The certified control weights are usually kept by the quality management department or by the head of the laboratory.

HOW TO PERFORM THE QUALITY ASSURANCE

There should be quality assurance at three levels:

- daily control by the laboratory personnel by measuring the control weights
- half yearly control by the laboratory personnel by measuring the control weight and the certified control weight of every balance
- if possible, a yearly servicing by the manufacturer of the balance.

DAILY CONTROL

Every day (when the balance is used), before the first use in the morning, the control weight should be measured and recorded on the daily control sheet. Do not touch the control weight with bare hands, always use gloves or forceps to handle it. Per balance and per control weight accepted weight limit values are set (according to the manufacturer of the control weight or on the basis of weight results gathered over a long period and the standard deviation on these results), see 'FORM-SOP-LABORATORY-

EQUIP-01–01 Daily control balance'. Indicate for each balance which control weight should be used (see 'FORM-SOP-LABORATORY-EQUIP-01–04-Balances and control weights'). Keep all information in the logbook.

If the accepted limit values are exceeded, take the following steps.

- Check if the balance is completely level and if all weight measurements are properly performed.
- If there is still a problem, perform an internal calibration of the balance (not all balances do have this feature).
- If still a problem, contact the supervisor and check the balance with a certified control weight.
- If still a problem, contact the manufacturer.
- If the limit values are still exceeded, the possible consequences on earlier weight measurements should be considered. All measurements after the last successful control weight measurement and before the exceeded measurement should be traced. Possible consequences on those laboratory results should be examined and rectified.
- The investigation of the problem and all corrective actions are registered on 'FORM-SOP-LABORATORY-EQUIP-01–02-Corrective actions'.

HALF YEARLY CONTROL

The balance is controlled half yearly by measuring the control weight as well as the certified control weight. Both weights are measured five times. The average difference between both weights has to be within a certain limit (see 'FORM-SOP-LABORATORY-EQUIP-01–05-Half yearly control of control weights'). This limit value is determined by each laboratory depending on the use of the balance. Balances used for preparing standard solutions have to be more accurate than balances used for weighing powders for bacteriology media preparation.

Always indicate (on a label) the due date of the half yearly control on the balance.

If the limit value is exceeded, take the following steps.

- Check if the balance is completely level and if all weight measurements are properly performed.
- If there is still a problem, use a different certified control weight.
- If still a problem, contact the manufacturer of the certified control weights.
- If the limit values are exceeded, the possible consequences on earlier weight measurements should be considered. All measurements after the last successful reference control weight measurement and before the exceeded measurement should be traced. Possible consequences on those laboratory results should be examined and rectified.
- The investigation of the problem and all corrective actions are registered on 'FORM-SOP-LABORATORY-EQUIP-01–02-Corrective actions'.

YEARLY SERVICING

If possible, the laboratory should have a contract for a yearly servicing on each balance. The maintenance contracts are stored with the quality manager. The maintenance company should be accredited and should have an ISO 9001 certificate.

During the yearly maintenance, the following tasks are performed:

- thorough cleaning of the balance
- calibrating the balance with certified calibration weights
- checking the accuracy, precision, linearity and eccentric weighing.

After the maintenance, a calibration report and certificate are delivered.

The yearly maintenance is registered on the form 'FORM-SOP-LABORATORY-EQUIP-01–03-Balance maintenance by maintenance company'. It is good practice to label each balance with the date of the yearly maintenance.

Water baths

Water baths (Figure 2.37) are used to maintain the temperature of liquids held in laboratory containers. The temperature range used for most purposes is 37°C to 60°C. There is a wide range of

Figure 2.37 Water bath.
Photo: Willy Schauwers, Provincial Institute for Hygiene, Antwerp, Belgium.

baths available, some made of plastic or glass (to allow observation). Most are made of a corrosion resistant metal, heated by an element and thermostatically controlled. A propeller may be used to distribute heated water evenly around the bath to maintain a set temperature. A thermometer (digital or analogue) is usually supplied with the bath. The temperature of a digital read-out water bath or thermometer should be recorded daily (see 'FORM-SOP-LABORATORY-EQUIP-05–02 Temperature registration thermometer'). Rotating or shaking water baths may also be used for some laboratory procedures but these are usually more expensive. Simple shelves or storage packs can be purchased for holding specimens in most types of water bath.

Use and maintenance

- Do not use a lid when open vessels are in the bath because condensate drips may contaminate the vessel contents.
- Water and electricity together are dangerous so regularly check electrical connections for signs of wear.
- Make sure that the water has reached the required temperature before starting incubation and that the water level in the bath is over the level of the incubating liquids.
- The water in the bath should be deionized, distilled, boiled or rain water and the recommended level maintained.
- Clean the water bath regularly, do not damage the heating element. If scale is present: add some concentrated acetic acid, heat up the water bath and leave the acid solution for a few hours until the scale is dissolved.

Water distillers and deionizers

Good quality water is essential in a laboratory for many tasks including preparation of reagents, solutions and media and for rinsing clean glassware. Good quality water contains little or no dissolved salts and gases, and it should have low electrical conductivity. It is essential that there is always a supply of stored pure water available. For laboratory use, water is usually stored in 5 or 10 l containers made of good quality glass or polyethylene with screw cap and stopcock.

Water should not be stored for too long before use and the containers should be cleaned regularly and checked visually for the presence of contaminants. Good quality water can be produced using either deionizers or distillers and maybe a combination of both (Figures 2.38 and 2.39).

Distillers

All distillers (stills) work on the principle that impure water is boiled and the resultant steam is passed over a cold surface where condensate is collected. This condensate is pure, or almost pure, water. The cold surface is maintained by running water and a distiller can use a lot of tap water in a working day. It is therefore important that this residual water is utilized either for washing glassware, since it is warm, or that it is diverted to a storage tank.

Stills are basically of two types, the long established 'Manesty' still,[6] which is made primarily of stainless steel (wall mounted), and glass distillers (like GFL water stills) which are usually bench based models. The SOP for the 'Manesty' distiller can be found in 'SOP-LABORATORY-EQUIP-08 Distiller'.

There are advantages and disadvantages to each type of still. When deciding which model is the right one, it is necessary to consider the relative robustness of the models, the price, running costs, and the quality and quantity of water produced. It is important that the model selected has automatic electricity cut out switch in case of overheating. This may occur as a result of power fluctuations or water supply failure. The cooling water supply often fluctuates so the tap should have two controls, one directly from the

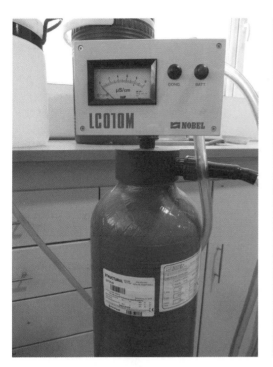

Figure 2.38 Deionizer. Photo: Dr Ziay Ghulam, Dr Wahidullah Bahaer, Central Veterinary Diagnostic and Research Laboratory, Kabul, Afghanistan.

Figure 2.39 Manesty water distillation apparatus. Photo: Dr Ziay Ghulam, Dr Wahidullah Bahaer, Central Veterinary Diagnostic and Research Laboratory, Kabul, Afghanistan.

water supply, which remains fully open (thereby providing at least the required minimum pressure), and another to further control water flow to the still. Cooling water should be adjusted so that residual water has a temperature of about 70°C. This way, the amount of cooling water used is minimal.

USE, CARE AND MAINTENANCE:
- Make sure there is enough supply of cool running water to feed the still.
- Check the boiling chamber regularly for build-up of scale.
- Cleaning and descaling: the day before the cleaning and descaling is planned one has to pour about 60 ml of concentrated acetic acid in the (warmed up) boiling chamber of about 5 l.

- After cleaning/descaling/rinsing, switch on the still and rinse it with the first 5 l of distilled water production. Then, the pH value of the distilled water should be around pH 6.0.

Deionizers

Deionized water is produced by using a unit containing a cation – anion exchange resin in a cartridge which can be refilled or exchanged. A built-in mains or battery-operated conductivity meter monitors the water quality and will indicate when resin cartridges should be refilled/replaced/regenerated. Sometimes charcoal is included, which removes contaminants other than mineral salts. Water from ion exchange is not sterile and may contain pyrogens. Deionizers have the advantage that large quantities of cooling water are not required.

If it is necessary to provide ultra-pure water (as in some central and reference facilities or research establishments), then water can be double distilled or, deionized first and then distilled. In laboratories with poor electricity supply, a deionizer is a good alternative for a distilling apparatus.

Automatic mechanical pipettes and dispensers

There are two types of automatic mechanical pipettes (Figures 2.40 and 2.41):

- air-displacement automatic mechanical pipettes:

  - recommended for aqueous samples and for general laboratory work
  - always have a cushion of air (dead volume) between the pipette piston and the liquid sample
  - the piston is a permanent part of the pipette

- positive displacement automatic mechanical pipettes:

  - recommended for problem samples (viscous, dense, volatile, radioactive, corrosive)

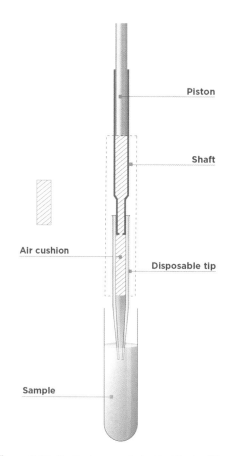

Figure 2.40 Air-displacement pipette. Photo: Gilson, Den Haag, courtesy of Gilson International BV.

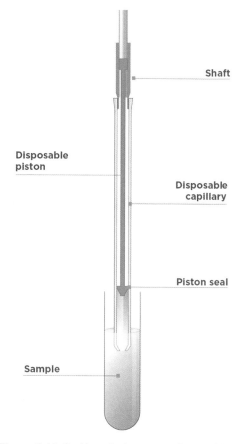

Figure 2.41 Positive displacement pipette. Photo: Gilson, Den Haag, courtesy of Gilson International BV.

- direct contact of the piston with the sample (no air cushion)
- disposable piston (not a permanent part of the pipette).

Automatic mechanical air-displacement pipettes

Methodologies in modern laboratories, in particular in biochemistry and serology, demand precise pipetting of volumes down to a few microliter (μl). In district laboratories, it is common to use glass graduated pipettes for volumes over 1 ml and for smaller volumes a micropipettor. The mechanically operated micropippettors use disposable plastic tips to hold liquids. In district facilities, the most commonly used automatic pipette is the hand-held mechanical plunger type (air-displacement). These pipettes work by means of a mechanical piston moving up and down on manual pressure that displaces a precise volume of air which, in turn, displaces a known volume of liquid. These 'displacement' pipettors can be used for a fixed volume ranging from 10 to 1000 μl or can be manually adjustable between set volumes, for example 5 to 50 μl, 50 to 250 μl and 200 to1000 μl. These pipettes are (usually available as single units or multi-channel units of four, eight or twelve) designed for work with microtiter plates. Manufacturers recommend that tips are not re-used. When not in use pipettes are stored upright in a stand (Figure 2.42).

It is important to follow the manufacturer's directions otherwise large errors occur, especially when dispensing small volumes.

HOW TO USE AN 'AIR-DISPLACEMENT' MICROPIPETTOR

For aqueous solutions use forward pipetting.

- Put the tip firmly on the pipettor.
- Press the plunger of the pipettor to the first stop.

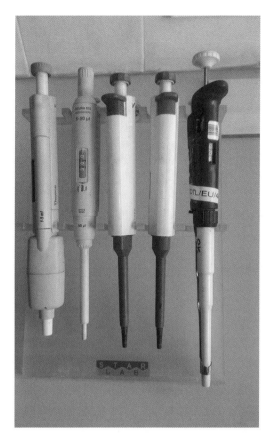

Figure 2.42 Micropipette. Photo: Dr Ziay Ghulam, Dr Wahidullah Bahaer, Central Veterinary Diagnostic and Research Laboratory, Kabul, Afghanistan.

- With the pipettor vertical and the tip 1–2 mm below the surface of the liquid, release the plunger smoothly and slowly to aspirate the fluid. Wipe the outside of the tip, do not touch the open end of the tip.
- To dispense, touch the tip at an angle against the side of the receiving vessel, press to the first stop and wait a few seconds and press the plunger to the second stop to expel all the fluid.

For high viscosity fluids (serum, plasma) and solvents, use reverse pipetting (or use positive displacement pipettes).

- Put the tip firmly on the pipettor.
- Press the plunger of the pipettor to the second stop.
- With the pipettor vertical and the tip 1–2 mm below the surface of the liquid, release the plunger smoothly and slowly to aspirate the fluid. Wipe the outside of the tip, do not touch the open end of the tip.
- To dispense, touch the tip at an angle against the side of the receiving vessel, press to the first stop and wait a few seconds. Do not press to the second stop (the liquid in the tip should be discarded).

Micropipettor tips can be packaged in bulk or in tip-boxes (racks, no hand contact with the tip needed). Tips can be sterile, non-sterile and with or without filter.

CARE, USE AND MAINTENANCE
- First organize your work station for maximum efficiency and minimum fatigue.
- Make sure the tip is properly mounted and fits well before you set the volume.
- Adjust the volume.
- Choose the mode of pipetting (reverse or forward mode) adapted to your sample.
- Eject the used tip and store the pipette in an upright position to avoid damage and cross-contamination.
- Do not leave your pipette lying on the workbench where it can come into contact with chemicals or fall off and break. Always store your pipette vertically to prevent liquids from running inside the shaft of the pipette.
- Use the right tip for the pipette, there is a very wide range of tips.
- Do not re-use tips.

QUALITY ASSURANCE
- Air-displacement pipette leakages can be tested using the BRAND PLT unit (Pipette Leak Testing unit).[7] The most frequent cause of inaccuracy in air displacement pipettes

is leakage. The leakages arise from damage either to the seals, pistons or tip cones. Often not detectable by the naked eye, leaks lead to significant volume errors. Air-displacement pipettes must be checked at regular intervals and the results must be compared with the error limits. However, a calibration certificate only reflects the results at the time of testing. The time between these calibrations is crucial, since leaks can occur at any time. While the PLT unit cannot replace regular gravimetric testing, daily pipette checks can provide a safeguard during the periods between calibrations. Even the smallest leaks are detected! Process reliability for the pipettes is thus significantly improved.
- Every 6 months the micropipettors should be checked for leakage, accuracy and precision. If leakage, accuracy and precision are within limits, the micropipettors can be used for another 6 months, if the micropipettor fails some parts have to be replaced and it has to be recalibrated. Some micropipettors can be re-calibrated; others have to be sent back to the manufacturer. Testing micropipettors requires the use of an analytical balance (sensitivity 0.0001 g, usually not available in a district laboratory) and some specialized training. See:

- 'SOP-LABORATORY-EQUIP-07 Quality assurance of mechanical pipettes'
- 'FORM-SOP-LABORATORY-EQUIP-07–01 Pipette Calibration Worksheet'
- 'FORM-SOP-LABORATORY-EQUIP-07–02 Gravimetric pipette test worksheet for Pipetman P-1000'.

Dispensers

Bottle top dispensers (Figure 2.43) are widely available for rapid repetitive dispensing of reagents. Some dispensers can be autoclaved. Dispensers are available for different volume

Figure 2.43 Dispenser. Photo: Willy Schauwers, Provincial Institute for Hygiene, Antwerp.

ranges and are supplied with a range of adaptors to fit commonly used reagent bottles. The dispensers must be used in such a way that no air-bubbles are dispensed while adding the reagent.

Follow the same quality assurance as for mechanical automatic pipettes.

Colorimetry equipment and spectrophotometer

Colorimeter and spectrophotometer are names given to instruments that produce and then measure light which passes through or is absorbed by a coloured solution representing the sample. Spectrophotometers use a diffraction grating prism to produce monochromatic light to pass through the sample whereas colorimeters use filters and the light that passes through the

specimen is polychromatic light but within a narrow range (for example, an Ilford blue filter has wavelength range of 450–480 nm). Most regional laboratories will have a simple spectrophotometer, which can operate at wavelengths in the visible spectrum between, for example, 345 and 750 nm (see Chapter 7 and Figure 2.44).

The majority of quantitative methods used in clinical chemistry result in a coloured product, which can be measured. In most cases the intensity of colour is directly related to the amount of substance in a solution, that is, the more intense the colour the higher the concentration of the substance in the sample. Measurements are made on either (1) the amount of light absorbed by the sample (absorbance) or (2) the amount of light passing through the sample (transmittance). The sample (coloured solution) is usually held in a cuvette. Cuvettes are small volume receptacles made of special glass or re-useable or disposable plastic. Most manufacturers guarantee that each batch is matched for consistency and precision. The manufacturer will provide details on how to use the spectrophotometer/colorimeter and the methodologies

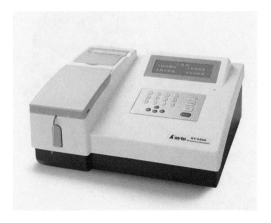

Figure 2.44 Semi-automatic photometer, Rayto RT-9200 (wavelengths in the visible spectrum between 330–800 nm, five standard filters, three optional filters, flow cell. Photo: Tom Gernaey, Tactilab RDC, Brussels, Belgium.

used for testing samples, as well as how to calibrate the machine for the calculation of results. All tests using the spectrophotometer or colorimeter should include positive and negative (blank) controls and the equipment should be calibrated prior to each batch of tests. Standard solutions are available for most procedures.

If a laboratory only needs a colorimeter or spectrophotometer to measure the haemoglobin content in blood, it might be better to purchase a dedicated haemoglobinometer. The EKF Diagnostics' DiaSpect Tm haemoglobinometer is one of the fastest haemoglobin analysers. DiaSpect Tm delivers laboratory quality results in just about one second of the microcuvette being placed into the analyser (no reagents needed) (Figure 2.45).

A few things to remember when using colorimeters and spectrophotometers.

- Cuvettes have opaque sides by which they should be held and clear sides through which light is directed.

Figure 2.45 Diaspect Tm handheld haemoglobinometer. Photo: Tom Gernaey, Tactilab RDC, Brussels, Belgium.

- The clear sides should not be allowed to become scratched or dirty.
- The clear side of the cuvette should be dry and free from finger marks.
- Before reading the absorbance or transmittance of a coloured solution, check if the solution is clear, there are no air bubbles in it, and that it is at room temperature to avoid condensation on the outside of the cuvette.
- Use a standard and blank with each test.
- Always keep spare lamps in stock.
- Protect the instrument with a dust cover.

Microscopes

The microscope is the most used tool in a laboratory therefore it is important that it is used properly. The cost of a microscope depends on a number of factors, but in most cases it is worth paying a little bit more to get the best quality optics and resolution possible.

A standard microscope consists of:

- a light source (usually an electric bulb, but 'field' microscopes use a mirror to reflect sunlight)
- a condenser to focus light on the specimen
- a stage with a bracket to move the specimen
- eyepieces through which the specimen is viewed and which contribute to the magnification of the specimen (usually 10×)
- objectives through which light is passed to the eyepieces and which also contribute to the magnification of the specimen (usually 10×, 20×, 40× and 100× magnification)
- a course and fine adjustment system to move the specimen up and down.

The compound light microscope

A compound microscope has a system of lenses. The lens system nearest the eyes is in the ocular (eyepiece). Another lens system is in the objective, which is nearest the object being viewed. See 'SOP-LABORATORY-EQUIP-09 Microscopes' and 'Human Parasitology in Tropical Settings Practical Notes ITG 171025' available on the website.

A microscope can be monocular but most are binocular with two eye pieces that can be focused separately to suit the vision of the viewer.

Parts of the microscope (Figures 2.46 and 2.47)

EYEPIECES

The eyepieces, located at the top of the microscope, contain magnifying lenses. The usual magnification is 10×, but eyepiece oculars of 15× and 20× magnification are also available.

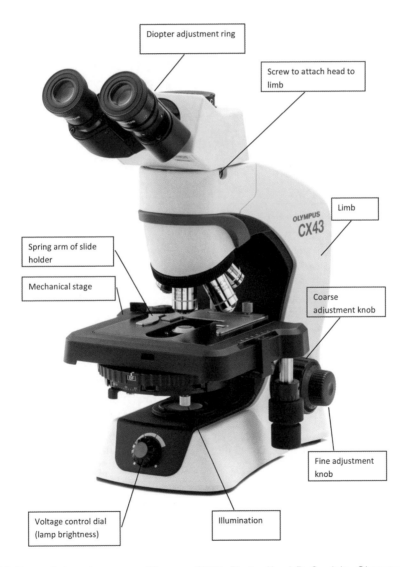

Figure 2.46 Parts of the microscope, Olympus CX43. Photo: Karel DuGardein, Olympus, Berchem, Belgium.

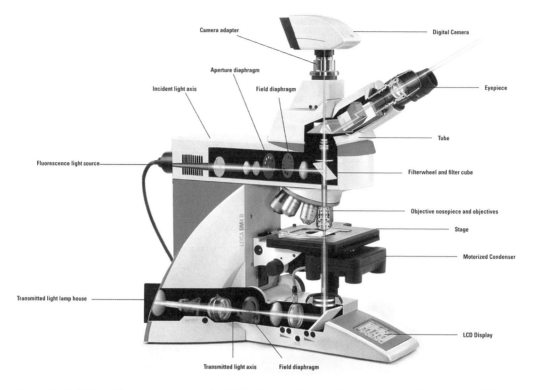

Figure 2.47 Parts of the microscope, Leica DM4. Photo: Manuela Jacobsen, Leica Biosystems, Wetzlar, courtesy of Leica Microsystems CMS GmbH, Wetzlar, Germany.

OBJECTIVES

The underside of the arm of the microscope contains a revolving 'turret' to which the objective lenses are attached. Most microscopes have four objective or magnifying lenses:

- the lower power objective, which magnifies 10×
- the high-power objective, which magnifies 40×
- the oil immersion objective, which magnifies 100×
- additional objectives (4×, 20×) can also be purchased.

Each objective is marked with colour-coded bands and the power of magnification.

To determine the degree of magnification in use, multiply the magnification mentioned on the eyepiece (usually 10×) by the magnification listed on the objective being used. For example, an object viewed using a 10× ocular and high power (40×) objective would be magnified 400 times. An object viewed with a 10× ocular and the oil immersion objective (100×) would be magnified 1000 times.

For the 'oil immersion objective' (100×) it is necessary to use special microscope oil. It is important not to get this oil on the other lenses as it will distort the image.

CONDENSER AND DIAPHRAGM

The arm of the microscope connects the objectives and eyepiece(s) to the microscope base, which supports the microscope. The base also contains the light source. The light source beam has a movable condenser and iris diaphragm located above it. The condenser focuses or directs the available light onto the objective as it is raised or lowered. The iris diaphragm located in the condenser unit regulates the amount of light that strikes the object being viewed (much like a shutter of a camera). The iris diaphragm is adjusted by a movable lever.

Currently, long-lasting LED-light-bulbs are used as a light source for microscopes. Quite often these LED-light-bulbs are not replaceable.

Coarse and fine adjustments

The course adjustment should be used to focus the low power objective only. The fine adjustment is required at higher power to provide a sharper image after the object is brought into view with the coarse adjustments at low power magnification. The working distance is the distance between the objective and the slide when the object is in sharp focus. The higher the magnification of the object, the shorter the working distance will be.

Stage

The stage of the microscope is supported by the arm and is located between the objectives/turret and the light source/condenser. The stage serves as the support for the specimen being viewed (usually a prepared microscope slide). Specimen slides can be moved by using knobs located just below the stage. These move the specimen slide left and right or backward and forward.

When the microscope is not being used, the stage should be centred so that it does not project from either side of the microscope. The microscope should be covered with a dust cover when not in use (preferably a cotton cover, plastic can cause condensation).

Using the light microscope

The eyepieces on binocular microscopes must be adjusted for each individual's eyes. The distance between the eyepieces is adjusted (as when using binoculars) until one image is seen. The object is then brought into sharp focus with the right eye. The right eye is then closed and the left eyepiece dioptric adjustment is used to bring the object into sharp focus while viewing through the left eye.

The low power objective is used with the coarse adjustment for initially locating specimens or slides and for viewing objects.

The high power (40×) is used when greater magnification is needed in procedures such as cell counts and viewing urine sediments. After focusing with the low power objective, the ×40 objective is rotated into position and then fine adjustment is used to bring the object into focus.

The oil immersion objective (100×) is used to view stained blood smears, histological sections and stained preparations of microorganisms. After initial focusing at low power magnification, a drop of immersion oil is placed on the slide. The oil immersion objective is then rotated into the drop of oil, taking care not to allow any of the other objectives to get into contact with the oil. At the end of the day, all oil should be cleaned from the objective preferably using lens paper.

Basic adjustments of the microscope

The condenser and diaphragm must be adjusted according to the objective being used and the type of specimen being observed. When viewing objects with the oil immersion lens (stained preparations), the condenser should be raised until it is almost touching the slide. The diaphragm should be completely open to give maximum light. When looking at objects with low power (when viewing unstained fluids, such as urine sediments or cell dilutions for counting) the condenser may need to be lowered to reduce the brightness of the light and increase the contrast between the constituents being viewed and the background.

Table 2.3 Basic adjustments of the microscope.

Objective	Condenser	Diaphragm	Immersion oil	Mirror	Optical limit and purpose
10×	Lowered	Open/closed	No	Flat	30–500 µm Search of eggs and larvae Search of Ciliata (Entomology)
40×	Halfway	Open/closed	No	Flat	5–30 µm Details of eggs and larvae Search of certain protozoans by means of direct wet smear (Counting of WBC and RBC)
100×	At the top	Open	Yes	Concave	0.25–5 µm Permanently stained smears

The basic adjustments are listed in Table 2.3. If the light source is a mirror, it is mentioned which side (flat or concave) should be used.

Precautions

- Use the coarse adjustment only with the low power objective.
- Use oil each time the oil immersion lens is used.
- Use immersion oil with the oil immersion objective only.
- Clean all oculars and objectives with lens paper after each use.
- Do not 'rack down' on a specimen slide, always 'rack up'.[8]
- Move or transport the microscope with one hand under the base and the other hand gripping the arm. For longer distance transport, it is advisable to place the microscope in its box (well-immobilized and padded), always avoid shocks.
- Lubricate the microscope with the correct lubricant and always clean it after use.
- Store the microscope covered and in a protected area.
- Remove the immersion oil from the immersion-lenses on a daily basis by wiping it off with a non-fluffy, soft and clean tissue or

with soft toilet-paper. Dried oil will cause the microscopic image to be foggy.

- If dried oil is sticking to the immersion-lenses, it can be removed, using a clean tissue, moistened with a mix of ether and alcohol (1/1 v/v). This way of cleaning should be kept to a minimum and should never be used on internal lenses and prisms. For this reason, one should always carefully check the manual of the microscope for suitable solvents. After cleaning, the lenses should be dried immediately with another clean and dry tissue or soft toilet paper.

- Other lenses should be cleaned with another tissue, which is not used for the immersion-lenses, as any contact between oil and non-immersion-lenses should be avoided. Also, plastic lenses exist, which are not resistant to organic solvents.
- Never leave the tube of the microscope open (that is, without eyepieces) as this can cause dust and fungal spores to enter the microscope.
- After every use, the microscope should preferably be covered with a cotton dust-sheet. Avoid plastic covers, as these will promote

fungal growth, especially in moist circumstances.

- Ideally, microscopes should be housed in a suitable acclimatized room. This is because of the risk of condensation forming within the optics of the microscope if temperatures fluctuate significantly. Local air humidity can be diminished by placing an electrical dehumidifier, or even more simply, by placing dried silica gel crystals, near the microscope if it is stored in a closed cupboard.
- Always make sure that some spare light bulbs for the microscope are in stock and keep the reference numbers of all parts and accessories for replacing them if necessary.
- Use a small soft brush to clean less accessible places.
- Do not mix immersion oils.
- Decontaminate the stage with 70% alcohol.
- Turning off: turn the lamp to its lowest setting, switch off.

The microscopic field

The microscopic field is the circular image one sees at a certain magnification.

The microscopic field is divided according to the plate of a clock. Doing so, we can locate any object in this microscopic field, starting from the centre (see Figure 2.48):

- between one and two o'clock we can see a thrombocyte between two red blood cells
- the object near two o'clock is a neutrophil
- the object between four and five o'clock is a monocyte
- between the centre and ten o'clock we can see a lymphocyte
- at the edge, near eight o'clock, we can see part of a lymphocyte.

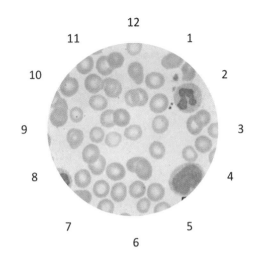

Figure 2.48 The microscopic field is divided according to the plate of a clock. Photo: Idzi Potters, ITG, Antwerp, Belgium.

Systematic examination

The search for eggs and larvae of helminths and of ciliates is performed with the 10× objective. The complete preparation is examined, leaving no parts missed out. To accomplish this, one should work systematically (Figure 2.49). Always start at a corner of the coverslip, for example, the upper-left corner and proceed by looking at the next microscopic field, with a small overlap. This means that each time when a field has been examined, an object in this field, for example, a crystal, at three o'clock is chosen, and is brought towards nine o'clock. This second field is examined and so on. In this way, one should go in a straight line from the upper-left corner towards the upper-right corner from the coverslip. Once we arrive there, we choose an object at six o'clock and move it towards twelve o'clock. This results in arriving at the row below the one that has just been examined (again with a small overlap). This time we work from right to left. This way the complete preparation should be examined within the edges of the coverslip, until arriving at the lower-right corner.

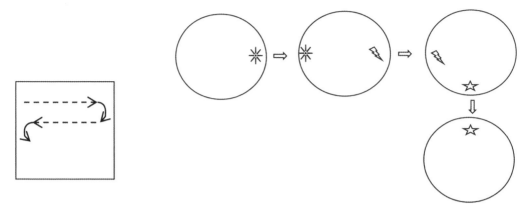

Figure 2.49 Systematic examination of the microscopic field. Photo: Idzi Potters, ITG, Antwerp, Belgium.

For searching some of the protozoans the 40× objective is used. In the same way as described above, a few overlapping rows (3 or 4) should be examined. A partial systematic examination is used for permanently stained smears as well. Also here, the examined part should consist of overlapping rows.

Immersion objectives

Rays of light that pass the optical system of the microscope pass through different kinds of materials (glass, air). Each one of these materials has a certain refracting capacity or refractional index n. When light passes through these materials (with different refracting indices), the rays of light will be bent and sometimes lost (when they are bent so strongly they leave the optical system of the microscope). The level and way of bending depend on the refracting capacity of the environment and on the order of the transition.

When light passes from an environment with low refractional index to an environment with higher refractional index, it will be bent to the normal NN' of the optical system (right half of the scheme) (Figure 2.50). When light passes from an environment with high refractional index to an environment with lower refractional

index, it will be bent, away from the normal NN' of the optical system (left half of the scheme).

High-magnification objectives (50×, 100×) need large amounts of light. To avoid loss of light by refraction, immersion oil with a sufficiently high refractional index is placed between the preparation and the objective.

Some examples of different refractional indices:

Air:	$n = 1.000$
Water:	$n = 1.330$
Normal glass:	$n = 1.500$
Immersion oil:	$n = 1.515$
Canada balm:	$n = \pm 1.5$.

Extra topics available on the website:

- calibration of the microscope
- Köhler illumination
- dark field illumination
- phase contrast microscopy
- fluorescent microscopes.

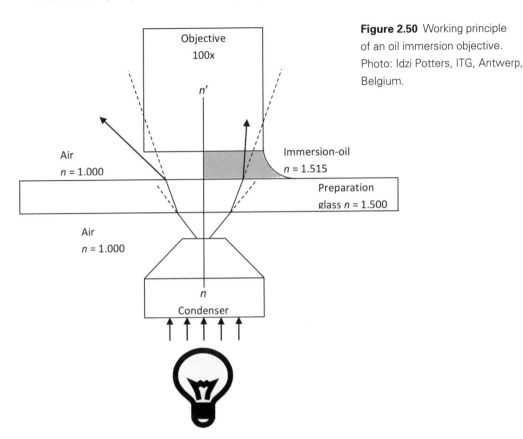

Figure 2.50 Working principle of an oil immersion objective. Photo: Idzi Potters, ITG, Antwerp, Belgium.

2.5 Specialized technology

ELISA

The technology associated with the ELISA (enzyme linked immunosorbent assay) has made diagnostic serology and antigen capture tests much more accessible for smaller veterinary laboratories. However, to perform a lot of tests it is necessary to use a semi-automated ELISA system. This may include a multiplate system, a photoelectric reader, an automated washer, shaker and specialist incubators. Many ELISA test-kits allow manual processing, but for multiple sampling, a reliable plate reader which can be set to the wavelength required for the test, is needed. The widespread availability of enzyme-linked reagents and suitable colour-linked enzyme systems has resulted in a wide

range of diagnostic tests-kits for veterinary disease diagnosis at the district and regional level.

The ELISA plate reader is a rather expensive but robust piece of equipment, which provides the opportunity to expand the technology significantly. It is basically a spectrophotometer that is able to rapidly read and record colour or turbidity in 96 wells in a microtiter plate (Figure 2.51 and Figure 2.52). The cheaper models of plate readers use filters to select a wavelength (each wavelength needs a specific filter), the more expensive readers use a monochromator (can select a wavelength from 200 to 999 nm) (Figure 2.53).

It is worth noting that dual-wavelength measurements can compensate for optical interference. Dual wavelength is used in many microplate-based applications to reduce optical interference caused by scratches, fingerprints or

Figure 2.51 ELISA reader using filters. Photo: Wouter De Sadeleir, BioSPX, Drogenbos, Belgium.

Figure 2.52 ELISA plate washing station. Photo: Wouter De Sadeleir, BioSPX, Drogenbos, Belgium.

other matter that absorb light equally at both wavelengths. For example, many investigators prefer to read microplate-based assays with lids or membrane seals in place to reduce biohazards, as well as evaporation. As a result of using lids, condensation may collect on the lid during the assay process. Experiments demonstrate

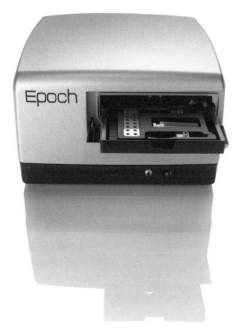

Figure 2.53 ELISA reader with a monochromator (filters no longer needed). Photo: Wouter De Sadeleir, BioSPX, Drogenbos, Belgium.

that when the plate is read at two wavelengths and the difference in optical densities is computed, this technique adequately compensates for these effects.

Polymerase chain reaction (PCR)

PCR is now a well-established and commonly used technology in reference and central laboratories. However, many rural centres may not be set up to perform PCR tests and technical staff would need specialized training to be able to interpret the results.

PCR is essentially an in vitro DNA amplification procedure. The normal PCR cycle involves a system that uses heat to denature the DNA and the use of primers which amplify fragments of the DNA segment in the presence of a thermostable polymerase enzyme (See Figure 2.54). The reaction is carried out using an excess of primers and polymerase enzyme and undergoes several reaction cycles. The equipment used is automated but due to the fact that the samples must not be contaminated by foreign DNA there needs to be a designated 'clean' room for the equipment to be set up and specialized training is necessary for staff required to perform PCR tests. The PCR is extremely sensitive and only requires very small amounts of clinical sample. There are currently a wide range of test-kits

Figure 2.54 RT PCR cycler, safety cabinet and no-break-system. Photo: Dr Ziay Ghulam, Dr Wahidullah Bahaer, Central Veterinary Diagnostic and Research Laboratory, Kabul, Afghanistan.

available for veterinary diagnosis at specialist centres and these may become more widely available in the future. The molecular techniques most commonly used in bacteriology and virology are discussed further in Chapter 4.

Automated haematology and biochemistry systems

Automated systems for testing blood samples (Figures 2.55 and 2.56) are popular for larger diagnostic laboratories where a large number of samples are processed daily. The Sysmex pocH-100iV Diff haematology analyser is designed for laboratories testing up to 25 samples per day. These are also mentioned briefly in Chapters 5 and 7.

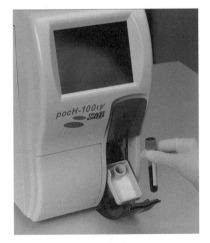

Figure 2.55 Sysmex pocH-100iV Diff haematology analyser. Photo: Jan Volkaert, Sysmex-Cyscope, Hoeilaart, Belgium, © Sysmex.

Figure 2.56 Sysmex pocH-100iV Diff screen. Photo: Jan Volkaert, Sysmex-Cyscope, Hoeilaart, Belgium, © Sysmex.

Most modern systems rely on flow cytometry, where particles (for example, blood cells) scatter the light directed at them as they pass through a flow chamber. Flow cytometers use the principle of hydrodynamic focusing to present cells to a laser (or any other light excitation source). The sample is injected into the centre of a flow chamber so that the stream of particles in the sample is very narrow allowing each particle to contact the laser individually. As the cells intercept the light source they scatter light and fluorochromes are excited to a higher energy state. This energy is released as a photon of light with specific properties (depending on the fluorescent dye used) and this can be recorded. Unlike spectrophotometry, which measures the percent absorption and transmission of specific wavelengths of light for the whole sample, flow cytometry measures the fluorescence per cell (or particle). Cytometry is a useful research tool and is widely used in more specialized veterinary diagnostic laboratories.

2.6 General laboratory ware

All laboratory procedures require the use of some type of container, often glassware (Figure 2.57). Basic laboratory glassware comes in many shapes and sizes and is used in specific procedures or preparation of reagents and solutions used in laboratory analysis. In recent years, it has become common to use laboratory containers made from plastic as well as glass. Because of this, the term laboratory ware or 'labware' is sometimes used to include glassware and plastic ware. Much of today's labware is designed to be disposable which eliminates the possibility of contaminating reagents as a result of inadequate cleaning. However, in some parts of the world, it is too expensive to use disposable plastic ware and a washing and sterilization service is required for re-using 'labware' (also ecological reasons might apply).

Glassware bottles and volumetric flasks are available with exterior amber coating, this feature is particularly useful when handling reagents that are light-sensitive. Some glassware bottles are also available with plastic coating to provide protection from mechanical impact and to help reduce leakage of the contents should the glass break. The maximum working temperature for these bottles is 135°C but long-term exposure (> 30 min) should be avoided. Avoid exposure to direct heat from a hotplate or a Bunsen flame.

Glass containers in laboratories are of two basic compositions.

Flint glass:

- flint glass has a low resistance to heat and chemicals but it is inexpensive.

Borosilicate glass:

- usually has a high thermal resistance and does not react with most chemicals, Pyrex is a brand of borosilicate glass commonly used for beakers, flasks and other labware
- contains a higher percentage of SiO_2 than flint glass – glass with a high percentage weight (> 80%) of silica (SiO_2) is less likely to be attacked by acids (with the exception of

Figure 2.57 Standard glassware store containing flasks and glass bottles. It is important to have a good system for washing and drying glassware to make sure that there is always sufficient glassware available for use. Lack of glassware and /or staff can hold up work and result in poor productivity. If funds are available it is useful to have glass fronted cases to store glassware so that it is easy to see what is available for use. Photo: Dr Ziay Ghulam, Dr Wahidullah Bahaer, Central Veterinary Diagnostic and Research Laboratory, Kabul, Afghanistan.

hot concentrated phosphoric acid and hydrofluoric acid). Glass is separated into four acid resistance classes and borosilicate glass corresponds to Class 1

- alkaline solutions attack all glasses and borosilicate glass can be classified as moderately resistant. The alkali resistance of borosilicate glass meets Class 2 requirements.

Plastic containers are useful in the laboratory because they are impact resistant and therefore less likely to break. Unlike some types of glass, plastics do not release ions. However, they may bind and release (leach) solutes. Plastics are unaffected by most aqueous solutions. Table 2.4 presents details of plastic ware's chemical resistance and possibility to autoclave.

The type of receptacle used for laboratory procedures may depend on:

- availability
- the nature of the test(s) to be performed.

Basic laboratory glassware includes bottles, beakers, flasks, test tubes, graduated cylinders and pipettes.

Table 2.4 Plastic ware properties: chemical resistance and possibility to autoclave.

Plastic ware	Chemical resistance	Autoclavable
Polypropylene (PP)	Good	Yes
Low-density polyethylene (LDPE)	Good	No
High-density polyethylene (HDPE)	Good	No
Polymethylpentene (PMP, TPX)	Good	Yes
Polycarbonate (PC)	Moderate	Yes
Polytetrafluoroethylene (PTFE)	Very good	Yes
Polymethylmethacrylate (PMMA)	Moderate	No
Polystyrene (PS)	Moderate	No
Polyvinylchloride (PVC)	Moderate	No

Volumetric accuracy

Volumetric glassware is manufactured and calibrated in accordance with international ISO standards to permit very accurate determination and measurement of specific volumes. All of the volumetric glassware is marked with a set of inscriptions in accordance with any specific standard associated with it.

Mandatory inscriptions include:

- manufacturer
- class:

  - class A/AS – highest level of accuracy (S stands for swift delivery of pipettes and burettes), glass or plastic
  - class B – general purpose work calibrated to a lower level of accuracy, glass or plastic

- nominal volume
- volume unit
- calibration temperature
- for pipettes:

  - IN: calibrated to contain, or
  - EX: calibrated to deliver
  - waiting time
  - blowout: indication that last drop should be blown out of jet.

The following information may also be added:

- country of origin
- error limit
- standard: the standard to which the product conforms
- certified glassware bears a serial number/ batch number for identification and traceability.

General guidelines for use include the following.

- Ensure that all volumetric glassware is kept scrupulously clean. Dirt, and especially grease, can distort the shape of the meniscus and also cause droplets of liquid to adhere to the vessel walls. Both seriously impair accuracy (good cleanliness is indicated by uniform wetting of the glass surface with distilled water).
- Never pipette by mouth. Always use a purpose designed pipette filler.
- Autoclaving at 121°C and cleaning in automatic dishwashers is acceptable and will not affect the accuracy of glassware.
- All items should be held in a vertical position when reading the meniscus. The meniscus should be at eye level to avoid parallax errors.

- When handling strong corrosive acids, and similar, select volumetric ware manufactured from chemically resistant borosilicate glass.
- Do not expose volumetric glassware to direct heat, for example, hotplates, Bunsen flame.

Volumetric glassware is available in two accuracy classes:

- class A/AS:

  - highest measurement accuracy
  - class AS has the same high tolerances as class A, but is designed to permit more rapid outflow on burettes and pipettes

- class B:

  - approximately half of the measurement accuracy of class A/AS.

Working with volumetric instruments

Always try to avoid significant temperature difference between the measuring instrument and the liquid.

Reading the meniscus

The volumetric instrument is held upright and the observer's eye must be at the same height as the meniscus. In this position, the ring mark will appear as a line. The meniscus will appear darker and more easily readable in front of a light background if a piece of dark paper is held behind the instrument immediately beneath the ring mark or graduation mark (Figure 2.58). The volume has to be read at the lowest point of the liquid level. The lowest point of the meniscus has to touch the upper edge of the graduation mark.

Figure 2.58 Measuring volumes, reading the meniscus. The volume has to be read at the lowest point of the liquid level. The lowest point of the meniscus has to touch the upper edge of the graduation mark. Photo: Willy Schauwers, Provincial Institute for Hygiene, Antwerp, Belgium.

Bottles

Reagent bottles are available in a variety of sizes and types. Plastic bottles should only be used for reagents that do not interact with plastic. The bottle size should not be much larger than the volume of reagent to be used. Brown bottles, glass and plastic, are available for storage of light sensitive reagents. The borosilicate reagent bottles with a blue pouring ring and screw cap (both polypropylene) are very convenient for most laboratory use (Figure 2.59). They are available in amber glass as well.

Figure 2.59 Glass reagent bottles. Photo: Dr Ziay Ghulam, Dr Wahidullah Bahaer, Central Veterinary Diagnostic and Research Laboratory, Kabul, Afghanistan.

Beakers

Beakers (borosilicate glass, polypropylene or polymethylpentene) are wide-mouthed, straight-sided containers (Figure 2.60) that have a pouring spout formed from the rim. Each beaker is labelled to indicate the approximate capacity in millilitres. Many beakers have additional markings to indicate volume increments. Beakers have many functions in a laboratory, but they should only be used for non-critical measurements or estimated measurements.

Figure 2.60 Beakers. Photo: Willy Schauwers, Provincial Institute for Hygiene, Antwerp, Belgium.

Flasks

Two commonly used flasks are the Erlenmeyer and the volumetric measuring flasks.

- The Erlenmeyer flask (Figure 2.61) has a flat bottom and sloping sides that gradually narrow in diameter so that the top opening is bottle-like. The opening may be plain, stoppered with a bung or it may have threads for a cap. Erlenmeyer flasks range from 10 ml capacity to 4000 ml capacity. They may be used to hold liquids, to mix solutions or to measure non-critical volumes. Markings on the side indicate the capacity in millilitres.
- The volumetric flask (Figure 2.62) is a pear-shaped flask used for making critical measurements that require accuracy. Volumetric flasks are manufactured to strict standards and are guaranteed to contain a specified volume at a particular temperature. The capacity in millilitres is marked on the flask. A line is etched in the neck of the flask to indicate the appropriate fill level. Usually a portion of water or other solvent is added to the flask before adding an exact amount of solute which has been weighed. The remaining solvent is then added to the flask until it approaches the 'fill' line. The last portion is added very slowly (use a wash bottle or a pipette) until the lowest point of meniscus, or curved liquid, is level with the marking on the neck of the flask when viewed at eye level. Close the flask and shake upside down to mix contents. Volumetric flasks are used to prepare solutions when the accuracy of the concentration is important.

Graduated cylinders

A graduated (measuring) cylinder (Figure 2.63) is an upright, straight-sided container with flared base to provide stability. It is used in the laboratory to make non-critical volume measurements and they are available in capacities ranging from 5 to 2000 ml.

Figure 2.61 Erlenmeyer flask. Photo: Willy Schauwers, Provincial Institute for Hygiene, Antwerp, Belgium.

Figure 2.62 Measuring flask. Photo: Willy Schauwers, Provincial Institute for Hygiene, Antwerp, Belgium.

Figure 2.63 Graduated (measuring) cylinder. Photo: Willy Schauwers, Provincial Institute for Hygiene, Antwerp, Belgium.

Markings on the side indicate the total capacity in millilitres. Liquids are measured in a graduated cylinder by pouring the liquid into the cylinder until the bottom (low point) of the meniscus is level with the desired volume mark.

Pipettes

Pipettes (Figure 2.64) are used in laboratory work to measure and are calibrated 'to deliver' or 'to contain' liquids.

Graduated and volumetric pipettes are TD = to deliver. These pipettes are designed to dispense the correctly measured volume, so there will be a minute amount of liquid left in the tip. Serological pipettes are TC = to contain. To accurately dispense the measured volume the last bit must be blown out.

Figure 2.64 Pipette volumetric bunch. Photo: Willy Schauwers, Provincial Institute for Hygiene, Antwerp, Belgium.

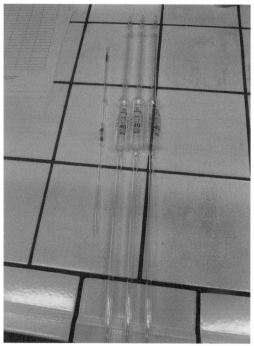

Figure 2.65 Pipette volumetric. Photo: Willy Schauwers, Provincial Institute for Hygiene, Antwerp, Belgium.

There are several types of pipettes:

- Bulb pipettes (calibrated to deliver, TD, Ex, volumetric pipettes) (Figure 2.65) are tubes with a wide opening on the upper end, a round or oval bulb in the centre, and a tapered tip on the other end. They are manufactured to deliver a precise volume with a high degree of accuracy. The volumetric pipette has usually a single graduation mark.

  - Calibration: Class AS: 'Ex + 5s' or Class B: 'Ex' (5s = 5 seconds waiting time).
  - Rinse the pipettes with tap water (by using a water jet pump) immediately after use.

- Graduated pipettes (calibrated to deliver, TD, Ex) (Figure 2.66) are long tubes with a total capacity marking near the top. The volume that can be measured and the volume between two consecutive graduation marks are mentioned at the top of the pipette. The most common graduated pipette has graduations to the tip: the volume that can be measured is contained between the 0 mark and the tip.

  - Less accurate than bulb pipettes.
  - Calibration: Class AS: 'Ex + 5s' or Class B: 'Ex'.
  - Usually the nominal (total) volume is mentioned at the top.
  - Rinse the pipettes with tap water (by using a water jet pump) immediately after use.

- Disposable capillary pipettes (calibrated to contain, TC, In).
- Mechanical, air-displacement pipettes.

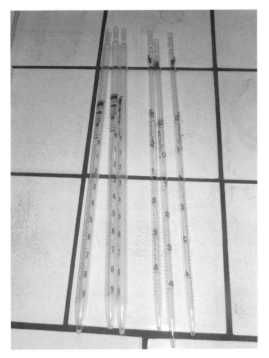

Figure 2.66 Pipette graduated. Photo: Willy Schauwers, Provincial Institute for Hygiene, Antwerp, Belgium.

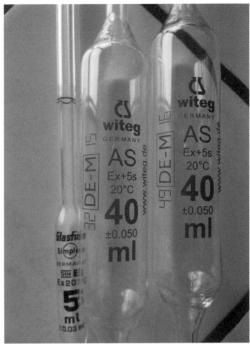

Figure 2.67 Pipette volumetric close up. Photo: Willy Schauwers, Provincial Institute for Hygiene, Antwerp, Belgium.

HANDLING OF PIPETTES

Pipettes calibrated 'to deliver' ('TD, Ex').

- Filling:

  - a pipette bulb or pipettor (manual pi-pump or electric/battery operated pipette-aid) is attached to the upper end and the liquid is suctioned up into the pipette ±5 mm above the marking on the stem above the centre bulb
  - wipe off any remaining liquid with a tissue at the outside of the tip
  - set the meniscus
  - wipe off any drop of fluid adhering to the tip.

- Delivering:

  - hold the pipette vertically and deliver the liquid with the tip of the pipette in contact with the inclining inner surface of the receiving vessel
  - the pipette is held nearly vertical, the tip is placed in contact with the inclining inner surface of the receiving vessel
  - when the meniscus comes to a rest in the tip, the waiting time of 5 s begins (class AS only)
  - after the waiting time, draw the pipette upwards along the inner wall of the receiving vessel through a distance of about 10 mm (Figure 2.67)
  - pipettes should be rinsed with tap water (by using a water jet pump) immediately after use.

Pipettes calibrated 'to contain' ('TC, In').

- Filling:

  - aspirate the liquid to the mark
  - hold the pipette horizontally and carefully wipe off with a tissue.

- Delivering:

  - to empty capillaries blow out the liquid with a pipetting aid, and rinse two to three times with the diluting medium (required due to calibration 'to contain').

Mechanical, air-displacement or positive displacement pipettes with disposable tips are frequently used to measure small volumes. They are available in a variety of volumes, ranging from 1 to 5000 μl. Used tips are disposed of directly into disinfectant using the ejector mechanism of the pipette. The micropipettes usually have two positions of the plunger operated by thumb. The first position is used to pick up the sample and the second to expel the sample from the tip into a tube or well. Micropipettes must be calibrated and maintained according to the instructions of the manufacturer. For how to use mechanical pipettes see section 'Automatic mechanical pipettes and dispensers'.

WORKING WITH PIPETTE AIDS

Pipetting by mouth is not allowed. The use of pipetting aids is mandatory. Pipetting aids can be manual or motorized.

Manual pipetting aids (pipette fillers) use a valve system to create pressure (Figure 2.68).

- The pipetting aid (pipette filler) is attached to the top of the pipette.
- Press 'A' to squeeze the ball and to create negative pressure.
- Press 'S' to fill the pipette just above the mark.
- Press 'E' to drain the pipette.

Motorized pipettors (Figure 2.69) are useful for longer pipetting series. The pipette aid is

Figure 2.68 Pipette filler. Photo: Willy Schauwers, Provincial Institute for Hygiene, Antwerp, Belgium.

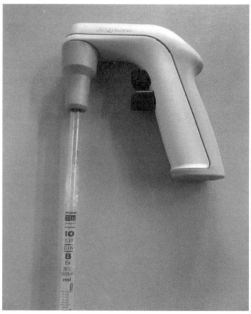

Figure 2.69 Pipette filler electric. Photo: Willy Schauwers, Provincial Institute for Hygiene, Antwerp, Belgium.

protected against penetrating liquids by a valve with a membrane.

- The pipetting aid is attached to the top of the pipette.
- Press the 'fill' button (the intake rate is variable by trigger pressure).
- Press the 'deliver' button. The discharge rate is variable by trigger pressure. One can choose between 'free delivery' and 'power delivery'.

2.7 Care, maintenance and cleaning of laboratory ware

Good quality labware is expensive and must be handled with care. Clean and sterile glassware should always be available in a laboratory and stored in a cupboard that offers protection from dust and from accidental breakage.

Care and maintenance of laboratory glassware

General precautions

To obtain maximum life and performance of your glassware, correct handling is essential. The following notes will serve as a guide to new users and to remind more experienced handlers of the recommended procedures.

- Before using any piece of glassware, take time to examine it carefully and ensure that it is in good condition. Do not use any glassware that is scratched, chipped, cracked or etched. Defects like these can seriously weaken the glass and make it prone to breakage while in use.
- Dispose of broken or defective glassware safely. Use a purpose-designed disposal bin that is puncture resistant and clearly labelled.
- Pyrex laboratory glassware should under no

circumstances be disposed of in a domestic glass recycling stream (for example, bottle banks), since its high melting point makes it incompatible with other glass (soda-lime glass) for recycling. The correct method of disposal is to include it in with general waste in accordance with the relevant guidelines, provided that the glass is free of any harmful chemical contamination.

- Never use excessive force to fit rubber bungs into the neck of glass aspirators, test tubes, conical flasks, and so on.
- Never use excessive force to connect a rubber hose or tubing.
- Lifting or carrying large glass flasks, beakers or bottles by the neck or rim can be very dangerous. It is best to provide support from the base and sides.
- When stirring solutions in glass vessels, such as beakers and flasks, avoid using stirring rods with sharp ends which can scratch and weaken the glassware.
- Do not mix concentrated strong acids with water inside a glass measuring cylinder. The heat of reaction can break the seal at the base of the cylinder.
- We recommend that all glassware is washed before it is first used.
- Wash glassware promptly after use to avoid hard dried residues. Use a biodegradable, phosphate free detergent, specially formulated for laboratory use.
- Do not use cleaning brushes which are badly worn and where the metal spine may scratch the glass.

Heating and cooling

- The maximum recommended working temperature for Pyrex glassware is 500°C (for short periods only). However, once the temperature exceeds 150°C special care should be taken to ensure that heating and cooling is achieved in a slow and uniform manner.

- Pyrex borosilicate glass is microwave safe. However, as with any microwave vessel, be sure it holds a microwave absorbing material, such as water, before placing in the oven. Plastic screwcaps and connectors manufactured from polypropylene or PTFE, are microwave safe.
- Heat vessels gently and gradually to avoid breakage by thermal shock. Similarly, allow hot glassware to cool gradually and in a location away from cold draughts.
- If you are using a hotplate, ensure that the top plate is larger than the base of the vessel to be heated. Also, never put cold glassware onto a hotplate which is already well heated. Warm up gradually from ambient temperature.
- When autoclaving Pyrex containers, for example, bottles with screwcaps – always slacken off the caps. Autoclaving with tightly screwed caps can result in pressure differences and consequent breakage.
- If you are using a Bunsen burner, employ a soft flame and use a wire gauze with ceramic centre to diffuse the flame.

Preparation of media

- Take great care when heating liquids that have a high viscosity. Viscous liquids can act as thermal insulators and can cause 'hot spots' leading to thermal breakage of the glassware. This is particularly important with media solutions as the viscosity usually increases considerably during preparation.
- Regularly stir the solution to assist even distribution of heat. If using a magnetic stirrer set the speed to ensure adequate agitation of the whole liquid.
- Do not use glass vessels with thick walls, for example, 'heavy duty ware' or standard beakers or flasks which have capacities of 5 l or greater.

Vacuum and pressure use

- The application of positive pressures inside glass apparatus is particularly hazardous and should be avoided if at all possible. Safety precautions should always be taken to protect personnel.
- If necessary, always use an adequate safety screen and/or protective cage.
- Under no circumstances use glassware that is scratched, cracked or chipped. Its strength will be seriously impaired.
- Do not use flat bottomed vessels, such as Erlenmeyer flasks and bottles, under vacuum as they are likely to implode. Exceptions are vessels with specially thickened walls such as Büchner filter flasks and desiccators.
- Avoid stress caused by over-tightening clamps. Support glassware gently where possible.
- Never subject glassware to sudden pressure changes. Always apply and release pressure gradients and vacuums gradually.

Care and maintenance of laboratory plastic ware

General precautions

The following guidelines are provided to ensure your plastic laboratory-ware is maintained in the best possible condition. These guidelines are not definitive and care must be taken as each polymer has its own unique properties.

Chemicals can adversely affect the performance of laboratory plastic ware resulting in cracking, loss of strength and flexibility. If in any doubt, note the type of polymer the product is manufactured from, the chemical that is to be used, then confirm compatibility by checking against a chemical resistance chart.

HEATING

- Never place plastic ware in direct contact with a flame or place onto a hotplate surface.

- Most plastics allow the transmission of microwaves. However, as with any microwave vessel, be sure it holds a microwave absorbing material, such as water, before placing in the oven.

STERILIZATION

- If the plastic ware is to be sterilized by autoclaving always pre-check that the polymer can withstand repeated exposure to temperatures of 121°C.
- When autoclaving bottles always ensure the caps are loosened or removed to prevent accidental collapse or deformation.

DISPOSAL

- If the disposal of an item of plastic ware is unavoidable, always follow local laws and regulations.

Cleaning of laboratory ware

Cleaning means the removal of dirt or of any other unwanted material (blood, food residues and so on). This removes visible contaminants. The goal of cleaning is to assure visible cleanliness.

Cleaning has the task of mechanically removing bacteria and fungi or of depriving them of their source of nutrients. The number of germs (bioburden) can be considerably reduced (50–90%) by cleaning thoroughly.

Factors decisive for effective cleaning:

- chemical action
- mechanical action
- time
- temperature.

If one, for example, wants to use fewer chemicals, then one must clean for longer or use greater mechanical action, which means having to scrub harder.

The raw materials used in detergents (cleaning agents, cleaners) are:

- surfactants (tensides)
- acids
- alkalis (lyes)
- water-soluble and water-insoluble solvents
- auxiliary ingredients.

Surfactants are the most important of these raw materials.

Aqueous detergents (as well as other aqueous liquids) are either neutral, acidic or alkaline. Commercially available cleaning products can be classified as follows:

- neutral detergents   pH approx. 5–9
- acidic detergents   pH < 5
- alkaline detergents   pH > 9
  (use concentration)
- solvent detergents.

Detergents that cannot be clearly assigned to any particular class.

- Neutral detergents, the principle ingredient of which is surfactants. Neutral detergents are, in general, much weaker than alkaline detergents (lyes), therefore alkaline detergents should preferably be used to clean surgical instruments.
- Acidic detergents remove lime and cement residues (acetic acid, citric acid, phosphoric acid), commonly are used as toilet cleaners, with addition of surfactants and as cleaners for sanitary fittings.
- Alkaline detergents remove incrustations in the kitchen as well as in industrial and hospital settings (caustic potash solution, soda [sodium carbonate], ammonia and so on). They are used concentrated as oven cleaners and are stronger than neutral detergents.

Use of detergents for cleaning glassware

The use of biological detergents for cleaning glassware in laboratories is essential. These detergents are available in either liquid or powder form. Detergents possess the following advantages over soaps.

- Their action is unaffected by the temperature of the water.
- They are equally efficient in either slightly alkaline or slightly acid water.
- They do not coagulate proteins.

The main disadvantage of detergents is that the slightest trace on labware causes contamination of samples and can result in haemolysis of blood. It is therefore extremely important in a washing procedure to ensure that all detergent is removed by thorough rinsing in changes of deionized water. Usually 20 ml of liquid soap per 1 l of water can be used for cleaning glassware.

Good laboratory technique demands clean labware, because the most carefully executed piece of work may give an erroneous result if dirty glassware is used. In all instances, glassware must be physically clean; it must be chemically clean; and in many cases, it must be bacteriologically clean or sterile. All labware must be absolutely grease-free. The safest criterion of cleanliness is uniform wetting of the surface by distilled water. This is especially important in glassware used for measuring the volume of liquids. Grease and other contaminating materials will prevent the glass from becoming uniformly wetted. This in turn will alter the volume of residue adhering to the walls of the glass container and thus affect the volume of liquid delivered. Furthermore, for pipettes and burets, the meniscus will be distorted and the correct adjustments cannot be made. The presence of small amounts of impurities may also alter the meniscus.

Insufficient cleaning and sterilization of glassware will lead to erroneous test results throughout the laboratory. Since untrained workers are often employed in the glassware/plastic ware washing sections of laboratories and are sometimes not well supervised, poor results can often be traced back to this section of the laboratory. For example, traces of soap and poor sterilization techniques will affect chemical analysis, haematological, serological and many microbiological assays.

General precautions

- Careful handling and storage should be used to avoid damaging glassware.
- Inspect the glassware before each use and discard if scratched on inner surfaces, chipped, cracked or damaged in any way.
- Use only plastic core brushes that have soft non-abrasive bristles or soft, clean sponges/rags. Use brushes to clean inside of deep glassware.
- Rubber sink and counter mats can help reduce the chance of breakage and resultant injury.
- Do not overload sinks or soaking bins.
- Do not place metal or other hard objects, such as spatulas, glass stirring rods or brushes with metal parts, inside the glassware. This will scratch the glass and cause eventual breakage and injury.
- Never use strong alkaline products and hydrofluoric acid as cleaning agents. These materials dissolve glass, leading to damage and eventual breakage.
- Do not use any abrasive cleansers, as these will scratch the glass and cause eventual breakage and possible injury. Scouring pads will scratch glass and should not be used.
- Do not use heat as a method to remove carbon residues. Heating glassware to temperatures > 400°C will cause permanent stresses in the glass and eventual breakage.
- Use proper drying racks for cleaned glassware.

- Cleaning of glassware is always simplified by rinsing dirty glassware with water immediately after use. If labware is not cleaned immediately, it may become impossible to remove the residue.
- Brushes with wooden or plastic handles are recommended as they will not scratch or abrade the glass surface.
- When washing, the water should be hot.
- During the washing, all parts of the glassware should be thoroughly scrubbed with a brush. This means that a full set of brushes must be at hand – brushes to fit large and small test tubes, burets, funnels, graduated cylinders and various sizes of flasks and bottles. Motor driven revolving brushes are valuable when a large number of tubes or bottles are processed.

- Do not use cleaning brushes that are so worn that the spine hits the glass. Serious scratches may be the result. Scratched glass is more prone to breaking during experiments. Any mark in the uniform surface of glassware is a potential breaking point, especially when the piece is heated.
- Do not allow acid to come into contact with a piece of glassware before the detergent (or soap) is thoroughly removed. If this happens, a film of grease may be formed.
- Grease is best removed by acetone.

Use of ultrasonic baths

An ultrasonic bath (Figure 2.70) is a metal vessel containing (mostly) water which produces high frequency sound waves. Contaminants are literally 'shaken off' the glassware. It is used for general cleaning, usually of small items, with detergent additive (acid and alkali are not used since these will attack the metal).

Cleaning method

Glass and plastic labware can be cleaned manually, in an immersion bath, in an ultrasonic bath or in a laboratory washing machine.

- Manual cleaning, wiping and scrubbing method: the generally accepted wiping and scrubbing method with a cloth or sponge soaked in cleaning solution is the most popular cleaning method. Labware must never be treated with abrasive scouring agents or pads, which might damage the surface. For gentle treatment of labware, clean immediately after use – at low temperatures, with brief soaking times and at low alkalinity. Glass volumetric instruments should not be exposed to prolonged immersion times in alkaline media above 70°C, as such treatment causes volume changes through glass corrosion, and destruction of graduations.

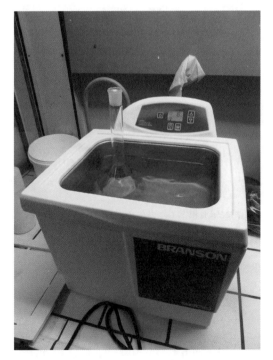

Figure 2.70 Ultrasonic bath. Photo: Willy Schauwers, Provincial Institute for Hygiene, Antwerp, Belgium.

- Immersion method: for the immersion method, labware is soaked in the cleaning solution for 20 to 30 min at room temperature, then rinsed with tap water, and finally with distilled water. Only for stubborn residues should the soaking time be extended and the temperature increased.
- Ultrasonic bath: both glass and plastic labware may be cleaned in an ultrasonic bath. However, direct contact with the sonic membranes must be avoided.
- Machine cleaning: machine cleaning with a laboratory washing machine is gentler to labware than cleaning in an immersion bath. The labware is only exposed to the cleaning solution for the relatively short flushing periods when sprayed by the jet or ejector nozzles. Lightweight objects will not be tossed and damaged by the jet if they are secured in washing nets. Labware is protected against scratching when the wire baskets in the washing machine are plastic coated.

Plastic labware items generally have smooth, non-wetting surfaces and can usually be cleaned effortlessly under low alkalinity conditions. Polystyrene or polycarbonate labware, for example, centrifuge tubes, must only be cleaned manually with neutral detergents. Prolonged exposure even to low alkaline detergents will impair their strength. The chemical resistance of these plastics should be verified in each case.

Washing and cleaning

- Most laboratory plastic ware is readily cleaned in warm water with a detergent and soft cloth or sponge. Avoid using abrasive cleaners or scouring pads which can result in surfaces becoming scratched.
- A low or non-alkaline detergent is suitable for cleaning most plastic ware. Note however that polystyrene and polycarbonate prod-

ucts are susceptible to attack by alkalis and a neutral detergent is recommended.
- If using an automatic laboratory washing machine to wash plastic volumetric ware, such as measuring cylinders, employ a wash temperature below 60°C. High temperatures can affect volumetric accuracy.
- Ultrasonic baths may be used for cleaning plastic ware. However, do take care that the products do not directly touch the transducer membrane.

Cleaning in trace analysis: To minimize metallic traces, laboratory equipment is placed into 1N HCl or 1N HNO_3 at room temperature for not more than 6 h (glass laboratory equipment can be boiled for 1 h in 1N HNO_3). It is then rinsed with distilled water. To minimize organic contamination, laboratory equipment can first be cleaned with alkalis, or a solvent such as alcohol.

There is extra material on the website detailing the cleaning of:

- new glassware
- dirty glassware
- pipettes
- dirty slides.

Disinfection and sterilisation of laboratory ware

Disinfection kills the disease-causing (pathogenic) bacteria. Bacterial spores are not killed. However, in many cases disinfection suffices. Disinfection means that one can no longer contract infection from the disinfected objects (*dis*-infection).

The goal of disinfection is to kill germs and reduce the number of germs such that the disinfected objects can no longer transmit infection.

Sterilization means the killing of all microorganisms, including bacterial spores. The goal of sterilization is to assure absolute absence of organisms.

Disinfection

Disinfection can be achieved with chemical (for example, alcohol) or physical (for example, temperature) methods.

CHEMICAL DISINFECTION (TABLE 2.5)

In chemical disinfection microorganisms are killed by certain chemicals. To that effect, several chemical disinfectants are available. To be deemed suitable for that purpose, they should:

- have as broad a spectrum of action as possible, that is, be able to kill several types of pathogens
- need only a short exposure time
- not be susceptible to any (or to only little) loss of efficacy in the presence of proteins
- not have any (or only very little) unpleasant odour
- not cause any (or only very little) irritation to skin or mucous membranes
- have a high material compatibility profile
- be environmentally friendly
- be economical.

Table 2.5 Classes of active substances used in chemical disinfectants.

Active substance	Spectrum of action	Fields of application	Pros	Cons
Aldehydes formaldehyde glutaraldehyde glyoxal	Covers virtually the entire spectrum	Surfaces Instruments	Biodegradeable Low use concentration	Unpleasant odour Allergenic
Alcohols ethanol n-propanol isopropanol	Bactericidal Fungicidal Virucidal to an extent	Hands Surfaces	Rapid onset of action Biodegradeable Dries quickly Generally good material compatibility	Risk of fire and explosion if used to disinfect large surfaces Skin-degreasing effect
Quaternary ammonium compounds (QUATs)	Depending on substance bactericidal fungicidal	Instruments Hands	Sustained effect Odourless	Adversely affected if used in combination with anionic surfactants (soap effects)
Halogens sodium hypochlorite povidone-iodine	Covers virtually the entire spectrum	Instruments Hands (mucous membranes)	Rapid onset of action	Poor biodegradability profile Corrosive to metals Irritant to mucous membranes
Per compounds hydrogen peroxide peracetic acid	Covers virtually the entire spectrum	Instruments Mucous membranes Water	Rapid onset of action Biodegradable	Unstable
Phenols and phenol derivatives	Bactericidal Virucidal to an extent	Surfaces Instruments	Few protein effects Good cleaning performance	Poor biodegradability profile Hazardous to health

Many chemical agents are referred to as disinfectants, a term that is applied to substances which destroy microorganisms. Other terms with a similar meaning are germicide and bactericide. A disinfectant that is non-injurious to human tissue is classified as an antiseptic and chemicals which are used to prevent organisms growing in a sterile medium are classified as preservatives. The action of a disinfectant is modified by several factors. For example, some disinfectants are very efficient in the absence of organic matter but are not so effective in its presence.

Chemical agents function as sterilizing agents in the following ways:

- interfering with the enzymatic system of the microorganism
- disruption of the cell membrane
- coagulation of protein
- oxidation.

In each case death of the microorganism occurs.

The following points must be borne in mind when handling disinfectants.

- To assure the right concentration of the solution used, the manufacturer's instructions must be observed. Measuring vessels or dosage systems must be used for dosing purposes. If too low a dose is used, the disinfectant will not work properly. If too much is used, the disinfectant action will not be any better, so this confers no advantage and simply damages the environment, is expensive, may damage materials and, not least, it is harmful to staff. A dosage table will make it easier to use disinfectants (see below).
- The disinfectants must be used only for the intended purpose. While that sounds logical, in practice this rule is not always followed.
- No detergents may be added (for example, all-purpose cleaners) since this could diminish the disinfectant efficacy.
- For their own protection, staff must always

wear protective gloves when handling disinfectants – except for hand disinfection.
- Staff must be trained.

ALCOHOL

Absolute alcohol (100%) is not a very effective sterilizing agent because the power of penetration into the microorganism is poor. When diluted with distilled water at a concentration of 70% however, it becomes effective as a skin sterilizer (disinfectant) and is often used prior to inoculations or venepunctures. It is also a useful preservative.

CHLORINE DERIVATIVES

Chlorine derivatives and the free chlorine present are useful for disinfecting contaminated glassware and some liquids. The disadvantages of these disinfectants are that they attack metal and rapidly deteriorate in the presence of organic material, dust and sunlight and are most effective only at a pH close to neutral. In an acid environment chlorine release is accelerated.

Chlorine releasing disinfectants include:

- sodium hypochlorite (bleach): usually contains 5 or 10% available chlorine, hypochlorites must be freshly prepared from stock solutions every day
- calcium hypochlorite granules or tablets: contains about 70% available chlorine, hypochlorites must be freshly prepared from stock solutions every day
- sodium dichloroisocyanurate (NADCC): contains about 60% available chlorine.

Which concentration of available chlorine to use for which purpose:

- a 0.1% available chlorine solution can be used to disinfect benches
- a 0.25% available chlorine can be used for waste containers (for discarding used pipettes, slides)
- a 1% available chlorine solution can be used for treatment of spills.

GLYCEROL

Glycerol in a 50% solution will kill contaminating microorganisms. It is used for the preservation of certain viruses and bacteria that are not affected by the glycerol.

PHENOLICS

- Lysol is a powerful phenolic antiseptic used mainly for disinfection of discarded culture plates, contaminated pipettes and other potentially infectious material. Solutions of 1 to 9% are generally used.
- Hycolin: a phenol derivative used at a 1% solution for all forms of laboratory disinfection.

Always read the manufacturer's instructions *before* using a chemical disinfectant. Used incorrectly, *all* disinfectants can be potentially hazardous to laboratory staff.

Sterilization

Laboratory ware and instruments may need to be sterilized but the procedure is also important for the preparation of culture media and some reagents. Sterilization can be achieved using a variety of methods and it is a case of choosing one that is appropriate for the situation.

PHYSICAL METHODS OF STERILIZATION

- Dry heat: The application of dry heat is a simple method of sterilization providing the material to be sterilized is not adversely affected by the heat. An example is nichrome wire loops used in bacteriology. These can be heated and sterilized in a Bunsen flame in excess of 300°C. For glassware/metal articles that must remain dry, a hot air oven is used at a temperature of 160°C. This temperature must be maintained for 1 h (2 h if the objects are heavy or bulky), then the glassware is allowed to cool slowly before removing it from the oven (when the temperature is around 40°C). Note that rapid cooling will cause contraction and unsterile air being sucked into the vessel. Prepare the objects to be sterilized in the same way as for the autoclave method. Cotton-wool plugs, if used, should not be too thick, otherwise the hot air cannot penetrate.

- Moist heat: A common method of wet sterilization is achieved by using an autoclave (steam sterilizer) at a temperature of 121°C (15 psi) for 20 min or at 110°C (10 psi) for 40 min. A temperature of 100°C (boiling) will kill non-sporing organisms within 10 min. Most spores will be killed in 30 min at 100°C but some spores can resist boiling for several hours. The addition of 2% sodium carbonate increases the bactericidal effect of boiling water and spores that resist boiling water for 10 h have been killed in 30 min this way. This method is suitable for sterilizing contaminated instruments following post-mortems so that they may be used again quickly. It is unsuitable if instruments are to be stored in a sterile condition.

- Biological fluids may be sterilized by heating them in a water bath at 56°C for periods of 1 h daily if necessary. The principle is called Tyndallization but at this low temperature more than three exposures may be necessary. This method of sterilization cannot be used when the fluid potentially contains resistant spores. It is a useful method to sterilize media containing protein or sugars which may be 'broken down' at higher temperatures and when filtration is not feasible.

STEAM STERILIZATION

Steam sterilization (autoclaving) is defined as the destruction or irreversible inactivation of all reproducible microorganisms under exposure to saturated steam at 121°C (15 psi).

The following points must be observed.

- Efficient steam sterilization is assured only if the steam is saturated and has unrestricted access to all contaminated areas.
- To prevent pressure build-up, containers or vessels must always be open.

- Dirty reusable labware must be cleaned thoroughly before steam sterilization. Otherwise, residue will bake on during sterilization and microorganisms may not be effectively destroyed if they are protected by the residue. Furthermore, any adhering chemical residues may damage the surfaces due to the high temperatures.
- Glassware that is contaminated with blood clots, such as serology tubes, culture media, petri dishes, and so on, must be sterilized before cleaning. It can best be processed in the laboratory by autoclaving or by placing it in a large bucket or boiler filled with water, to which 1–2% soft soap or detergent has been added, and boiled for 30 min. The glassware can then be rinsed in tap water, scrubbed with detergent, rinsed again.
- You may autoclave glassware or sterilize it in large steam ovens or a similar apparatus. If viruses or spore-bearing bacteria are present, autoclaving is absolutely necessary.
- Not all plastics are resistant to steam sterilization. Polycarbonate, for example, will lose its strength. Polycarbonate centrifuge tubes cannot be steam sterilized.
- During sterilization (autoclaving), plastic equipment in particular should not be mechanically stressed (for example, do not stack). Thus, to avoid shape deformation, beakers, flasks, and graduated cylinders should be autoclaved in an upright position.
- Culture tubes which have been used previously must be sterilized before cleaning. The best method for sterilizing culture tubes is by autoclaving for 30 min at 121°C (15 psi). Media that solidifies on cooling should be poured out while the tubes are hot. After the tubes are emptied, brush with detergent and water, rinse thoroughly with tap water, rinse with distilled water, place in a basket and dry.

If tubes are to be filled with a medium that is sterilized by autoclaving, do not plug until the medium is added. Both medium and tubes are thus sterilized with one autoclaving. If the tubes are to be filled with sterile medium, plug and sterilize the tubes in the autoclave or dry air sterilizer before adding the medium.

RADIATION

- Ultraviolet (UV) radiation: sunlight can destroy bacteria due to the action of UV light which acts as a sterilizing agent by producing peroxides which are oxidizing agents. The most highly germicidal rays are those having wavelengths between 296 and 210 nm. However, since the penetrating power of UV light is low, its practical use is limited. UV light is often used in safety cabinets to sterilize the bench when the cabinet is not in use.
- Ionizing radiation: gamma rays have more penetrating power than UV light and can pass through materials. They affect organisms by ionizing the cell constituents. This method is used in commercial laboratories and industry to sterilize disposable polystyrene Petri dishes, syringes, vials and so on, while they are in their packaging.

FILTRATION

There are several types of filters made from diatomaceous earth, porcelain, asbestos, sintered glass and membranes, which have been used in laboratories. Membrane filters are gradually superseding other types as they are easy to handle and can be discarded after use thus minimizing cleaning. They produce little loss of solution and are available in various levels of porosity. Diatomaceous earth and porcelain filters will not be described since they are not used in veterinary laboratories.

Membrane filter units utilize membranes made from a cellulose ester which has pores that range down to 5 μm. After use and sterilization the membrane is discarded and a fresh membrane is inserted. In veterinary laboratories pore sizes of 22 μm or 45 μm are commonly used for sterilization of biological fluids. Membrane filters are supplied in various forms and sizes. Some are disposable and can be attached to a syringe for filtration of small volumes of biological fluids.

Information

There is a range of text books which outline the general use of laboratory equipment but the most up to date information and advice on the use of equipment or supplies can be obtained from the manufacturer.

Endnotes

1 http://www.surechill.com/. Youtube movie on the Sure Chill technology: https://www.youtube.com/watch?v=FKjLvcwt7M0.

2 https://www.scilogex.com/scilogex-dilvac-portable-dry-ice-maker.html, US$1325.

3 https://www.dometic.com/en-gb/uk.

4 http://www.dixons-uk.com, Dixons Surgical Instruments, UK.

5 https://www.steris.com/healthcare/products/sterility-assurance-and-monitoring/biological-indicators/verify-assert-self-contained-biological-indicator/.

6 http://www.impact-test.co.uk/products/2894-manesty-water-still/.

7 https://www.brand.de/en/products/liquid-handling/plt-unit-pipette-leak-test/.

8 Rack down: to move the objective towards the slide (or, if the stage is movable: to move the stage towards the objective). Rack up: to move the objective from very close to the slide in an upward direction away from the slide (or, if the stage is movable: to move the stage from very close to the objective away from it).

9 More information can be found at: https://www.biotek.com/resources/technical-notes/dual-wavelength-measurements-compensate-for-optical-interference/.

10 For more information see 'Scilabware-Overview', http://www.obrnutafaza.hr/pdf/scilabware/Scilabware-Overview.pdf.

Bibliography

Biotek, Biotek Catalog_2017_2018, https://www.biotek.com/products/detection-microplate-readers/800-ts-absorbance-reader/?Product_Interest_Source=Email%20Mktg%20-%20BioTek&source=Email%20Mktg%20-%20BioTek&Lead_Campaign_Source=7011O0000001x2Kw&Campaign_Source=7011O0000001x2Kw, https://www.biotek.com/products/liquid-handling-microplate-washers/50-ts-washer/?Product_Interest_Source=Email%20Mktg%20-%20BioTek&source=Email%20Mktg%20-%20BioTek&Lead_Campaign_Source=7011O0000001x2Kw&Campaign_Source=7011O0000001x2Kw

Brand, Brochuere_Volumenmessung (n.d.) https://www.brand.de/fileadmin/user/pdf/Information_Vol/Brochuere_Volumenmessung_EN.pdf.

Cheesbrough, M.C. (2005) Medical Laboratory Manual for Tropical Countries. Volume I Cambridge University Press. Cambridge.

Fleming, D.O., Hunt, D.L. (2006) Biological Safety. Principles and Practices, 4th edn. ASM Press. Washington, DC.

Gilson (n.d.) Gilson Guide To Pipetting Third Edition, https://www.gilson.com/pub/static/frontend/Gilson/customtheme/en_US/images/docs/GuideToPipettingE.pdf.

Michigan State University (n.d.) https://msu.edu/course/lbs/159h/Spectrophotometry04.pdf.

Rainin (n.d.) Procedure for Evaluating Accuracy and Precision of Rainin Pipettes, https://www.mt.com/dam/RAININ/PDFs/TechPapers/ab15.pdfRainin_PipetteAccuracyandPrecision.

Unpublished laboratory manual: 'Human Parasitology in Tropical Settings – Practical notes' – Edition 2017 – Idzi Potters, Philippe Gillet, Jan Jacobs (Institute of Tropical Medicine – Antwerp).

WHO (2003) Manual of Basic Techniques for a Health Laboratory. WHO, Geneva.

WHO (2004) Laboratory Biosafety Manual, 3rd edn. WHO, Geneva.

Wikipedia: https://en.wikipedia.org/wiki/Fluorescence_microscope.

World Federation for Hospital Sterilisation Sciences, http://wfhss.com/wp-content/uploads/wfhss-training-1-03_en.pdf.

PART II

SPECIALTIES

chapter 3

Parasitology

Susan C. Cork and Mani Lejeune

3.1 Introduction

Parasitology is an important part of any routine health monitoring programme and often forms the largest component of the work in regional and district veterinary laboratories. Gastrointestinal worms, and other helminths, protozoa and ectoparasites are responsible for reduced productivity and a range of clinical problems in young and adult livestock. Faecal examination is the most common diagnostic approach employed to detect endoparasitic infection in a host. It must be noted that parasites that have predilection to various organ systems such as central nervous system, circulatory, musculoskeletal, respiratory and, to some extent, urinary system may also shed their infective stages (for example, eggs, cysts, oocysts, trophozoites, larvae and so on) in the host's faeces. A comprehensive faecal evaluation can provide very useful information about the parasitic status of the host. Government and commercial veterinary diagnostic laboratories may offer a wide range of parasitological tests and have the ability to handle large numbers of samples using semi-automated systems. However, in resource poor settings, and at the district level, more basic manual tests and field adapted protocols are commonly used.

Classical parasitology testing performed in regional and district laboratories does not require cost prohibitive resources but it is worth investing in good quality microscopes. The basic laboratory protocols for parasitology are generally easy to perform but interpretation of the results requires adequate knowledge of at least general parasitology. In this chapter, we will outline the most frequently performed parasitology tests, along with a brief mention of some more specialized techniques. The emphasis will be on techniques that can be performed using simple protocols yet produce reliable results. The following sections will outline key parasitic diseases and the relevant diagnostic procedures used in veterinary helminthology, protozoology and entomology. Further information on the biology of a wider range of parasitic diseases can be found in the references provided at the end of the chapter.

3.2 General helminthology

Helminths (worms) are an important group of organisms. There are both free-living and parasitic species but in veterinary medicine the term generally refers to the latter.

There are two main groups or phyla:

1 nemathelminthes: nematodes (roundworms), currently, they are classified as phylum: nematoda
2 platyhelminthes: trematodes (flat worms) and cestodes (tapeworms).

There is also a group of helminths of lesser veterinary importance, the acanthocephala (thorny headed worms).

Nematodes are the most abundant of the worms but not all are parasitic, there are many free-living forms and some parasitize plants. There are a wide range of parasitic nematodes, the adults of which have a predilection for a particular organ in the host. While the majority reside in the gastrointestinal tract (for example, *Haemonchus* sp., *Trichostrongylus* sp., *Ascaris* sp., *Trichuris* sp.), some live in the lungs (for example, *Dictyocaulus* sp., *Protostrongylus* sp.), peritoneal cavity (*Setaria* sp.), kidney (*Stephanurus* sp.), eye (*Thelazia* sp.), connective tissue (*Onchocerca* sp.) and meninges (*Paraelaphostrongylus tenuis*). The life cycle of nematodes can be simple; with the eggs hatching in the environment and developing into infective larvae that are ingested by the host and reach the gastrointestinal tract to develop into adults, for example, *Haemonchus contortus*. Others have complex life cycles with larval stages migrating through organs such as the liver or lungs before reaching gastrointestinal tract to reside as adults, for example, *Ascaris suum*. The life cycle can also be direct or indirect. The latter requires development of early parasitic stages in a suitable intermediate host (for example, houseflies for *Habronema muscae*). With the majority of helminths having an environmental component to their life cycle the diseases that they cause can be seasonal and are effected by climatic conditions. In addition, some parasitic helminths have specific host and habitat preferences limiting their distribution to specific parts of the world.

Trematodes and cestodes are considered to be more primitive than nematodes. They are largely hermaphrodites (that is, the sexes are not separate) with the notable exception being the schistisomes. Most trematodes and cestodes that are parasitic to animals have an indirect life cycle requiring an intermediate host. For example, the majority of trematodes use snails as intermediate hosts; whereas, most cestodes exist by cleverly exploiting the predator–prey relationship between the intermediate and definitive hosts (for example, *Taenia* spp.). Adult trematodes invariably reside in the organs of the gastrointestinal system (for example, *Paramphistomum* sp., *Fasciola* sp. and so on) with one exception (*Schistosoma* sp. in blood vessels). Similarly, adult cestodes generally reside in the gastrointestinal tract of the definitive host (usually a predator), however, the larval forms tend to have a predilection for various organ systems in the intermediate hosts (usually prey species).

Some 'parasitic' worms appear to live 'in balance' with the host causing little apparent harm (for example, *Cooperia* sp. of sheep). However, some are highly pathogenic and cause poor production and ill health in livestock (for example, *Haemonchus contortus* of sheep). The disease caused by helminths may be subclinical or clinical. Clinical signs depend on (1) the number and species of parasite present, (2) the location of the parasite(s) in the body and (3) the general immune status of the host animals (which can be determined by age, previous exposure, concurrent disease, nutrition and so on). Clinical disease is also exhibited if the host animal becomes physiologically stressed as a result of lack of feed, harsh climate, concurrent disease, pregnancy or during seasonal migration. Not all animals in a herd or flock will be equally infected with a particular parasite species because some animals may be genetically susceptible, whereas others may have a degree of resistance. Parasites residing in resistant animals will often shed fewer eggs than those in more susceptible animals. This is exemplified by the fact that in a horse herd with strongyle infection only 20–30% are heavy egg shedders, the remaining animals are often either low or moderate shedders. For this reason, eggs counts are often used to determine which animals in a herd need to be given treatment (anthelmintic).

In the last few decades it has become more widely accepted that helminthic infection is not always harmful. The beneficial role of helminths for host health has been documented through the 'hygiene hypothesis' developed to explain the steep increase in inflammatory bowel disease and allergic response in hosts that are devoid of parasitic infections. Nevertheless, treatment or control measures are warranted when helminth infection is perceived to be the cause of poor animal health or is documented to significantly decrease the production of a livestock unit.

> A sound helminth control strategy requires breaking the life cycle(s) of the species implicated. The majority of helminth parasites have a life cycle that has an environmental component (for example, egg development and hatching of larvae), which may require a specific range of temperature and humidity. Environmental conditions will, therefore, influence the number of infective stages of the parasite which can successfully develop to infect the host and this must be taken into consideration when implementing any sort of helminth control program in livestock.

Collection and preservation of faecal samples

As with any samples sent to the laboratory, it is important to submit full details of the animal(s) sampled (that is, species, age, sex, breed) and a description of any clinical findings, recent use of anthelmintic(s), ectoparasite control or other treatment. The date of collection and the name of the submitting animal health officer or veterinarian should be clearly marked on the submission form. For detection of parasite eggs, and lungworm larvae, it is important that fresh faecal samples are collected rectally. In situations where fresh faecal samples need to be picked up from the ground, care must be taken to avoid environmental contamination as free-living helminths may appear similar and complicate the diagnosis. If possible, at least 10 g of faeces should be submitted for laboratory examination. All samples should be collected into plastic screw cap containers or in securely sealed, labelled plastic bags (Figures 3.1 and 3.2), filling the container to the brim and securing it. Containers should be air tight to create anaerobic conditions and prevent egg hatching and subsequent larval development. If a delay of 1–4 days is expected, samples should be stored at 4°C. For long-term storage, samples can be fixed in a preservative such as 70% alcohol or 10% formal saline. If molecular testing is planned, samples are required to be saved in 70% alcohol. The animal(s) identification, date of sample collection and the name of the owner (or submitter) should be marked on the container label. At the laboratory, if it is necessary to make an accurate assessment of the number

Figure 3.1 Various plastic bottles and bags may be suitable for collecting faecal samples. Regardless of shape or size, specimen bottles should have a wide neck to allow easy access and a thread cap for secure closure. If plastic bags are used for sample collection these should be sealed and placed inside another bag to ensure that leakage does not occur. Label all containers to provide a clear identification and record relevant details (that is, reference number, animal identification and specimen details and so on) at the time of sample collection.

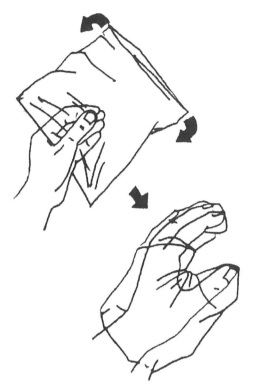

Figure 3.2 The use of a plastic bag to protect hands from direct contact with faecal material if water-proof gloves are not available.

of parasite stages (eggs/larvae) present in the samples, it is important to measure and record the weight of the sample and the volume of preservative fluid if added (see later).

Processing

The preparation of a standard suspension of faeces is the first step in the examination of a faecal sample. Make a note of the colour, smell, consistency and texture of the sample. Thorough mixing of the faecal sample is required before suspension because parasite eggs (especially tapeworms) may not be evenly distributed in faecal samples. If possible make a large volume (150 ml) of suspension with the ratio of 1 g

faeces to 15 ml of water and take the required aliquot (15 ml) for further processing.

Preparation of a standard suspension

1 Break up the faecal specimen with a fork.
2 Weigh 3 g of faeces and place in a bowl.
3 Pour about 30 ml of water into the bowl and macerate the faeces.
4 Sieve the sample through a tea strainer to remove fibrous/coarse material.
5 Wash the suspension with water into a beaker (or open necked bottle) previously marked to reach an indicated volume of 45 ml, thus the sample is suspended in a final ratio of 1 g in 15 ml water.
6 Thoroughly mix the suspension and dispense 15 ml into one or two 20 ml round-bottomed centrifuge tubes.
7 Centrifuge[1] briefly (low level [1500–2000 rpm] setting for 10 min) and decant the supernatant.

Methods to detect helminth life stages (eggs, larvae, and so on) in faeces

The choice of method for parasite detection depends on the purpose for which the test is performed. A direct faecal smear can be preferred for a quick parasitic disease diagnosis of a terminally sick animal. In contrast, more comprehensive parasitology testing is performed for disease assessment, either in an individual animal or in population. In general, comprehensive testing utilizes protocols that concentrate parasite stages in faecal samples. This can be performed in a qualitative or quantitative manner. Qualitative techniques are those that can be used to determine the relative number and variety of parasite eggs present in a sample, these are often simple to perform and do not require a lot of laboratory equipment. Methods which allow a more accurate assessment of parasite numbers are known as quantitative techniques. Quantitative techniques are important for routine screening of livestock and can also be used to determine whether or not a parasite treatment or control program is required, or to determine if an anthelmintic treatment has worked. These quantitative methods will be discussed later.

Qualitative techniques

DIRECT SMEAR

For a quick qualitative assessment, a direct smear can be made by taking a few drops of a standard suspension of faeces and adding a few drops of water to make a smear on a glass slide, adding a cover slip and examining the preparation under the microscope. This is a simple method but it is not very sensitive, that is, only a very small sub-sample of faeces is examined so many eggs may be missed. The method is usually adequate to detect heavy infections.

A SIMPLE FLOTATION TECHNIQUE

Life stages of parasite species possess specific buoyant densities. The flotation technique works on the principle that a flotation fluid with specific gravity greater than the buoyant density of a parasite stage will allow it to float in that solution. Based on this, it is possible to selectively segregate parasite eggs (and larvae) from coarse faecal matters. A basic flotation solution includes a saturated (super concentrated) salt (NaCl) solution (1.2 specific gravity [SpG]). This flotation fluid can be added to the sediment from a standard faecal suspension. In most cases a standard 15 or 20 ml test tube is used but smaller volumes can be prepared, for example, using 2.0 ml Eppendorf tubes (Figure 3.3). Add a small volume of flotation solution and mix thoroughly. Top it up with enough solution to reach the top of the test tube to form a positive meniscus. Allow it to stand for about 10–15 min and using a loop or a glass rod scoop the fluid at the meniscus and place on a cover-

slip for microscopic examination. Alternatively, a cover slip can be positioned over the meniscus for the required period and then placed on a microscope slide. Note that centrifugation steps can help speed up the process and also increase the sensitivity of flotation technique.

Nematode eggs, most cestode eggs and coccidial (protozoan) oocysts will float to the surface of the meniscus and adhere to the cover slip. Heavy trematode eggs such as *Fasciola* sp. may not float in saturated salt (NaCl) solution and cysts of *Giardia* sp. (protozoan) may be distorted. Sheather's sugar solution with 1.32 SpG may float fluke eggs but will distort the delicate egg wall. Zinc sulphate ($ZnSO_4$) with 1.18 SpG is preferred for detecting lungworm larvae and *Giardia* sp. cysts. Strategically, a faecal sample can be processed simultaneously by both Sheather's sugar and $ZnSO_4$ methods to recover the whole range of parasite species present in the sample. The appearance of helminth eggs in faecal samples is illustrated in Plates 2a, 2b and 3. If the sample is not fresh (> 12 h old) some parasite eggs may have hatched and there will be larvae in the sample. It is possible to identify the larvae of different helminth species but this is time consuming and requires skill and experience (see Figure 3.4). As outlined earlier, larval development

Figure 3.3 A temperature controlled centrifuge with an Eppendorf rotor. Eppendorf tubes hold up to 2.0 ml of fluids and usually have a well-fitting lid which reduces the risk of contaminating the rotor.

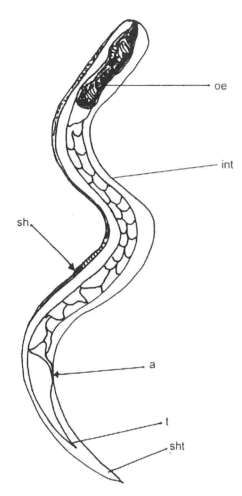

Figure 3.4 Larval identification is difficult but some common species may be recognized by the characteristic length and shape of the tail as well as the number of cells in the intestinal tract. Free living larvae may be stained using iodine solution which may help distinguish them from potentially pathogenic species. Try to get specialist training before attempting to identify helminth larvae. Oe= oesophagus; int = intestine; sh = sheath; a = anus; t = tail of the larvae; sht = tail of the sheath.

can be prevented by keeping samples in air tight container or refrigerated (< 4°C) or by adding a fixative (10% formal saline or 70% alcohol) to the sample.

See Plate 1 for nematode larval culture.

A SIMPLE SEDIMENTATION TECHNIQUE

Sedimentation relies on the principle that parasite eggs will settle to the bottom of a suspension after a period of time, this is best facilitated by prior centrifugation (1500–2000 rpm, 3 min) of the sample and then leaving the sediment to settle for at least 10 min. A direct smear can then be made (see above). There may be some difficulty in identifying parasite eggs using this method if there is a lot of debris present and centrifugation can distort the structure of some eggs, for example, *Fasciola* sp. trematodes.

Quantitative techniques

For the purposes of survey work and diagnosis it is usually necessary to determine the number of parasite eggs per gram (EPG) of faeces, this is because some parasites are only pathogenic in high numbers and therefore low counts may not be of any clinical significance (for example, *Moniezia* sp. in cattle, cyathostomes in horses). Most animals will carry a low number of a range of helminth parasites without showing any clinical signs. However, some parasites are pathogenic even in low numbers (for example, *Fasciola hepatica* in sheep) and some produce more eggs than others (for example, *Haemonchus contortus* in sheep). For these reasons, it is important to understand the life cycle and epidemiology of parasite infections before making decisions about the significance of the results.

Quantitative assessment is done using various methodologies such as the differential centrifugal flotation technique, McMaster technique (flotation based, see Figure 3.5) or Stoll counting method (sedimentation based). Another commonly used method is the Wisconsin double centrifugation flotation technique. As indicated earlier, flotation relies on the principle that parasite eggs will float in a solution of higher specific gravity (SpG). Some of the most commonly used techniques for the determination of EPG in a sample are outlined below. More detailed descriptions of less frequently used or more complicated techniques can be found in the bibliography at the end of the chapter.

DIFFERENTIAL CENTRIFUGAL FLOTATION TECHNIQUE

Although saturated salt solution can be used to perform this protocol it is preferable to use Sheather's sugar (1.32–1.27 SpG) and $ZnSO_4$ (1.18 SpG) solution simultaneously as this can help in quantitation of a whole gamut of parasite stages that could be present in a faecal sample. The steps involved are outlined below:

1 Prepare a faecal suspension as indicated previously.

2 After discarding/decanting the supernatant, place the tube(s) in a test tube rack and re-suspend the sediment in a small volume of flotation solution by mixing thoroughly. Advisable to process two tubes of same faeces simultaneously, one with Sheather's sugar (see box) and the other with $ZnSO_4$ solution.

3 Gently top up with flotation solution to form a small convex meniscus at the top of the tube.

4 Carefully lower a cover slip(s) vertically onto the top of the tube(s). Avoid trapping large air bubbles. Leave to stand for 15–20 minutes.[2] Alternatively, centrifuge at a slow speed (1000–1500 rpm) for 15 min.

5 Vertically lift the cover slip off the tube(s), together with the fluid adhering to it and place it on a microscope slide.

6 Examine under the medium (10×) or high (20×) power of a microscope and count all parasite stages on the slide. Lungworm larvae and *Giardia* sp. cysts are counted in $ZnSO_4$ float, whereas the rest of all are counted in sugar float.

Calculation: If the faecal suspension is prepared as described previously, 1 g of faecal sample would be present in each tube. Therefore,

> ## Recipe for Sheather's sugar solution
>
> 454 g granulated sugar, 355 ml tap water, 6 ml full-strength (37%) formaldehyde
>
> Heat the tap water to near boiling. Add the granulated sugar, and stir until the sugar is dissolved. Allow the mixture to cool to room temperature, and then add the formaldehyde. Check the solution's specific gravity, and adjust it to 1.27 by adding water or sugar.
>
> Source: Blagburn, B.L., Butler, J.M. (2006) Optimize intestinal parasite detection with centrifugal fecal flotation. Veterinary Medicine 101(7): 455–464.

parasite stages counted during microscopic examination is easily expressed as number of eggs present per gram of faeces.

These calculations allow for the fact that some eggs will not be recovered on the cover slip. There is considerable error in many of the standard counting techniques but the results are adequate for clinical diagnostic purposes. For greater accuracy repeating the procedure for each sample is advised (that is, run duplicates).

MODIFIED MCMASTER COUNTING TECHNIQUE (FLOTATION)

See Figures 3.5 and 3.6 for an illustration of the required equipment and procedure.

1 Complete steps 1–7 (as described previously) and re-suspend the deposit in saturated salt (NaCl) solution or Sheather's sugar in the same volume (that is, 15 ml).

2 Mix the sample and then remove a small volume of fluid from the saturated salt suspension and fill the counting chamber of a ruled McMaster slide (usually calibrated to hold 0.15 ml). For the greatest accuracy fill two chambers and count the eggs in

both and divide by two to get the average number.

3 Allow the slide to stand for 3–4 min and then count all the eggs under the ruled square.

4 The total number of eggs counted × 100 indicates the number of eggs per gram of faeces.

(Calculation: This depends on the volume of the counting chamber under the grid [usually 0.15 ml for the McMaster slide]. If in doubt, check with the manufacturer of the slide used.)

If there was 3.0 g of faeces in 45.0 ml of the standard suspension, then in 0.15 ml there is 0.01 g of faeces. Thus the number of eggs in 0.01 g × 100 = the total number of eggs in 1.0 g of faeces. The faecal egg count = eggs per gram of faeces so if there were an average 4 eggs counted under each grid of the McMaster slide then this × 100 = 400 eggs per gram (EPG).

This technique may not pick up all eggs, therefore to ensure detection where there are only a few eggs in a sample a more sensitive technique such as the differential centrifugal flotation technique is recommended.

MODIFIED STOLL COUNTING TECHNIQUE (SEDIMENTATION)

Use a calibrated pipette or pre-determine the volume of a number of drops from a vertically held Pasteur pipette (usually about 0.03 ml/drop). Calibration can be done by weighing the drops. Once this has been done the pipette can be used for several counts.

1 Prepare a standard suspension (as outlined previously), with water as the suspension fluid.

2 Thoroughly mix the standard suspension and fill the pipette, place two to three pools of

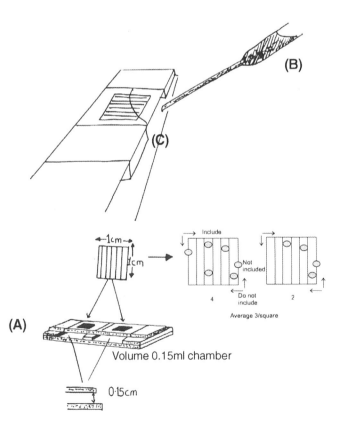

Figure 3.5 Filling the chambers on a McMaster slide (A) for a worm egg count (see text). A fine tipped pipette (B) is used to fill the counting chamber(s) of the slide (C) with faecal suspension. The McMaster slide is then viewed under the microscope. Parasite eggs will float to the surface of the grid. When counting the eggs in a grid it is convention to include eggs which touch the line on the left and top sides but not eggs which lie on the lines at the right or bottom sides.

suspension (2–3 drops each) on a slide and cover with a cover slip.

3 Examine under low power and count all the eggs or larvae seen.

4 For identification of some eggs it will be necessary to use high power.

Calculation: eggs per gram of faeces = no. eggs counted × 15/*nv*

where *n* = no. of drops counted, for example, 5 and *v* = volume of each drop, for example, 0.03 ml (30 μl)

Therefore if 10 strongyle eggs were counted under low power:

EPG = 10 × 15/ 5 × 0.03 = 10 × 100 = 1000

The appearance of helminth eggs commonly seen in the faeces of domestic animals is illustrated below in Figures 3.15 to 3.19 and in colour in Plates 2a, 2b and 3.

Culture, recovery and identification of larvae

Larval extraction for lungworm larvae

Unlike many helminths of veterinary importance, lungworm eggs tend to hatch before they are coughed up and swallowed to be passed out in faecal material (see Figure 3.25). This means that even when the faecal sample is fresh, lungworm eggs may not be present. The following method can be used to harvest lungworm larvae from a sample to allow assessment of the number present. Preferably, fresh faeces should be used for this technique to avoid confusion between lungworm larvae and the larvae of other parasites and free-living nematodes. The following technique is sensitive for the diagnosis of lungworm infections in sheep and donkeys but

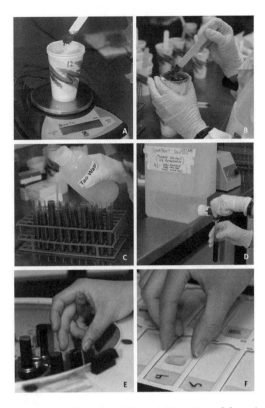

Figure 3.6 Flotation: (A) weighing 4 g of faecal material; (B) dilution of faecal material in water and filtering through a gauze pad into centrifuge tube; (C) filling of centrifuge tubes to equal level; (D) diluting faecal pellet in Sheath solution after centrifugation; (E) filling tubes enough to form meniscus, cover tubes with cover slips; (F) remove cover slip from tube and transfer to labelled microscope slide. See also Plate 5. Photo: Dr Regula Waeckerlin, University of Calgary, Canada.

for detection of cattle lungworm (*Dictyocaulus viviparus*) half saturated salt solution should be used instead of water.

1 Weigh about 1 g of faeces and place the sample in a small piece of tissue paper to form a small cylindrical 'bag'. Fill a conical centrifuge tube with water and place the paper 'bag' containing the sample inside. Try to avoid trapping air beneath the bag.

2 Refill the tube with water and leave overnight at room temperature (25°C) under light.

3 The next day remove the paper 'bag' containing the faecal material from the tube and discard it.

4 Centrifuge the tube containing the larvae and water and discard the majority of the supernatant.

5 Suspend the sediment in the small amount of fluid remaining in the tube and add 1 drop of Lugol's iodine.

6 Using a Pasteur pipette transfer the suspension on to a microscope slide in 3-drop pools. Apply a cover slip to each 'pool' and examine under the microscope using the low-power lens. Identify and count all the lungworm larvae present.

Alternatively, the Baermann method can be used, this is essentially the same as the above method but the first three steps are performed using the Baermann equipment. The last steps (4–6) are the same (Figure 3.7).

A modified Baermann larval extraction technique is now commonly used because it is simple and requires less space to perform. It is sensitive and comparable to traditional Baermann technique. The method is summarized below.

1 Make an 8 × 8 cm envelope of cheese cloth/kim wipes sandwiched between two window screens.

2 Place 2–5 g of faeces inside the envelope on the side of cheese cloth.

3 Submerge the faecal envelope into 200 ml

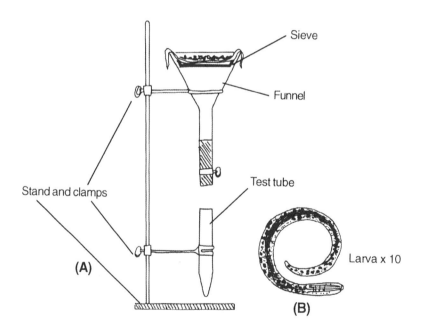

Figure 3.7 The Baermann equipment (A) used for extracting lungworm and an illustration of a lungworm larva (B). A sieve (250 μm) is placed in the wide neck of a glass funnel held in a retort stand. A rubber tube is attached to the bottom of the funnel which has been partially filled with water. Gauze is placed in the sieve and faeces are added. The apparatus is left overnight after filling the funnel with water. Larvae migrate out of the faeces and are collected in the neck of the funnel to be released into the test tube below. Next progress with steps 4–6 as outlined for larval extraction. See also Plate 6.

of lukewarm water kept in a 250 ml glass beaker.

4 Leave the setting overnight under light.

5 Discard the envelope and decant the supernatant without disturbing the sediment (15–20 ml).

6 Collect the sediment in a centrifuge tube and centrifuge at low speed (1000–1500 rpm) for 10 min. Carefully discard the supernatant and examine under low or medium power objective for larvae present in 1 ml of the sediment.

LARVAL CULTURE

Owing to the fact that many important helminth eggs look similar (that is, the common Trichostrongyles) it is often necessary to hatch out the larvae to enable identification of the genus and species of parasite. In cases of anthelmintic resistance and/or assessment of treatment regimens in experimental and field trials this becomes very important. The following simple method is easy to perform but requires patience and some practice to allow accurate measuring of larvae and recording of data. Some of the features used to identify parasite larvae are given in Figure 3.4.

Procedure

1 Weigh out 10 g of the faecal sample and mix with charcoal or a desiccant (calcium carbonate) to form a mixture with a dry consistency.

2 Place the faeces in a culture jar (wide necked jar, 0.5 to 1 l volume) and place a Petri dish on top.

3 Incubate the jar for 7–8 days at 28°C and check the jar regularly. If necessary, add extra water to prevent the culture drying out.

4 After 8 days fill the jar with water, and allow to stand for 2–3 h. The larvae will migrate into the water which can then be poured into a cylinder for sedimentation.

5 Allow the sediment to settle and pour off the supernatant or pour the suspension of larvae into several test tubes and follow

the steps 4–6 as for extraction of lungworm larvae.

6 Add 4 drops of 10% formalin, to kill the larvae, and 3 drops of iodine to stain them.

7 Examine the sediment using the low or medium power lens (4× or 10/20×) of the microscope.

8 Examine morphological features including the tail shape and measure the length of the larvae using a calibrated eyepiece (see Chapter 2 – microscope section), make detailed drawings so that these can be compared with those in text books (or consult a local expert). See Figure 3.4.

One drawback of the above technique is the failure to obtain a clean culture without faecal debris. Recent advancements in the field of molecular testing for anthelmintic resistance requires relatively pure larval culture and the following protocol may yield a better result.

Modified nematode larval culture

1 Weigh 30 g (minimum 10 g) faeces and place it into a 250 ml wide mouthed glass/plastic beaker containing up to 1.5 cm of vermiculite at the bottom.

2 Add same amount of vermiculite to the top and spray with tap water.

3 Mix faecal sample with vermiculite and push the mixture down to the bottom of the beaker and make a hole in the middle to increase aeration to the culture.

4 A plastic Petri dish is used as a lid for the beaker but a rolled-up paper strip is placed between the rim of the beaker and the Petri dish to facilitate aeration.

5 Leave at room temperature (25°C) for 10–14 days but check regularly every other day to moisten the culture by spraying with water.

6 For harvest, gently add lukewarm water on top of the culture up to the rim of the beaker.

Place the plastic Petri dish on top and avoid trapping air bubbles between.

7 In one go, carefully invert the beaker and add 12 ml of lukewarm water to the Petri dish and leave the setting under light for 4 h allowing larvae (L3) to swim from the edge of the beaker into the clean water in Petri dish.

8 Collect the content from petri dish and examine under microscope for L3. Identified L3 can be saved further for molecular testing.

Other helminthological techniques

Post-mortem (necropsy)

Standard helminthological techniques allow an estimate of parasite burdens in livestock but cannot accurately represent the parasite burden in each and every animal sampled. Some parasites may produce few eggs or produce them only intermittently, as a result not all infections will be identified during routine faecal screening. In addition, the animal may develop clinical signs of disease during larval migration before adult worms reach maturity and produce eggs (that is, patency). In some of these situations the extent of parasite infestation may only be appreciated at necropsy. The total worm count is a laborious procedure but can provide valuable information, especially in field trials, to assess the efficacy (effectiveness) of new anti-parasitic drugs (anthelmintics) or in cases where anthelmintic resistance is suspected.

Total worm count

NECROPSY (POST-MORTEM) WORM
COUNTING TECHNIQUES
This procedure is performed at post-mortem in cases where an estimate of the total parasite load of an animal is required. The following procedure is recommended for sheep and cattle but can also be modified for pigs, horses and other livestock. Several clean 10 l buckets and 200 ml lidded jars, plenty of water, a clean knife and some string will be required.

1 Open the abdomen as for a routine post-mortem (see Chapter 8). Tie off the rectum and oesophagus, in two places, at each end using string and cut between the tied portions to release the entire gastrointestinal tract. Tie off and remove the colon, caecum, abomasum and small intestine (ileum, jejunum). These components should be placed in separate dissecting trays (see Figures 3.8–3.10 for the anatomical structures of the ruminant gut).

2 For each portion of the intestinal tract open up the tract wall to examine the mucosal surface and look for the presence of parasitic worms and/or lesions suggestive of parasite activity, that is, evidence of haemorrhage, thickening or sloughing. Record any unusual findings. If coccidiosis or other protozoal infections are suspected take mucosal scrapings from the intestinal wall for examination under the microscope.

3 Wash the contents of each portion of the intestinal tract into a 10 l bucket flushing the gut lining with clean water (about 5 l). Mix the contents of the bucket and remove 100 ml from each sample, this should be placed in a lidded jar. Take a small amount of the contents from each jar and examine under a dissecting microscope (10× magnification) in a ruled Petri dish (Figure 3.11). If a microscope is not available stain the larvae and adult worms with iodine solution (2–3 min) and decolourize with sodium sulphate (30%). The adult worms can be seen by the naked eye and can be counted.

> Calculation: Number of worms counted (note also the number of each species identified) × 100 = total number of worms present.

4 Sieving could be done to remove particulates that hinder examination. This is achieved

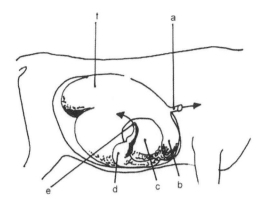

Figure 3.8 The general topography of the bovine abdomen to show the location of the gut in situ. (a) oesophagus; (b) reticulatum; (c) omasum; (d) abomasum; (e) duodenum; (f) rumen. The intestinal tract is tightly coiled within the mesentery. For a total egg count remove the entire intestinal tract which will be full of fluid and therefore quite heavy. Tie off the oesophagus in two places and cut between these two areas so that each cut end is sealed thereby avoiding leakage of fluids. Tie off the rectum (see Figure 3.9) or end of the large intestine in two places and make a cut to remove the entire gastrointestinal tract. The general plan of the ruminant intestinal tract is illustrated in Figure 3.9.

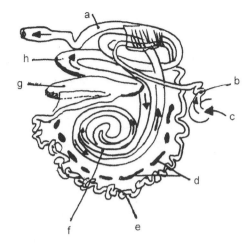

Figure 3.9 Simple plan of the ruminant intestinal tract (the omentum has been removed to show the general layout of key structures). (a) descending colon; (b) duodenum; (c) abomasum; (d) mesenteric lymph nodes; (e) small intestine; (f) colic spiral; (g) caecum; (h) ascending colon. See Figure 3.10 for more detailed information.

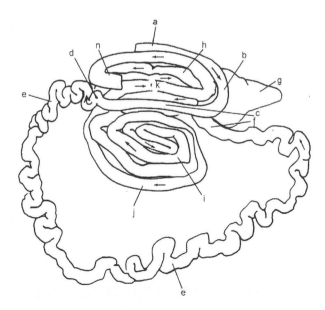

Figure 3.10 Detailed anatomical outline of the intestines of a cow. (a) descending duodenum; (b) caudal flexure of the duodenum; (c) ascending duodenum; (d) duodenal flexure; (e) jejunum; (f) ileum; (g) caecum; (h) proximal loop of the colon; (i) centripetal gyri; (j) centrifugal gyri; (k) distal loop of colon; (n) descending colon.

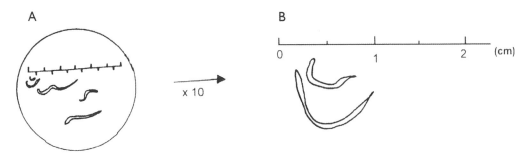

Figure 3.11 Relative size of adult worms which may be seen during a total worm count at post mortem examination. (A) Calibrated petri dish, (B) view under dissecting microscope.

by running the aliquots through a sieve of 400 μm mesh. The sieve size is selected based on the helminths that is expected in the sample. For example, a 37.5 μm mesh is preferred for abomasal specimen (for L4 of *Haemonchus* and *Ostertagia*); whereas, a larger sieve is required for large intestinal specimen (for adult *Trichuris*). Descriptions and measurements of the more common intestinal helminths are given in Table 3.1.

5 Encysted larvae (L4) on the mucosa of abomasum can be recovered by either digesting the whole or a portion of the organ.
6 For complete morphological examination helminth specimens should be appropriately preserved. Live nematodes collected at necropsy should be dropped into warm 70% ethanol so they stretch and present a wrinkle free cuticle to aid morphological identification. Similarly, live cestodes/trema-

Table 3.1 Approximate size and location of common helminth parasites.

Location	Size	Gross characteristics
Abomasum		
Haemonchus sp.	2 cm long	Females are red when fresh (Barber's pole appearance)
Ostertagia sp.	1 cm long, slender	Brown when fresh
Trichostrongylus axei	< 0.5 cm long	Grey when fresh, not readily seen on the abomasal wall
Small intestine		
Bunostomum sp.	2 cm long	Stout white worms with a bent head
Nematodirus sp.	2 cm long	Slender worms, often twisted in clumps
Trichostrongylus sp.	0.5 cm long	Grey slender worms
Strongyloides sp.	0.5 cm long	Grey slender worms
Cooperia sp.	0.5 cm long	Grey and comma shaped
Large intestine		
Chabertia sp.	1.5–2.0 cm long	Large bell-shaped mouth
Oesophagostomum sp.	< 2 cm long	Head bent, tapered buccal capsule
Trichuris sp.	< 8 cm long	Whip like, long filamentous anterior part longer than posterior part

Table 3.2 Examples of tapeworm species.

Tapeworm	Final host	Larvae**	Intermediate host/site of cyst
Taenia saginata *	Human	*Cysticercus bovis*	Cattle/muscle
Taenia solium *	Human	*Cysticercus cellulosae*	Pig, human/muscle ('pork measles')
Taenia multiceps (*Multiceps multiceps*)	Dog, wild canids	*Coenurus cerebrallis*	Sheep, cattle, yaks/ Central nervous system ('Gid' disease)
Taenia hydatigena	Dog, wild canids	*Cysticercus tenuicollis*	Sheep, cattle, pig/peritoneum
Taenia ovis	Dog, wild canids	*Cysticercus ovis*	Sheep/muscle ('sheep measles')
Taenia pisiformis	Dog, wild canids	*Cysticercus pisiformis*	Rabbit/peritoneum
Taenia serialis	Dog, wild canids	*Coenurus serialis*	Rabbit/connective tissue
Taenia taeniaformis	Cat, wild felids	*Cysticercus fasciolaris*	Mouse, rat/Liver
Echinococcus granulosus *	Dog, wild canids	Hydatid cyst	Ruminants, horses, pigs, human/liver, abdominal cavity
Echinococcus multilocularis *	Dog, wild canids, cat	Hydatid cyst	Rodents, other small mammals, human/liver, abdominal cavity

Notes: *These parasites are zoonotic, that is, they can be transmitted from animals to humans. *Taenia* sp. infection in humans occurs due to ingestion of cysts in undercooked meat, humans are the definitive host. In some cases *Cysticercus cellulosae* infection can occur in humans when *T. solium* eggs are accidentally ingested. *Echinococcus* sp. infection (or hydatids) in humans occurs when parasite eggs from infected dogs (and other carnivores) are accidently ingested, this may happen if contaminated salad greens and vegetables are harvested and consumed raw. Humans are a dead end host for this parasite. The common canine and feline Taeniid worms are also transmitted via the food chain.

**Tapeworm larvae often have different names from their adult stage. In some cases this was because the larvae were named before the full life cycle of the tapeworm was fully understood.

todes must be placed in water so they relax and then transferred to appropriate preservatives. It is advisable to save relaxed specimens in 70% ethanol that further aid molecular analysis.

TAPEWORM CYSTS SEEN IN LIVESTOCK SPECIES

During a routine post-mortem the presence of cystic structures may be noted in the liver, lungs, peritoneum, skeletal muscles, brain and other organs in livestock. Many of these cysts are the intermediate stages of tapeworm parasites for which the main final host is often the dog or other predator. Table 3.2 lists the common tapeworms and their definitive and intermediate hosts.

Other tapeworms may be transmitted by vectors such as insects or mites, for example, *Dipylidium caninum*, the dog/cat tapeworm is transmitted by the dog/cat flea (*Ctenocephalides* sp.), *Anoplocephala perfoliatia* the equine tapeworm is transmitted by an oribatid mite. In these cases the cystic stage will be found in the arthropod vector and the adult worm is found in the intestine of the dog/cat or horse respectively.

TRICHINOSCOPY AND TRICHINELLA SP. DIGEST

Meat samples are examined for the presence of *Trichinella* sp. as members of this genus are known to be zoonotic. This organism is a nematode parasite which is found in the muscles of pigs, wild boar, wild carnivores, including bears, rodents, walrus and other marine mammals. It can cause disease in humans who ingest under-cooked infected meat or who accidentally ingest cyst material while handling meat products. The disease in humans may be serious and so it is advisable to wear gloves when handling potentially infected samples. Samples to collect: 35–50 g of muscle especially the masseter (cheek), tongue, intercostal muscles and the diaphragm.

Tests:

1 Trichinoscopy

This technique involves taking small sections of tissue and making a squash preparation by squeezing the tissue between two slides. The slide is then examined under 10× magnification using a stereoscopic microscope.

The appearance of the larvae is illustrated in Figure 3.12.

2 Pepsin digest

1 Trim the fat and fascia from the muscle sample and weigh out 25 g.
2 Homogenize for 1 min in 250 ml of digestion solution (1% HCl and 1% pepsin).
3 Incubate the homogenate overnight at 37–40°C (or 4–6 h with stirring). Allow to settle and decant about 200 ml of the supernatant.
4 Centrifuge the sediment at 200×g for 3–5 min and remove the supernatant. Re-suspend the sediment in 30–50 ml of saline.
5 Take successive aliquots and place in a Petri dish. Count the number of larvae present under the microscope (stereoscope 12.5× magnification).

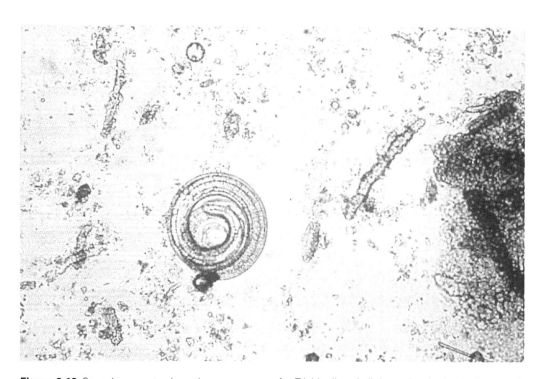

Figure 3.12 Squash smear to show the appearance of a *Trichinella spiralis* larva developing in the muscle tissue of a pig (20×). Note that smoking and curing pork does not kill the *Trichinella* larvae.

See Figure 3.13 for the life cycle of *Trichinella spiralis*.

Owing to its significant public health consequences any trade of meat products that might contain *T. spiralis* requires an appropriate level of testing. A detailed protocol is outlined in the OIE Manual of Diagnostic Tests and Vaccines for Terrestrial Animals which is available online (see text box). Regional food safety regulations will usually stipulate what testing is required for retail of pork and high-risk meat, such as wild boar and other wild game (see also Figure 3.13). Adequate cooking of meat will kill *T. spiralis* but some traditional forms of curing meat will not.

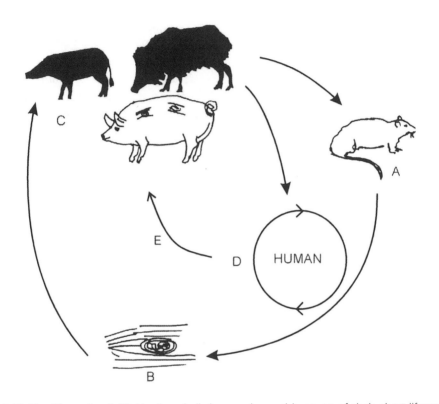

Figure 3.13 The life cycle of *Trichinella spiralis* is complex and because of their short lifespan, the adult worms are rarely found in natural infections. (A) the adult nematode lives in the small intestine of a rodent (or other host) and liberates larvae which encyst in muscle tissue. The infected tissues of the host are ingested by omnivorous or carnivorous species such as domestic and wild pigs or bears. (B) The encysted larvae encyst in the ingested are liberated tissues liberated to become adult worms in the intestine of the pig. (C) The fertilized female worms produce larvae which encyst in the tissues of the pig. (D) Pork (wart hog or bear meat) is ingested by humans and released larvae become adults in the intestine and then further produce larvae which enter the circulatory system and encyst in muscle. The migration of *Trichinella* sp. larvae in human tissues may result in severe anaphylactic (allergic) reaction and death several days after eating the infested meat. Infection in humans (Trichinosis) can be prevented by ensuring that pork meat and that of wild pigs and/or bears is well cooked before consumption. (E) Pigs may become infected when fed waste food. Infection in pigs is prevented by not feeding waste food and controlling rodents in piggeries.

The OIE Manual of Diagnostic Tests and Vaccines for Terrestrial Animals is a compilation of diagnostic procedures and a ready source of information for any veterinary diagnostic laboratory. The Terrestrial Manual has been designed for practical use in the laboratory setting and contains over 110 chapters on infectious and parasitic diseases of mammals, birds and bees. http://www.oie.int/en/standard-setting/terrestrial-manual/access-online/

Interpretation of parasitology results

Table 3.3 provides a rough guide to the interpretation (clinical significance) of some examples of worm egg counts in cattle and sheep. It is generally accepted that a low parasite burden is 'normal' in grazing animals with low levels of helminth parasites inducing natural resistance in older livestock. Complete elimination of parasitic helminths may therefore not be beneficial. However, at certain times of the year, especially when young animals are born, or at other times when additional nutritional or other 'stresses' may occur, previously subclinical infections may become clinical with resultant rapid weight loss, diarrhoea and other clinical signs. Helminth parasites may reproduce rapidly in non-immune younger stock and reach high levels very quickly; this is especially common where rotational grazing is not practiced and young stock graze areas which may have previously been heavily contaminated with worm eggs. The infective larval stages of most helminths tend to develop rapidly when the temperature and humidity are at a suitable level for larval development in the environment, for example parasitic disease tends to be more prevalent following the rains and in areas where the temperatures are mild. Owing to factors such as variation in land topography and climate, differences in herd/flock age structure

and management, the recommendations for parasite control based on worm egg counts will vary from one situation to the next. The life cycle, pathogenicity and fecundity of the parasite must also be considered. However, some general rules can be applied. In most cases, helminths of low pathogenicity such as the tapeworm *Moniezia* sp. must be present in high numbers (that is, > 2000 EPG) before the count would be considered significant. In contrast, low numbers of the nematodes *Ostertagia* sp. (EPG < 200) can cause clinical disease. The life cycle of some helminth parasites can be complex, resulting in a long pre-patent period (time between infection and the presence of eggs in the faeces), for example, the liver fluke (*Fasciola hepatica*). In cases of fasciolosis the animal usually develops clinical signs before eggs are found in the faeces so the presence of any *F. hepatica* eggs is always considered significant.

3.3 Helminths of veterinary importance

The term helminth means 'worm' and the drugs used to control worm infestations are 'anthelmintics'. Helminth life cycles may be simple, with direct infection of the final host by ingestion of infective eggs or larvae, or complex, involving one or more intermediate hosts and several developmental stages. The life cycles of helminths of particular significance will be described in the relevant subsection (see index).

Nematodes (roundworms)

Nematodes, for example, *Nematodirus* spp., *Ostertagia* spp., *Haemonchus* sp. also known as roundworms, are a common cause of ill thrift and diarrhoea in young animals. Roundworms vary in size from thread-like worms < 5.0 mm long (Trichostrongyloids) to large worms

Table 3.3 A guide to the significance of faecal egg count (EPG – eggs per gram) in faecal samples collected from sheep and cattle.

Worm species	Degree of infestation – cattle			Degree of infestation – sheep		
	Light	Moderate	Heavy	Light	Moderate	Heavy
Nematodes (round worms)						
Mixed		200–700	> 700			
Mixed without *Haemonchus* sp.				< 500	500	> 1000
Haemonchus sp.	200	200–500	> 500	100–2500	2500–8000	> 8000
Ostertagia (Teladorsagia) circumcinta				50–200	200–2000	> 2000
Ostertagia ostertagi	150		> 500			
Trichostrongylus axei	50	50–3000	> 300			
Trichostrongylus spp.			> 500	100–500	500–2000	> 2000
Nematodirus sp.				50–100	100–600	> 600
Strongyloides sp.						> 10,000
Bunostomum sp.	20	20–100	> 100			
Cooperia sp.	500	500–3000	> 3000			
Cooperia punctata	50	200	> 200			
Oesophagostomum radiatum	50–150	150–500	> 500			
Oesophagostomum columbianum				100–1000	1000–2000	> 2000
Chabertia ovina						> 1000
Cestodes (tapeworms)*						
Moniezia sp.			>10.000			> 10,000
Trematodes**						
Fasciola hepatica	> 0		> 50	50–200	200–500	> 500

Notes: Information collated from a number of sources. Note that it is difficult to differentiate species of many of the strongyle worms using egg examination. Faecal egg counts should be assessed in view of the clinical condition of the animals sampled. The number of eggs produced by the parasite depends on the species and fecundity of the worm(s) and the degree to which the host's immunity interferes with the parasites' ability to compete for resources.

*In most cases host-adapted adult ruminant cestodes are only pathogenic in very high numbers, however, the cyst form of carnivore tapeworms may cause clinical disease in the ruminant intermediate host.

Trematodes (Flukes)

**Schistosoma* sp. – demonstrate variable levels of pathogenicity, depends on species of fluke, host age and immune status. Low levels (10–100) of some species are considered pathogenic in young stock. Need to assess significance in view of clinical signs seen, usually recommend treatment.

5.0–40.00 cm long (Ascarids). Although most adult nematodes reside in the gastrointestinal system, some species have become specialized to live in other organ systems, for example, lung-worm (*Dictyocaulus* sp.) and the kidney worm (*Stephanurus* sp.). Owing to the fact that some animals will have mixed helminth infections, the severity and nature of the clinical signs seen will vary depending on the number, location and species of nematodes present. Clinical signs during the early stages of infection will depend on the route of migration taken by the larval forms of the nematode, for example, *Strongylus vulgaris* infection in the horse may result in inflammation and blockage of abdominal blood vessels that results in colic and *Parascaris equorum* infection in foals may be associated with pneumonia as larvae migrate through the lungs before maturing in the intestine. The identification of gastrointestinal nematodes may be possible by the examination of eggs in faecal samples but in some cases, because some eggs (for example, the strongyles) are difficult to distinguish in closely related species, it is necessary to examine the larval or adult stages (see Figures 3.4

and 3.14). Common genera (groups) are listed in Table 3.4.

The life cycle of most nematodes involves 4–5 larval stages (L1–L5), three of which may develop in the external environment (see Figures 3.21 and 3.23). The L3 stage is often the infective stage and will go on to develop into the mature parasite within the host species if ingested from pasture (direct life cycle). L2 within egg is the infective stage for most ascarid worm. In some cases the L3 develops in an intermediate host, for example, invertebrates such as snails, earthworms and flies, and the final host becomes infected by either ingesting the intermediate host (indirect life cycle, Figure 3.22) or ingesting L3 shed by the intermediate host or L3 transmitted through bite of

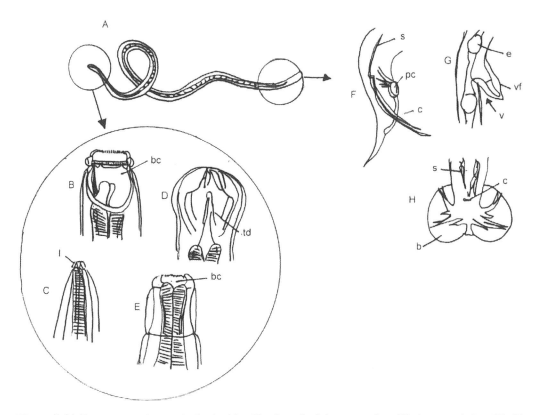

Figure 3.14 Features used to assist in the identification of adult nematodes. (A) shape and size; (B)–(E) morphology of the head; (F)–(H) morphology of the tail and reproductive organs.

Table 3.4 The classification of nematodes of veterinary importance.

Superfamily	Genus	Comments
Trichostrongyloidea (Thread like worms– 'strongyles')	*Ostertagia /Teladorsagia*	Abomasal parasite of ruminants.
	Haemonchus	Abomasal parasite, may cause anaemia in sheep.
	Trichostrongylus	Intestinal parasite, causes enteritis in herbivores.
	Cooperia	Enteritis in ruminants.
	Nematodirus	Enteritis, young sheep, occasionally cattle.
	Hyostrongylus	Stomach worm of pigs, gastritis.
	Dictyocaulus	Lungworm of ruminants and horses, coughing.
Strongyloidea	*Strongylus*	Small and large 'red worms' in horses, enteritis colic and other complications.
	Cyathostomum (Trichonema)	Caecal worm of horse and ruminants.
	Chabertia	Colitis in ruminants.
	Oesophagostomum	Caecal/colon worm of ruminants.
	Syngamus	Tracheal worm in birds, causes 'gapes', may result in asphyxiation.
	Stephanurus	Kidney worm in pigs, urinary tract infection.
Ancylostomatoidea	*Ancyclostoma, Uncinaria*	Hookworms of carnivores, these species live in the small intestine, infective larvae penetrate the skin causing dermatitis, ZOONOTIC (cutaneous larva migrans).
	Bunostomum	Hookworm of ruminants, enteritis.
Metastrongyloidea	*Metastrongylus*	Lungworm in pigs, indirect life cycle, causes coughing.
	Muellerius	Lungworm in sheep, indirect life cycle (mollusc intermediate host), coughing.
	Filaroides, Aelurostrongylus	Lungworm in dogs and cats respectively, complex life cycle requires mollusk or annelid intermediate host and can include other hosts.
Rhabditoidea	*Strongyloides*	Small intestine of mammals, enteritis.
Ascaridoidea (large nematodes, cause, blockage of intestine)	*Ascaridia, Heterakis*	Poultry, enteritis & death.
	Ascaris, Parascaris	Ascarids, young animals, enteritis.
	Toxocara	Weight loss and death. *Toxocara canis*, zoonotic (Visceral larva migrans).

Table 3.4 *continued*

Superfamily	Genus	Comments
	Toxascaris	Gastrointenstinal parasites of dogs, cats, foxes and related host species.
Oxyuroidea	*Oxyuris*	Pinworm in foals, infective larvae on mares' teats, causes irritation and rubbing of anal area.
Spiruoidea	*Habronema, Draschia*	Gastritis in equids, muscid flies act as intermediate host.
	Thelazia	Eye worm, several species of parasitize in birds and mammals including humans.
Filaroidea	*Dirofilaria*	Heart worm, intermediate host – mosquitoes, adult worms live in the pulmonary artery and right ventricles, microfilariae circulate in the blood stream. Diagnosis by blood smear or Knott's test.
Trichuroidea	*Trichuris*	Whip worms, caecum and large intestine of mammals.
	Capillaria	Enteritis in mammals and birds.
	Trichinella	Trichinosis in humans (usually infection by ingestion of undercooked pork) ZOONOTIC, parasite of mammals.

Notes: This table summarises some of the genus in each main group but is not comprehensive. Additional information may be found in Urquhart *et al.* (1996) and Taylor et al. (2007).

intermediate host. Once the nematodes develop to maturity eggs will be produced, this is then termed a 'patent' infection. The time taken from infection with the L3 to the presence of eggs in the faeces is known as the 'prepatent' period. The length of the 'prepatent' period is variable (usually 10–21 days) but depends on the species of parasite and the immune status of the host species. Most nematodes are fairly specific to a particular host species but may infect other hosts, these 'aberrant' infections rarely become patent. Owing to the environmental phase of the parasite life cycle, the duration of the total life cycle may depend on the prevailing weather conditions. Most environmental stages of nematode development have specific requirements for humidity and temperature to allow development to the infective stage, for this reason the build-up of worm infections in animals is often seasonal. Effective control and prevention of nematode infestation requires (1) identification of the species and number of parasite(s) present and (2) knowledge of the life cycle. Some examples are discussed in the text and further information may be found in Taylor et al. (2007) and other texts listed in the bibliography. The microscopic appearance of helminth eggs commonly seen in the faeces of domestic animals is illustrated in Figures 3.15–3.19. The microscopic appearance of a nematode (*Ancylostoma caninum*, the canine hookworm) is illustrated in Figure 3.20b.

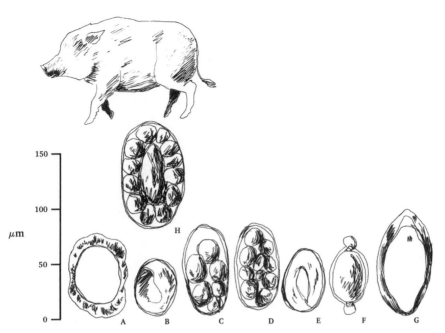

Figure 3.15 The microscopic appearance of parasite eggs commonly found in the faeces of pigs. (A) *Ascaris* sp., (B) *Strongyloides* sp., (C) *Oesophagostomum* sp., (D) *Hyostronglyus* sp., (E) *Metastrongylus* sp., (F) *Trichuris* sp., (G) *Macracanthorhynchus* sp., (H) *Stephanurus* sp. found in urine. Illustration: Louis Wood. See also Plates 2a, 2b and 3.

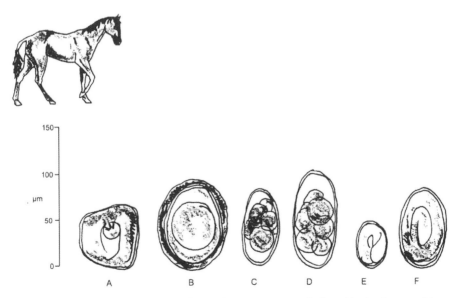

Figure 3.16 The microscopic appearance of parasite eggs commonly found in the faeces of horses and donkeys. (A) tapeworm eggs (*Anoplocephala* sp.), (B) *Parascaris* sp. (young horses), (C) Strongyle type (large and small strongyles, *Strongylus* sp.), (D) *Trichostrongylus* sp., (E) *Strongyloides* sp. (young horses), (F) *Dictyocaulus arnfieldi* (lungworm). See also Plates 2a, 2b, 3 and 4.

Figure 3.17 The microscopic appearance
of parasite eggs commonly found in the
faeces of ruminants. (A) *Nematodirus*
sp., (B) Strongyle type (*Osteragia* sp.,
Trichostrongylus sp. etc.), (C) *Chabertia* sp.,*
(D) *Oesophagostomum* sp., (E) *Strongyloides*
sp., (F) *Neoascaris* sp. (*Toxocara vitulorum*),
(G) *Trichuris* sp., (H) *Fasciola* sp.
(liver fluke), (I) *Paramphistonum* sp., (L)
Capillaria sp., (M) *Moniezia* sp. (tapeworm
with six hooked oncosphere), (N) *Dicrocoelium*
sp. (fluke containing miracidium). Illustration:
Louis Wood. See also Plates 2a, 2b, 3 and 4.

Note: * *Gaigeria pachysalis* which closely resembles
Bunostomum sp., is found in sheep and goats and
is a voracious blood sucker, as few as 100–200 adult
worms are sufficient to produce death in sheep
within a few months of infection.

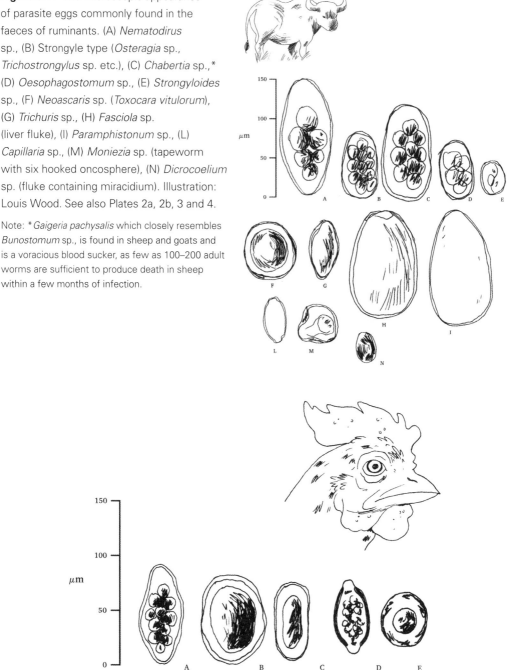

Figure 3.18 The microscopic appearance of parasite eggs commonly found in the faeces of poultry.
(A) *Syngamus trachea* (gapeworm), (B) *Ascaridia* sp., (C) *Heterakis* sp., (D) *Capillaria* sp., (E) tapeworm
eggs (*Davainea* sp., *Raillietina* sp.). Coccidial oocysts are also frequently found in poultry faeces and are
usually numerous but smaller than helminth eggs. Illustration: Louis Wood. See also Plates 2a, 2b and 3.

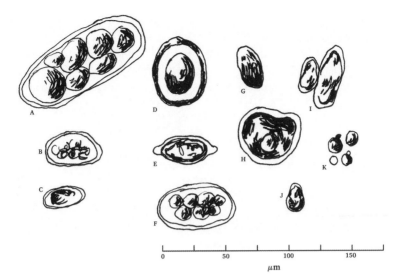

Figure 3.19 The relative size of helminth ova, coccidial oocysts and artefacts. (A) *Nematodirus* sp., (B) *Cooperia* sp., (C) *Capillaria* sp., (D) *Ascaris* sp., (E) *Trichuris* sp., (F) *Chabertia* sp. (strongyle type), (G) mite egg, (H) *Moniezia* sp. (tapeworm), (I) free-living nematode ova, (J) coccidial oocyst, (K) yeast. It is common to see larvae in specimens which have not been examined soon after collection and not preserved or stored in the fridge. Some larvae may be free-living species, identification of helminth larvae and interpretation of results takes some skill and experience. To avoid larvae hatching make sure that samples are collected and stored correctly. Illustration: Louis Wood. See also Plates 2a, 2b, 3 and 4.

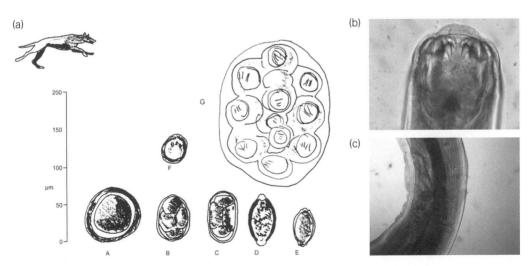

Figure 3.20 (a) The microscopic appearance of parasite eggs commonly found in the faeces of carnivores. (A) *Toxocara* sp., (B) *Ancylostoma* sp., (C) *Uncinaria* sp., (D) *Trichuris* sp., (E) *Capillaria* sp. (*Capillaria plica* may be found in the urine), (F) tapeworms (eggs, but may also see segments), for example, *Taenia* sp. and *Echinococcus* sp., (G) *Dipylidium caninum*. (b and c) Photomicrographs (40×) of the canine hookworm (*Ancylostoma caninum*) demonstrating the structure of the nematode mouth parts and the gut. Photo: Stephano Catalano, University of Calgary, Canada. See also Plates 2a, 2b, 3 and 4.

Cestodes

Cestodes, or tapeworms, range in size from small worms < 1 cm long (*Echinococcus* sp.) to long tape-like structures 1–2 cm wide and several metres long (*Taenia* sp.). The typical 'tapeworm' is the mature stage of a life cycle, which results in patent infections in the final host. Eggs or worm segments may be passed in the faeces of the definitive host. In many cases the segments are still mobile and result in irritation around the anus (for example, *Dipylidium caninum* in dogs and cats). Most cestodes have an indirect life cycle which involves a final host (predator) and an intermediate host (prey). The intermediate host may be another mammal, a bird, reptile, fish or an invertebrate. Cestodes rarely cause clinical disease in the final host but the intermediate (cystic) stages of tapeworms may cause clinical signs related to the size and location of the cyst. Some important examples are discussed further in other parts of the text and are outlined in Table 3.2. The life cycles of *Taenia saginata* and *Echinococcus* sp. are outlined in Figures 3.27 and 3.29. Control and prevention of tapeworm cysts in the intermediate host (for example, 'gid' in ruminants) requires the use of anthelmintics to kill the adult tapeworm in the final host and also the prevention of re-infection of the final host by making sure that the cystic stage is not ingested (that is, do not feed raw offal to dogs).

Trematodes (flukes)

Trematodes, or flukes, generally have indirect life cycles which involve an invertebrate intermediate host (snails). Flukes range in size from small flat organisms 2–3 mm long to large specimens of *Fasciola gigantica* which may reach 5–6 cm in length. Trematodes can cause significant clinical disease, for example, liver fluke (*Fasciola hepatica*), but may only be common in particular regions due to the environmental needs of the intermediate host. The common flukes of veterinary importance are discussed in pages 157–160. Control and prevention requires a good understanding of the life cycle of the trematode involved and may require control of the intermediate host.

Effects of parasitic helminths on their hosts

There are many species of parasites that are relatively harmless. However, there are also many species that produce pathological changes leading to ill health and/or the death of the host. Parasites cause damage to the host in following ways.

1 Parasites can compete with the host for food and nutrients in the intestinal tract resulting in emaciation and poor condition, for example, *Diphyllobothrium* sp.
2 By mechanical obstruction or compression of organs, for example, ascarids in the intestine of young animals, *Syngamus trachea* causing 'gapes' in birds or cystic stages of tapeworms, for example, *Echinococcus* sp. hydatid cysts.
3 Parasites feed on the tissues of the host, for example, worms such as *Chabertia* species cause irritation of the gut lining (enteritis).
4 They suck blood from the walls of the intestinal tract and cause anaemia, for example, wire worms (*Haemonchus* sp.) and hookworms (*Ancylostoma* sp.).
5 Migration of the larval stages of helminths may cause damage to the lungs, liver and other organs (*Ascaris suum*). In some cases the clinical signs are more severe in the prepatent period than when the adult worms are present in their final location. Due to the fact that these clinical signs occur before eggs appear in the faeces it may be difficult to confirm the diagnosis unless additional assays are available to detect antibody or antigen.

6 Meat quality may be impaired by the presence of cystic stages of tapeworms for example, 'pork measles' associated with *Taenia solium*, this also has implications for public health.

Diseases caused by nematodes (roundworms)

Gastrointestinal nematodes

These are probably the most economically important of the internal parasites of livestock and commonly occur throughout the world.

LIFE CYCLE

The life cycles for most of the common gastrointestinal nematodes in livestock are fairly similar. In most cases, infection is by ingestion of the infective larval stage (free living L3 or L2/3 in the egg) from the pasture, except for species such as *Ancylostoma* sp. (canine hookworm) and *Bunostomum* sp. (cattle hookworm) in which L3 may enter the host through the skin and some (for example, various *Capillaria* sp.) which involve a transport host. The adult worms live in the stomach and the large or small intestine of the definitive host where the female worms

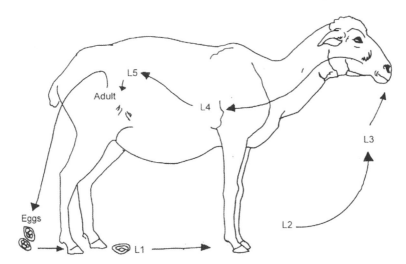

Figure 3.21 Basic gastrointestinal nematode life cycle. Nematode adult worms are either male or female, the females are usually larger and require plenty of nutrients. After mating (usually in the intestine of the host) female worms lay eggs or larvae which are passed out in the faeces of the host. During development the nematode moults at intervals shedding its cuticle (outer coat) at each stage. In the complete life cycle there are usually four moults, the successive larval stages being designated L1, L2, L3, L4 and finally L5, which is the immature adult. Some development occurs in the faecal pat of the host (L1 to L3) if the life cycle is direct or in an intermediate host (for example, a free-living invertebrate) in an indirect life cycle. The L3 is usually the 'infective stage' and can be ingested by the final host while grazing. The L4 and L5 larvae develop within the definitive host. The time period between ingestion of the L3 and the appearance of eggs in the faeces is known as the 'prepatent' period. Within the animal the larvae often migrate through the liver and/or lungs on their way to the intestinal tract where they mature to the adult form and mate. When eggs appear in the faeces the infection has become 'patent'. In some cases the migrating larvae may cause clinical signs in the animal before patency and so the absence of parasite eggs in faecal samples may not necessarily preclude the possibility that nematode infestation is the cause of the problem.

produce large numbers of eggs, which pass out with the faeces. Given appropriate environmental conditions the eggs hatch out in the faecal mass to become first stage larvae. The larvae moult and pass through a number of changes (L1–L2) before they become infective L3 (3rd stage larva), which migrate onto the grass near the faeces. Animals grazing on the contaminated pasture ingest the L3, which develop further to the L4 and L5 adult stages during migration through the host's tissues (Figure 3.21). The entire life cycle takes from 3 to 8 weeks depending on the species of worm. See also Figures 3.22 and 3.23.

CLINICAL SIGNS

Clinical signs are rarely specific for a particular parasite and in most cases there will be a mixed population of helminths in the gastrointestinal tract, which may include cestodes as well as nematodes and possibly trematodes as well. A heavily infected animal may become weak, lose weight, develop a scruffy coat and, in chronic cases, may become anaemic. Mild to heavy diarrhoea is commonly, but not always, apparent. Very heavy burdens of gastrointestinal parasites may cause acute disease (diarrhoea, dehydration, metabolic imbalance) and rapid death whereas moderate burdens may result in chronic weight loss eventually leading to cachexia with death after 2 to 3 months unless the animal is treated with anthelmintic or is placed on good quality clean pasture and develops immunity. Mild infections may cause few clinical signs and

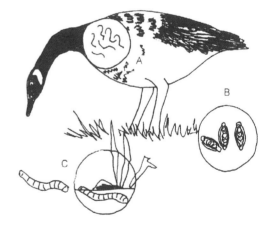

Figure 3.22 Life cycle of a nematode (*Capillaria contorta*) with an indirect life cycle. *Capillaria* sp. are nematodes which infect mammals and birds. Some species require an intermediate host to complete the life cycle, for example, *C. contorta*. This species infects birds and utilizes an earthworm as the intermediate host. The adult parasite lives in the oesophagus and crop of poultry and wild birds (A), eggs are passed out in the faeces (B) and the developing larvae are ingested by earthworms (C). Birds are re-infected when they ingest earthworms during feeding. Other species of *Capillaria* (*C. obsignata* and *C. caudinflata*) also occur in birds. Both of these species live in the small intestine but *C. obsignata* has a direct life cycle and *C. caudinflata* has an indirect life cycle. The prepatent period for all these species of avian *Capillaria* is 3–4 weeks; the clinical signs observed in infected birds depend on the level of infection and the immune status of the bird. Control of *Capillaria* sp. infection requires identification of the parasitic species involved. Prevention of infection is generally more difficult for species with an indirect life cycle (*C. contorta* and *C. caudinflata*) than for those with a direct life cycle (*C. obsignata*) due to the fact that it is usually not possible to prevent access of extensively reared birds to potentially infected intermediate hosts. Other nematodes with indirect life cycles may use other invertebrates such as a free living or parasitic mite as the intermediate host. All parasite life cycles have evolved to utilize the hosts' normal feeding patterns so that transmission is ensured.

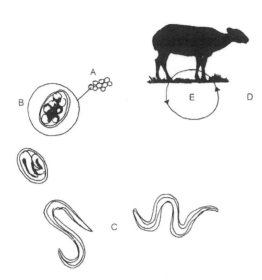

Figure 3.23 Life cycle of a typical ruminant round-worm (for example, *Haemonchus contortus*). (A) Roundworm eggs are shed in the faeces at the morula stage of development (B), (B)–(C) The L1 stage develops and the eggs hatch in the faeces in a day or two (if the conditions are favourable). The second moult (L2) starts and continues complete, in the environment. (D) The infective L3 remains within the L2 cuticle until it is ingested by the host (sheep). (E) The L2 sheath is cast off in the aboma-sum and the L3 parasitic stage undergoes a moult to the L4. The L4 moults to the L5 within the host and develops into the sexually mature adult. (A) Eggs are passed in the faeces and the cycle starts again. The time scale for the life cycle depends on the environmental conditions but is commonly 2–3 weeks. The number of eggs produced by the para-site will depend on the immune status of the host and the fertility of the parasites present.

are only indicated by lower than expected weight gain or increased susceptibility to other diseases. Mild infection is often termed 'sub clinical'. It should be noted that animals which develop high faecal egg counts may do so because they have limited immunity to the parasites; this may be a result of a lack of previous exposure in young livestock, but in adult animals it is more likely

to be the result of poor nutrition and/or poor husbandry (for example, high stocking rate, lack of grazing rotation and so on) which limits the immune capability of the host animal and also increases the degree of pasture contamination. Good animal husbandry is a vital component of parasite control.

PATHOLOGICAL FINDINGS
Infected animals may develop watery or slightly liquid diarrhoea, often with mucus (mucoid enteritis) and sometimes with blood (haemor-rhagic enteritis), but these signs may also be seen in other diseases. At necropsy there is pres-ence of nematode parasites in the digestive tract (some may not be easily seen by the naked eye so a magnifying glass is useful). The intestinal wall and abomasal mucosa may be thickened and swollen (oedematous). Sometimes there is nodule development in the mucosa of the intes-tine and enlargement (hypertrophy) of the mesenteric lymph nodes.

DIAGNOSIS
• Clinical signs.
• Microscopic examination of the faeces to find the parasite stages (eggs, larvae). If larvae are present in faeces it is an indication that the faecal sample is not fresh (unless lungworm is suspected, see page XX).
• Faecal culture may be necessary to allow dif-ferentiation of strongyle eggs. This is done by hatching the eggs and examining the morphology of the larvae. Details of larval identification can be found in Taylor et al. (2007).
• Post-mortem and total worm count with identification of adult parasites.
• Note that regular examination of faecal sam-ples for the presence of parasite eggs is very important for good livestock husbandry. This is because even low parasite burdens can limit production and apparently healthy ani-mals may contaminate pasture with a large

number of parasite eggs. In a good rotational grazing system, the stock person usually ensures that pasture is spelled after young livestock have been grazing because young animals often pass large numbers of parasite eggs in their faeces until they gain some resistance.

CONTROL MEASURES

There are a wide range of effective drugs available for the treatment and prevention of parasitic disease but it cannot be overemphasized that the use of anthelmintics is only effective if concurrent husbandry practices are also implemented to reduce the burden of infective larvae on pasture.

- Periodic use of anthelmintics (chemoprophylaxis): regular treatment of animals in order to keep them free of the most pathogenic (harmful) worms. There are a wide range of products available for treatment and/ or prevention of parasitic disease and advice should be sought from the regional veterinary officer or animal health advisor with regard to the best choice and use of available anthelmintics. Many of the pour-on and injectable preparations used to control ectoparasites may also be effective against some, but not all, helminths.
- Avoid overcrowding: overcrowding (high stocking rate) tends to increase the concentration of worm eggs on pasture and should therefore be avoided.
- Rotation of pasture: pasture rotation allows the animals to graze on 'clean pastures' at intervals while the infected paddocks can be left to become 'clean'. It is, however, important that pasture should be left without grazing livestock for sufficient time (this depends on climatic factors and land use) so that part of the population of infective larvae die.
- Young stock: calves, lambs and other young stock are the most susceptible so they should

not be grazed on pasture contaminated by last year's young stock.
- Wet or moist pastures: wet places like the edges of pools of water are suitable for the development and maintenance of nematode larvae or trematode metacercariae, and if possible should be fenced off.
- Stabled animals: raised hay-racks ensure that feed is not contaminated by bedding or faeces so that infective larvae do not migrate into it from the floor.

A note on anthelmintic resistance

In recent years, anthelmintic resistance has been spreading at an alarming rate. It is widespread in nematode parasites of ruminants, pigs and horses. Resistance can develop wherever anthelmintics are frequently used and this has been seen with commonly used formulations of benzimidazole, levamisole/morantel and ivermectin. In South America, 100% resistance to almost all known anthelmintics on the market has been reported in small ruminant populations. The risk of resistance developing is greatest where animals have not been given a sufficient dose of the anthelmintic and in cases where the same product is used repeatedly over a period of time. The issue also appears to have emerged due to the unnecessary usage of anthelminthic for 'blanket treatment' of all animals in a herd.

Improvements in our understanding of the biochemistry and molecular genetics of parasite-anthelmintic interactions has helped us to better understand what leads to resistance, for example, levamisole/morantel resistance appears to be associated with alterations in cholinergic receptors in resistant nematodes whereas benzimidazole resistance appears to be associated with an alteration in beta-tubulin genes which reduces or abolishes the high affinity binding of benzimidazoles for tubulin in these parasites. Further discussion on the mode of action of anthelmintics and anthelmintic resistance is beyond the scope of this book but in cases where a population of animals has a heavy worm infestation despite having been treated this needs to be reported to the veterinarian in charge so that a suitable parasite management plan can be developed and alternative treatment provided.

It is generally best to alternate the products used. Advice on which products will work best in an area can be obtained from the regional animal health authorities. Worm count (at necropsy) or faecal egg count examination after treatment (Faecal Egg Count Reduction Test – FECRT) are used to determine the efficacy of anthelmintic treatment. In the laboratory, other tests can also be used to identify possible anthelmintic resistance, for example, the egg hatch assay and larval development assays.

It should be noted that helminth infections in a normal flock or herd usually follows a binomial distribution; wherein, ~20% are heavily infected and a similar percentage are infected with very few to no parasites. The rest of the flock or herd will have a moderate to low level helminth infection. This suggests that we only need to treat a proportion of the flock or herd, for example, the heavy egg shedders, and leave the more resistant animals (that is, moderate and low shedders) so they form a refugia of 'anthelmintic susceptible' helminths in pasture and slow down the development of anthelmintic resistance in the flock/herd. An overall assessment of the whole flock/herd to identify the heavy egg shedders and targeting them for subsequent anthelmintic treatment and monitoring drug efficacy by FECRT is a reasonable way to move forward in tackling anthelmintic resistance.

Specific nematodes of veterinary significance

The next section is a brief summary of some of the important nematode parasites that are commonly seen in tropical and temperate climates. The examples selected illustrate key concepts that should be taken into account when dealing with parasite problems. We have provided the 'common' name where appropriate but have also used the scientific (Latin) name as the 'common names' vary from region to region. For additional information, there are a good range of reference texts available that provide a more comprehensive coverage of helminthology (see references at the end of the chapter).

Nematodes found in the stomach or abomasum

- Wire worms (*Haemonchus* spp.) are also called 'barbers pole' worms and are red in colour, the males are 10–20 mm long and the females are 18–30 mm long, red striped and are easily seen at necropsy. In sheep, the worms (in sufficient numbers) can suck a significant amount of blood resulting in anaemia and eventually oedema, due to fluid leakage associated with protein loss and debility. Fluid may specifically collect under the jaw, hence the term 'bottle jaw' for haemonchosis in some areas. Acute disease may result from heavy infestation (2000–20,000 worms) in young sheep (*Haemonchus contortus*), adult sheep may develop a degree of immunity. The life cycle is illustrated in Figure 3.23.
- Brown 'stomach worm' (*Ostertagia* spp.). These are smaller than wire worms but can cause severe clinical disease in ruminants. *Ostertagia (Teladorsagia) circumcincta* and *O. trifurcata* cause inflammation of the abomasum in sheep and goats. *O. ostertagia* is responsible for significant losses in cattle causing abomasitis, diarrhoea and weight loss. Ruminants

are infected by ingestion of the L3 from pasture, exsheathment of L3 occurs in the rumen and further development occurs in the lumen of an abomasal glands. The entire life cycle of the parasite may take as little as three weeks if conditions are favourable. In some cases the L4 may become arrested in development and only progresses to the L5 adult stage after 4–6 months (hypobiosis). Not all anthelmintics are effective against the arrested stages of *Ostertagia* sp. Diagnosis is based on identification of typical eggs in the faeces but blood samples[3] may also be useful to assess the degree of abomasal damage.

Owing to the fact that ovine ostertagiasis is not caused by *O. ostertagia* it is generally safe to rotationally graze cattle and sheep together. Mixed grazing reduces the burden of species specific parasites. Periodic use of fields for crops may also be beneficial to avoid overutilization of 'problem' areas where previous egg contamination has occurred.

- Stomach or 'hair worm' (*Trichostrongylus* sp.). These are the smallest of stomach worms. This parasite is a common cause of enteritis and weight loss in various host species (cattle, sheep, goats, horses and rabbits). *Trichostrongylus axei* lives in the abomasum of ruminants and in the stomach of horses and pigs. *T. colubriformis* occurs in most ruminants and *T. capricola* and *T. vitrinus* occur in sheep and goats. Mixed grazing of horses and ruminants may lead to the buildup of *T. axei*.
- 'Stomach worm' (*Hyostrongylus* sp.) – *H. rubidus* causes severe gastritis in pigs.
- 'Stomach worm' (*Habronema* spp.) – causes gastritis in horses and donkeys. The life cycle involves Muscid flies as the intermediate host.

Nematodes found in the small intestine

- 'Bankrupt worm' (*Cooperia punctata, C. oncophora, Cooperia* spp.). These very small worms are hardly visible to the naked eye. They cause diarrhoea, loss of body condition and infected animals go off their feed. Disease associated with *Cooperia* spp. is commonly seen in young ruminants.
- 'Hookworms' (*Ancylostoma* sp. and *Bunostomum* sp.). Red in colour with teeth or cutting plates in the ventral margin of buccal capsule. Hookworms are very vigorous blood suckers causing anaemia, loss of condition and hypoproteinaemia with associated swelling under the jaw (oedema) due to the collection of fluid. *Uncinaria* sp. and *Ancylostoma* sp. are the canine hookworms. Infection occurs by penetration of the skin by the infective larvae (L3). These hookworms may infect humans causing a skin rash but the infection does not become patent. Control requires attention to hygiene. *Bunostomum* sp. occur in the intestine of cattle, sheep and other ruminants.
- 'Long necked worm' (*Nematodirus* spp.). Similar in size to the hookworm but with a curled up neck. *Nematodirus battus* is especially important as a cause of severe diarrhoea in young sheep. Other species, for example, *N. spathiger*, can infect cattle as well as sheep and goats.
- *Capillaria* sp. Occur in the small intestine (and oesophagus) of mammals and birds. The worms can cause severe diarrhoea and weight loss. Effective control requires identification of the parasite (Figure 3.22). Recent taxonomic classification has led to members of this genus being placed under some newly created genus.
- *Trichinella spiralis.* This nematode worm is zoonotic and is the cause of trichinosis. There are both sylvatic and domestic life cycles. The adult worms can occur in the small intestine of rodents and other omnivores. The worms migrate to muscle tissues. The infective larvae present within cysts in muscle tissue may then be consumed by pigs, bears, humans and other omnivorous hosts. Death in humans

may occur as a result of an acute allergic response if ingested larvae migrate through body tissues. Infection may be prevented by ensuring that meat (especially pork and wild game meat) is very well cooked. The life cycle is illustrated in Figure 3.13.[4]

- 'Roundworm' – this term is usually used to refer to the large roundworms which are readily seen at necropsy or in the faeces of puppies and other animals following dosing with anthelmintics. Most of the worms in this group belong to the super family Ascaridoidea (that is, *Ascaris* sp., *Toxocara* sp. and so on). and are discussed further in the next section.

Nematodes found in the large intestine and caecum

- Large mouthed worm (*Chabertia* sp.) has a large mouth and sucks the mucosa of the intestinal wall causing severe irritation and diarrhoea, which is sometimes blood stained. It is found in the colon of sheep, goats, cattle and other ruminants.
- Nodular worm (*Oesophagostomum* sp.). This is the same size as the large mouthed worm but it is white in colour. The larvae of this parasite penetrate the intestinal wall, sometimes causing small abscesses (called 'pimply gut' condition). It can also cause diarrhoea. These worms are found in sheep, goats and other ruminants.
- 'Whip worm' (*Trichuris* sp.). This worm is found in the caecum and large intestine of a number of species. The adult worms have a long tail like a stock whip. *Trichuris ovis* occurs in sheep, *T. vulpis* occurs in dogs and may cause mucoid diarrhoea and abdominal pain.
- 'Equine red worms' (*Strongylus* sp.). These are large worms found in the large intestine of horses (for example, *S. vulgaris*, *S. equinus*, *S. edentatus*). *Strongylus vulgaris* is particularly pathogenic as it migrates via the mesenteric arteries and may cause damage to the blood

vessel walls with resultant changes to the blood supply to the intestine. In severe cases colic and death may follow. The cyathastomes or 'small red worms' the larval forms of which may also cause significant disease in horses.

Large 'roundworms' or ascarids

The large roundworms are known as ascarids (family Ascaridoidea) and occur in most domestic animals. The adult stages in the intestine cause un-thriftiness in young animals and occasional intestinal obstruction. Another important feature of the group is the harmful effect caused by the migratory behaviour of the larval stages of some ascarids. Ascarid parasites are distributed worldwide and occur in ruminants, pigs, horses, poultry, dogs and humans, all are comparatively large with an elongated, cylindrical and un-segmented body. The common name used is 'roundworm' and adult specimens may be easily seen in the faeces after anthelmintic dosing. The canine 'roundworm' *Toxocara canis* is common in puppies and the life cycle involves transmission of infective larvae from the pregnant bitch via the placenta to unborn puppies and also via milk and from the environment (Figure 3.24). Larvae within the adult dog may become encysted as the host's immunity develops. However, encysted larvae in the bitch are often activated during pregnancy due to inhibition of the bitch's immunity. Infection of young puppies may occur due to continued exposure to contaminated bedding where larvae develop or by suckling from the bitch. If eggs containing larvae (L2) are ingested by human babies or young children, larvae may migrate to the eye causing defects although patent infections do not occur. *T. cati* and *T. leonina* occur in kittens and the life cycle involves transmission from cat to kittens in the milk as well as direct via environmental contamination (but not via the placenta). The clinical signs include pneumonia, ascites and severe emaciation in severe

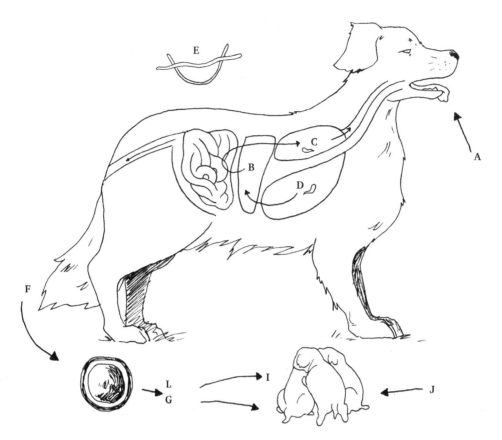

Figure 3.24 Life cycle of the dog roundworm (*Toxocara canis*). This worm has a fairly complex life cycle that is specialized to ensure infection of young puppies. (A) Ingestion of larvated (L2) roundworm eggs from the environment. (B) The eggs hatch in the dog's stomach and the larvae invade the intestinal wall to spread in the blood stream to the lungs. (C) Larvae pass from the circulation into the alveoli of the lungs or, depending on the immune status and/or age of the animal, travel back to the heart in the circulation (D) and are carried away to the tissues. Larvae may then encyst in the tissues as an arrested infective stage. This commonly occurs in adult animals and in a bitch these arrested stages may reactivate during pregnancy to infect the developing puppies. (E) Larvae that do enter the lungs migrate to the trachea and travel to the small intestine to become adults. (F) Adult worms mate and produce eggs which pass out in the faeces. (G) Eggs embryonated in the environment to form the infective stage. (H) Larvated eggs may be ingested by an accidental or 'paratenic' host in which the larval parasites survive but do not develop (for example, rodents, humans, sheep). (I) Puppies may ingest infective eggs from the environment or by ingestion of larvae in the bitch's milk (J), or before birth if dormant larvae reactivate in the pregnant bitch and cross the placenta. Illustration: Louis Wood.

cases or bloating and abdominal discomfort in moderate infections.

Toxocara vitulorum is the roundworm found in cattle and buffaloes (syn. *Neoascaris vitulorum*), which causes 'ascariasis'. It is especially common in warm areas. *Ascaris suum* is the 'roundworm' of piglets and is common in extensively reared young stock. *Parascaris equorum* is the 'roundworm' of foals (it is rare in adult horses due to the development of immunity) and causes diarrhoea and ill-thrift. *Ascaridia galli* occurs in poultry and other birds. Other species which may be of clinical significance in birds include *Heterakis* sp., *Porrocaecum* sp. and others.

Disease process and clinical signs

Larval and adult ascarids may cause significant damage to the lungs, liver and intestine. The clinical signs may include weakness and stunted growth in young animals. Heavy infections result in a potbellied appearance, poor coat and diarrhoea, which alternates with constipation. In dogs, vomiting may also occur.

Diagnosis

1 Clinical signs.
2 Faecal examination for parasite eggs.

Treatment

There are a wide range of anthelmintic drugs that can be used and guidance should be sought from the local animal health advisor or veterinary surgeon. Hygiene and good husbandry are important in order to prevent re-infection of treated young animals.

Nematodes that live in the lungs

The clinical disease associated with lungworm infection in animals must be distinguished from other causes of pneumonia, such as viral (for example, Parainfluenza, infectious bovine rhinotracheitis) and bacterial (for example, *Pasteurella* sp.), diseases. In many cases a combination of infectious agents may be involved and the first step in treatment should be to identify the pre-disposing factors and improve husbandry (housing, nutrition, appropriate stocking rates and so on).

- 'Lungworm' – *Dictyocaulus* sp., the adult worms live in the trachea and bronchi of the host and cause bronchitis with coughing and loss of weight. The clinical signs may vary depending on the presence of concurrent disease(s). In host adapted infections, where the animal has some immunity, the disease may be mild but in some cases the animal may develop an allergic response to migrating larvae and develop severe acute pneumonia. *Dictyocaulus viviparus* is the species seen in cattle and deer, *D. filaria* occurs in sheep and goats and *D. arnfieldi* occurs in horses and donkeys. Donkeys may have sub-clinical lungworm infections resulting in contamination of pasture which may lead to clinical disease in horses grazing in the same locality. Generally, in areas where lungworm is a problem, horses and donkeys should not be grazed together. Diagnosis is made by examining the faeces for the presence of lungworm larvae which hatch from the egg during passage through the gastrointestinal tract after being coughed up and swallowed by the host (see Figure 3.25). A modified Baermann technique can be done to identify the larvae (L1) in the faeces. Blood samples may also be collected for serological examination (CFT and ELISA, see Chapter 6). The disease is treated by removing infected animals to clean grazing after treatment with anthelmintics. There are also vaccines available for cattle which are derived from irradiated larvae.
- Ovine lungworm – *Muellerius* sp., the adults live in the large bronchi and bronchiolar tissue resulting in respiratory signs. The parasite has an indirect life cycle with molluscs (for example, snails) acting as the intermediate host. A modified Baermann is preferred for detecting L1.
- Porcine lungworm – *Metastrongylus* sp., the adults live in the bronchi resulting in respiratory signs. This lungworm has an earthworm intermediate host. The eggs rather than the larvae can be detected in faecal float.

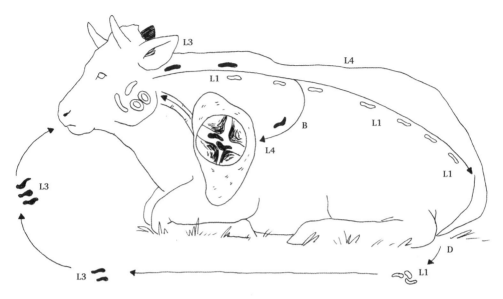

Figure 3.25 The life cycle of *Dictyocaulus viviparous* (bovine lungworm). The adult bovine lungworm is found in the trachea and bronchi resulting in bronchitis and pneumonia with coughing and respiratory distress. The disease associated with the parasite is known by many names, for example, husk, parasitic pneumonia and dictyocauliasis, and typically affects young cattle in their first grazing season. The disease process can be divided into four phases (1) penetration phase (days 2–7, ingested larvae make their way to the lungs), (2) prepatent phase (days 8–25, larvae arrive in the lungs, signs of bronchitis develop and heavily infected animals may die), (3) patent phase (days 26–60), adult worms produce eggs, lung tissue fills with eosinophils, the extent of the lung damage depends on the number of worms present and the immune response of the animal and (4) the post patent phase (days 61–90, recovery period as the immune system clears the infection and debris is coughed up and swallowed). Complications may occur if secondary bacterial infection results in severe pneumonia. (A) The infective L3 stage is ingested and passes into the intestinal tract. (B) The L4 stage develops and passes in the blood and lymph to reach the lungs. (C) Adult worms mature in the lungs and eggs hatch, the L1 migrates to the trachea. The L1 is then coughed up and swallowed to pass out in the faeces. Larvae and not worm eggs are passed in the faeces (see Figure 3.7). (D) The prepatent period is 3–4 weeks (this is the time between ingestion of the infective stage and the presence of larvae in the faeces), note that clinical signs may develop before the infection is patent. (E) The L1 develops into the L2 and L3 in the environment, this takes about 5 days under optimal conditions. The larvae are motile which allows them to move away from the faecal pat. In some cases, a fungus (*Pilobolus* sp.) assists the transfer of larvae from the faecal pat to pasture or bedding material. Illustration: Louis Wood.

- *Oslerus osleri*, *Crenosoma vulpis* (dogs) and *Aelurostrongylus abstrusus* (cats) can occur in the respiratory tract of carnivores. A modified Baermann is preferred for detecting L1.

Nematodes that live in other organs

There are several other nematodes of veterinary importance that live in the kidneys (for example, *Stephanurus dentatus* in pigs) and the heart and blood vessels (for example, *Dirofilaria immitis*) of animals. *Dirofilaria immitis* is a common parasite of dogs in areas where the intermediate hosts (mosquitoes) are prolific. Adult worms live in the large blood vessels and chambers of the heart. Clinical signs include coughing and congestive heart failure in the later stages of the disease. Treatment may be difficult due to the location of the parasites and the risk of blockage of important blood vessels if the worms die. The control of the disease requires routine preventative treatment with an anthelmintic and the control of the mosquito intermediate host. *Parelaphostrongylus tenuis* is a meningeal worm of white tailed deer and camelids/horses are aberrantly infected with this helminth. *Setaria digitata* commonly dwells in the peritoneal cavity of sheep but their larval migration can cause cerebrospinal nematodiasis. *Onchocerca cervicalis* is found in the ligamentum nuchae of horses and mules. *Brugia phangi* resides in the lymphatic system of cats and dogs. *Stephanofilaria* spp. have predilection to skin of cattle.

THE THORNY HEADED WORMS

The Acanthocephala (thorny headed worms) include *Macracanorhynchus hirudinaceus* which inhabits the small intestine of the pig. The parasite has an indirect life cycle using the millipede as the intermediate host. Members of this phylum are pathogenic particularly to water fowls.

Diseases caused by cestodes

Adult tapeworms are usually found in the small intestine of a 'final' host but require an intermediate host, which may be a vertebrate or an invertebrate, to complete their life cycle. Tapeworm eggs hatch when they are swallowed by the intermediate host. They penetrate the

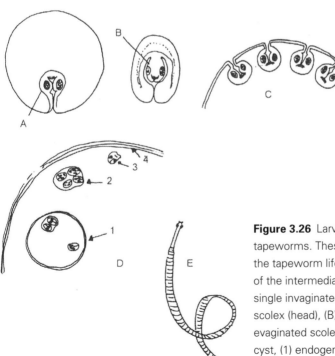

Figure 3.26 Larval (cyst) forms of some tapeworms. These are the immature stages of the tapeworm life cycle that occur in the tissues of the intermediate host. (A) Cysticercus with a single invaginated (inverted into the cyst space) scolex (head), (B) cysticercoid with a single evaginated scolex, (C) coenurus, (D) hydatid cyst, (1) endogenous cyst, (2) brood capsule, (3) protoscolex, (4) cyst wall, (E) Strobilocercus.

intestinal wall and migrate to suitable parts of the body such as muscle, lung or liver where they develop into a cyst. The cysts of some tapeworm (Taeniidae) contain a fluid filled cavity with one or more heads, this is called a bladder worm (Cysticerci, Coenurus, hydatid cyst). These larval forms are illustrated in Figure 3.26. When the definitive host eats part (or all) of the intermediate host in which the bladder worm or cysts are situated, the heads of the larval worm attach themselves to the intestinal wall of the final host and develop into an adult tapeworm. The intermediate hosts of *Anoplocephala* sp. tapeworms are invertebrates such as free-living mites that are ingested accidentally while horses are feeding or grazing. In dogs and cats the tapeworm *Dipylidium caninum* is transmitted by the flea which may be ingested accidentally by dogs and cats while grooming. Most species can be infected with tapeworms without showing obvious clinical signs. In cattle, adult *Moniezia* sp. tapeworms may occur in fairly high numbers without causing any significant clinical disease but in horses, adult *Anoplocephala* sp. tapeworms are occasionally reported to cause significant damage with occasional perforation of the intestinal wall. There are a range of products available to kill adult tapeworms but not all of the products used to treat roundworms contain suitable active compounds. The local veterinary officer or animal health advisor can usually provide advice on selecting a suitable product.

The following are some of the important diseases associated with tapeworms or, more commonly, their cysts that have been frequently reported in livestock. In some cases confusion has arisen due to the fact that the adult tapeworm and the cyst stage have had different names, this was initially due to the fact that it was not known that both were part of the same life cycle.

Cysticercosis

Taenia sp. tapeworms are distributed worldwide. Although the final host is usually a carnivore or omnivore there is usually little sign of disease caused by the adult tapeworm in these species. The main clinical concern is the cystic stage which develops in the intermediate herbivore or omnivore host, for example: *Taenia saginata* adult parasite occurs in the intestine of humans and the cysts occur in cattle (Figure 3.27). The adult of *T. solium* occurs in the intestine of humans and the cysts occur in various organs of swine. Other species cause cysts in rabbits and rodents and the adult parasite occurs in the intestine of carnivores such as dogs and cats (see Table 3.5).

LIFE CYCLE

The adult tapeworm lives in the small intestine of a definitive host (for example, human for *T. saginata*). The mature worms can be up to 10m long and have a life span of more than 20 years. Each tapeworm can produce 400 segments each containing 8000–50,000 eggs. The eggs, or the segments containing the eggs, are passed with the faeces of a definitive host and contaminate soil, pastures, hay or drinking water and can survive in the environment for 4 to 6 months. If ingested, the egg hatches in the stomach of the intermediate host releasing embryo that cross the intestinal wall, enter the blood stream and then localize in the muscles or internal organs throughout the body to form cysticerci. It can take 10 to 12 weeks for cysticerci to develop in the intermediate host and become infective. In some cases, cysticerci can also develop in the final host (for example, *Taenia solium* in humans) by accidental ingestion of eggs in contaminated food. A final host usually develops infection by eating raw tissues which contain tapeworm cysts (Figure 3.27). In the case of *T. solium*, the principal location of the cysticerci is the muscle tissue of the pig (or rodent, bear and so on), but the cysticerci may also develop in other organs

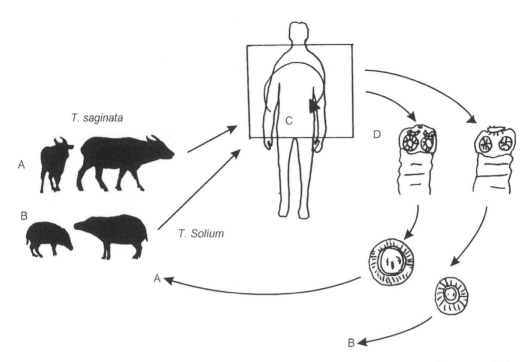

Figure 3.27 Life cycle of *Taenia solium* and *T. saginata*. (A) *T. saginata*, the beef tapeworm, forms cysts in the tissues of the bovine (cattle) intermediate host. (B) *T. solium*, the pork tapeworm, forms cysts in the tissues of the porcine (pig) intermediate host. (C) Humans are the definitive host for the beef and pork tapeworms. Infection occurs in humans following ingestion of undercooked beef or pork meat. (D) Adult tapeworms develop in the human small intestine. In some cases humans may also act as the intermediate host for *T. solium* and develop cysts in the body, occasionally these cysts occur in the brain.

such as lungs, liver, kidney or brain. This type of cysticercus is technically called *cysticercus cellulosae*. The cystic stages and hosts of other *Taenia* sp. tapeworms and the hydatid tapeworm are outlined in Table 3.5.

CLINICAL SIGNS AND DISEASE PROCESS
There are often no clear clinical signs of the disease in the intermediate host unless the cyst becomes very large or, due to its location, it starts to compress surrounding tissue resulting in clinical signs related to local dysfunction. If cysts rupture there may be a generalized allergic reaction to released tapeworm larvae and fluids. Adult tapeworms may cause diarrhoea and abdominal discomfort in the final host.

DIAGNOSIS
In the definitive host tapeworm segments are often passed in chains but individual eggs can also be observed in faeces. Diagnosis in the intermediate host depends primarily on the inspection of carcasses for cysticerci in the heart, liver, lungs, masticator muscles and tongue. Cysts can range in size from 5 mm to 6 cm long. The preferred sites for cysticerci of different *Taenia* sp. are variable but meat inspection procedures are usually based on the compromise between detection of cysticerci and the preservation of economic value of the carcasses. Ante-mortem diagnosis of cysticercosis in the intermediate host is difficult. In humans, it is presently carried out by laboratory detection of cysticerci by CAT (computerized axial tomography)

Table 3.5 Final and intermediate hosts of common tapeworms.

Definitive host			Intermediate host		
Adult	Host	Site	Larvae	Host	Site
Taenia solium	Human	Small intestine	*Cysticercus cellulosae*	Pig, human	Muscle, liver, lungs, kidney
Taenia saginata	Human	Small intestine	*Cysticercus bovis*	Cattle, other ruminants	Muscle
Taenia ovis	Dog	Small intestine	*Cysticercus ovis*	Sheep	Muscle
Taenia (Multiceps) multiceps	Dogs, foxes	Small intestine	*Coenurus cerebralis*	Sheep, goat, cattle, pig	Central nervous system
Taenia hydatigena	Dog, wild canids	Small intestine	*Cysticercus tenuicollis*	Sheep, cattle, pig	Peritoneum, liver capsule
Taenia pisiformis	Dogs, foxes	Small intestine	*Cysticercus pisiformis*	Rabbit	Peritoneal cavity, liver and kidney
Taenia taeniaeformis	Cat	Small intestine	*Cysticercus fasciolaris* (strobilocercus)	Mouse, rat	Liver
Taenia serialis	Dog	Small intestine	*Coenurus serialis*	Rabbit	Connective tissue
Echinococcus granulosus	Dog or wild canidae	Small intestine	Hydatid cyst (unilocular)	Sheep, cattle, horse, human, other wild and domestic mammals	Liver, lung, other visceral organs, muscle and brain
Echinococcus multilocularis	Dog or wild canidae	Small intestine	Hydatid cyst (multilocular)	Arvicoline Rodents, dogs, human, other ruminants	Liver, other organs

scanning techniques for cerebral cysts and on finding antibody to cysticerci in the cerebrospinal fluid.

CONTROL MEASURES
- Health education, construction and maintenance of latrines (human sanitation), treatment of all human cases.
- Compulsory meat inspection and thorough cooking of meat. Cysticerci can be destroyed at 57°C.
- The infected carcasses could be frozen at –10°C for at least 10 days, which would kill the cysticerci, although this method does reduce the economic value of the meat. In heavy infection, where more than 25 cysticerci are detected, the carcass should be destroyed and discarded.
- The use of human faecal sludge as fertilizer should be confined to cultivated fields or to those on that pigs and cattle will not be grazed for at least 2 years.
- Safe disposal of infected carcasses, that is, the cyst material in the carcass of intermediate hosts should not be made available to foraging scavengers such as domestic dogs.

Coenurosis or gid

Coenurosis is caused by the intermediate stage (*Coenurus cerebralis*) of the tapeworm *Taenia (multiceps) multiceps* which occurs in the small intestine of dogs and other canid species such as the fox and jackal. The adult tapeworm is 40–100 cm long and the gravid segments measure 8–12 mm by 3–4 mm. *Coenurus cerebralis* is found in sheep, cattle and other ruminants in many parts of the world. The cyst varies in size from a pea to a hen's egg and is composed of a thin transparent wall on the inner side of which are a number of small white irregular white spots 400 to 500 in number each representing an invaginated larval tapeworm head (see Figures 3.26 and 3.28).

LIFE CYCLE

The intermediate stage (*Coenurus cerebralis*) develops in the brain and spinal cord of sheep, yak, cattle, other ruminants and has also been found in man. Infection occurs as a result of ingestion of ripe segments of *Taenia (multiceps) multiceps* usually in grass or water contaminated with infected dog faeces. The embryos hatch in the intestine and are carried in the blood stream to the liver (where many die) then to the heart and finally to the systemic circulation. At about 8 to 14 days after infection the embryos reach the brain. The young cysts wander about in the brain before settling down and become fully developed in 7 to 8 months. The fully grown cyst can measure 5 cm or more in diameter and has a delicate, translucent wall. On the inner surface of the cyst

Figure 3.28 Life cycle of *Taenia multiceps*, the tapeworm parasite that has a cyst stage in the brain causing 'gid' in the ruminant intermediate host. (A) Adult tapeworm in the small intestine of the dog (final or definitive) host. The adult worm may reach up to 1 m in length. (B) Mature tapeworm segments are passed in the dog's faeces and release eggs into the environment. (C) The ruminant intermediate host ingests the tapeworm eggs while grazing. (D) The larval cyst (*Coenurus cerebralis*) develops in the brain causing circling. (E) The dog becomes infected by ingesting infected brain tissue. The parasite is controlled by (1) treating infected dogs with a drug to kill the adult tapeworm and (2) preventing the dog eating infected ruminant tissues.

are a number of heads which resemble the scolex of the adult worm. The final host (that is, dog) becomes infected by ingesting the cyst.

CLINICAL SIGNS AND DISEASE PROCESS

The adult tapeworm in dogs causes no obvious clinical disease. In the intermediate host, there may be signs of nervousness and excitability (meningo-encephalitis) as the embryos wander about in the brain tissues. Post-mortem examination at this stage indicates the presence of numerous narrow tunnels in the brain. Yellow streaks of dead tissue and small blisters are apparent on the brain surface. However, this tends to occur when several parasites invade the brain simultaneously and in many cases these preliminary signs do not appear. The primary acute stage is followed by a latent period of 4 to 6 months. This is followed by a chronic stage associated with the growth of the cyst and the production of pressure on the brain. The animal may develop stereotypic movements depending on location of the cysts in the central nervous system. Most cysts tend to develop in a region on the surface of the cerebral hemisphere resulting in an animal that holds its head to one side and turns in a circle. Movement is usually towards the affected side of the brain and in many cases such animals are blind in the eye on the opposite side. When the parasite is lodged in the front part of the brain the head is held high and the animal steps high (trotters) or may walk in a straight line until it meets an obstacle. When the cyst is found in the cerebellum the animal is easily frightened and it has a jerky and staggering gait in the hind legs which gradually grows worse and finally leads to collapse. In advanced cases, there may be actual deformity of the skull with softening of the skull bone. Post-mortem examination usually reveals only one large vesicle, 5 cm or more in diameter, in the brain. The embryos arrested in other organs degenerate and either disappear or become a spherical encapsulated body.

DIAGNOSIS

- From the clinical signs coenurosis must be differentiated from other conditions which affect the brain such as tumours, abscesses, listeriosis, poisoning by certain plants, acute septicaemia and so on.
- Very often it is not possible to arrive at a definite diagnosis except during the post-mortem examination.
- Clinical epidemiology, previous history of cases in the area.

TREATMENT

- When the cyst is located on the surface of the brain surgical removal can be carried out.
- In some instances, trephining or removal by means of special trocar and cannula has given good results.
- Recently some anthelmintics have been tested for efficacy against coenurosis in sheep, for example praziquantel may be effective (seek advice from the local veterinary officer for current recommendations).
- Regular anthelmintic medication for the definitive host (usually domestic dogs), for example, praziquantel is often very effective.

CONTROL MEASURES

- Prevent the final host (dog) from eating carcasses potentially infected with coenurosis by destroying the carcass.
- Regular treatment of the definitive host (dog) with anthelmintics. With some products the dog should be confined for 48 h after treatment to facilitate the collection and proper disposal of infected faeces.
- Education for the public with regard to the implementation of hygiene measures.

Hydatidosis or echinococcosis

Echinococcosis is caused by the tapeworm *Echinococcus granulosus* or *E. multilocularis*, which are some of the smallest cestodes of domestic carnivores. The adult tapeworm consists of a scolex (head) and three to six segments. The parasite is distributed worldwide and occurs in the small intestine of the dog and many wild canids. The intermediate stage is called the 'hydatid cyst' and is found in a wide range of species including domestic ruminants, pigs, horses and humans. An adult *E. granulosus* has a life span of up to 20 months and can produce 6000–12,000 eggs/month. A dog may host a hundred or even thousands of these worms. Hydatids are transmissible to humans and so are of public health importance. Children who have close contact with infected dogs are exposed to the greatest risk of infection. Human infection, in this case, does not occur through ingestion of cysts in meat but can occur through ingestion of soil, water or any other vegetable material contaminated by dog faeces.

LIFE CYCLE

The adult tapeworm is found in the small intestine of dogs (Figure 3.29). The eggs are voided in faeces which may be dispersed over a wide area in dust and contaminated herbage and pasture. This provides a means of infection for herbivores. The embryonated eggs are capable of survival on the ground for about 2 years. After ingestion by the intermediate host, activated oncospheres (embryos) pass through the intestinal wall and are distributed to other organs via the blood. The liver filters out most of them but others can lodge in the lungs, kidney, bones and brain.

Oncospheres (Figure 3.26) that are not destroyed in the tissues develop into cysts (hydatid cyst) which can attain a diameter of 1 cm in 5 months. At this stage the cyst contains a number of secondary cysts or brood capsules each containing about 40 protoscolices or tapeworm heads. Brood capsules may detach themselves from the cyst wall to float free in

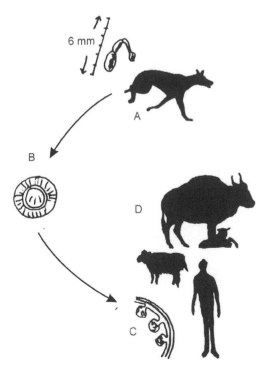

Figure 3.29 Life cycle of the hydatid tapeworm (*Echinococcus granulosus*). (A) The adult tapeworm lives in the small intestine of the dog and many wild canids, which are the definitive hosts. The mature worm is very small, about 6 mm long. (B) The eggs are passed in the faeces of the dog to contaminate food and water supplies. Eggs are accidentally ingested by grazing animals and humans. (C) The hydatid cysts develop in the liver, lungs and brain of the intermediate hosts. (D) In humans, the hydatid cysts may reach a very large size. The definitive host (dog) becomes infected by ingesting the cyst stage in the tissues of the intermediate host. This commonly occurs where dogs and wild canids have access to the raw visceral organs of animals which die and are not buried. Infection can be controlled by (1) treating infected dogs for the adult tapeworm and (2) preventing access of dogs and wild canids to dead livestock by burying carcasses or by cooking any meat and offal fed to dogs.

6 mm

the vesicular fluid forming hydatid 'sand'. The cyst progressively increases in size, displacing organs. Not all the cysts contain protoscolices but those that do are infective to dogs after 5 to 6 months. Infectious scolices are transmitted to dogs when they eat raw infected meat. Adult tapeworms reach maturity 6 to 7 weeks later.

CLINICAL SIGNS

The adult tapeworm is not pathogenic and thousands may be present in a dog without clinical signs. Domestic animals with hydatid cysts in the liver or lungs may not demonstrate clear signs of the disease and the majority of the infections are only revealed at post-mortem inspection. However, the rupture of a cyst may cause sudden death due to the development of anaphylactic shock. Pressure by the growing cyst may cause a variety of clinical manifestations if hydatid cysts are located in the kidney, pancreas, nervous system or marrow of the long bones.

DIAGNOSIS

Diagnosis in the intermediate host is generally made by post-mortem examination of viscera and other organs. The protoscolices and their hooks inside the cysts can be identified under the microscope. In live animals, specific diagnosis of hydatid infection is rarely possible. Percussion of infected liver and lungs may lead to detection of hydatid cysts but these must be distinguished from abscesses or other cystic structures.

- In humans, serological tests such as ELISA, complement fixation or immunoelectrophoresis are commonly used. Scanning techniques may also be used to locate the cysts in organs.
- In dogs, adult tapeworms are difficult to identify because the segments are small. Faecal examination for the presence of eggs can be attempted but *Echinococcus* eggs are difficult to distinguish from those of other tapeworm belonging to family: Taeniidae.

TREATMENT

- The hydatid tapeworm can be more difficult to remove than the *Taenia* sp. but several anthelmintics, notably praziquantel, are now available which are highly effective. After treatment it is advisable to confine dogs for up to 48 hours to facilitate collection and disposal of faeces containing parasite eggs.
- In humans, hydatid cysts may be excised surgically. In some cases, hydatid cysts have been found incidentally when a patient is undergoing surgery or radiography for other conditions.

CONTROL

- Health education.
- Control of livestock slaughter with confiscation and destruction of affected viscera.
- Regular treatment of dogs to eliminate the adult tapeworms and prevention of infection in dogs by exclusion from their diet of animal material containing hydatid cysts. This can be done by preventing the access of dogs to abattoirs and by proper disposal of carcasses.
- In some countries, incidental benefit from the destruction of stray dogs for rabies control has seen a great reduction in the incidence of hydatid infection in humans.

Diseases caused by trematodes

Flukes of veterinary importance include *Fasciola* sp., *Paramphistomum* sp., *Dicrocoelium* sp. and *Schistosoma* sp. Effective fluke prevention and control strategies require accurate and early diagnosis and an understanding of the life cycle of the parasite involved. As with other helminth diseases the clinical signs are variable and depend on the level of infection as well as the immunity of the host.

Fasciolosis

Fasciolosis has a worldwide distribution and is common in rice growing areas. Liver fluke infection can be caused by *Fasciola hepatica* (common liver fluke or sheep liver fluke) or *F. gigantica* (liver fluke). *F. hepatica* is usually found in temperate areas whereas *F. gigantica* is seen in tropical/sub-tropical zones.

SUSCEPTIBLE DOMESTIC SPECIES

Cattle, sheep, buffaloes, goats and pigs are susceptible. The disease is rare in other species although the number of reported cases in humans has been rising in some parts of the world.

LIFE CYCLE OF *FASCIOLA HEPATICA*

Fasciola sp., in common with other flukes, are hermaphrodite (both sexes present in same individual), flat, un-segmented, leaf like parasites. They are greyish-brown in colour. The life cycles of *F. hepatica* and *F. gigantica* are similar and complex involving an invertebrate interme-

diate host (Figure 3.30). Adult flukes are found in the bile ducts of the liver of the host and lay eggs which pass into the intestine to be voided with the faeces. Eggs develop into minute immature worms (miracidia). The miracidia penetrate certain species of snail (*Lymnaea* spp.) which act as the intermediate host. The immature flukes amplify within snail tissue and develop for 3 weeks or more until they are released into the water (as cercariae) to encyst (as metacercariae) on aquatic vegetation. Each egg can potentially develop into 1260 metacercariae; each worm can shed around 4000 eggs per day; and each animal could be infected with, on an average, 100 adult worms.

Infection of sheep (and other hosts) takes place in damp, marshy and swampy areas where snails are found. Livestock become infected when metacercaria encysted on vegetation are consumed. In sheep, the larval stages burrow through the wall of the intestine into the abdominal cavity from where they migrate to

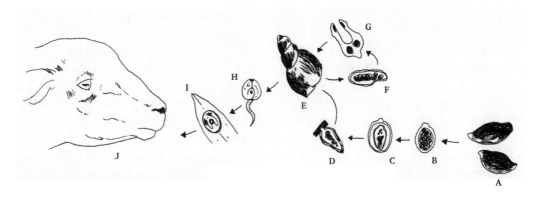

Figure 3.30 The life cycle of the liver fluke (*Fasciola hepatica*). (A) The adult liver flukes live in the bile ducts of the liver. (B) Eggs are passed in the host's faeces (most common in sheep). (C) Egg containing the first immature stage, the miracidium. (D) Miracidium penetrating the snail intermediate host (*Lymnaea* sp.). Without the snail the life cycle cannot continue. (E) The snails prefer wet areas and boggy land, this is why liver fluke disease is more common in these environments. (F) The sporocyst forms within the snail and develops into the next stage, the redia (G). (H) Redia develop into cercaria which are shed from the snail. (I) The cercaria encyst on the vegetation as metacercaria where they are ingested by the definitive host (J) sheep host. Cattle and other grazing animals can also be infected but liver fluke disease is less commonly seen in these other species. Illustration: Louis Wood.

and through the liver. During this stage, the infection may cause serious damage to the liver. The mature stage of the fluke migrates to the bile ducts to lay eggs. It should be noted that clinical signs may be apparent before eggs can be seen in faecal samples. The prepatent period lasts from 10–12 weeks for *F. hepatica* and from 13–16 weeks for *F. gigantica*. It is not possible for the liver fluke to complete the life cycle without the snail intermediate host.

GREATEST DANGER OF INFECTION

As outlined earlier, infection is common in marshy, swampy zones along water ways or lakes where the intermediate hosts (snails) are present. It is not possible for the liver fluke to complete the life cycle without the snail. The liver fluke may take 17–18 weeks to complete the entire life cycle (depending on suitable environmental conditions).

CLINICAL FINDINGS

Acute disease is seen when large numbers of immature forms penetrate and migrate through the liver. This occurs 2–6 weeks after ingestion of large numbers of metacercariae. Chronic liver fluke disease occurs 4–5 months after ingestion of moderate numbers of metacercariae and is more common in cattle. The clinical signs seen in acute liver fluke infection reflect the degree of liver damage and depend on the number of migrating fluke present. Young sheep often die in the acute phase following severe diarrhoea with development of dehydration. Some survive heavy infections and develop signs of chronic liver damage, for example, anaemia and oedema in the dependent parts of the body (especially in the sub-mandibular and sub-sternal regions). Adult sheep and cattle usually have milder infections resulting in a reduction in growth rate, weight loss and reduced milk yield or poor wool production. Chronic liver fluke infection in cattle is characterized by thickening of bile duct which appears like a 'clay-pipe'. This is due to fibrous

changes of bile duct mucosa due to chronic irritation caused of the spines on the tegument of *Fasciola hepatica*.

DIAGNOSIS

- Diagnosis can be based on post-mortem findings, for example, fibrous hardening of the liver with thickening of the bile ducts in which flukes can be seen. In acute cases haemorrhage is seen in the liver parenchyma, often with severe haemorrhage into the abdominal cavity.
- Ante-mortem diagnosis is primarily based on clinical signs, previous history of fasciolosis in the area and identification of snail habitats.
- Diagnosis of bovine fasciolosis is sometimes difficult. In this context, routine haematological examination of blood for evidence of eosinophilia and examination of faeces for fluke eggs is essential.
- Blood biochemistry such as estimation of plasma levels of enzymes released by damaged liver cells can also be used for diagnosis (as well as a predictor for recovery). Two enzymes are normally used, glutamate dehydrogenase (GLDH) is used to estimate damage caused in liver tissue and gamma glutamyl transpeptidase (GGT) is used to indicate the extent of damage to the epithelial cells lining the bile ducts. For more information about biochemical assessment of liver function see Chapter 7.
- Serological tests may be used for the detection of antibodies against components of the immature flukes, for example, ELISA and the passive haemagglutination test (see Chapter 6).

CONTROL AND PREVENTION

A. Reduction of the intermediate host (snail) population

1 The snail intermediate host is found in marshy, wet places and near rivers and livestock should be kept away from these areas.

2 Practice land reclamation (drainage) and fencing around infected areas.

3 Chemicals such as copper sulphate have been used to kill the snail but they have the potential to kill other aquatic organisms and to damage the environment, so widespread use is not advised.

B. Reducing land contamination and the use of anthelmintics

The prophylactic use of fluke anthelmintics should be aimed at reducing pasture contamination by fluke eggs. Note that not all drugs routinely used for nematode control are effective against liver fluke.

1 Remove the fluke population in the host before the development of heavy infection.

2 Avoid grazing animals in areas previously contaminated.

3 Always check the faeces of recently purchased livestock for fluke infection before introducing them to new land.

Paramphistomosis Stomach fluke or conical fluke infection (Paramphistomomum sp.)

There are several species of ruminal (stomach) or conical fluke of which *Paramphistomum cervi* and *P. microbothrium* are the most common. They are distributed worldwide but are more common in sub-tropical and warm areas. The adult flukes are found in the rumen and reticulum of cattle, sheep and goats, and immature stages are found in the duodenum. The life cycle of Paramphistomes is similar to that of *F. hepatica* and involves several species of fresh water snail, which serve as the intermediate host. The pathogenic effect in the final host (ruminant) occurs during the intestinal phase of infection. The young flukes feed on intestinal epithelium and cause severe erosions in the duodenal mucosa characterized by oedema, haemorrhage and ulceration. In heavy duodenal infections, the most obvious sign is diarrhoea accompanied by anorexia and thirst. The adult parasites in the fore-stomach rarely have a harmful effect even when many thousands are present and feeding on the wall of rumen or reticulum. Diagnosis is usually based on clinical signs with disease more common in the young animals of the herd especially those with a history of grazing areas around snail habitats. As with *Fasciola* sp. clinical disease often occurs during the prepatent period so faecal examination for eggs may be of limited use. The paramphistome eggs are colourless and can be easily distinguished from the yellow *Fasciola* sp. eggs which are similar in shape. Control and preventive measures are similar to those for fasciolosis.

Dicrocoeliosis (Dicrocoelium dendriticum)

Dicrocoelium dendriticum is distributed worldwide. This parasite lives in the bile ducts of many herbivores, particularly sheep, cattle, goats, rabbits and wild ruminants, such as deer. In heavy infections, production losses may occur as a result of chronic damage to the liver. The life cycle is fairly complex and involves three hosts. These eggs, when excreted through the faeces of the definitive host contain fully developed embryos. They hatch when ingested by the first intermediate host (many types of land snail) in which they develop into cercaria. These are passed in a mass cemented together by slime (slime balls). These are then ingested by ants (*Formica* sp.) in which they develop into metacercaria. Infected ants exhibit a peculiar behaviour of climbing up the grass blades during cooler parts of the day when animals are generally expected to graze. The final host becomes infected when these ants are ingested along with grass. In the final host, the metacercaria hatch in the small intestine and the young flukes migrate to the main bile ducts in the liver. There is no migration in the liver tissue. The flukes can survive in the final hosts for several years. The prepatent period is 10 to 12 weeks. In many instances, clinical signs are not present in the final host but anaemia, oedema and emaciation may occur in heavy infections. Diagnosis is made by finding the characteristic small, thick and dark brown eggs during faecal examination.

PREVENTION AND CONTROL

Control is difficult because of the wide distribution of intermediate hosts and the longevity of *D. dentriticum* eggs. Regular anthelmintic treatment of livestock using a flukicide anthelmintic will reduce pasture contamination.

*Schistosomosis (*Schistosoma *sp.)*

Schistosomes generally parasitize the blood vessels of the gastrointestinal tract and other organs. The parasites of this group are not hermaphrodites (that is, the sexes are separate) but the male and female are permanently combined. Most domestic animals are capable of acting as a final host for some *Schistosoma* sp. Some species infect humans (*S. mansoni*, *S. japonicum*). Schistosomosis is more common in tropical and subtropical zones but has also been reported in southern Europe. As with other flukes, an intermediate host is essential for the completion of the life cycle. The male adult schistosome is about 2 cm long (the female is smaller) and the eggs are 100 μm long with a characteristic spindle shape and a terminal spine. Eggs are usually passed out in the faeces of the host and hatch in the water. Miracidia hatch and infect the intermediate host. Water snails are particularly important in the life cycles of species which infect sheep and cattle (*S. bovis*, *S. mattheei*). The final host is infected by the penetration of skin or the gut lining by motile cercaria (there is no metacercarial stage). In sheep, acute disease results in severe diarrhoea containing blood and mucus and the development of dehydration

7–8 weeks after infection. The development of anaemia and emaciation follows. Diagnosis can be difficult but infection can be confirmed at necropsy or by examination of faecal smears for the characterizitc eggs which usually contain a developing miracidium. The egg shell is so delicate that the flotation process may distort them. Faecal smears prepared in saline rather than water are preferred.

The eggs of some species of Schistosome are released in the urine causing bladder irritation and haematuria (*S. haematobium*). Nasal schistosomiasis occurs in the nasal mucosal veins of cattle and horses in Asia (*S. nasalis*). Infected animals may develop polyps on the nasal mucosa resulting in blockage of the air passages, this condition is known as 'snoring disease'. Examination of nasal swabs smeared in saline on a glass slide will help identify the characteristic 'palanquin/boomerang' shaped eggs of *S. nasalis*.

3.4 Protozoa

General protozoology

Protozoa are single-celled organisms, which are largely free living but there are some important parasitic species such as the coccidia which are primarily intestinal parasites of animals (for example, *Eimeria* sp., *Cystoisospora* sp.). Coccidia are transmitted from one host to another mainly by the faecal–oral route. There are some species specific coccidia that usually cause disease in young animals (for example, *Eimeria* sp. in calves and lambs) or when the host's immune system is compromised. Coccidiosis tends to become a serious cause of production loss in extensively managed poultry especially when husbandry is poor (for example, lack of hygiene, high stocking rates and so on). For the diagnosis of disease caused by coccidia and other gastrointestinal protozoa it is important that fresh faecal samples are collected. If there is to be a delay in examination these samples should be stored at 4°C. In most cases, coccidial oocysts ('eggs') present in the faeces will be detected during routine examination for helminths using flotation techniques. Most oocysts are easily seen using medium magnification (20×) under the microscope but are smaller than most helminth eggs. Identification of the coccidial species present may require sporulation (development to the next stage of the life cycle, see Figure 3.31a) before identification can be completed, this requires freshly collected material. Trophozoites of *Entamoeba* sp. and *Giardia* sp. may be seen in fresh smears taken from rectal scrapings or from mucus on the surface of faecal material from infected animals; whereas, cysts can be easily observed using a $ZnSO_4$ float.

Haemoparasitic protozoa are another important group that can have serious effects on livestock health, these include *Babesia* sp., *Theileria* sp. and trypanosome species which are transmitted by vectors such as ticks and/or biting insects. *Anaplasma* sp. are not protozoa but are often considered along with this group as they also parasitize blood cells and are transmitted by ticks and biting insects. *Anaplasma* sp. are rickettsia and are briefly mentioned in Chapter 4.

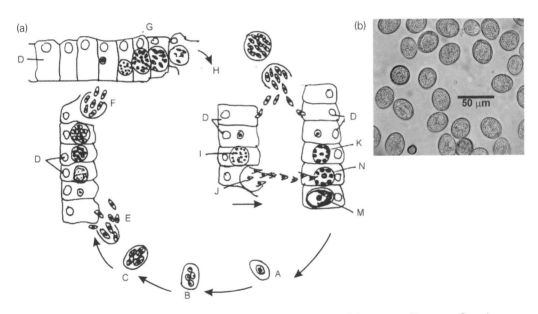

Figure 3.31 (a) The life cycle of *Eimeria* sp. Protozoal organisms of the genus *Eimera* or *Cystoisospora* cause 'coccidiosis' in a number of species, in most cases the species of coccidia is host specific. The life cycles of protozoal species tend to be quite complex with various names for different stages of the life cycle. (A) Oocyst released in the faeces of the definitive host. (B) Sporulation of oocyst and formulation of sporocysts. (C) Each sporocyst divides to form two banana shaped sporozoites. (A)–(C) is known as 'sporogony'. (D) Intestinal epithelial cells with nuclei illustrated. (E) Release of sporozoites and invasion of intestinal cells. (F) Formation of first stage schizont and release of merozoites. (G) Schizogony (asexual reproduction). (H) Formation of second stage schizont. (H)–(M) is known as 'gametogony'. (I) Microgametocyte (male), (J) release of microgametes, (K) macrogametocyte (female), (N) microgametes (male) enter a macrogamete (female) and a zygote (M) is formed. (b) Photograph of *Cystoisospora* sp. Photo: Dr Susan Kutz, University of Calgary, Canada.

Table 3.6a Overview of the classification of some protozoa of veterinary significance.

Phylum	Disease – characteristic*	Genus	Genus/species – Examples
Sarcomastigophora	Trypanosomiasis/Trypanosomosis (Nagana, surra, human sleeping sickness, Chagas disease, etc.)	*Trypanosoma*	*Trypanosoma congolense, T. evansi, T. cruzi*, etc.
	Leishmaniasis/Leishmanosis (visceral and cutaneous forms)	*Leishmania*	*Leishmania donovani, L. infantum, L. mexicana*, etc.
	Trichomoniasis/Trichomonosis – causes infertility in cattle	*Trichomonas*	*Trichomonas foetus, T. vaginalis, T. gallinae*
	Histomonosis (Blackhead in turkeys – associated with *Heterakis* sp. nematode infection)	*Histomonas*	*Histomonas meleagridis*
	Giardiasis, Giardiosis –causes diarrhoea, ill thrift, malabsorption	*Giardia*	*Giardia duodenalis* (Assemblages A-H)
	Amoebiasis/Amoebiosis – major cause of serious gastrointestinal disease in humans and primates	*Entamoeba*	*Entamoeba histolytica, E. dispar, E.coli.* Some species are commensals
Apicomplexa	Coccidiosis – diarrhoea, ill thrift	*Eimeria*	*Eimeria* spp. (there are many species with varying degrees of host specificity) *Cystoisospora* spp., *Isospora* spp., *Atoxoplasma* sp. (in birds)
	Cryptosporidiosis – can be zoonotic	*Cryptosporidium*	*Cryptosporidium* spp. (around 26 valid species)
	Toxoplasmosis – cause abortion in sheep, zoonotic	*Toxoplasma*	*Toxoplasma gondii*
	Neosporosis – cause abortion in cattle	*Neospora*	*Neospora caninum*
	Sarcocystosis – cysts in meat	*Sacrcocystis*	*Sarcocystis* spp.
	Babesiosis (Redwater, Tick Fever) – hemoglobinuria, fever	*Babesia*	*Babesia bovis, B. bigemina, B. divergens*
	Theileriosis (East coast fever) – fever, enlarged lymph nodes	*Theileria*	*Theileria parva, T. annulata*
	Malaria – fever in humans, severe and mild forms	*Plasmodium*	*Plasmodium falciparum*, etc.
	Leucocytozoonosis	*Leucocytozoon*	*Leucocytozoon* spp.
	Avian malaria – mild and severe forms	*Plasmodium, Haemoproteus*	*Plasmodium relictum, Plasmodium* spp. *Haemoproteus* spp.
Ciliophora	Balantidiasis/Balantidiosis, Buxtonellosis – enteritis	*Balantidium, Buxtonella*	*Balantidium coli, Buxtonella sulcata*

Note: *In the medical world the disease naming convention is -iasis and in the veterinary world -osis. Some in vet world still use -iasis to indicate subclinical infection. For this chart, we will retain both -iasis and -osis as we have both human and animal parasites listed.

Table 3.6b Classification of Rickettsiae. Rickettsia are non-motile, gram-negative, highly pleomorphic bacteria that can be present as cocci or rods and are often transmitted by arthropod vectors.

Order	Family	Genus
Rickettsiales	Rickettsiaceae	*Rickettsia*
		Coxiella
		Ehrlichia/ Cowdria
	Bartonellaceae	*Bartonella*
	Anaplasmataceae	*Anaplasma*
		Eperythrozoon

Sample collection for haemoparasites is outlined in Section 3.6.

In some protozoal diseases, for example, toxoplasmosis (see Figure 3.32), the oocysts are only found in the faecal material of the definitive host (that is, the cat). To detect infection in intermediate (rat, sheep) or abnormal hosts (humans) serological tests are used to detect antibodies to *Toxoplasma gondii*. In most cases two blood samples taken two weeks apart would be necessary to distinguish between previous exposure and current infection (that is, a rising titre for IgG antibodies indicates current infection, see Chapter 6). Alternatively, checking for IgM antibodies can indicate an ongoing infection. Some protozoal organisms are highly motile (for example, *Trichomonas* sp. that can cause infertility in cattle). These organisms can be seen in fresh smears prepared from scrapings of infected tissues such as vaginal mucus or preputial washings.

Classification, development and life cycles

In common with other living organisms, protozoa are classified according to the Linnaean system characterized by a species and a genus name. Recent developments in genetic profiling and scientific studies of parasite life cycles have resulted in a number of changes to the names given to many protozoal organisms. Table 3.6a provides an overview of the classification of some protozoal species of veterinary importance. The life cycles of some common protozoa are illustrated in Figures 3.31–3.34; the terminology is quite complex and in some cases the predominant clinical signs seen in the infected host will depend on the presence of concurrent disease as well as the immune response of the host to the parasite.

3.5 Protozoal diseases

There are a variety of protozoal species (*Cryptosporidia* sp., *Cystoisospora* sp., *Eimeria* sp., *Giardia* sp., *Entamoeba* sp.) that may be associated with diarrhoea in animals, some are obligate pathogens but many are relatively harmless in healthy adults and only become pathogenic in severely debilitated individuals. Clinical signs, for example, blood tinged mucoid diarrhoea associated with *Eimeria* sp., may be severe in young animals and non-immune adults especially where stocking rates are high and hygiene is poor. Other protozoa such as *Trypanosoma* sp. and *Leishmania* sp. can cause severe systemic disease which may be fatal in some cases. Early diagnosis and treatment is important. In this section the cause, prevention and control of some common protozoal diseases will be discussed. It is not within the scope of this book to cover all of the important protozoal diseases in detail so we have chosen some common organisms which illustrate the key approach to diagnosis and prevention. Diseases caused by 'blood protozoa' and other haemoparasites of ruminants will be considered in Section 3.6.

Coccidiosis

Coccidiosis is a general term used to refer to clinical disease caused by a number of protozoal organisms in the class Coccidia. The species of veterinary importance fall into two distinct families, the Eimeridae and the Sarcocystidae. The term 'coccidiosis' in poultry and other live-stock is usually reserved for diseases caused by *Eimeria* sp. and/or *Isospora* sp. *Eimeria* sp. occur in poultry, ruminants, pigs, horses and rabbits, and typically cause disease following invasion of intestinal epithelial cells, see Figure 3.30. Some species also invade the epithelial cells in the kidney and liver. Most species are host specific. Some of the most common species are outlined in Table 3.7. The life cycles of *Isospora* sp. and *Eimeria* sp. have many similarities, but differ in a number of respects. The organisms can be identified in faecal samples by the morphology of their oocytes. *Isospora* sp. have sporulated oocysts which contain two sporocysts each with four sporozoites whereas sporulated oocysts of *Eimeria* sp. tend to contain four sporocysts with two sporozoites. Some species of *Isospora* have extra-intestinal stages occurring in the liver, spleen and lymph nodes which then may re-infect the intestinal mucosa. In some cases, rodents may ingest oocysts from carnivore *Isospora* sp. and become infected with the asexual stages of the parasite, the rodents then act as reservoir hosts and dogs and cats are re-infected when ingesting infected rodents. These species of Isospora that involve a reservoir/transport/paratenic hosts are now classified under the genus *Cystoisospora*. The genus *Isospora* once included species now in the genera *Toxoplasma*, *Besnotia* and *Sarcocystis*. The nomenclature for many protozoal organisms has changed over the

past few decades – this is partly the result of molecular phylogenetics but it also reflects our better understanding of the complex life cycles of protozoal parasites. Table 3.6b outlines one current taxonomic list of protozoa of veterinary importance.

The life cycle of *Eimeria* sp. is often representative of the coccidial group and other life cycles are compared with it (Figure 3.31a). Typically there are two main phases of reproduction in the life cycle: asexual and sexual. Asexual stage occurs following ingestion of infective sporulated oocysts. The sporozoites from oocysts emerge

Table 3.7 Some coccidial protozoa and their hosts.

Species	Host
Eimeria tenella, E. necatrix, E. brunetti, E. maxima, E. acervulina, E. praecox, etc.	Chickens
Eimeria meleagrimitis, E. adenoeides, etc.	Turkeys
Eimeria zuerni, E. bovis, E. canadensis, E. auburnensis, E. alabamensis, E. ellipsoidalis, E. cylindrica, E. bukidnonensis, etc.	Cattle
Eimeria ovinoidalis, E. ahsata, E. bakuensis, E. crandallis, E. parva, E. intricata, E. granulosa, E. faurei, etc.	Sheep
Eimeria ninakohlyakimovae, E. arloingi, E. christenseni, E. caprina, E. hirci, E. parva, E. aspheronica, etc.	Goats
Eimeria intestinalis, E. steidae (liver)	Rabbits
Eimeria leuckarti	Horses
Eimeria debliecki, E. perminuta, E. spinosa, E. scabra, E. porci, E. neobeliecki, Cystoisospora suis	Pigs
Cystoisospora felis, C. rivolta, Toxoplasma gondii, Sarcocystis sp., etc.	Cats
Cystoisospora canis, C. ohioensis, Neospora caninum, Sarcocystis sp., etc.	Dogs

in the intestinal lumen and penetrate epithelial cells to initiate schizogony by developing into a large, first generation schizonts. They undergo merogony to produce first generation merozoites which then invade other cells to develop as second generation schizonts and subsequently as second generation merozoites. There may be several cycles of this asexual reproduction with damage to the gut wall resulting each time. The sexual stage is initiated when merozoites from last cycle of merogony invade new epithelial cells and develops as gamonts. Some gamonts increase in size (macrogamont) and develop as female gametocyte while others undergo series of division to produce hundreds of microgametocytes (male gametocytes). Microgametocytes that are released into the lumen will seek cells containing female gametocyte to fuse with and complete fertilization. Following fertilization, a zygote develops in the host's intestinal cell. A wall forms around the zygote until it develops as an oocyst which is then released by rupture of the host cell. Sexual multiplication results in the production of oocysts which pass out in the faeces. Within each oocyst sporogony (sporulation) ensures development of eight infective sporozoites. These are infective for the next host but the process of sporulation may take a day or two depending on the ambient environmental temperature and humidity. Owing to the fact that sporulation occurs in the environment this is the stage at which good hygiene can prevent infection by ensuring all potentially infective faeces are removed and that the stocking rates are not too high.

The location of the reproductive cycle in the intestinal tract depends on the species of *Eimeria* involved. In poultry, the location of intestinal lesions at necropsy may assist in the identification of the *Eimeria* species involved. The severity of the damage depends on (1) the number of organisms present, (2) the species of protozoa and (3) the immunity of the host. As the host gets older there will be some degree of immunity, which will result in inhibition of the reproduction of the protozoa. Developing trophozoites compete with the host for ingested nutrients and damage to the intestine results in a reduced absorptive capacity of the intestinal lining. Severe diarrhoea in heavy infections can be rapidly fatal in young animals.

Disease diagnosis

Identification of oocysts in a faecal sample (see Figure 3.31b) or examination of intestinal smears and typical lesions at necropsy. Treatment and prevention will depend on the veterinary products available in the area (for example, in feed anti-protozoal drugs, sulphonamides) but the most important factor for the prevention of coccidiosis is good husbandry and hygiene.

Prevention

1 Good hygiene.
2 Ensure good husbandry.
3 Vaccination (oral vaccines containing attenuated forms of *Eimeria* sp. are available).
4 Coccidiostats in the feed (must be accompanied by 1 and 2).

Treatment

1 Antiprotozoals mainly coccidiostats (for example, Amprolium, Decoquinate, Lasalocid, Monensin, Sulphonamides and so on).
2 Supportive care (for example, fluid therapy for dehydration).

Toxoplasmosis

Toxoplasma gondii is an intestinal parasite of young cats and may cause transient diarrhoea. Cats are the definitive hosts in which oocysts are produced. More than 250 mammals and birds are known intermediate hosts in which tissue cysts containing bradyzoites develops. The life cycle is complex. The definitive host (cats) gets infected by consuming the tissue cysts in intermediate hosts (prey animals). Intermediate hosts get infected through various routes of transmission but mainly through environment contaminated with sporulated oocysts. Cats may become infected by ingestion of sporulated oocysts from cat faeces or by ingestion of cystic stages (bradyzoites) in an intermediate host such as a rodent. The life cycle of *Toxoplasma gondii* is illustrated in Figure 3.32. Humans may

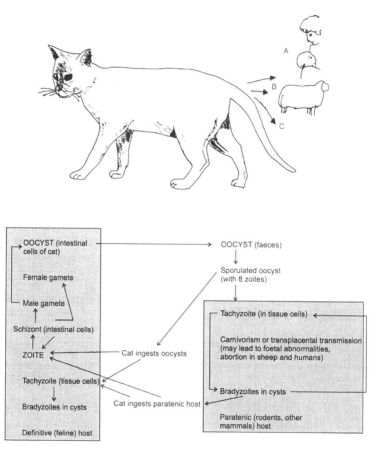

Figure 3.32 The life cycle of *Toxoplasma gondii*. The cat or other wild felid species are the definitive host for this protozoal organism. It is only in the definitive host that infective oocysts develop and are shed in the faeces to infect other animals. (A) Humans can become infected by accidental ingestion of sporulated (infective) *Toxoplasma* sp. oocysts in food or from inanimate objects contaminated with cat faeces. Healthy adults rarely develop clinical disease but if a pregnant woman is infected the organism migrates to the placenta and can cause damage to the developing foetus. (B) Healthy adult sheep and other animals may also ingest sporulated oocysts while grazing. Pregnant sheep may abort if they become infected during gestation unless the ewe has developed immunity, that is, she has had previous exposure to the organism (natural infection) or has been vaccinated. (C) Cats and kittens can become infected by ingestion of sporulated oocysts or by ingestion of paratenic hosts. The protozoal organisms undergo asexual reproduction in the cells of the cat but only cause clinical disease in young animals or if the number of organisms is very high.

become infected but as the life cycle of the organism cannot be completed in this species their role is that of a 'paratenic host', that is, a host in which the parasite may grow or multiply but that is not required for the parasite to complete its life cycle. If a woman is pregnant at the time of infection abortion may follow and congenital defects in newborns are common. Similarly, toxoplasmosis may result in high neonatal losses and abortion in sheep. In non-pregnant animals and humans there may only be a transient fever or 'flu like' signs with no further development although cystic stages may settle in the tissues.

Diagnosis

1 Identification of oocysts in cat faecal samples.
2 Serology (paired samples).
3 Abortion: typical lesions in placenta (Figure 3.33).
4 In neonatal animals there may be congenital defects such as microphthalmia (abnormally small eyes).

Prevention and control

1 Keep cats away from ruminant grazing areas.
2 Hygiene and hand washing for humans. Pregnant women should not clear out cat/kitten litter trays and should wear gloves when gardening.
3 Vaccinate young ewes before breeding or mix non-pregnant animals with aborting adults to induce immunity and to reduce future lamb losses.

Sarcocystosis

Many species of *Sarcocystis* have an obligatory two host life cycle. Asexual reproduction (schizogony) occurs in the intermediate host and sexual reproduction (gametogony) including sporogony occur in the definite host. There is

Figure 3.33 Aborted foetus which was lost near to full term. A healthy lamb was born to the same ewe. Note the white calcified cotyledons on the placenta of the aborted lamb, this is indicative of infection with *Toxoplasma gondii*, Scotland, UK.

generally no development in the environment as oocysts are sporulated by the time they are shed in the faeces. The oocysts wall is so delicate to hold two developed sporocysts; therefore, individual sporocysts containing four sporozoites are appreciated in faecal examination. Infection of the intermediate herbivore host occurs following ingestion of fully sporulated oocysts or the sporocysts in faecal material from omnivores and carnivores. The definitive host becomes infected by ingestion of bradyzoites in sarcocysts in muscle tissue of the intermediate hosts (Figure 3.34). Host relationships of some species of sarcocysts are outlined in Table 3.8.

Sarcocystis rarely causes clinical disease in the definitive host but schizogony in the tissues of the intermediate host may result in an elevated temperature, muscular pain, enlarged lymph nodes and other signs. Some species of *Sarcocystis* may cause neurological signs (*Sarcocystis neurona* of horses). For more information see the text list at the end of the chapter.

Diagnosis

Cysts may be found in muscle tissue at necropsy or at meat inspection and can result in carcass

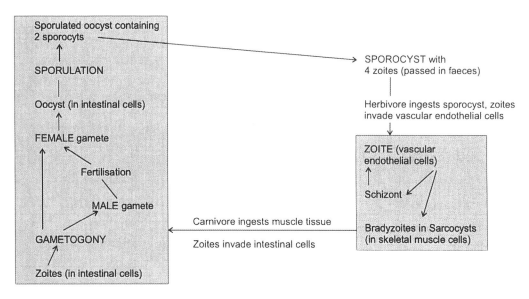

Figure 3.34 Life cycle of *Sarcocytis* sp.

Table 3.8 Host relationships for some species of *Sarcocystis*.

Intermediate host	Definitive hosts		
	Human	Dog	Cat
Cattle	S. hominis	S. bovicanis (cruzi)	S. bovifelis (hirsuta)
Water buffalo		S. levinei	S. bovifelis (fusiformis)
Sheep	-	S. ovicanis (tenella)	S. ovifelis (gigantea)
Goat	-	S. capracanis	S. hircifelis
Horse	-	S. equicanis (bertrami), S. fayeri	?
Pig	S. suihominis	S. suicanis (miescheriana)	S. suifelis (porcifelis)
Rabbit	-	?	S. leporum, S. cuniculi
Llama, alpaca	-	S. aucheniae	

condemnation or the need for severe trimming of parts of the carcass.

Prevention and control

1 Hygiene and good management.
2 Control concurrent disease(s).

Neosporosis

Neospora caninum is a member of the family Sarcocystidae. The dog is the definitive host for this parasite with ruminants (cattle) commonly acting as the intermediate host. The dog can also act as an intermediate host for this parasite in prenatal infections. The life cycle of the parasite has only recently been elucidated. *Neospora caninum* occurs worldwide. Oocysts are passed in dog faeces from 8–23 days after infection. In cattle, infection may result in abortion, this is usually observed at 5–6 month gestation although it can occur any time between 3 months and full term. Abortions may be sporadic or clustered. Foetuses can be born alive (often underweight and with neurological signs) or they may die *in utero* in which case they often undergo mummification.

Diagnosis

Diagnosis is confirmed based on histological examination of freshly aborted foetuses. There are also serological tests; IFA and ELISA available (see Chapter 6). Cows infected with *N. caninum* are likely to infect more than 95% of their calves transplacentally.

Prevention and control

Prevention and control require efficient disposal of any aborted material to prevent dogs gaining access to potential sources of infection and also ensuring that dog faeces does not contaminate areas where cattle feed.

Cryptosporidiosis

Cryptosporidium sp. are very small (4–8 μm diameter) organisms that are often associated with diarrhoea in young or debilitated animals. They are now considered to be gregarines rather than true coccidia, this may explain their refractoriness to commonly used coccidiostats. The organisms undergo schizogony, gametogony and sporogony in vacuoles formed in the epithelial cells lining the intestinal, and sometimes the respiratory tract, renal epithelium and gallbladder. Cross infection among a wide range of vertebrate hosts is common and auto-infection by ingestion of food contaminated with an animal's own faecal

material is relatively easy. An infected animal may show no clinical signs if the infection is light but concurrent disease or environmental stress may tip the balance resulting in rapid build-up of infective organisms and the development of diarrhoea, dehydration and other complications. There are 26 valid species of *Cryptosporidium* reported of which not all are zoonotic. Some that are of zoonotic potential includes, *C. parvum*, *C. meleagridis*, *C. ubiquitum*, *C. muris*, *C. canis*, *C. felis*, and so on.

Diagnosis

1 Floating fresh faecal samples on 1.33 SpG sugar solution and examining under medium and high power (20× and 40×) would help. It appears with a pinkish tinge specifically on sugar float. The organism floats just beneath the coverslip and so on a plane of focus slightly above than where we usually expect parasites stages to be seen.

2 Examination of fixed, Giemsa stained faecal smears. Owing to the fact that the organisms are very small it is necessary to gain some familiarity with the appearance of the stained organism to be confident of identifying *Cryptosporidia* in samples.

3 Phase contrast microscopy may be helpful for examination of fresh specimens.

4 Antigen capture ELISA (see Chapter 6).

Prevention and control

1 Hygiene and good management.

2 Control concurrent disease(s).

Trypanosomosis and leishmanosis

Trypanosoma sp. (Figure 3.35) and *Leishmania* sp. are broadly classified as 'flagellates' due to many of their life stages possessing flagella that allow effective motility. Trypanosomes are transmitted by biting/blood-sucking insects (for example, flies of the genus *Tabanus* sp., *Stomoxys* sp. and *Hippobosca* sp.), tsetse flies (*Glossina* sp.), kissing bugs (Reduviidae) or mechanically, and cause a range of diseases in humans (for example, sleeping sickness in Africa and Chagas disease in Latin America) and animals (for example, nagana disease, Surra, and so on). Nagana, caused by *Trypanosoma congolense congolense*, is the most important trypanosome of cattle in tropical Africa. In many cases there may be mixed infections with *T. brucei* and *T. vivax*. *Trypanosoma congolense* is widely distributed between 15°N and 25°S. Hosts include cattle, sheep, goats, equids, camels, dogs and pigs. The wildlife reservoirs include antelope, giraffe, zebra, elephants and warthogs. A similar condition is caused by *T. brucei brucei*, which can infect a similar host range. It currently occurs in sub-Saharan Africa between the latitudes 14°N and 29°S. Trypanosomosis and other diseases associated with haemoparasites cause significant economic loss in ruminants and other livestock, including horses, in tropical countries and are discussed in more detail in Section 3.6.

Leishmania sp. protozoa cause a range of clinical signs in humans as well as wild and domestic animals and are transmitted by insect vectors. *Leishmania donovani* is responsible for human visceral leishmaniasis (kala-azar), which causes enlargement of the spleen and other internal organs. *Leishmania tropica* causes mucocutaneous leishmaniasis. Both forms occur in the tropics but have also been seen more recently in the Mediterranean region. In domestic animals, *L. infantum* is the most important which causes potentially fatal disease in dogs. This parasite is endemic in many continents.

Other flagellates

Trichomonads and *Giardia* sp. are mucosoflagellates and are motile on mucosal surfaces. *Giardia* sp. are frequently associated with diarrhoea in a wide range of animals but may often be present in healthy animals with no evidence of disease. Severe enteritis may occur following heavy infection with *Giardia* sp. especially where the host has concurrent disease. In some cases, cysts may remain in the gut wall long after recovery or treatment but may re-activate when the host's immunity is compromised.

Diagnosis

Diagnosis is by identification of trophozoites in faecal smears. Faecal flotation tests for cysts (using $ZnSO_4$ as flotation fluid) and antigen capture ELISA are used to diagnose *Giardia* sp. infection. Trophozoites of Trichomonads (*T. foetus*, *T. blagburni*) can be cultured using specific in pouch.

Prevention and control

Prevention and control involves improved hygiene and prevention of faecal contamination of feed and water supplies.

Members of the group Trichomonads may cause a range of diseases in various host species. *Tritrichomonas foetus* can cause abortion in cattle and may be isolated from infected foetal material, the uterus and vagina of the cow and the prepuce, penis and epididymis of infected

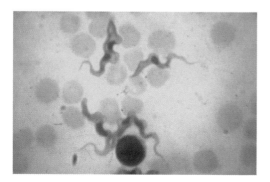

Figure 3.35 Trypanosome (T) as seen in a Giemsa stained blood smear from a horse (1000× oil immersion).

bulls. The disease is sexually transmitted so it is always important to check the bull for the presence of *Tritrichomonas foetus* following a suspected *Tritrichomonas* abortion outbreak. *Trichomonas gallinae is* associated with mouth and crop infections in song birds and raptors. Recent studies confirm the role of *Trichomonas blagburni* in the intestinal infections of cats. *Control and prevention of Tritrichomonas foetus* requires testing all bulls to be used for breeding and possibly changing to artificial insemination if it is not sure that a bull is clear following treatment. Treatment with antiprotozoal drugs is effective but must be accompanied by improved hygiene and husbandry practices.

Other protozoa

A wide range of other protozoa may cause significant clinical disease in wildlife, humans and livestock but these will not be discussed further here. However, before we go on to consider haemoparasites there is one more enteric protozoan that will be considered briefly.

Balantidiosis

Balantidium coli is a cilate parasite found in the colon and caecum of pigs, rats and other mammals. It occurs worldwide and is the only member of the ciliate phylum known to be pathogenic to humans. Humans can become infected via the faecal-oral route, usually from the normal host, the pig, which is usually asymptomatic. Contaminated water is the most frequent mechanism for transmission. Transmission between species requires a period of adaptation. Infection occurs when a host ingests a cyst (that is, via water or contaminated food). Once ingested, the cyst passes through the digestive system to the small intestine. Once in the small intestine trophozoites are produced which colonize the large intestine and feed on intestinal flora. In acute disease, especially where many trophozoites invade the wall of the colon, a severe diarrhea can develop. In severe cases perforation of the colon has been reported. In the lumen of the intestine some trophozoites will undergo encystations and are passed out in the faeces to infect another host. Balantidiosis is common in the Philippines and other places where there is close contact between humans and pigs and where sanitation is poor.

Disease prevention

Prevention requires the application of basic sanitation measures and educating the public about mitigating potential health risks by avoiding contaminated water and food sources.

3.6 Haemoparasites of ruminants

Haemoprotozoa are of clinical importance worldwide, especially in tropical countries where the vectors that transmit them may be prolific. The economic losses due to haemoprotozoal parasites in ruminants can be significant with high mortality in some areas due to diseases such as East Coast Fever (ECF, *Theileria parva*). However, in many cases these diseases are endemic so local livestock, especially the indigenous breeds, develop immunity and the consequent losses are less dramatic. There is typically a high level of subclinical infection with an associated loss in production. Two of the most important haemoparasites that infect ruminant livestock are *Babesia* sp. and *Theileria* sp. These are transmitted by ticks. Although not strictly 'parasites', *Anaplasma* sp. and other rickettsia are also dealt with in this section (see Table 3.6b). These organisms have some similarities with bacteria and may 'parasitize' red blood cells to cause significant clinical disease with red cell destruction and the subsequent development of anaemia. *Anaplasma* sp. may be transmitted by ticks as well as by biting flies and mechanical means (for example, contaminated needles). Many animals carry low numbers of haemoprotozoa or *Anaplasma* sp. without apparent clinical disease.

Diagnosis, control and prevention

Diagnosis of haemoparasitic disease is usually made on clinical and epidemiological grounds with confirmation of the cause obtained through collection and examination of stained blood smears. The appearance of the parasites in blood smears is usually characteristic allowing ready identification, although in some cases an animal may be infected with more than one organism at the same time. In cattle and sheep, smears can easily be made from blood collected from the ear vein but smears can also be made from whole blood collected in EDTA for concurrent haematology tests. However, peripheral (capillary) blood is more likely to contain haemoparasites than blood from the central veins. In many cases the number of haemoparasites in the circulation may be low so it is a good idea to prepare thick smears as well as thin smears of peripheral blood. Where possible, ensure that all glass slides are labelled using a glass marker because most normal marker pens will be washed off when the smears are fixed. In addition to the examination of blood smears there are molecular and antigen detection assays and a range of serological screening tests (predominantly ELISA based) to detect the presence of, or exposure to, a range of haemoparasites. These will not be considered further here but are well described in a number of texts listed at the end of this chapter.

The prevention and control of haemoparasitic disease requires an understanding of the life cycle of the parasite and that of the vector. A successful control programme will require concurrent ectoparasite control (see Section 3.7). There has been significant research into the development of preventative vaccines to reduce the impact of haemoparasitic disease in ruminant species and these seem to be effective in some parts of the world. It is also possible to select animals that are genetically more resistant to the vectors which spread the diseases and therefore have a reduced challenge. Details about the treatment, prevention and control of specific haemoparasites are provided later in this section and more information is available in the bibliography at the end of the chapter.

Preparation and examination of blood and tissue smears for the presence of haemoparasites

The preparation of a good blood film requires the use of clean grease free slides (storage of slides in 70% alcohol will remove grease but allow these to dry before use). The film should be made from a single drop of peripheral blood, for example, from the ear vein. Peripheral (capillary) blood is more likely to contain haemoparasites than blood from the major veins but smears can also be prepared from an EDTA sample submitted in a vacutainer if a capillary sample is unavailable. The prepared film should be smooth and even, allowing random distribution of white cells throughout the film; the erythrocytes should be distributed in a single layer (see Chapter 5). A spreader can be used for preparation of smears, this should have smooth even edges with the corners cut to give a side 1.5 cm wide (slightly narrower than the microscope slide). For examination of some haemoparasites it is a good idea to prepare thin smears as well as thick smears, the latter especially where the number of parasites is low. The following techniques (see Figure 3.36) may be used to make thin and thick smears.

Methods

PREPARATION OF A THIN FILM
Place a small drop of blood 1.5–2 cm from one end of a clean slide. Hold the spreader at an

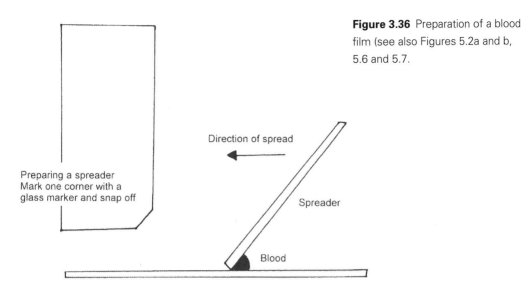

Figure 3.36 Preparation of a blood film (see also Figures 5.2a and b, 5.6 and 5.7.

angle of 30° in front of the drop of blood and bring back to touch the blood allowing it to spread along the edge. Make the film by pushing the spreader forward in a direct and even movement. The film should be about 3–4 cm long. Air dry rapidly by waving the film in the air (this avoids crenation of the erythrocytes). Label the film, either by writing with a pencil in the film itself or, if in the laboratory, etch the slide using a diamond tipped pen. Fix the film in methanol for 2 min as soon as the blood film is completely dry (the film will not fix to the slide if it's wet at the time methanol is added). If fixing is delayed the film can deteriorate rapidly therefore it is advisable to carry a small bottle of methanol for this purpose when out doing fieldwork.

PREPARATION OF A THICK FILM

Place a drop of blood at one end of the slide. Touch the drop of blood with a spreader and make a thin film starting at 3 cm from the end of the slide (Figure 3.36). Spread the remaining blood in the shape of a cube (3 × 20 mm). If the film is too thick it will peel off the slide. Air dry completely and draw a line between the two films. Haemoglobin can be removed from the thick film by inverting the slide and placing

it at an angle in distilled water (2 min). Fix the slide in methanol. In the laboratory, the slides can be stained with Giemsa or Leishman stain.

There are a number of smear preparation (apart from the two mentioned above) and staining methods recommended but the following have also been found satisfactory in our experience.

Smear preparation

DIRECT EXAMINATION OF A FRESH WHOLE BLOOD

For some blood parasites, especially that live outside of blood cells, such as *Trypanosoma* sp. and microfilaria of helminths, it is a good idea to initially examine fresh whole blood. This can be done by placing a drop of blood on a slide and examining under a microscope with or without a cover slip on top of blood. The organisms can be appreciated based on their motion, if they are still viable and motile. A parasite-free blood film appears standstill. However, the corkscrew motion of trypanosomes cause tumbling movement of adjacent red blood cells. Similarly, snakelike motion of some microfilariae is reflected by a wavy movement of red blood cells. This method is not sensitive and it is also hard to differentiate parasite species. Protozoa in faeces and *Trichomonas* sp. in vaginal mucus can also be identified by direct examination.

BRAIN CRUSH SMEAR

Cut a small piece of brain tissue containing capillaries. Put this on a clean slide and then crush between two slides. Slide these together gently and then move the slides apart. Allow the smear to dry in air. Rapid fixation is absolutely essential.

LYMPH NODE SMEAR

Locate the position of the lymph node of interest and hold it firmly. With sterile precautions, and by means of a syringe-needle pierce the gland and move the needle tip around. Pull the plunger of the syringe to aspirate a sample of the contents of the gland and withdraw the needle. Expel the contents on to a clean slide and make a smear (as for a thick and/or thin blood film) and stain. This method is good for detecting stages of *Theileria* sp. in lymphocytes.

SPLEEN SMEAR

At necropsy, cut open the spleen using a knife. With the edge of a slide scrape the cut surface and spread the material across the slide using a spreader to draw the spleen tissue behind it. Allow to dry in air. It is important to protect prepared smears from dust and rain by covering them or putting them into a box immediately after they are prepared.

FIXATION

Fixation of smears should be carried out as soon as possible after smear is air dried and is a preliminary step before staining can be carried out. The fixative[5] usually used is methanol. Smears are fixed for at least 1 min (could be more, but not less) and again air dried.

Stains and staining

Stains and staining techniques vary according to individual preference. The stain usually used is Giemsa diluted 1 : 15 and the usual staining technique is as follows. (Note that timing, dilution and pH may vary slightly with different batches of stain.)

1 Using a pipette, measure 6 ml of concentrated (stock) Giemsa stain into a clear empty bottle (the stock stain may need to be passed through a filter to remove particles).
2 Add 90 ml of water at pH 7.2 (the ideal pH may vary with batches of stain).
3 Mix well and pour into a staining jar containing the smears.
4 Allow to stain for 20–30 min (in some cases 10 min may be sufficient).
5 Wash the smears in distilled water or tap water.
6 Dry in air and examine.

The Leishman stain may also be used (see below). Other stains are available in kit test form to allow quick staining for fieldwork (for example, Diff Quick™[6]). Most of the stains are based on Romanowsky stains (see Chapter 5).

PURPOSE OF STAINING

Staining helps in the differentiation of cells. Cellular constituents stained by the basic components of the stain appear blue, purple or violet in colour and are said to be basophilic; while those constituents stained by the acidic components of the stain appear red, pink or orange in colour and are said to be acidophilic or eosinophilic. Those cellular constituents staining between the two extremes in colour are said to be neutrophilic (see also Chapter 5).

MICROSCOPIC EXAMINATION

It is good technique to examine smears using low magnification before going on to examine specific areas under high magnification using oil immersion. This is especially important when examining brain smears or smears for trypanosomes.

Stain preparation

GIEMSA STAIN

Preparation of a stock solution: Add 1g of Giemsa powder to 66 ml Glycerol and mix thoroughly. Heat to 56°C for 90 minutes and add 66 ml methanol. Label the stock solution and mix thoroughly. Leave to stand for 7 days before filtering and storage in glass-stoppered bottles.

To prepare the buffer, add pH tablets to distilled water. pH 7.2 is usually recommended but some laboratories have found that the stain will work in the range of pH 6.8–7.2.

LEISHMAN STAIN: PREPARATION OF A STOCK SOLUTION:

Add 0.15 g Leishman powder to 100 ml of methanol. Dissolve by grinding and stirring for 2 h. Keep the stock solution in a dark glass dropper bottle.

To prepare the buffer, add pH tablets to distilled water. pH 7.2 is usually satisfactory for staining blood smears and some protozoa.

Fix air dried smears in Leishman stain for 2 min. Add an equal amount of buffered distilled water. Mix by gently rocking the slide to allow even distribution. Leave for 10–15 min. Rinse in buffer and immerse until the smear appears pink (1–2 min). Blot or air dry.

A note on the examination of blood smears for the presence of microfilariae

Microfilariae are the pre-larval forms of helminth parasites which may occur in the blood, for example, the microfilaria of the canine heart worm, *Dirofilaria immitis*. The parasite is transmitted by mosquitoes (*Aedes* sp.). The following technique is also useful for detecting trypanosomes in fresh blood.

CONCENTRATING MICROFILARIAE AND TRYPANOSOMES:

1 Take about 3–5 ml of fresh blood and add it to 0.5 ml of 3.8% sodium citrate to stop it clotting. Solid sequestrene can also be used as an anticoagulant, but a liquid anticoagulant solution is better.

2 Centrifuge the blood sample for 10 min. If looking for trypanosomes, centrifuge at fast speed. If looking for microfilariae centrifuge the blood at about 1000 rpm. The 'buffy' coat of a prepared and centrifuged microhaematocrit can be examined for the presence of trypanosomes.

3 After centrifuging there will be three layers – a layer of plasma, a layer of white cells (buffy coat) and a layer of red cells.

4 Using a Pasteur pipette take off the supernatant plasma to the white cell layer or, if a haematocrit tube has been used, break off the haematocrit tube at the white cell layer.

5 Take as much of the white cell layer as possible and prepare a smear on a slide. Examine under a coverslip. If looking for microfilariae take some of the red cell layer as well and examine this under the coverslip.

6 Scan the entire coverslip using the low power (10× or 20×) objectives for microfilariae and a high power (100×) objective for trypanosomes (see Figure 3.35).

CONCENTRATING MICROFILARIAE (MODIFIED KNOTT'S TEST)

1 Measure 1 ml of anticoagulated (EDTA/heparin treated) whole blood in a 15 ml centrifuge tube.

2 Add 9 ml of 2% formalin.

3 Cap the centrifuge tube with rubber stopper and mix by inversion.

4 Centrifuge for 10 min and pour out the supernatant.
5 Add a drop of Methylene Blue stain to the sediment and mix gently.
6 Place two drops of this stained sediment on a glass slide and examine under microscope for microfilaria.

Immunological tests

There are a number of serological tests (for example, complement fixation [CFT], agglutination, FAT and ELISA – see Chapter 6) currently available for the diagnosis of haemoparasitic diseases, however, the interpretation of the test results requires experience and knowledge of the epidemiological pattern of the disease(s) in a given area. Many animals will have antibodies to haemoparasites without developing clinical disease and low titres may imply a level of immunity from past exposure. If the disease is endemic in an area it is important to gather information about the 'normal' titres expected from healthy animals. This allows the laboratory to determine a 'cut off' level above which serological titres will be considered positive for animals in that area. The results of any serological tests need to be evaluated with reference to the clinical signs exhibited in the animals sampled. There are also tests available to look for antigenic material in tissues or smears (for example, immunofluorescence, see Chapter 6). Survey work to assess the prevalence of haemoparasites in livestock usually includes detailed evaluation of the number and type of arthropod (that is, tick or insect) vectors in the area. The species of arthropods found locally can have some predictive value for the risk of exposure to arthropod borne diseases although the mere presence of a potential vector does not always imply that the disease can be effectively transmitted (see Chapter 14 – vector borne diseases).

Diseases caused by haemoparasites in ruminants

Theileriosis

Theileriosis has been recorded in Africa, Asia, North America and parts of Europe and is transmitted by a number of tick vectors (for example, *Hyalomma* sp., *Rhipicephalus* sp., *Haemaphysalis* sp.). *Bos indicus* cattle tend to be more resistant than European breeds, although this may be due to resistance to the tick vectors. There are a number of species of *Theileria* (*T. parva*, *T. orientalis*, *T. mutans*, *T. sergenti* and so on) some of which cause a non-persistent and mild clinical disease. *Theileria mutans* (benign bovine theileriosis) can be transmitted by *Amblyomma* sp. and *Haemaphysalis* sp. ticks. *Theileria annulata* (syn. *T. dispar*) causes tropical theileriosis in cattle (also known as Mediterranean Coast fever), this is transmitted by *Hyalomma* sp. ticks. In Africa, *Theileria parva*, transmitted by *Rhipicephalus appendiculatus* ticks, is the cause of ECF. ECF is characterized by lymph node enlargement, high temperature, weakness, emaciation and a high death rate. *Theileria* sp. parasitize red and white blood cells and can be found in lymph nodes and the spleen. The life cycle of *Theileria* sp. involves development in the tick vector. The incubation period may take up to 40 weeks depending upon the species and environmental conditions, but for most it is 1–3 weeks.

In sheep and goats *Theileria hirci* causes a disease similar to ECF in cattle. *Theileria hirci* causes moderate disease and is transmitted by *Hyalomma* sp. ticks. *Theileria ovis* causes mild disease and is transmitted by *Rhipicephalus* sp. and *Haemaphysalis* sp. ticks.

LIFE CYCLE
Theileria infected ticks may remain infected for a number of months. Developmental stages of parasites occur in the tick and these pass through the stages (larva, nymph and adult) of the tick life cycle (trans stadial) but there is no transovarial

(that is, from adult to egg) transmission. The parasites localize in the salivary gland of the tick ready for injection into the host animal.

The risk of infection depends on the distribution of vector-ticks and in tick-infested areas all animals become infected. In areas where animal reservoirs of the infection exist, for example, wild bovids, the disease is difficult to eradicate. Imported breeds and immulonogically naïve cattle of all ages will be susceptible to the development of clinical theileriosis in areas where the disease is endemic.

CLINICAL DISEASE

The clinical findings will depend on:

1 the number and species of parasite present
2 the level of general health and immune status of the host.

Clinical cases typically have a high temperature, the animal appears depressed and refuses to feed. Anaemia may develop following haemolysis of red blood cells and this is followed by the development of jaundice (yellow mucous membranes) as the breakdown products of haemoglobin build up in the tissues. The lymph nodes may swell and superficial ones can be palpated (see Figure 8.8). In the later stages of a severe infection the animal may develop swelling of the eyelids and ears, nasal discharge and diarrhoea with mucus and blood leading to rapid emaciation followed by death within 8–15 days. Mortality may reach 100% in imported and immunologically naïve animals. Occasionally cerebral infections can occur resulting in nervous signs such as circling.

DIAGNOSIS

Some of the clinical signs associated with *Theileria* sp. may be similar to those seen in clinical babesiosis and severe cases of anaplasmosis. Diagnosis can usually be confirmed at necropsy; typically, the carcass is anaemic and yellow

(that is, jaundiced as a result of intravascular haemolysis), the liver is enlarged and brown or yellow in colour, the spleen is enlarged and soft and the lymph nodes are enlarged and haemorrhagic. The kidneys are often pale in colour and there may be ulceration of the abomasum with haemorrhages. Diagnosis can be confirmed by finding the parasite in stained smears taken from the lymph nodes or spleen. Ante-mortem diagnosis can be confirmed by the examination of a stained blood smear (see Figure 5.6) or lymph node needle biopsy and by serological testing (that is, complement fixation and indirect agglutination, see Chapter 6). Antigen detection and immunofluoresence can also be used to confirm a diagnosis.

CONTROL AND PREVENTION

Control is primarily achieved through:

1 control of the tick vector, for example, treating animals with acaricide through dipping, spraying and dusting (see section 3.8)
2 periodic changing of pasture and burning abandoned pastures to destroy juvenile ticks.
3 the use of vaccination
4 selection for genetic resistance. E.g. some indigenous cattle breeds are highly resistant to ticks.

The control of cattle movement can also be considered.

Babesiosis (Piroplasmosis) = red water fever

Babesiosis is another tick borne disease of domestic animals with a worldwide distribution. Cattle and buffaloes are most often affected but sheep, goats, horses, donkeys, dogs and rodents are also susceptible. There are a number of species in the genus (*B. bovis, B. divergens, B. bigemina* and so on) and some cause only mild or subclinical disease. In warm humid regions *Babesia bovis* and *B. bigemina* are the commonest species. In cool damp countries, *B. divergens* and *B. major* predominate. *Babesia* sp. protozoa are transmitted by various species of ticks such as *Rhipicephalus (Boophilus)* sp., *Ixodes* sp., *Haemaphysalis* sp. and *Dermacentor* sp. Unlike the tick cycle of *Theileria* sp., adult ticks transmit some *Babesia* sp. in the eggs to their progeny (transovarian transmission) and therefore may reinfect new hosts from generation to generation. Mechanical transmission through contaminated instruments and needles is also possible.

CLINICAL SIGNS AND DIAGNOSIS

The typical clinical signs include a high temperature (41°C/106°F) usually 8–17 days after the bite of an infected tick, this may be associated with sudden loss of appetite. Anaemia (mucous membranes become pale and later may become yellow) develops as the course of the disease continues and some animals develop nervous signs (may be confused with rabies). Animals tend to become weak as the anaemia becomes more pronounced and the urine is often reddish or brown in colour, hence the name 'red water fever'. After 2–3 days untreated animals may die or the disease may become chronic with metabolic disturbance and eventual death. In chronic babesiosis the animal appears dull, emaciated and has a poor coat. Diagnosis can be confirmed at necropsy although the post mortem signs can be confused with Theileriosis. The spleen is characteristically swollen and the carcass is anaemic, sometimes yellowish (jaundiced). The bladder contains red or pink urine. The kidneys are enlarged and congested and there may be haemorrhages in the heart, stomach and intestine. In areas where tick borne diseases are common more than one type of blood parasite may be present in blood smears but some may not be very pathogenic. To check for the presence of *Babesia* sp. and other blood parasites, samples of blood, lymph nodes, spleen smears

and brain crush smears should all be taken. If rabies is suspected the brain should also be sent to a laboratory equipped for rabies diagnosis.

Ante-mortem diagnosis can be confirmed by clinical signs along with blood smears which show *Babesia* sp. organisms in red blood cells. Thick and thin smears should be made and stained with Giemsa. Immunological tests may also be used to diagnose infection, the most frequently used tests being the complement fixation test (CFT), fluorescent antibody tests (FAT) and the gel precipitation test (GPT). For details see Chapter 6. In areas where laboratory services are not available an animal health officer may try 'response to treatment' to distinguish Babesiosis from other haemoparasitic diseases. *Babesia* sp. are readily killed by a number of drugs (for example, Diminazene, Imidocarb) and this helps to differentiate the disease from others such as anaplasmosis.

Anaplasmosis – gall sickness

This is a disease caused by rickettsial organisms of the genus *Anaplasma*. It was previously thought to be a protozoan and is often described along with other haemoparasitic diseases. The organism is seen in the red blood cells on infected mammals and is transmitted by tick vectors. *Anaplasma* sp. are found worldwide both in tropical and temperate areas. The main species of veterinary importance are *Anaplasma marginale*, which causes significant clinical disease, and *A. centrale*, which is usually associated with a milder form of disease. The incubation period for the development of clinical anaplasmosis is variable and can range from several days to 5 weeks (usually 30–40 days). The source of infection is usually ticks (*Dermacentor* sp., *Rhipicephalus* sp. and others) which have fed on sick or convalescent cattle, asymptomatic carriers or possibly, infected wild ruminants. Horse flies (Tabanidae), stable flies (*Stomoxys* sp.) and other biting insects can also transmit anaplasmosis and occasionally the organism can be transmitted mechanically during veterinary procedures for example, by use of contaminated syringes and other equipment during vaccination or surgical procedures.

CLINICAL SIGNS AND DIAGNOSIS

Clinical signs are characterized by fever (temperature > 40°C), loss of appetite, marked anaemia (mucous membranes are pale and may then become yellow as jaundice develops), haemoglobinuria (urine darkish yellow to brown) and enlargement of the lymph nodes. Animals usually lose condition rapidly and are susceptible to the development of concurrent disease. Diagnosis may be confirmed at necropsy although the gross findings resemble babesiosis except that the urine in the bladder is usually brown, not red. Ante-mortem diagnosis is based on the clinical signs, epidemiological pattern, the examination of stained blood smears for the presence of *Anaplasma* sp. and the use of serological tests.

CONTROL AND PREVENTION

The disease may be treated by using antibiotics (for example, tetracycline). Prevention is possible by implementing a vaccination program. Low-pathogenic *A. centrale* has been used effectively as a vaccine, as have killed vaccines. Vaccination of all stock is rarely practical but calves exposed to infection can develop 'premunity' that protects them in later life although these calves may remain carriers. The vaccination, along with administration of hyperimmune serum, has also been used in treatment. Convalescent animals must be well fed to ensure full recovery. Adult animals recovering from an attack of the disease have a strong resistance to it which lasts for years. As with other tick-borne diseases, tick control is important and should include the regular use of repellents, acaricides (dipping, spraying, dusting) and the periodic rotation of pasture or burning of overgrown or abandoned pasture. Additional strategies would include the control of livestock movement and keeping stock away from infected wildlife. Insects may also need to be controlled to prevent mechanical transmission.

Trypanosomosis

Trypanosomes are flagellate protozoa that live in the blood, bone marrow, lymph nodes and sometimes the nervous system of affected animals and in the digestive tract and salivary glands of the insect vector. Trypanosomiasis occurs throughout Africa, South America and many Asian countries. Unlike the other haemoprotozoan diseases discussed in this section, Trypanosomosis is primarily transmitted by biting flies (tsetse and other flies of the family Glossinae) and not by ticks. Trypanosomes come from a group of protozoan organisms which usually undergo part of their reproductive cycle in an intermediate host. Many domestic (cattle, sheep, goats, dogs, swine, mules, donkeys, camels and buffalo) and wild animals may become infected with one or more of the trypanosome species. The following information refers mainly to African Trypanosomosis in cattle ('nagana' or tsetse fly disease). Wild and domestic bovids (cattle, buffalo, eland, antelope) may be infected following a bite from an infected vector (tsetse fly). New outbreaks of disease may develop if cattle in a herd move into an infected tsetse fly area and (as long as vectors are present) if wild (reservoir) hosts and/or asymptomatic carrier animals come into the same range as an immunologically naïve herd.

LIFE CYCLE AND CLINICAL SIGNS

There are many species of Trypanosome (for example, *Trypanosoma congolense*, *T. vivax*, *T. brucei*, *T. evansi*, *T. equiperdum*) and a number of specific disease conditions caused by each type. The clinical presentation depends on the species of Trypanosome involved, the mode of transmission, the host species and the host immune response. In Africa, the Trypanosomes most commonly infecting cattle include *Trypanosoma congolense*, *T. brucei* and *T. vivax*. Infections with more than one species of Trypanosome are common. The clinical disease associated with Trypanosomes is often referred to as 'nagana'. Nagana is a zulu word meaning 'to be depressed'. In general, the incubation period is 1–3 weeks but it can be up to 2 months for less virulent strains. The clinical signs observed are variable. In cattle, chronic Trypanosomosis presents as an elevated temperature, weakness and a loss of appetite followed by a reduction in milk production in lactating animals. Anaemia may develop along with diarrhoea and swollen lymph nodes. In the final stages of the disease emaciation may become severe and fluid swelling may appear along the belly and dewlap (oedema). In pigs and horses the disease is often acute and animals usually die shortly after a period of high temperature.

Other trypanosomes

Trypanosoma evansi causes a serious disease known as Surra in horses and camels. Although *T. evansi* can infect other species it is often asymptomatic but in horses and camels it can be fatal with fever, anaemia, oedema and neurological signs developing prior to death. Surra occurs in North Africa, Central and South America, Southern Russia and parts of Asia. It is thought that *T. evansi* is transmitted by biting flies and also vampire bats (in Central and South America). Another protozan, *Trypanosoma equiperdum* (Dourine) causes an acute venereal disease in horses and donkeys, treatment is rarely successful.

DIAGNOSIS

Diagnosis may be confirmed at necropsy although the gross findings may be similar to those of other chronic diseases in cattle and acute septicaemia in horses/pigs. The carcass is usually emaciated and anaemic. The lymph nodes are enlarged and the spleen and liver are enlarged and soft. The muscles appear pale and the blood watery with evidence of haemorrhages on the pleural surfaces and pericardium. Smears of lymph nodes or spleen may be made for indirect fluorescent antibody tests or a brain crush smear may be submitted to the laboratory. Antemortem diagnosis can be made on clinical signs and the epidemiological profile but will need to be confirmed by examining stained (thick and thin) blood smears (Figure 3.36). These should be taken from the ear vein of animals in the early stages of the disease when a fever is present. Whole blood may be collected for laboratory animal tests (sub-inoculation of blood in rats (*T. congolense, T. brucei*) or in other laboratory animals) or for isolation of trypanosomes. A haematocrit centrifuge is used to concentrate the trypanosomes in the buffy layer of the blood (the white cells).

CONTROL AND PREVENTION

Trypanosomosis requires the control of the insect vector (regular spraying with insecticides, controlled clearing of bush and use of fly traps). If possible, the movement of susceptible livestock into infected zones should be prevented. Treatment of exposed livestock with anti-protozoal drugs can be effective in preventing clinical disease but these usually need to be repeated (consult the nearest veterinary officer for current recommendations). In some areas disease-resistant cattle have been bred (or have evolved), for example, the West African N'Dama breed which seems to be resistant to ticks and other parasites.

In some cases, it is possible to predict which vector borne diseases may be present in an area by evaluating the climate, vegetation and topographical features. Figure 3.37 outlines the broad global climatic zones and zoogeographic areas. These factors, along with the presence of suitable hosts (wild or domestic animals), determine which arthropods will thrive in an area and

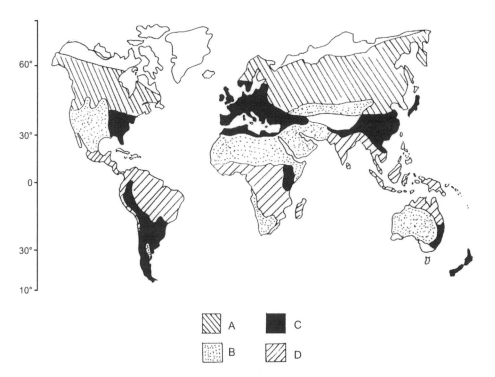

Figure 3.37 Climatic zones. The climate and topography of a country will determine the range of arthropod vectors present and therefore allows some prediction of the likely presence or absence of specific vector-borne diseases: (A) cool humid zones (continental summer and subarctic); (B) dry zones (desert and steppe); (C) warm humid zones (Mediterranean, humid subtropical and west coast regions); (D) tropical humid zones (rain forest and savannah). There are a wide range of variations within zones depending on the factors such as altitude and vegetation. The map illustrates climatic zones and zoogeographical regions.

therefore allow an assessment of which protozoal diseases and other arthropod transmitted diseases are likely to occur in a given region. The role of wildlife reservoirs in livestock disease outbreaks is discussed further in Chapter 9.

Trypanosomiasis in humans and community engagement

Trypanosomiasis causes significant human morbidity and mortality in many parts of the developing world. Chagas disease is caused by *Trypanosoma cruzi*, a parasite spread by bloodsucking reduviid bugs (for example, *Rhodnius prolixus*), and is one of the major health problems in South America. Due to immigration,

the disease also affects people in the USA. The symptoms of Chagas disease vary over the course of an infection. In the early, acute stage, symptoms are often mild and usually produce no more than local swelling at the site of infection. After 4–8 weeks, individuals with active infections enter the chronic phase of Chagas disease that is asymptomatic for 60–80% of chronically infected individuals through their lifetime. There are various treatment options. Antiparasitic treatments given in the acute phase of the infection may delay or prevent the development of disease symptoms during the chronic phase of the disease, but 20–40% of chronically infected individuals may still go on to develop life-threatening heart and digestive system disorders. Prevention is based on vector control and avoiding insect bites.

In Africa, human trypanosomiasis takes two forms depending on the parasite involved. *Trypanosoma brucei gambiense (T.b.g.)* is found in west and central Africa and currently accounts for over 95% of reported cases of sleeping sickness. In this chronic infection a person can be infected for months or even years without major signs or symptoms of the disease. When symptoms emerge, the patient is often already in an advanced disease stage where the central nervous system is affected. *Trypanosoma brucei rhodesiense (T.b.r.)* is found in eastern and southern Africa. According to the WHO, this form now represents less than 5% of reported cases and causes an acute infection. First signs and symptoms are observed a few months or weeks after infection. The disease develops rapidly and invades the central nervous system. The parasite is transmitted by tsetse flies (*Glossina* sp.) which are found in sub-Saharan Africa although only certain species transmit the disease. Rural populations living in regions where transmission occurs and which depend on livestock agriculture or hunting are the most exposed to the tsetse fly and therefore to the disease although not all areas where tsetse flies occur have cases of sleeping sickness. The disease develops in areas ranging from a single village to an entire region. There are various treatment options available depending on the stage of the disease. Prevention is based on vector control and avoiding insect bites.

Chagas disease and African sleeping sickness are both diseases in which the parasitic agent causing the disease also occurs in animal species. Although the transmission from animals to humans is not direct, the arthropod vectors that transmit these diseases are not host specific so interspecies transfer of the parasites is possible. For this reason, in areas where trypanasomes and their vectors are endemic, vector control and monitoring of animal health is important for maintaining community health.

The 'One Health' approach

In rural areas, throughout the world, there are often concurrent human and animal health issues to deal with. This provides an ideal opportunity for medical and veterinary teams to work together with other experts to provide technical support for district and regional human and animal health extension services. The latter are responsible for delivering health advice to the local community and may often be required to visit villages where human and animal disease outbreaks are occurring concurrently. In the case of vector borne zoonotic diseases there may be a link between human and animal disease outbreaks but even though this is often not the case there can be significant benefit to taking a collaborative approach to ensure that good hygiene and sanitation practices are implemented and that potential disease vectors are effectively controlled. Where possible, the focus should be on disease prevention with the promotion of good animal husbandry practices and support provided for community education on topics such as food and water safety and how to prevent disease transmission between humans and between humans, animals and the wider environment. The latter is often best achieved by engaging a local leader who is well placed to involve schools and community groups in educational activities.

Zoonotic diseases endanger people's livelihoods by affecting their livestock as well as directly compromising their own health and survival. A list of common zoonotic diseases is provided in the Appendix 1. The effective control of any zoonotic disease, including vector borne parasitic diseases such as sleeping sickness, enteric protozoal infection and helminth parasites such as *Taenia saginata* and *T. solium*, has the potential to alleviate illness and poverty, particularly in marginalized rural and peri-urban communities living in close contact with animals.

3.7 Introduction to veterinary entomology

Veterinary entomology is the study of arthropods of veterinary importance. This includes ectoparasites, that is, arthropods which parasitize animals directly, and arthropod vectors of diseases such as West Nile virus, Bluetongue and Babesiosis, that is, mosquitoes, midges and ticks respectively. Ectoparasites include arachnids (that is, mites and ticks) and insects (that is, mosquitoes, flies and lice). Some are obligate parasites but many spend much of their life cycle in the environment.

Ectoparasites cause disease in a number of ways: (1) feeding directly on the blood of the host; (2) burrowing under the skin feeding on tissue proteins; (3) they may lay eggs on the host that develop into larvae that feed on host tissues; and/or (4) acting as vectors for other diseases. Vector surveillance methods, and a summary of some important vector borne diseases, is provided in Chapter 14 (and Tables 14.1 and 14.2). Some ectoparasites live for most of the time on the host (for example, lice), some feed intermittently on one or more hosts during the life cycle (for example, ticks) and others feed periodically (for example, fleas). To effectively control arthropods it is important to be familiar with their life-cycles. This is because anti-parasitic treatment of the host alone is unlikely to be effective if ongoing re-infection from environmental sources is not prevented.

Ectoparasites, and free-living arthropods, are responsible for transmitting a wide range of pathogens (that is, viruses, bacteria and protozoa) capable of causing significant disease in livestock as well as in companion animals, wildlife and humans. Some ectoparasites are fairly host specific (for example, lice) but many are able to parasitize a wide range of hosts. In large numbers ectoparasites can be a direct cause of

Table 3.9 The classification of ticks of veterinary importance and the diseases they transmit.

Family	Genus	Diseases transmitted
Ixodidae	*Dermacentor*	*Anaplasma* sp. (anaplasmosis) and other rickettsia sp. (Q fever), *Babesia* sp. (babesiosis)
	Hyalomma	*Babesia* sp. *Theileria* sp. (tropical theileriosis), tick typhus, haemorrhagic fever
	Ixodes	*Babesia* sp., *Borrelia* sp. (Lyme disease), tick pyaemia (*Staphylococcus* sp.). Q fever, tick typhus
	Rhipicephalus (Boophilus)	*Theileria* sp. (*T. parva*, East coast fever), Nairobi sheep disease (*Ehrlichia* sp.), various viral encephalitis, *Anaplasma* sp. (anaplasmosis) and other *Rickettsia* sp. *Babesia* sp. (babesiosis)
	Haemaphysalis	Rickettsial diseases (spotted fever), encephalitis (viral e.g. Kyasanur forest disease)
	Ambylomma	Rickettsial diseases (*Cowdria* sp. heartwater) , Q fever, tularemia and relapsing fevers
Argasidae	*Argas*	*Aegyptionella pullorum, Borrelia anserina*
	Ornithodorus	African swine fever, *Borrelia* sp.

Notes: Ticks are not species specific and some multiple host species may feed on different hosts during the life cycle. Diseases may be transmitted from animal to animal and to humans. Some ticks feed on birds as well as mammals.

Table 3.10 The classification of mites of veterinary importance.

Family	Genus	Species	Host species	Clinical signs
Astigmata (scab mites)				
Psoroptidae	*Psoroptes*	*ovis*	Sheep	Sheep scab, wool damage, severe inflammation and exudation with scab formation
	Psoroptes	*cuniculi*	Rabbits and other mammals	Ear mites, irritation, head shaking, excessive wax
	Chorioptes	*bovis*	Cattle, sheep, goats and horses	Scabs on skin, irritation, lesions around tail base and legs spreading over back and neck, Loss of condition
	Otodectes	*cyanotis*	Cats, dogs	Ear mites, shaking head, excessive wax
Sarcoptidae	*Sarcoptes*	*scabiei*	Various spp. including humans	Intense irritation, itching (pruritis), hypersensitivity may occur, self trauma and secondary bacterial infection
	Notoedres	*cati*	Cats, dogs, rabbits	Similar to sarcoptic mange but predominantly around the external ears
Cnemidocoptidae (or Knemidokoptidae)	*Cnemidocoptes* (or *Knemidocoptes*)	*mutans*	Birds	Scaling on head and legs in poultry and other bird species
		gallinae	Birds	'Depluming itch' in poultry
		pilae	Birds	Parrots and cage birds, scaling on beak and legs, may cause deformity
Dermanyssidae	*Dermanyssus*	*gallinae*	Birds	'Red mite' causes irritation and self trauma. The mites reside in the environment when not feeding
Prostigmata (follicle mites)				
Demodicidae	*Demodex*	*canis* *bovis* *gatoi* *injai* *folliculorum*	Dogs, cats, cattle, humans	Demodecosis, may be associated with hair loss, not usually pruritic. Underlying disease can worsen clinical signs. May be aymptomatic
Cheyletidae	*Cheyletiella*	*blakei* *parasitivorax* *yasguri*	Cats, rabbits	Infects humans (zoonotic), often few clinical signs. Humans can develop a rash around mid abdomen
Psorergatidae	*Psorergates*	*ovis*	Sheep	Itch mite in sheep, wool damage. May develop a hypersensitivity response

debility as well as being an annoyance (Tables 3.9 and 3.10). Healthy animals usually develop some immunity against common ectoparasites but if animals become debilitated (due to undernourishment or concurrent disease) ectoparasite numbers may increase. With this in mind it is important to look for other underlying cause(s) of 'ill health' as treating the ectoparasite problem.

can require specialist knowledge but some basic keys are available to help identify the most common ectoparasites to the level of genus. Where possible, when selecting specimens of ticks and mites for examination, it is a good idea to collect representative adult males and females as juveniles are often harder to identify. The trapping of free living arthropods is discussed in Chapter 14.

Classification and identification

The classification of ectoparasites of veterinary importance is outlined in Figure 3.38 and is described in more detail under each section heading. Definitive identification of arthropods

Samples for ectoparasite examination

Ectoparasites (see Figure 3.43) collected from individual animals can be submitted as either whole fixed or fresh specimens (for example, ticks, lice) or in skin scrapings (mites), hair samples (lice, mites) and ear wax samples (ear

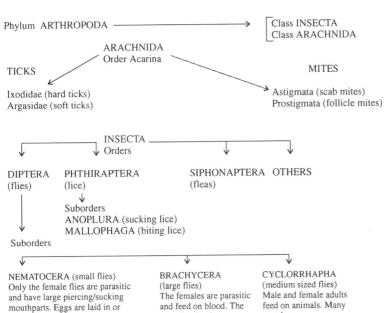

Figure 3.38

The general classification of arthropods of veterinary importance.

mites). Samples should be placed into sealed labelled containers. If there is to be a delay in examination it is preferable to fix the specimens using preservatives such as 70% alcohol. If the specimens are to be used for museum collections, or for display, it is advisable to consult an entomologist to determine the best way to fix the specific parasites which are collected because some fixatives alter the colour and/or size of the specimen.

Skin and hair samples may also be examined for the presence of fungal hyphae. Fungal species such as *Trichophyton* sp. and *Microsporum* sp. may cause 'ringworm'. Ringworm is not a 'parasite' but it is a skin infection which can resemble ectoparasite infestation causing irritation, red skin and hair loss especially if there is a secondary bacterial infection (see Chapter 4, section 4.6 and Chapter 10, Tables 10.7 and 10.8).

3.8   Ticks and tick-borne diseases

Ticks are common worldwide and are responsible for transmitting a wide range of micro-organisms including haemoprotozoa (for example, *Babesia* spp. and *Theileria* spp.), Rickettsia (for example, *Anaplasma* spp.), bacteria (for example, *Borrelia* spp.) and viral diseases (louping ill, various encephalidities) (see Table 3.9).

Ticks can be collected directly from the animal host and fixed in 70% alcohol or in 10% formal saline prior to examination under a dissecting microscope. Try to make sure that the head and mouthparts of the tick are removed along with the body. Swabbing the skin where the tick is attached with alcohol and waiting a few minutes may help. Collected ticks should be stored in labelled glass or plastic vials and should be accompanied by information about the host (species, breed, age, sex, recent ectoparasite treatment, general health and so on), the location (altitude/latitude of the area) and the date when the ticks were collected.

The distribution of ticks and other arthropods depends on a number of environmental and climatic factors. Owing to changes in land use and climate the distribution of some arthropod species, including ticks, is changing. Studies on tick distribution and abundance can give good predictive information about the likely occurrence of specific tick-borne diseases in an area. Tick survey work should be done using strict guidelines with an agreed sampling protocol. The latter would usually include whole body tick counts along with pasture tick counts to assess the number of ticks present in the environment. Data from such studies can be used for the development of vector distribution maps and for risk assessments. Disease modelling, which engages the expertise of epidemiologists, climatologists, disease specialists and entomologists, has become a very useful tool for establishing surveillance and disease prevention programs for vector borne diseases in many parts of the world.

There are two main types of tick, the Argasidae (soft ticks: no scutum) and the Ixodidae (hard ticks: scutum present). These are distinguished by the presence or absence of a hard outer scutum on the dorsal aspect of the body. In the Ixodid species the scutum covers most of the dorsal surface of the male tick but not of the fed female tick (Figure 3.39). Ticks exist as eggs, larval, nymph and adult forms. The larval forms have three pairs of legs like insects (Figure 3.40). The nymphs and adults have four pairs of legs. Some tick species have a single host and others have several different hosts during their life cycle. Ticks feed on blood and swell to many times their original size as they feed. Heavy tick infestations in young animals may cause anaemia. In ticks which feed on multiple hosts, the larval, nymph and adult ticks drop off each host after feeding and then begin the next stage of development in the environment (Figure 3.41). In ticks which feed on a single host the tick progresses through each stage of

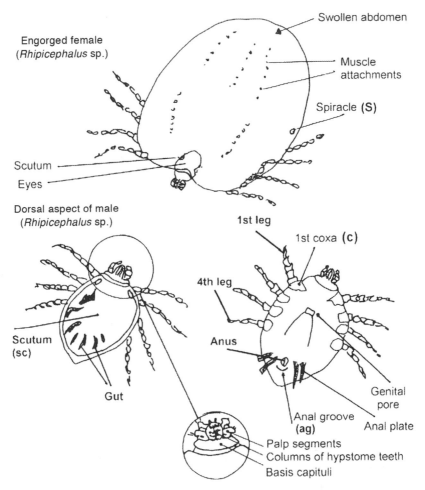

Figure 3.39 Morphological characteristics used to identify ticks. The identification of ticks requires careful examination, usually of adult males, under a dissecting microscope. Fully engorged females are difficult to examine due to the distortion of their anatomical features. On the dorsal (upper) surface the shape and size of the scutum (sc), the colour of the body and legs and the type of mouth parts may also assist in the identification to genus level. Identification to species level requires experience but may be achieved using detailed keys. Adult males should be placed on their back and the ventral (under) side examined for the shape and size of the anal groove (ag), spiracles (s) and the coxae (c).

development on the same host (for example, *Rhipicephalus [Boophilus] annulatus*) (Figure 3.42).

Tick identification

The identification of ticks takes time and experience. Figure 3.39 outlines the main mor-

phological features that need to be identified using a dissecting microscope with a good light source. Identification to genus level may be done using basic classification guidelines but identification to species level may need referral to a specialist centre. It is important to identify the type of tick present as this allows more targeted and effective control measures to be developed.

Figure 3.40 Typical tick life cycle. (A) Adult male and (B) female mate, (C) the female takes a large blood meal and lays her eggs in the environment. (D) A larva develops from each egg, (E) this feeds and moults to form a nymph. The nymph feeds and moults to form an adult (A or B).

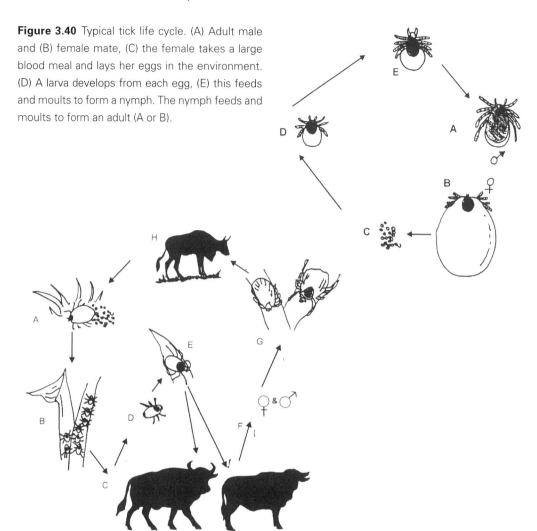

Figure 3.41 Life cycle of a three-host tick (for example, some *Rhipicephalus* sp., *Ixodes* sp.). These ticks can feed on three individual hosts separated by long periods in the vegetation where moulting to the next stage of the life cycle and egg laying occurs. Unfed ticks appear flat but when fed the abdominal area expands to 5–10 times the original size. Feeding at each stage may take a few days to several weeks. Females mate once and take an enormous meal before detaching to die after laying eggs in the vegetation. In most cases the aim of tick control is to kill the adult females before they detach and contaminate the environment with eggs. Adult males tend to remain on the host for several matings. It is often easier to identify the species of adult male specimens versus adult females due to the excessive engorgement of the females (see text). (A) Engorged females lay eggs off the host in long grass. Eggs hatch when conditions are favourable (that is, suitable temperature and level of humidity). (B) Larvae emerge and cluster on grass to infect the first host (C). (D) Engorged larvae moult off the host to form the nymph stage. (E) Nymphs attach to the next available host which may be a different species of mammal or bird. (F) Engorged nymphs moult off the host and emerge as adult ticks. (G) Adult male and female ticks mate and attach to the next host. The female engorges and detaches into the environment where she lays her eggs.

Figure 3.42 Life cycle of a typical one-host tick (for example, *Rhipicephalus* [*Boophilus*] *microplus*, *R. annulatus*). All stages of the life cycle, except for egg laying, occur on the same host. (A) Eggs laid by an engorged female off the host in the vegetation. (B) the larvae hatch and cluster on long grass before attaching to the chosen host. (C) Larvae feed and moult to form the nymph stage. Note that larvae have three pairs of legs as compared with four pairs of legs in the nymph and adult stages. Some people may confuse larval ticks with lice which is why it is important to collect samples of ectoparasites for correct identification in the veterinary laboratory. (D) Nymphs feed and moult to form the adult. (E) Mature male and female ticks mate and the female engorges (often > 10× unfed size) and drops off to lay her eggs in the vegetation. The eggs survive best in long vegetation and when the climate is warm and fairly humid. It is for this reason that ticks are more common in some regions and tend to have different seasonal prevalence (see text).

For example, single host ticks (for example, *Rhipicephalus* [*Boophilus*] *annulatus*) are easier to control than two (*Hyalomma* sp.) or three host ticks (*Rhipicephalus sanguineus*, *Ixodes* sp.) due to the fact that the latter are present in large numbers in the environment for much of their life cycle. Where animals are hosting a number of ticks, even if they appear to be healthy, it is useful to collect blood smears to check for the presence of haemoparasites (see Figure 3.36). Examination of blood smears from apparently healthy animals in an area provides a useful comparison for samples collected from sick animals where haemoparasites are suspected to be the cause of disease.

Owing to the fact that each stage in the life cycle of multiple host ticks (for example, *Ixodes* sp.) spends some time in the environment their distribution and abundance is determined by environmental conditions, that is, the larval and nymph stages require suitable levels of humidity and ambient temperature to ensure survival to infect the next host. In one-host ticks, typically only the larvae need to search for a host, the remainder of the life cycle occurs on the same animal. Heavy infestation of ticks can result in 'biting stress' with resultant irritation and damage to the skin. In addition, ticks can transmit a range of clinically important diseases. For example, *Rhipicephalus* (*Boophilus*) *microplus* and

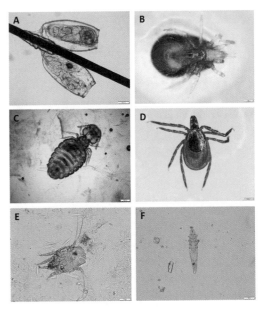

Figure 3.43 Photos of common ectoparasites: (A) eggs of lice; (B) *Dermanyssidae* (mite) – cat; (C) *Damalinia* sp. (lice) – goat; (D) *Ixodes scapularis* (tick) – dog; (E) *Chorioptes bovis* (mite) – cattle; (F) *Demodex canis* (mite) – dog. See also Plate 7. Photos: Dr Mani Lejeune, Cornell University, USA.

R. (Boophilus) annulatus transmit *Babesia bigemina* and *B. bovis. R. (Boophilus) decoloratus* ticks transmit *B. bigemina* and rickettsial species (*Anaplasma marginale* and *A. centrale*), see Table 3.9.

Tick control

Animal health authorities will develop their own tick control guidelines but some basic principles are outlined below.

1 Identify the species of tick(s) present.
2 Treat infested livestock with a suitable product (acaricide).
3 Move livestock to a 'clean' (that is, tick free) pasture after treatment.
4 Clear overgrown pasture/scrub to remove the environmental stages of tick development.
5 If 'clean' land is not available, acaricide treatment may need to be repeated every 2–3 weeks.[7]

3.9   Mites and mange

Parasitic mites are small arachnids which can cause mild or severe skin damage depending on the species of mite, the level of infestation and the underlying health status of the host. Mite infestations are associated with skin irritation and hair loss; severe infections are usually clinically obvious and cause discomfort and itching. The latter often results in secondary trauma, the formation of scabs and concurrent bacterial infections. Mite infestation is sometimes referred to as mange. Some species, such as the sheep scab mite (*Psoroptes ovis*) may cause significant morbidity in livestock and economic loss due to wool or pelt damage and poor body condition. Many mites, for example, *Chorioptes* sp., are not strictly host specific, for example, *Chorioptes bovis* may infect cattle, sheep, goats and horses. *Sarcoptes scabiei* infects pigs, dogs

(Figure 3.45), camels, sheep, goats and horses as well as humans but specific subspecies are not commonly zoonotic.

A summary of the mites of veterinary importance is given in Table 3.10 (above) and two species are illustrated in Figure 3.44. Some non-parasitic, or free-living, mites such as members of the Trombiculidae (that is, harvest mites) may also cause irritation in individual animals but the infestation is usually transient. Mites may also be found in stored food (for example, *Glycyphagus* sp. and *Acarus* sp.) and in human dwellings, for example, the house dust mite (*Dermatophagoides pteronyssinus*), which feeds on flakes of loose epidermal tissue and may result in allergy in certain individuals.

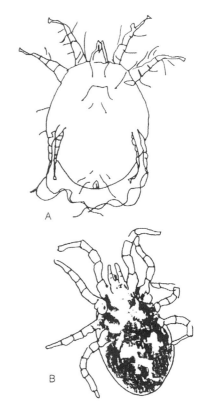

Figure 3.44 Ventral view of two species of adult mites. (A) *Psoroptes* sp. mite; (B) poultry 'red mite' *Dermanyssus gallinae*.

Figure 3.45 Stray dog with advanced sarcoptic mange. Kathmandu, Nepal. The mite burrows under the skin causing irritation and, in most cases, stimulates a hypersensitivity response associated with intense irritation. The host animal loses hair and develops raw areas, especially on the legs and ear tips, as a result of self-mutilation. Secondary bacterial infections commonly complicate diagnosis and treatment.

As mentioned earlier, the clinical signs of parasitic mite infestation are variable and depend on (1) the species of mite involved, (2) the number of mites present, (3) the immune status of the host as well as (4) environmental factors. In some cases, a host can develop a severe hypersensitivity response to the presence of mites resulting in severe irritation and self-trauma, in these cases there may only be a few parasites present. This response is often seen in cases of sarcoptic mange (*Sarcoptes* sp.) in dogs, camels and pigs. Skin scrapings taken from animals with mange should be carefully collected using a sharp sterile scalpel blade to remove a sample of the superficial and deeper epidermis. It is advisable to swab the area with alcohol prior to sample collection. If the skin condition is severe, the wounds can be bathed in an antiseptic wash afterwards. In severe cases of sarcoptic mange the animal may have secondary bacterial skin disease but, as there may be few active mites present, it might be necessary to take several skin scrapings to isolate the mites. In addition to skin scrapings, hair samples are often collected at the same time for other tests.

To find ear mites (for example, *Otodectes* sp., *Psoroptes* sp.) take samples of ear wax. Typical signs of ear mite infection in dogs include red sore ears due to repeated scratching, head shaking and the subsequent development of aural haematomas due to self-trauma. Treatment of ear mites requires removal of the wax and other debris before acaricide is added. Supportive care to prevent further self-trauma may also be required. A veterinary professional should thoroughly examine the animal before ear drops are given to make sure that the ear drum is intact and that there are no other complications.

Demodectic mange (*Demodex* sp.) may affect many species but in debilitated cattle and dogs it can become very severe and may result in thickened skin, hair loss (especially around the eyes) and deep skin lesions. In dogs, there may also be an underlying disease problem (for example, hypothyroidism) and in these cases *Demodex* sp. may be difficult to treat. Infections can be diagnosed by examining skin scrapings. Animals may need antibiotic treatment for secondary bacterial infections.

Some mite infections are superficial (for example, *Cheyletiella* sp.) and in these cases treatment is usually quite simple. The mites can be identified in hair or fur samples. Superficial mites may be caught on Sellotape strips by placing the sticky side of the tape on the animal's skin and hair and removing it quickly. The mites and their eggs remain stuck to the tape which can be placed on a microscope slide for examination. Note that *Cheyletiella* species mites may also affect humans causing a red rash around the abdomen or other areas of soft thin skin that have come into contact with infested animals.

Skin scrapings and tissue digestion

Examination procedure

DIRECT METHOD

Place the sample (skin/hair) on a slide and mix with a little oil or water. Examine the slide with the microscope using low power (4×). For large parasites (ticks, some lice) examination may be done using a dissecting microscope.

ALKALI DIGESTION METHOD

Digestion methods are used to allow a more 'clear' view of parasites taken from skin scrapings. Place a portion of the sample in a test tube and add 5 ml of 10% KOH (or NaOH). Heat the sample gently to boiling point (about 5 min) until a homogenous solution remains in the tube. Do not allow the solution to boil as it will splash and KOH is corrosive. Allow the sample to cool and then centrifuge for 5 min at low speed. The sediment at the bottom of the tube can then be examined under the microscope for mites and fungi. For examination of fungal hyphae, fix the sample in 70% alcohol and stain with lactophenol cotton blue stain (LPCB) (see also Chapter 4, section 4.6).

3.10 Lice and other insects

Lice

There are two main suborders of lice, the anoplura (sucking lice) and the mallophaga (biting lice). In general, lice are fairly host specific with each species of animal having its own range of lice. In cattle, the most common species include the biting louse, *Damalinia* sp., and the sucking lice, *Linognathus* sp. and *Haematopinus* sp. (Figure 3.46). Disease caused by lice is known as pediculosis. Heavy infestations are more common in housed, over-crowded and debilitated individuals. Infestations are fairly easy to treat but re-infection is often a problem unless animals are moved to a 'clean' environment. Pediculosis

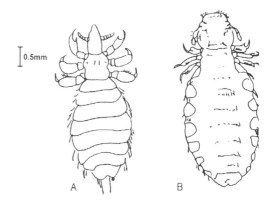

Figure 3.46 Dorsal view of two species of adult louse (insects – three pairs of legs). (A) *Linognathus* sp. (sucking louse), (B) *Damalinia* sp. (biting louse). Most species of lice are fairly host specific.

may cause some irritation but this is usually less severe than that seen in mite infestation. Sucking lice largely infest mammals whereas biting lice can be found on mammals and birds. Heavy infestations of sucking lice can cause severe anaemia. Both sucking and biting lice can cause irritation and self-trauma, which may lead to secondary bacterial infection and damage to the skin/hide. The control and prevention of louse infections requires an understanding of the louse life cycle and attention to animal husbandry and environmental conditions.

Lice are known to transmit some bacterial and rickettsial diseases such as typhus (for example, *Rickettsia typhi*). Diagnosis of louse infestation is confirmed by collecting samples of hair, this may be done by removing loose hair by the roots or by cutting with a scalpel blade at the base of the shaft. In this way, adult lice and eggs may be collected. Such samples can be placed in a labelled and sealed plastic bag and submitted for examination under the dissecting microscope (Figure 3.46).

Lice common in livestock

Anoplura (sucking lice)

Haematopinus sp. are found on cattle, pigs and horses. *Linognathus* sp. are found on cattle, sheep, goats and dogs (Figure 3.46). *Solenoptes* sp. are found on cattle.

Mallophaga (biting lice)

Damalinia sp. are found on cattle, sheep, goats and horses (Figure 3.46). *Felicola* sp. are found on cats (usually young or debilitated individuals). *Trichodectes* sp. are found on dogs and can be a vector for the tapeworm (*Dipylidium caninum*). *Heterodoxus* sp. are found on dogs. *Lipeurus* sp., *Menopon* sp. and *Menacaullus* sp. all affect poultry but do not occur worldwide.

Fleas

There are estimated to be over 2500 species of fleas worldwide. Fleas are small flightless insects that spend much of their life cycle in the environment. The parasitic adults live by consuming the blood of avian and mammalian hosts. Adult fleas are usually about 3 mm long, brown in colour with flattened bodies (Figure 3.47) to enable them to move through their host's fur or feathers. They lack wings but have long hind legs adapted for jumping and mouthparts adapted for piercing skin and sucking blood. The eggs and larvae develop in the environment with the length of the life cycle depending on environmental conditions and temperature. Larvae are maggot-like and have chewing mouthparts to feed on organic debris. Each species of flea is adapted for a specific host species but, most will bite a range of other hosts, for example, dog and cat fleas (*Ctenocephalides* sp.) often bite humans. Fleas are principally a nuisance to their hosts, causing itching and over grooming.

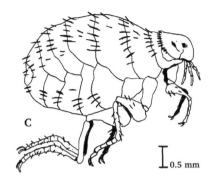

Figure 3.47 Lateral view of a flea. (C) (*Ctenocephalides* sp.) cat and dog fleas. The fleas are specialized for jumping. Most fleas are not host specific and will readily infest a wide range of hosts. Illustration: Louis Wood.

However, flea infestation can result in significant skin irritation, especially in hosts that become hypersensitive to the flea saliva. The latter is commonly seen in cats and dogs and may be associated with a significant amount of hair loss, because of over grooming. Fleas are vectors for a number of viral, bacterial and rickettsial diseases in humans and animals (Table 3.11). The oriental rat flea, *Xenopsylla cheopis* was considered to be a key vector for *Yersinia pestis* (the cause of bubonic plague) in the middle ages.

Other insects

There are a wide range of other insects which are capable of causing disease in domestic and wild animals, some of these are outlined in Table 3.11. The morphology and life cycles of some insects of veterinary importance are given in Figures 3.48a and 3.48b. There are some species of fly which spend part of their normal life cycle inside live animal hosts, for example, the bot flies which are members of the family Oestridae. The life cycle of the horse bot is illustrated in Figures 3.49a and 3.49b. Other flies, especially the blow flies, may lay eggs in

Table 3.11 Some insects of veterinary importance and the diseases they transmit (for additional information see Taylor et al., 2007).

Family	Species	Main hosts	Disease and comments
Diptera (flies)			
Tabanidae	*Tabanus* sp. *Chrysops* sp. *Haematopota* sp.	Predominantly large mammals	Causes annoyance, can transmit trypanosomes, tularemia, *Anaplasma* sp, equine infectious anaemia
Glossinidae	*Glossina* sp. (Tsetse fly)	Large mammals	*Trypanosoma* sp. (sleeping sickness in humans, Nagana)
Muscidae	*Musca* sp. (house fly), *Stomoxys* sp. (stable fly)	Many	Cause annoyance, may mechanically transmit bacteria, cause wound infections, Surra (*T. evansi*), keratoconjunctivitis (pink eye)
Oestridae	*Hypoderma* sp. (bot fly), *Oestrus* sp. (warble flies), *Gasterophilus* sp. (nasal bot)	Many	Local irritation, may cause secondary problems due to location of lesions (i.e. *Oestrus ovis* larvae migration to brain)
Calliphoridae	*Calliphora* sp. (blue bottles), *Lucilia* sp. (green bottles = blow flies)	Many	Myiasis, wound infection with larvae (maggots) = 'fly strike'
Hippoboscidae	Forest flies and keds, e.g. *Melophagus ovinus*, (wingless fly)	Several species	Cause 'biting' distress, annoyance
Ceratopogonidae	*Culicoides* sp. (biting midge)	All domestic and wild mammals	May transmit Blue tongue virus, epizootic haemorrhagic disease (deer) and African horse sickness as well as filarial worms
Psychodidae	*Phlebotomus* sp. (sand flies)	Many wild and domestic mammals and birds	May transmit Leishmaniasis
Culicidae	Mosquitoes* (*Aedes* sp. *Anopheles* sp. *Culesita* sp. *Culex* sp. etc.)	Many wild and domestic mammals and birds	Breed in damp humid areas, some species may transmit heart worm (*Dirofilaria* sp.), malaria, viral encephalitis and rift valley fever
Phthiraptera (lice)	*Trichodectes* sp.	Dog	Can transmit the *Dipylidium* sp. tapeworm
	Felicola sp.	Cat	Chewing louse of cats, can cause irritation, usually seen in debilitated animals
	Pediculus sp.	Human	Can transmit Typhus/Relapsing fever

Family	Species	Main hosts	Disease and comments
Siphonaptera (fleas)	*Ctenocephalides felis, Ct. canis* and others	Cat, dog, most mammals have host adapted fleas	Can transmit *Dipylidium* sp. tapeworm
	Pulex sp.	Humans, pigs and other mammals	May transmit a range of diseases including typhus, tularemia, and tapeworm
	Xenopsylla sp.	Rats and other rodents	May transmit *Yersinia pestis*-a cause of the bubonic 'plague'

Notes: *There are many species of mosquitoes, not all are competent to transmit disease, to be an effective vector the pathogen must be able to multiply in the insect (or tick) and the environmental conditions must be favourable so that the life cycle of the vector can accommodate sufficient replication of the virus, parasite or other potential pathogen.

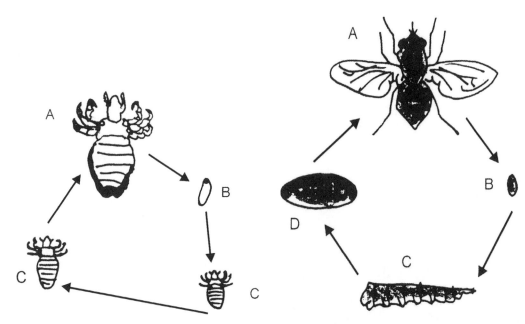

Figure 3.48a Typical louse life cycle in which there is no metamorphosis. (A) adult (insect – three pairs of legs), (B) egg, (C) nymphs.

Figure 3.48b Typical fly life cycle characterized by metamorphosis. (A) adult, (B) egg, (C) larva (maggot), (D) pupa.

Figure 3.49a The life cycle of the equine stomach bot (*Gasterophilus intestinalis*). (A) Adult fly lays eggs on the horse's hair (usually the lower limbs). (B) The horse ingests the bot eggs while grooming. (C) The eggs hatch in the horse's mouth or throat and the larvae migrate to the stomach to pupate (D) (see also figure 3.49b). The pupae pass out of the horse's system in the faeces (E/F). Pupae hatch and adult flies emerge when the weather conditions are appropriate. Illustration: Louis Wood.

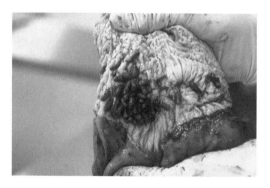

Figure 3.49b Horse stomach opened out to illustrate the appearance of the larvae of the horse bot fly (*Gasterophilus* sp.) which develop and overwinter in the horse's stomach before passing out in the faeces to emerge as adult flies. Small numbers of bots probably do very little harm but high numbers may damage the gastric mucosa.

diseased tissue or fresh wounds of animals leading to maggot infestation or myiasis. Myiasis may lead to severe tissue damage if maggots are left to invade healthy tissue so wounds should be regularly checked for the presence of maggots. Identification of insects takes skill and experience and it is necessary to have access to a good dissecting microscope and identification keys.

Identification of insects

Using identification keys, it is possible to identify most of the common ectoparasites to genus level but for species identification samples may need to be stored for later submission to a specialist institute. The classification of lice, flies and other insects of veterinary importance is outlined in Figure 3.37 and key species are listed in Table 3.11. Molecular methods of insect identification are also gaining popularity.

There are a lot of important animal diseases linked to arthropods but these are beyond the scope of this book. The tables and illustrations in this chapter are provided to give an overview of some of the main features of arthropod life cycles and morphology but additional information may be found in the texts listed in the reference section (see also Chapter 14). For further information on this subject it would be useful to visit an institute which offers specialist training and expertise in entomology and vector borne diseases.

Endnotes

1   If a centrifuge is not available prepare a standard suspension as outlined above (without step 7). Allow the sample to settle by gravity for 15–20 min and discard the supernatant. Note: this step may affect the sensitivity of detection.
2   Use a shorter time for fragile thin walled parasite eggs such as fluke eggs. The latter will collapse but should still be recognizable. Collapsed eggs may sink so examination should be done within a reasonable time after flotation.
3   Plasma pepsinogen levels are elevated in any disease which causes abomasal crypt damage and this is an indirect indication of the presence of *Ostertagia* sp. in cattle. In severe cases there is usually evidence of severe abomasal ulceration at necropsy.
4   https://www.cdc.gov/parasites/trichinellosis/biology.html.
5   Fixatives are reagents that solidify cells and prevent them from undergoing autolytic degeneration.
6   Diff Quick™ is a commercial stain manufactured by B.M. Brown Ltd, Reading, UK.
7   Depends on the type of tick and the duration of the life cycle which may in turn depend on the climate.

Bibliography

Bowman, D.D., Georgi, J.R. (2009) Georgi's Parasitology for Veterinarians, 9th edn. Saunders, Elsevier Health Sciences, Philadelphia, PA.

Goddard, J. (2008) Infectious Diseases and Arthropods, 2nd ed. Humana Press, Springer Nature, Cham, Switzerland.

Kasai, T. (1999) Veterinary Parasitology. Butterworth Heinemann, Oxford.

Merck (2000) The Merck Veterinary Manual, http://www.merckvetmanual.com/mvm/index.jsp. accessed June 2011.

Mullen, G., Durden, L. (2009) Medical and Veterinary Entomology. Elsevier, Academic Press,. Cambridge, MA.

Soulsby, E.J.L. (1982) Helminths, Arthropods and Protozoa of Domesticated Animals. Bailliere Tindall, London.

Taylor, M.A., Coop, R.L., Wall, R.L. (2007) Veterinary Parasitology, 3rd edn. Blackwell Publishing, Oxford.

Urquhart, G.M., Armour, J., Duncan, J.L., Dunn, A.M., Jennings, F.W. (1996) Veterinary Parasitology, 2nd edn. Longman Scientific and Technical, Harlow.

Zajac, A., Conboy, G.A. (2006) Veterinary Clinical Parasitology. Blackwell Publishing, Ames, IA.

chapter 4

Microbiology

Susan C. Cork and Roy Halliwell

4.1 General sample collection, preparation and handling

The discipline of microbiology includes bacteriology, virology and the study of fungi (mycology). Most regional veterinary laboratories and small district laboratories are equipped to perform basic bacteriology and some mycology work. Traditional virology is more specialized requiring cell culture and controlled conditions for viral growth. However, many viral diseases can initially be diagnosed based on clinical and epidemiological characteristics and confirmed using serological or antigen capture tests (see Chapter 6). The emergence of commercial 'kits' for molecular testing (see Table 4.3) has also made targeted molecular screening for microorganisms more feasible with some regional laboratories now able to offer a limited range of molecular tests. The type of samples submitted for microbiological examination is varied and includes faecal samples, milk, urine, tissues, exudates, transudates, pus and mucus. In stained preparations and wet smears submitted for microbiological examination, protozoal organisms may also be present, for example, *Coccidia* and *Cryptosporidia* spp. (see Chapter 3, section 3.5).

The correct collection and handling of microbiology samples prior to laboratory testing is important because poor technique may lead to contamination and overgrowth of the initial causative organism. Where possible, specimens

for microbiology testing should be kept chilled (but not frozen). Owing to the difficulty in getting suitable diagnostic samples to the laboratory from the field, many of the samples submitted to a microbiology unit may be of limited value and a cautious approach must be taken when interpreting culture results. To assist laboratory staff in the selection of suitable media and culture conditions for microbial growth it is important that the submitter completes a laboratory submission form to accompany the samples. On the submission form, the submitter should outline the nature of the disease observed (clinical signs/epidemiology) and note the species, age and sex of the animal(s) sampled as well as the type of sample (skin, faeces, milk, tissue swab, urine and so on). The submitter's name and contact details should be clearly noted along with the date of sample collection. Where possible the submitting veterinarian or extension officer should list the likely differential diagnoses on the form to ensure that the pathogens suspected can be specifically tested for, that is, some bacteria will only grow under anaerobic (oxygen free) conditions (for example, *Clostridium* spp.) and others require special nutrients in the media for growth (for example, *Leptospira* spp.). If suitable culture media or conditions are not provided culture will be unsuccessful even in cases where the pathogen was initially present in the specimen.

This chapter outlines the basic principles of microbiology and is divided into separate sections for bacteriology, virology and mycology. There is also a supplementary section on molecular techniques used in the microbiology laboratory. The emphasis is on the practical aspects of sample handling, testing methods commonly used in the district laboratory and the preliminary interpretation of results.

4.2 Introduction to bacteriology

Two hundred years ago factual knowledge of microorganisms and their importance in disease was limited. The smallest living organisms known were those that could be seen by the naked eye (lice, fleas, maggots and so forth). In the 17th century, a Dutch merchant called Leuwenhoek (1632–1723) was the first to get an insight into the 'invisible' world of microorganisms. Leuwenhoek's hobby was to grind down and produce lenses from glass and by placing one on top of another he produced a primitive microscope. Leuwenhoek, with his insatiable scientific curiosity, used this microscope to examine things around him, such as oil, drops of water, yeast, plants and body fluids and was

one of the first people to describe cells, bacteria, yeasts and a range of parasites. During the latter part of the 19th century the French scientist Louis Pasteur (1822–1895) demonstrated the role that microorganisms played in the processes of disease, fermentation and putrefaction (rotting). Numerous outstanding scientists have added to our knowledge since then.

The majority of bacteria vary in size between 0.003 mm and 0.005 mm (most viruses are much smaller, Figure 4.1). To observe bacteria it is necessary to use a microscope with a magnification of 500× or 1000× (we usually use an oil immersion lens with a 100×10× magnification). Bacteria are single-celled organisms which multiply by binary fission and, under optimum growth conditions, are capable of dividing approximately every 20 min. Theoretically that means, after 24 h, one cell can produce a mass equal to many millions of times its original weight. This rarely occurs naturally but it can happen under ideal laboratory conditions.

The majority of bacteria isolated from animals are harmless or even beneficial. These are referred to as normal flora or commensal bacteria. It is important to be familiar with the

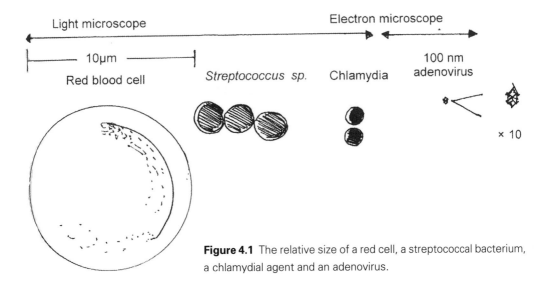

Figure 4.1 The relative size of a red cell, a streptococcal bacterium, a chlamydial agent and an adenovirus.

normal flora that might be isolated from samples such as faeces and skin. (see also Appendix 3). Although these are not usually considered significant, some of these (usually harmless) bacteria can invade tissues and cause disease if the animal is debilitated or injured, that is, they can be opportunistic pathogens. It is important to make the distinction between primary pathogens and opportunistic pathogens because in the case of the latter, treatment would need to focus on determining the underlying cause(s) of the disease as well as dealing with the bacterial infection.

Clinical bacteriology involves the isolation and identification of bacteria that cause disease. Pathogenic bacteria cause disease either by the direct effect of the bacteria and the associated host response (for example, inflammation, pus formation and so on), or due to the toxins they produce (for example, Clostridial diseases and some enteric pathogens). Disease causing organisms are called pathogens. Only a small proportion of bacteria present in and on the body and in the environment are pathogenic.

Anatomy of the bacterial cell

The principal structures of a bacterial cell are shown in Figure 4.2. The cell is bound peripherally by a very thin, elastic semi-permeable cytoplasmic membrane. Outside and covering the membrane is the rigid supporting cell wall, which is porous and relatively permeable. The cytoplasm consists of a watery fluid packed with large numbers of small granules including ribosomes and other structures. Other intra-cellular and extra-cellular structures may be present in some types of bacteria and this is dependent on the presence of favourable growth conditions. For example, a protective gelatinous covering layer called a capsule may be produced (as for some *Clostridium* spp.). Some bacteria (for example, Salmonellae) have flagella and fimbria (pili) which facilitate adhesion and motility. Differences in the latter are used to subtype strains of *Salmonella enterica*.

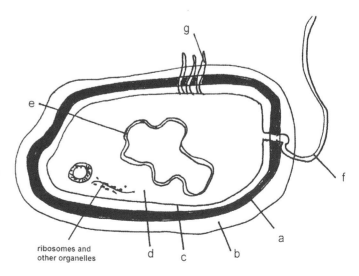

Figure 4.2 The principal structures of a bacterial cell. (a) Cell wall; (b) capsule of slime layer; (c) cytoplasmic (cell) membrane; (d) cytoplasm; (e) chromosome (nucleoid); (f) flagellum; (g) pili or fimbriae. See also Figure 4.3.

Brief description of cell components and their function (Figure 4.2)

Cell wall

The porous cell wall covers the whole organism. Division of one bacterium into two (by binary fission) is initiated by a constriction of the cell wall in the centre. This splitting does not always occur completely, and may result in characteristic clumps of bacteria (Staphylococci), chains (Streptococci) or pairs – diplococci.

Cytoplasmic membrane

Between the cell wall and the cytoplasm is a very thin cytoplasmic membrane. The cytoplasmic membrane is permeable allowing water and nutrients to pass into the cell and waste products to leave the cell. The nature of the membrane present in different groups of bacteria determines the uptake of dyes and therefore influences the result of Gram staining.

Cytoplasm

Bacterial cytoplasm is a gelatinous substance that contains soluble metabolites and other material. Lying within the cytoplasm is the bacterial chromosome.

Genetic information

Bacteria do not have a true 'nucleus'. The genetic information, deoxyribonucleic acid (DNA) is contained in the chromosome (nucleoid) which does not have a membrane cover. The nucleoid divides into two during bacterial multiplication. Some bacteria also have an additional short piece of circular extra-chromosomal DNA (plasmid) which can be exchanged with other bacteria. The transfer of plasmids can be important in the development of antibiotic resistance.

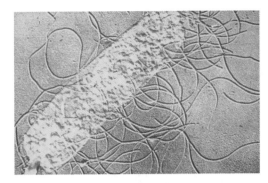

Figure 4.3 Scanning electron micrograph (SEM) of a gram-negative bacterium (*Yersinia* sp.). SEM allows the detailed examination of the external structures of bacteria. The elongated structures are fimbriae which form part of the 'antigenic' component of the bacteria (surface antigens).

Capsule

Many, but not all, bacteria have a gelatinous layer of material on the outer surface of the cell called a 'capsule'. The thickness of this layer varies considerably but it is usually visible when viewed under the light microscope. Well-developed capsules are often slimy and form a covering for the organism which may protect it from the environment or from attack by the defence mechanisms of the body.

Flagella

Bacterial flagella originate from basal granules inside the cell membrane and are usually about 0.02 μm thick and can be longer than the body of the bacterium. There may be one single flagellum or several. Flagella allow movement (motility) which enables bacteria to come into contact with nutrients. Motility can sometimes be observed when preparations are examined under the microscope or when cultures are inoculated into specialized media.

Fimbriae or pili

Fimbriae or pili are small filaments found on the surface of some bacterial cells. They measure up to 1.5 μm in length and 0.5–0.8 μm in width. Most fimbriate bacteria have the property of adhering to red blood cells and to tissues which may increase their pathogenicity. A mixture of fimbriate bacteria and red blood cells, especially red cells derived from the chicken or the guinea pig results in easily visible haemagglutination (see Chapter 6).

Capsules, flagella and pili are antigenic (that is, they stimulate an immune response) and if inoculated into an animal will stimulate the production of specific antibodies. This is the principle used to produce antisera for bacterial sero-typing tests (see Chapter 6).

Endospores

When some species of bacteria (that is, *Clostridium* spp., *Bacillus anthracis*) are in an environment unfavourable for multiplication they will develop endospores. These spores are dormant forms of the organism capable of survival for long periods of time under adverse conditions. Each bacterium develops one endospore

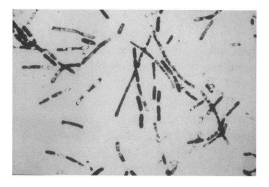

Figure 4.4 Microscopic appearance of rod shaped bacteria as seen using an oil immersion lens (1000×). Gram stain. *Clostridium novyi*. Gram-positive rods with oval subterminal spores.

that may be located centrally, sub-terminally or terminally. The size of the endospore may be sufficiently large to cause the walls of the bacterium to bulge (Figure 4.4). Spores are relatively resistant to extremes of physical and chemical environments including high temperatures, desiccation and disinfection. Under dry conditions or in soil, spores may remain viable for many years. When conditions improve, each spore will develop into one viable bacterium. The production of resistant spores allows *B. anthracis* (the cause of anthrax) to contaminate land for prolonged (> 100 years) periods.

Morphology and classification

The original names given to bacteria often reflected their microscopic appearance, that is, the shape of individual organisms and their distribution in relation to one another after division (fission) takes place. Bacteria can be ovoid or spheroid (coccus), rod like or cylindrical (bacillus), curved (vibrio), spiral-shaped (spirillum) or coil-shaped (spirochaete). The shape and the staining characteristics of bacteria can assist identification but experience is required to distinguish bacteria visually and in most cases further tests are necessary. When bacteria divide, the daughter cell may remain attached to the parent cell (but separates before fission occurs again) resulting in pairs of cocci (diplococci). If fission continues while they remain attached they form chains (streptococci) but if the division is not in one direction, random clumps of cocci occur (staphylococci). Sometimes the cocci remain in pairs for one further division and a regular aggregate of four cocci is formed (tetracocci). Most cocci are Gram-positive. The bacilli, or rods, do not form as many groupings as the cocci, but include the diplobacilli and streptobacilli (pairs and chains). After fission, rods will take different positions – the daughter cell, for example, can remain attached to the parent cell and 'wing away' at varying angles to give the appearance of Chinese lettering (*Corynebacterium* spp.). Sometimes a cuneiform bundle is the characteristic form (that is, *Mycobacterium* spp.). Spirilla are rigid rods with a helical (corkscrew) shape. They are motile by means of a tuft of flagella and are generally Gram-negative. Vibrios are short, curved, rigid rods shaped rather like a comma. They are motile usually by means of a single flagellum and are generally Gram-negative. Spirochaetes are also motile, and possess an axial fibre around which the body is twisted in a helical manner. Their length is usually 10–20 μm and thickness 0.2–0.4 μm. The number of spirals varies with the species. Silver impregnation is the staining method of choice for many of these organisms.

Modern bacterial nomenclature (naming) is based on genetic analysis of the chromosomal material as compared with observational studies of the phenotypic characteristics. Molecular characterization has resulted in changes in the classification of some bacteria with subsequent re-naming at the genus level, for example, *Pasteurella pestis*, (the cause of bubonic plague or the 'Black Death') is now known as *Yersinia pestis*.

4.3 Sample collection, preparation and submission

As outlined earlier, the type of sample submitted for microbiological examination is varied. Correct collection and handling of samples under field conditions is often difficult. As mentioned earlier, to improve the chance of isolating the microorganism responsible for a disease samples should be kept cool and transported directly to the laboratory accompanied by a completed sample submission form. An example of a sample submission form is provided in Appendix 2. Although different laboratories will have their own submission form, most of these have a space where the submitter can describe

the nature of the disease observed (clinical signs, epidemiology), list the causative agents suspected and make a note of any previous or current treatment. This information is important because antibiotic treatment provided to the animal before samples have been collected, for example, may interfere with the culture of a pathogen even if the antibiotic therapy has not been successful in eliminating it.

Aseptic techniques should be used when collecting and submitting specimens for microbiological examination. To avoid contamination of samples they should be placed in suitable sterile containers or transport media immediately after collection. The skin and mucous membranes of animals (and people) has a normal flora of microbes which are usually of no clinical significance but which may overgrow pathogenic microbes if mixed with the sample. To reduce the risk of sample contamination, sterile gloves can be worn. The sample collection procedure used will depend on the nature of the sample to be collected (see below). In each case use sterile or very clean, dry equipment. If there is going to be a delay between collection and submission of samples they should be kept chilled in the refrigerator (4°C).

Samples commonly collected

Prior to collecting samples the appropriate personal protective equipment (PPE) should be worn especially in cases where zoonotic diseases are suspected (see Chapter 1).

Hair and skin samples are usually collected for parasitology and fungal examination but may also be submitted for bacteriology. These samples can be sent in a clean labelled plastic bag, plastic pot or in a labelled paper envelope.

Faecal samples for microbiological examination should preferably be collected (with care) directly from the rectum of the animal and submitted in labelled plastic bags (double sealed), sterile glass or plastic bottles. Requests for special tests should be made on the submission form, that is, Ziehl–Neelsen stain for faecal smears where *Mycobacterium avium paratuberculosis* sp. are suspected or micro-aerophilic culture for *Campylobacter* spp. Note the colour and consistency of the sample as well as any unusual odour. If there is blood or mucus present in the sample it should also be examined for the presence of parasites such as schistosomes, amoebae or coccidia (see Chapter 3).

Swabs are used to sample pus and fluids from abscesses, skin wounds and so on. These are preferably placed in a transport medium, which contains nutrients for bacterial survival but not proliferation. If transport media are not available the swab can be kept moist by placing it in a bottle with a little sterile distilled water. Special moisture retaining swabs and kits containing transport media are available commercially but these can be expensive.

Milk samples should be collected from each quarter of the udder, after cleaning and disinfecting the teats, into sterile containers. Note the colour, smell and consistency of the milk samples. If clinical mastitis is present (that is, swollen painful teats or udder, discoloured milk) samples can be cultured directly and a smear made for later staining. Where subclinical mastitis is suspected simple indicators of inflammation are useful and screening tests such as the Whiteside test and California Mastitis Test (CMT) can be carried out. For routine screening of milk, somatic cell counts are often carried out.

Urine samples can be collected directly into sterile labelled bottles. However, free caught urine samples can be heavily contaminated with normal flora from the gut and reproductive tracts so it is preferable to collect mid-stream samples to minimize this. In some cases, it may be necessary for an attending veterinary professional to use a sterile catheter to get a representative sample.

Nasal secretions, vaginal discharge and eye discharges can be collected using a dry swab which is moistened using sterile water. It is important to ensure that the swab does not dry out so, where possible, it should be placed in transport medium to maintain good conditions for isolation of pathogens. It can be useful to prepare smears from these samples on microscopic slides to submit along with the swab(s). For example, motile trichomonads and protozoa may be readily observed in smears of vaginal mucous.

Necropsy examinations (post-mortem samples): Tissue samples can be collected during post-mortem examinations along with heart blood/tissue impression smears and body fluids (pericardial fluid, pleural fluid, ascitic fluid). These samples should be submitted in sealed containers along with details of the gross findings observed (see necropsy report form Appendix 2). Submitted tissue samples (for example, liver, muscle and so on) should be at least 2 cm³ in size. When tissue samples are ready to be cultured the surface is seared with a flame and a clean section exposed using a sterile scalpel blade. A portion of this tissue is then ground up and mixed with sterile water or extracted with a sterile loop and inoculated onto standard culture media. Body fluids can be collected aseptically using a syringe and needle, the volume and appearance of the fluid present should be recorded. Tissue impression smears can be made by running a clean microscope slide along the freshly cut edge of an organ or tissue, for example, liver, spleen, bone marrow cavity and stained for cytology. See Chapter 8 for more explanation.

Laboratory examination of specimens

When samples arrive at the laboratory, the macroscopic appearance of the submitted specimen(s) should be described along with the colour, smell and consistency. The presence of mucus, blood, pus or parasites should also be noted. If indicated, smears of specimens submitted for culture can also be prepared for microscopic examination. Prepared smears will usually need to be fixed in methanol or heat fixed and then stained to allow good visualization of any microorganisms present. The staining techniques commonly used in microbiology are selected to differentiate and highlight specific cell features (see Table 4.1). Motility is also used as a means of identifying bacteria and can be studied by examining an unstained wet[1] preparation of the organism suspended in a fluid.

Table 4.1 The staining characteristics of some bacteria of veterinary importance. The list is not comprehensive but provides an overview of the sort of differences to look out for when stained smears or fixed bacterial colonies are examined.

Stain	Specimen and some of the organisms that might be seen	Appearance in stained sections
Gram stain	Pus or exudate	
Gram +ve (purple – dark blue)	*Staphylococcus* spp.	Gram +ve cocci* (usually in clumps)
Gram –ve (pink)	*Streptococcus* spp.	Gram +ve cocci* (usually in chains)
	Corynebacterium spp.***	Gram +ve rods**
The Gram stain is used widely for routine staining of bacteria in smears. Gram +ve bacteria retain crystal violet in their cell wall so appear blue, Gram –ve bacteria to not retain the crystal violet stain and are counterstained red	*Pasteurella* spp.	Gram –ve rods**
	Scabs	
	Dermatophilus spp.	Gram +ve filamentous and branching with coccal zoospores
	Aspirated material from a lump	
	As above for exudates	
	Actinobacillus spp.	Gram –ve rods
	Actinomyces spp.	Gram +ve filamentous and branching
	Feacal sample (enteritis)	
	Escherichia coli	Gram –ve rods
	Salmonella spp.	Gram –ve rods
	Enterotoxaemia, deep infections, cellulitis	
	Clostridium spp.	Gram +ve fat rods, with or without spores
Carbol fuchsin Pink-purple colour	Faecal samples (enteritis)	
	Campylobacter spp.	Curved rods, 'seagull' forms
Especially used for recognizing *Campylobacter* spp.	Foot rot exudates	
	Bacteriodes spp.	Rods with knobs at one or both ends Long slender filaments, irregular staining

Table 4.1 *continued*

Stain	Specimen and some of the organisms that might be seen	Appearance in stained sections
Modified Ziehl-Neelsen (or Ziehl-Neelsen acid-fast) Purple-pink to red colour (acid fast = red) Carbol fuchsin stain penetrates the bacterial cell wall and is retained after decolorisation leaving *Mycobacteria* spp. characteristically red	Vaginal exudate, placental material	
	Brucella spp.	Small red coccobacilli in clumps
	Chlamydia spp.	Small red coccobacilli in clumps
	Lymph node, intestinal and generalized	
	Mycobacterium paratuberculosis	Fairly short, red 'acid fast' rods in clumps
	Mycobacterium bovis	Long thin bright red rods, may appear beaded, usually few present
	Mycobacterium tuberculosis	Long thin bright red rods, usually few present
Giemsa or methylene blue Pink-purple or Purple-blue Polychrome methylene blue is often used to stain slides with suspected anthrax – *Bacillus anthracis* stains characteristically blue with distinctive pink capsules	Blood smear from ear vein or body fluids	
	Bacillus anthracis	Purplish or blue square ended rods in short chains surrounded by a pinkish or red capsule. May also see spores.
	Skin preparations and tissue smears	
	Dermatophilus congolensis, rickettsiae and *Borrelia* sp. stain blue with Giemsa	
Unstained wet preparations	Hair, skin and scabs	
	Ringworm fungi **** (*Trichophyton* sp.; *Microsporum* spp.)	Chains of refractile, round arthrospores on hairs
	Other fungi (high dry objective)	Fungal mycelial elements or budding yeast cells
	Eye fluid, urine	
	Leptospira spp.	Elongated thin, helical structures with hooked ends, difficult to see without dark field microscopy
	Vaginal smear	
	Trichomonas spp.	Protozoan with undulating membrane and four flagella

Notes: *Cocci = circular. **Rods = elongated (some may be cocco-bacilli, these are elliptical in shape). ***Corynebacterium spp. can be quite variable in shape (pleomorphic) and may be rods, cocci, club shaped or filamentous. **** The Lactophenol cotton blue stain can also be used; *****The Azur B stain can also be used to identify *Bacillus anthracis* in blood smears. The main advantage compared to the Polychrome methylene blue staining is that Azure B stain (Aldrich 227935-5G) will be ready for use immediately after preparation. Storage can be at room temperature in a dark container or dark area. The shelf life is at least 12 months.

Routine and special stains used for microbiology smears

A Gram stain is routinely performed on bacteria from cultures and on smears submitted for microbiological examination. This is because the Gram stain reaction is an integral part of the routine bacterial identification procedure. For effective staining, it is important to make sure that smears prepared from specimens and cultures are not too thick and that cells are evenly distributed. A Bunsen flame can be used to kill the bacteria and fix them to the slide.

Gram stain

GRAM STAIN REAGENTS

1 Crystal violet (stock solution: crystal violet 1 g, water 200 ml. Dissolve by mixing and filter before use).
2 Gram's iodine (iodine 1 g, potassium iodide 2 g, water 300 ml).
3 Acetone or methylated spirits (industrial methyl alcohol).
4 Dilute carbol fuchsin (ZN carbol fuchsin diluted 1/10 with distilled water). As an alternative to carbol fuchsin, safranin (0.5%) or neutral red (0.1%) can be used.

It is important to avoid constant re-use of staining jars for a series of slides because bacterial organisms may be transferred from one slide to another giving false results. Most commercial Gram stain kits come with four small labelled bottles of reagents – a dropper bottle or a lidded bottle with a pipette are preferred.

PROCEDURE

1 Prepare a smear (from tissue swabs, sediments or from a culture plate) by mixing with 1–2 drops of sterile water using a 'flamed' wire loop.
2 Heat fix the smear preparation on to the microscope slide.
3 Flood the slide with crystal violet solution and leave for 1 min.
4 Wash the slide, drain and flood with Gram's iodine, leave for 1 min.
5 Wash the slide, drain and add acetone or alcohol. Hold the slide under running tap water until no further colour is removed (approximately 30 s). Do not over-decolourize).
6 Counter-stain with dilute carbol fuchsin (leave for 1 min).
7 Wash the slide again in water and leave to dry in a slide rack.
8 When dry the slide can be examined under the microscope using the oil immersion lens (1000×).

Gram-positive organisms stain blue/purple and Gram-negative organisms stain pink in colour. It is useful to have reference cultures of Streptococci (Gram +ve) and *Escherichia coli* (Gram −ve) to check that the staining reagents/technique is working well (see page XX).

MODIFIED ZIEHL–NEELSEN STAIN.

Modified Ziehl–Neelsen stain is used to detect 'acid-fast' bacteria, that is, Mycobacteria in faecal samples and in smears made from tissues. The stain may also be used to detect cryptosporidia in faecal samples.

REAGENTS

1 Dilute 1/10 carbol fuchsin (basic fuchsin 1 g, absolute alcohol 10 ml, 5% phenol in 100 ml of water).
2 Acetic acid (0.5%).
3 Methylene blue (1%).

PROCEDURE

1 Heat fix the smear preparation (prepared from tissue, faecal swabs or fixed bacterial colonies).
2 Flood the slide with carbol fuchsin and leave for 10 min.
3 Wash the slide in water and drain.

4 De-colourize with acetic acid for 30 s.
5 Wash the slide in water.
6 Counter-stain with methylene blue (leave for 30 s).
7 Wash the slide again in water and leave to dry.

When dry, the slide can be examined under the microscope using the oil immersion lens ($10\times100\times$). Acid-fast bacteria will stain pink to red, the background will be blue.

4.4 Culture and identification of bacteria

The successful culture of microorganisms on artificial culture media (see Figure 4.5) is dependent upon a number of very important factors including the following:

1 the condition (and degree of contamination) of the specimen before attempting culture
2 the nutritive elements available in the culture medium
3 level of available oxygen
4 the degree of moisture present
5 the pH of the medium
6 the incubation temperature
7 the sterility of the medium.

Laboratory animals have been used for the isolation of bacteria but better techniques that do not use laboratory animals are now preferred for diagnostic testing. Although laboratory animals may still be used in research and for the production of biological products this will not be discussed further here.

A point to consider when performing microbiological procedures is that they can be time consuming and that 'turnaround' and reporting times for culture results are often longer than for tests in other sections of the laboratory. Before, or at the time of, sample submission the submitter should be made aware of the likely turnaround times for the different diagnostic tests provided by the laboratory.

Media and special requirements

Oxygen requirements

Most bacteria grow well under ordinary conditions of oxygen tension (aerobes). However, some bacteria grow better without oxygen (anaerobes). Organisms that can grow in reduced oxygen tension are classified as microaerophilic. This is usually achieved in the laboratory by replacing oxygen in the incubating atmosphere with carbon dioxide.

Anaerobic conditions can be achieved in a number of ways including the following:

1 adding an oxygen reducing substance to the medium
2 inoculating the specimen into the deeper layers of solid media or under a layer of oil which covers a liquid media
3 incubation in an atmosphere devoid of free oxygen, for example, anaerobic jars.

For anaerobic culture, it is common practice to use an anaerobic jar along with the 'GasPak' system (Figure 4.6). To achieve anaerobic conditions the GasPak[2] system comes with a foil packet containing a mixture of chemicals to which water is added. The packet of chemicals is then placed in the anaerobic jar along with an anaerobic indicator to check that true anaerobic conditions have developed. Full instructions and information are given with the GasPaks. GasPaks to provide an enhanced carbon dioxide level are also available commercially.

The addition of a catalyst (for example, palladiumized alumina) may also be required to initiate the reaction. The resulting reactions are as follows:

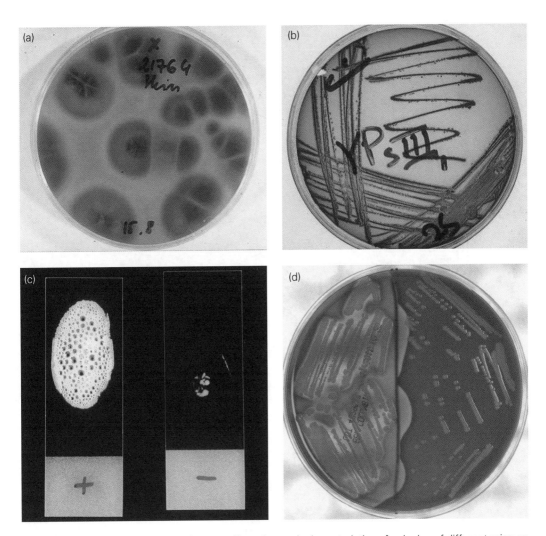

Figure 4.5 Agar plates showing culture media and growth characteristics of colonies of different microorganisms. (A) Agar plate with a culture of a fungus (*Paecilomyces* sp.) obtained from the skin of a reptile. Fungal identification depends on the colour and type of growth on specialized agar media (for example, sabouraud dextrose agar) and on the morphology of the fruiting bodies or Hyphal structure (see also Figures 4.14a and b). (B) Specialized media (CIN) with a culture of *Yersinia pseudotuberculosis*. This agar is selective for some enteric bacteria, *Yersinia* spp. colonies appear small and pigmented (dark grey). CIN = Cefsulodin-Irgasan-Novobicin agar (Difco). (C) This is an example of a positive (left) and negative (right) catalase test. Catalase is an enzyme which breaks down hydrogen peroxide (H_2O_2) into H_2O and O_2. When a drop of 3% H_2O_2 is added to a colony of catalase positive bacterium, the reaction produces bubbles. Photo: Paul Gadja. (D) This is an example of the appearance of *Staphylococcus* spp. grown on blood agar. On the left side of the plate is *S. aureus* showing characteristic golden colonies surrounded by clear zones of beta haemolysis. On the right side of the plate is *S. xylosus*, which is coagulase negative and does not produce haemolysis. Photo: Karen Liljebjelke. See also Plate 8a–d.

(a)　　　　　　　　　　　　　　　　　　　(b)

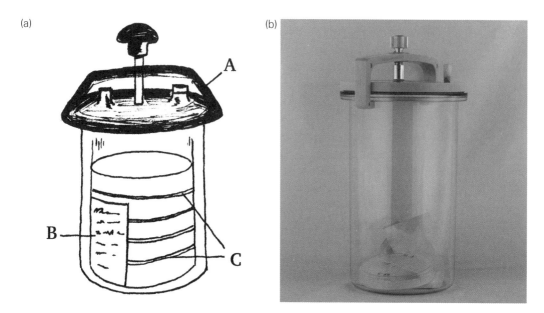

Figure 4.6 (a) Anaerobic jar with sealed lid (A) and a GasPak (B). A new pack is used prior to re-incubation after the lid is removed to examine the culture plates (C). The jar is usually placed in a normal incubator at 37°C and the plates are examined after 24 and 48 hrs of culture. Illustration: Louis Wood. (b) This is an example of a chamber used for growth of anaerobic bacteria. The blue pouch in the chamber contains a palladium catalyst which when water is added generates carbon dioxide and hydrogen gases, reducing the oxygen concentration to 0–5 ppm, and creating water vapour.

Table 4.2a Media used for the isolation and identification of common bacterial pathogens.

Medium	Comments
Blood agar	An enriched medium used to support the primary isolation of a wide range of bacteria. Also allows detection of haemolysis characteristics e.g. haemolytic Steptococci and Staphylococci
MacConkey agar	Selective medium containing bile to select for Enterobacteria and some other Gram –ve bacteria. Also allows demonstration of lactose fermentation (+ve colonies are pink)
Nutrient agar	A basic medium useful for non fastidious bacteria. Often used for colony counting and to demonstrate pigment production
Brilliant green agar	A useful indicator medium for the presumptive identification of Salmonellae. *Salmonella* spp. appear pink
Chocolate agar	Heat treated blood agar with added growth factors (X/V) to support isolation of *Haemophilus* sp and *Taylorella* sp.
Selenite broth	Selective enrichment media used for the isolation of Salmonellae
Edwards medium	A special blood agar based medium used for the isolation and identification of Streptococci

$$NaBH_4 + 2 H_2O = NaBO_2 + 4 H_2$$

$$C_3H_5O(COOH)_3 + 2 NaHCO_3 + [CoCl_2] = C_3H_5O(COONa)_3 + 3 CO_2 + 3 H_2 + [CoCl_2]$$

$$2H_2 + O_2 + [Catalyst] = 2H_2O + [Catalyst]$$

Essentially, the chemicals react with water to produce hydrogen and carbon dioxide along with sodium citrate and water as by products.

Media

A satisfactory microbiological culture medium must contain available sources of carbon, nitrogen, inorganic salts and in certain cases, vitamins or other growth promoting substances (see Table 4.2a). These can be supplied in the form of meat infusions which are still widely used in culture media. Beef extract can replace meat infusion and the addition of peptone provides

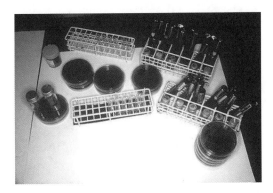

Figure 4.7 Classical microbiology requires a wide range of reagents and relies on the technical expertise and experience of the laboratory technician for the successful culture and identification of disease causing agents. This photograph illustrates the range of media and biochemical tests required to identify a strain of *Yersinia pseudotuberculosis* isolated at necropsy from a spleen abscess in a parrot. See also case study example (collibacillosis in poultry) and diagnostic flow charts Appendix 5, A5.1 and A5.2.

a readily available source of nitrogen and carbon. Some bacteria require the addition of other nutritive substances such as serum or blood. Carbohydrate supplements such as dextrose may also be desirable along with salts of calcium, manganese, magnesium, sodium or potassium. Dyes and chemicals may be added to media to (1) indicate specific metabolic activity, or (2) to aid or inhibit the development of certain types of bacteria (these are then known as 'selective media').

Sterility

The media used for isolating bacterial pathogens must be sterile, that is, free from all other organisms whose development might influence or prevent the normal growth of the bacterial inoculum. Autoclaving is the usual method for the sterilization of culture media. Excess sterilization or prolonged heating should be avoided as this will change the composition of the medium, for example, agar in the medium may form a precipitate. Culture media which may be damaged by autoclaving can be sterilized by discontinuous or intermittent heating at lower temperatures or by filtration (check the instructions which come with specific media or contact the manufacturer for more detailed guidance).

Storage

Media should always be stored in a cool moist atmosphere to prevent evaporation and drying. Prepared agar plates can be sealed and stored in a refrigerator but prolonged storage (> 10 days) of sterile media is not recommended. Once a laboratory can gauge how many plates are going to be required it is easier to plan ahead, however, when starting out it is advisable to make up batches of media on a regular basis as plates are needed.

General types of media and terminology used

Solid and liquid media can be used to culture bacteria. Most are commercially available in a dehydrated compound form that can be stored. Liquid media are especially useful for propagating anaerobes and allow motility. Liquid media are also useful for maintaining cultures and for biochemical tests. They have the disadvantage of not allowing assessment of colony characteristics which are important for identification.

Solid media are used for most primary culture work. The consistency of the medium is modified by addition of agar, gelatine or albumin in order to change it into a solid or semi-solid state. Agar (derived from a seaweed) is commonly used because it is relatively inert. For solid media agar is used at a concentration of 1%, it melts at about 95°C and sets at 42°C. Heat sensitive nutrients such as blood and some carbohydrates can be added at temperatures below 50°C before the agar sets.

Basal media are media or compounds which can be used alone or as a basis for other media, for example, nutrient broths (peptone and/or meat infusion) or digest broths (digested with proteolytic enzymes to release protein). The addition of agar to a broth at 1% will produce nutrient agar which is often used as a base agar to make media such as blood agar.

Enriched media contain ingredients which will facilitate bacterial growth (see Figure 4.5). Many pathogenic bacteria require additional nutrients such as blood or specific salts, sugars or proteins and would grow very slowly or not grow at all in basal media. Blood and serum are usually added at 5 to 10% and glucose at 1%.

Selective media contain ingredients which inhibit the growth of certain microorganisms and allow others to grow. For example, MacConkey agar contains bile salts which inhibit the growth of most Gram-positive bacteria but allow enteric (gut) organisms to grow. Selective media may also contain a colour indicator and specific sugars. For example, when a lactose fermenting organism grows in media containing lactose there is acid production (fermentation) and consequently pH is lowered. This can be detected by a change in the colour of an indicator in the media (for example, from colourless to red on MacConkey agar). Growth of a non-lactose fermenting organism will not result in acid production and there is therefore no colour change. Media such as these are sometimes referred to as 'differential media' (Table 4.2b).

Table 4.2b Cultural characteristics of some common bacteria found in veterinary medicine.

Bacterial species	Blood agar**	MacConkey agar*	Comments
Beta haemolytic Streptococci	Translucent round colonies, 0.5–1 mm diameter, glistening. Clear area around colonies (haemolysis)	No growth	The size of the zone of haemolysis varies with the Lancefield group type and species. Catalase –ve
Alpha haemolytic Streptococci	White, smooth and round colonies, 0.5–1 mm diameter	No growth	Greenish or partial zone of haemolysis. Rarely pathogenic. Catalase –ve
Staphylococcus aureus or *S. intermedius*	White or yellow, smooth round colonies, 2–3 mm diameter, may have a double zone of haemolysis	No growth (depends on strain)	Some strains do grow on MacConkey agar, Catalase +ve. Some strains of *S. aureus* show a characteristic yellow pigment

Bacterial species	Blood agar**	MacConkey agar*	Comments
Haemolytic *Escherichia coli*	Grey, smooth, shiny round colonies, 2–3 mm, clear zone around colonies	Good growth, colonies pink indicating lactose fermentation	Characteristic 'coliform' smell, bright pink colonies on MacConkey the same size as on blood agar
Non-haemolytic *Escherichia coli*	Grey, round and shiny colonies, 2–3 mm, no clear zone	Good growth, colonies pink indicating lactose fermentation	As above
Salmonella spp.	Grey, round and shiny colonies, 2–3 mm, no clear zone	Good growth	Not lactose fermenting, pale colonies on MacConkey agar. No smell
Yersinia spp.	Grey, round and shiny colonies, 2–3 mm, no clear zone	Good growth	Not lactose fermenting
Pseudomonas aeruginosa	Blue-green, flat round colonies, 2.5–4.0 mm, may have clear zone	Growth but pale colonies	Not lactose fermenting, some strains have metallic sheen on blood agar
Corynebacterium renale	Grey-white, moist colonies, 0.5–1.0 mm, no clear zone	No growth	Colonies may appear dry over time
Pasteurella multocida	Transluscent, round shiny colonies, 1–2 mm, no clear zone	No growth	Colonies may appear pinkish on blood agar, characteristic sweet odour
Campylobacter fetus	Small opaque colonies, 0.5 mm diameter, no clear zone	No growth	May require 2–3 days for culture, requires reduced oxygen conditions to grow (Micoaerophilic)
Clostridium perfringens	Grey, flat and often an irregular edge to colonies, 2–3 mm diameter, some have a clear zone	No growth	Requires anaerobic conditions to grow. Double zone of haemolysis is possible
Bacillus spp.	Grey, dry granular appearance, 3–5 mm diameter, may have clear zone	No growth	Many haemolytic but not *Bacillus anthracis*. Dry colonies as no capsular material is produced on lab. media

Notes: *Colony characteristics may vary but the descriptions given are for cultures read at 48 h. **The type of haemolysin present will determine the appearance of the zone around the colonies. Note that alpha haemolytic staphylococci appear different to alpha haemolytic streptococci.

pH requirements

The pH of a culture medium is extremely important for the growth of microorganisms. The majority of bacteria prefer culture media which are neutral (pH = 7.0) while some may require a medium which has an acid pH (for example,

Brucella spp.) and others require an alkaline pH (for example, some Salmonellae). Essentially, the optimum pH for bacterial growth in the laboratory is the same pH as that in the animal host from which the sample was collected. Optimum pH for bacterial growth varies and therefore this is often used in media as a means of 'selecting' the desired organism from a sample while 'inhibiting' the growth of others (acid, pH < 7.0; alkali, pH > 7.0).

Temperature

The temperature range suitable for the growth of most microorganisms lies between 15°C and 43°C. Some specialized microorganisms can grow at 0°C (and below) and others can grow at temperatures of over 80°C. However, most of the organisms pathogenic to animals have a narrow preferred temperature range based around 28–37°C.

Routine approach to samples submitted for bacteriological examination

Goals of primary inoculation

1 To cultivate the causative (pathogenic) agent(s) but minimize the growth of contaminants.
2 To obtain discrete colonies of organisms to allow selection of the pathogen(s) for subculture and identification.

To culture bacteria from a specimen inoculate the edge of an agar plate with the sample material using a swab or sterile wire loop. Wire loops can be bought or made and are useful making smears and for 'streaking' culture plates. Most loops are made of platinum alloy or nicrome wire which withstands heating/cooling. The loop should be 1.5 × 3 mm in diameter and the standard thickness of the wire is standard gauge 26 or 27. Streaking out, thereafter, depends on the anticipated load of bacteria in the inoculum. For example, for a faecal sample that will have a heavy load of bacteria, a loop is used to allow dilution of the sample. After each stage, the loop must be re-sterilized by passing the tip through a flame. This is repeated for each step so that the inoculum is diluted to a point when individual and separate colonies will develop (Figure 4.8). The idea is to spread the bacterial inoculum across the largest distance possible on the surface area available. For anticipated light loads of organisms, for example from a milk sample, the inoculum can be plated out without flaming the loop between each stage while streaking out. Alternatively, a culture swab or plastic loop could be used. On well prepared plates individual bacteria will grow into discrete colonies and the growth characteristics of these colonies can then be described. If streaking is not well done, non-pathogenic contaminants may overgrow the more fastidious pathogenic organisms which may then not be seen.

The plating out method for solid media is illustrated in Figure 4.8. This method is designed to ensure that isolated bacterial colonies will be present to facilitate subculture and bacterial identification.

Identification of bacteria

After primary culture and incubation, the plates are examined and colonies can be described (see characteristics to record below). If, after 24 h, bacterial growth is not readily observed the plates should be replaced in the incubator and read again after a further 24 h. The number of colonies of each type of bacteria may not directly relate to the number of bacteria present in the sample nor do they indicate the relative significance of each type of bacteria isolated. Pure or predominant growth of one type of organism is usually a significant result.

Characteristics to record

1 Is there heavy, light, pure or mixed growth? Is there a predominant colony type?
2 What is the size and shape of the colonies

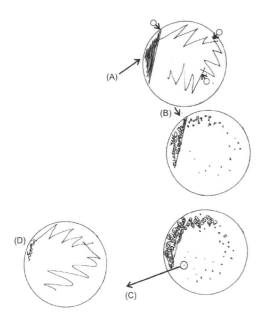

Figure 4.8 Initial 'streaking' of a culture plate and preparation of a subculture (or 'purity' plate). Initial plating out is usually done on blood agar and MacConkey agar unless special media are requested. (A) Initial zone (if this is from a faecal sample the initial growth will probably be mixed and prolific therefore streaking out should be carefully done to avoid contamination of the rest of the plate). Flame the loop between each set of streaks (O ->). Incubate the plate at 37°C and check the plate at 24 h and 48 h for growth and describe what is seen. (B) If the procedure has been done carefully there should be individual colonies to select from for the subculture. Select one or two of any colonies which may be pathogenic (disease causing) for Gram staining. (C) The subculture is prepared by selecting an isolated distinctive colony, this can then be grown overnight in broth before plating out. There is no need to flame the loop after the first stage. (D) Check that there is a pure growth of bacteria before harvest at 24 h (or 48 h if slow growing) for biochemical tests.

present (that is, 2–3 mm, 1–2 mm, pinpoint colonies, round, rough edge, entire edge, and so on)?

3 Describe the colour and consistency of the colony types present (that is, cream coloured, white, flat, convex, dry, mucoid and so on).

4 Describe the growth on blood agar and comment on the degree of haemolysis around the colonies (for example, beta, alpha or no haemolysis[3])

5. Record the growth on MacConkey agar; are the colonies pink (that is, lactose fermenting) or colourless (non-lactose fermenting)?

In most cases bacterial cultures are plated directly onto blood agar. MacConkey agar may also be used for the initial plating of enteric pathogens. The majority of bacteria of veterinary importance grow reasonably well on blood agar but, only the enteric bacteria (gut flora) will grow well on MacConkey agar, this is because this media contains bile salts that are inhibitory to the growth of many other bacteria. Many commensal enteric organisms ferment lactose and, due to the presence of an indicator dye, these bacteria produce pink colonies after overnight growth on MacConkey media.

To interpret microbiological findings, it is important to know the characteristics of the normal commensal bacteria found in different species of animal and to have a good case history. The latter should be provided by the submitter on the submission form. Culture results from faecal samples and samples from the skin are very difficult to interpret without knowledge of the case. Microbiological examination of samples collected aseptically from a catheterized bladder, correctly submitted necropsy samples or aseptically collected fluid aspirates (for example, joint fluid, cerebrospinal fluid) is usually more straightforward because the degree of contamination is considerably reduced and the chances of growing a good culture of the causative agent much greater.

Once a bacterial culture has grown, it is routine practice to sub-sample and stain bacteria from individual colonies using Gram stain (or other stains) to assist in the identification of bacterial isolates of interest. Next, preliminary biochemical tests will need to be performed. Further biochemical tests and specialist media are usually necessary to identify the organisms to family and genus level. Before this can take place, a sub-culture is made of each colony of interest to produce a 'purity' plate (see Figure 4.8). These plates are incubated further until a good pure growth of bacteria is present. From the purity plate(s) a suspension of bacteria can be prepared for inoculation of secondary media and biochemical reagents. Detailed biochemical tests should not be performed on mixed cultures as the results will be impossible to interpret.

Broths made from purity plates can be used for a wide range of biochemical tests and can also be used to determine the antibiotic sensitivity of the suspect pathogen(s). Table 4.2b

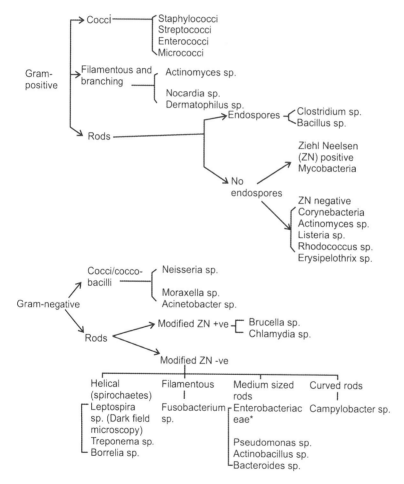

Figure 4.9 Classification chart according to staining reaction and cellular morphology of common bacteria of veterinary importance.

Note: *The Enterobacteriaceae include *Escherichia coli*, Salmonellae, Yersinae and other bacteria that are frequently isolated from the intestinal tract of healthy and sick animals. The staining characteristics and appearance under the microscope are similar for most bacteria in this group so it is usually necessary to perform a range of biochemical tests to identify the bacteria present in a sample.

outlines the cultural characteristics of some bacteria of veterinary importance and Figure 4.9 outlines the staining characteristics that can be used to help identify bacteria to the family or genus level. However, because several species of bacteria have a similar appearance on agar, and share similar staining characteristics, additional tests are required. It is usually necessary to perform a series of six or more preliminary biochemical tests to allow confirmation of the genus and further tests to allow identification to species level.

As outlined earlier, the number of colonies of each type of bacteria grown on a culture plate does not directly relate to the number of bacteria present in the sample submitted nor does it indicate the relative importance of each type of bacteria grown from the sample. Until a technician is experienced in deciding which colonies to follow up it is important to note down everything that is seen on the plate and to seek advice from a senior technician. Owing to the time and expense involved in the culture and identification of bacteria it is also helpful to discuss the case with the submitter so that a targeted approach can be taken to ensure all likely pathogens are considered.

Routine stains used in microbiology

Gram staining of bacterial colonies grown on culture media is part of the routine bacterial identification procedure. See Figure 4.10 (staining station in a microbiology laboratory).

The Gram stain

As outlined earlier, the Gram stain is an extremely important stain and is routinely used in bacteriology tests (see Figures 4.11a and b). The Gram staining characteristic of bacteria is based on the fact that some bacteria will retain a specific dye (methyl violet) and others will not.

Figure 4.10 Sink in a small microbiology laboratory at a district veterinary diagnostic facility. This illustrates the location of staining racks and commonly used reagents. Each laboratory technician will have his/her own preferred reagents which are usually placed in an area where they can be readily used. However, it is important to make sure that standard operating practices are followed. Photo: Regional Veterinary Laboratory, Khaling, Bhutan.

This is believed to be associated with the 'porosity' of the cell wall and cytoplasmic membrane.

METHOD
A small colony of the bacteria to be examined is mixed with a drop of water on a microscope slide and heat fixed using the cool part of the Bunsen flame (or air dried). The staining procedure is as follows:

Step 1. Apply methyl violet
All bacteria pick up the dye

Step 2. Fix (or mordant) the dye with iodine
A change takes place in the dye and/or the bacteria

Step 3. Apply decolouring agent (acetone/ alcohol)
Dye retained by some bacteria, not others

Step 4. Apply counter-stain of different colour
Counter-stain picked up by bacteria not retaining methyl violet

Gram-positive organisms (dye retained) will be blue/purple and Gram-negative organisms (dye not retained) will be pink in colour due to the counter-stain.

Gram stain reaction of some groups of organisms

Gram-positive	Gram-negative
Staphylococci	Coliforms
Streptococci	Neisseriae
Corynebacteria	*Campylobacter* sp. (Vibrio)
Clostridia	Salmonellae
Bacillus spp.	Shigellae
	Haemophilus group
	Brucellae
	Pasteurellae

Many other stains are used in microbiology (see page XX for the modified Ziehl–Neelsen stain) but will not be covered here. For more information see the bibliography at the end of the chapter, see also Figure 4.9.

Biochemical tests

Regional veterinary laboratories will usually have enough microbiology samples submitted to warrant the maintenance of supplies for a number of routine biochemical tests. The following are a few examples of simple tests which could also be considered for busy regional and district laboratories. For more detailed descriptions of tests consult a text book or, preferably, attend a bacteriology training course. There are also some useful online resources available that illustrate many of the following tests.

Beta-galactosidase test

Suspend a loopful of the bacterial colony to be tested in 0.25 ml saline in a tube. Add one drop of toluene. Put the tube in a water bath at 37°C for 2–3 min. Add 0.25 ml of ONPG[4] (O-Nitrophenyl-Beta-D-Galactopyranoside) reagent and mix. Return the tube to the water bath at 37°C for 24 h. A yellow colour change is a positive reaction. ONPG is a spectrophotometric and colorimetric substrate for the detection of B-galactosidase activity. If the bacteria under test contains galactosidase the ONPG reagent turns from colourless to yellow. This reaction is linked to the ability of the bacteria to ferment lactose.

Oxidation fermentation test

Two tubes of Hugh-Liefson media[5] are inoculated with broth containing the bacteria to be tested. One is covered with a layer of sterile liquid paraffin. The tubes are incubated at 37°C overnight. This test differentiates bacteria which utilize glucose by oxidation from those which utilize it by fermentation. The uninoculated medium is green (pH 7.1), if the pH becomes acidic, as a result of glucose utilization, it turns yellow.

- Oxidation (aerobic). The colour indicator turns yellow in open tubes and remains green in covered tubes, for example, *Pseudomonas* sp.
- Fermentation (anaerobic). The colour indicator turns yellow in both tubes for example, *Aeromonas* sp. Bromothymol[5] blue is the indicator (pH 6.0 yellow – pH 7.6 green/blue).

Catalase test

For this test 3% hydrogen peroxide (H_2O_2) is added to colonies taken from a pure culture of bacteria on a glass slide. Immediate release of bubbles of gas indicates a positive reaction. The test can also be done on an agar slope or directly on a plate culture.

The enzyme catalase, which is produced by some bacteria, is responsible for the release of oxygen from hydrogen peroxide.

Citrate test

For this test inoculate a tube of Koser[6] citrate medium with the test bacterium and incubate at 37°C for 96 h. Growth will only occur when a bacterium can use a citrate compound as a carbon and energy source and an ammonium compound as a source of nitrogen. These two chemicals, but no other sources of carbon and nitrogen, are in Koser citrate medium. Growth is assessed by the degree of turbidity.

Coagulase test

The enzyme coagulase is produced by the vast majority of pathogenic Staphylococci and its presence therefore is taken as an indication of pathogenicity. Coagulase has the ability to clot or coagulate blood plasma.

Emulsify a colony of the bacteria to be tested in sterile distilled water on a slide and add a loopful of rabbit plasma.[7] A positive test is indicated by clumping or agglutination of the bacteria. Occasionally a bacteria that is coagulase positive will be negative to this test so 'negative' tests should be confirmed by a tube test. To do this, dilute rabbit plasma 1 in 10 with normal saline. Place 0.5 ml of diluted plasma and five drops of an overnight broth culture of test bacteria in tubes approximately 80 × 12 mm in size. Incubate at 37°C and examine for clotting at intervals up to 5 h.

Note: Use a known coagulate positive *Staphylococcus* sp. as a positive control and use the diluent without the organism as a control for auto-agglutination. Human (but not bovine) plasma can be used if rabbit plasma is not available.

Gelatine liquefaction

Inoculate a tube of nutrient gelatine medium with test bacteria using a straight wire stab. Incubate at room temperature for up to 7 days and observe for liquefaction. Liquefaction occurs due to breakdown of protein in the medium.

Hydrogen sulphide production

Inoculate a tube of TSI[8] (triple sugar iron) agar with test bacteria by stabbing the butt and streaking the slope; incubate for 24–48 h and observe for blackening due to the production of hydrogen sulphide (H_2S), see Figure 4.11c. TSI agar contains three sugars, glucose (0.1%), lactose (1.0%) and sucrose (1.0%). Phenol red is the pH indicator. Unreacted and alkaline TSI slopes are red, and an acid pH results in a yellow colour change. Ferrous sulphate or ferric ammonium citrate with sodium thiosulphate react to form a black deposit to indicate hydrogen sulphide production.

Methyl red (MR) reaction

Inoculate MR-VP medium (glucose phosphate peptone water) with test bacteria and incubate at 35–37°C for 2 days. Add two drops of methyl red[9] solution, shake and examine. A red colour change indicates a positive reaction and a yellow colour indicates a negative reaction. A positive reaction occurs when glucose is fermented producing acidic conditions (pH < 4.5) which result in a change of colour of MR dye in the medium.

Voges-Proskauer (VP) reaction

After completion of the MR test, add 0.6 ml of 5% α-naphthol[10] solution and 0.2 ml of 40% potassium hydroxide (KOH) aqueous solution to the sample, shake, slope the tube and examine after 15 min (or up to 4 h) later. A positive reaction is indicated by the development of a pink colour. To intensify and speed up the reaction add a few crystals of creatine[10] or two drops of 0.5% creatine solution. The pink/red colour is due to the formation of acetyl methyl carbinol.

Nitrate reduction test

Inoculate nitrate broth with test bacteria and incubate for 24 h. Add two drops of sulfanilic acid[11] reagent (0.8% in 5 M acetic acid) and two drops of α-naphthol reagent (0.5% in 5M acetic acid). An orange/red colour reaction indicates a positive test. Bacteria that produce nitrate reductase enzyme will produce nitrites in the medium which result in a pink colour.

Oxidase (cytochrome oxidase) test

Add several drops of oxidase[12] reagent to the bacterial colonies to be tested. If the organisms produce oxidase the colonies will turn pink, red, then black. This can be done on a culture plate or, if colonies are picked off the plate, it can be done on filter paper. Oxidase reduces the dye in the reagent to a blue/black colour. Note that the reagent is not particularly stable so it must be prepared fresh. It also destroys the bacteria after a few minutes so if subculture is required it needs to be done quickly.

Urea hydrolysis test

Inoculate a slope of Christensen's urea[13] medium with a heavy load of test bacteria. Incubate for 1–2 days at 37°C. A red colour change indicates decomposition of urea in the medium by urease producing bacteria. Urease splits urea with the formation of ammonia which is alkaline. Bacteria which inhabit the renal tract are often urease positive (see Figure 4.11c – biochemical screening tests).

Motility test

Inoculate tubes of motility medium[14] by stabbing the medium to a depth of about 5 mm. Incubate at the appropriate temperature (28 or 37°C), note that motility can be temperature dependant and some organisms are not motile at 37°C. Motile organisms migrate through the medium which becomes turbid; growth of non-motile organisms is confined to the stab. 'Motility' medium is 'sloppy' or semi-solid with an agar concentration

of only 0.4% which enables the organism to move and movement to be traced. Motility can also be checked by the 'hanging drop' method (see below). True motility must not be confused with Brownian movement (vibration caused by molecular bombardment) or convection currents. A motile organism is one which actively changes its position relative to other organisms present and is seen in some bacteria and many protozoal organisms.

To assess motility using the 'hanging drop' method.

1 Use a clean slide and a square coverslip.
2 On the slide place small pieces of modelling clay to fit the corners of the coverslip or use a 'well slide' with a depression in the centre.
3 Transfer a loopful of culture to the centre of the coverslip.
4 Gently press the Plasticine on to the coverslip, ensuring that the 'drop' of culture is in the centre of the circle, and does not come into contact with the slide.
5 With a quick movement, invert the slide, so that the coverslip is uppermost and the drop of culture is underneath, 'hanging'.
6 Examine the drop of culture under the microscope, focusing first with the 10× objective, and when in focus swing round to the 40× objective to investigate motility.
7 Discard the preparation into disinfectant.

There are a number of other tests which are commonly used in some laboratories as well as variations on those described. For more details consult Quinn et al. (2000) or, preferably, attend a bacteriology training course.

Commercial identification systems for bacteria

Although the conventional method of identifying bacteria by phenotype is accurate and reliable, it

(a)

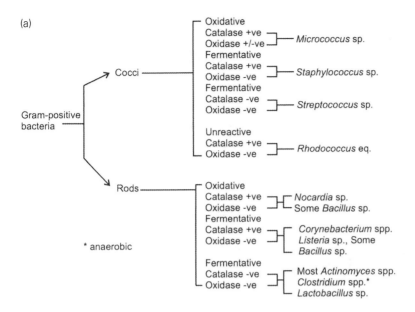

Figure 4.11 Primary
biochemical tests for
(a) Gram +ve and
(b) Gram −ve bacteria.

(b)

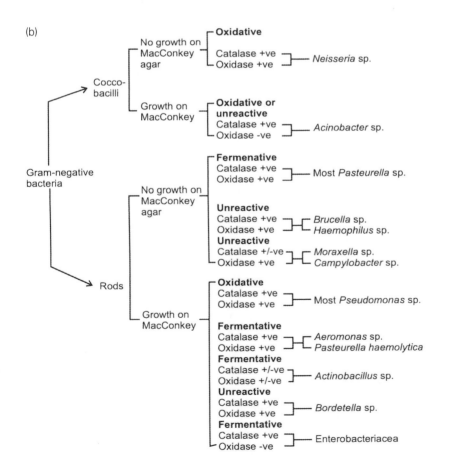

Figure 4.11c Simple biochemical screening test (for example triple sugar iron agar slopes) can be used to determine the species of selected bacteria during survey work but in most cases a series of 20–30 tests will be required. Illustrated are the Urease test, lysine iron agar (LIA) and triple sugar iron agar (TSI) slopes used to help identify Enterobacteriaceae. See also Plate 9.

involves time-consuming preparation of media and inoculation procedures. In addition, considerable expertise is needed to interpret the results. Although miniaturized 'kit' systems are now available it is still necessary to do a lot of preliminary work (primary and secondary culture, Gram stain, +/- selective media/growth conditions, preparation of 'purity' plates) and to ensure good technique. In some laboratories, traditional methods of bacterial identification have been superseded by molecular techniques. However, the latter are less useful when dealing with uncommon microorganism and where a clear list of differential diagnoses is not available,

for this reason the traditional methods are still required for most samples submitted for diagnostic purposes.

Much clinical microbiology now relies heavily on rapid, miniaturized systems which can be used to identify microorganisms in 4–24 h. Test results can be read and interpreted by eye or the microbiologist can use computer aided interpretation. Key advantages of using API kits include the use of small amounts of media (in small cupules or chambers), simple inoculation and incubation techniques, and easy and quick interpretation of test results based on colour changes. However, the kits can be expensive to maintain and, if shelf life is a constraint, use may not be justified unless there is a high throughput of samples.

At present, the most commonly used commercial identification system in veterinary diagnostic microbiology are the API (Analytical Profile Index) systems, of which there are a number designed to identity different groups or species of bacteria (Figure 4.11d). For example, specific kits are used for 'anaerobes' and 'enterobacteria' (Analytical Profile Index 20 enterobacteria or API 20E). Clear instructions for the use are provided with each kit. A range of available kit tests are outlined in Table 4.3.

Criteria to be considered when selecting a bacterial identification system for a particular laboratory include reliability, versatility, the time required for preparation of reagents, incubation time, relative difficulty in determining positive and negative reactions, safety factors for laboratory personnel, shelf life of the test kits and the price of the system. Owing to cost factors, the use of commercial identification kits is not sustainable for most of the smaller district laboratories.

Serological typing

A limited capacity for 'sero-typing' strains of bacteria might be developed in a regional veterinary

Table 4.3 A summary of available miniaturized biochemical test kits. All of these systems have been used in large diagnostic facilities. To some extent the use of these systems will be dictated by the funding source available.

System	Comments	Manufacturer
API	This system claims to have over 90% agreement with conventional procedures	Analytab Products Inc., Plainview, New York
API 20E	Primarily used for Enterobacteriaceae	BioMerieux Inc.
Rapid AE	Similar to the API 20E, results are read after 4 hr of incubation	
API 20A	Used for the identification of anaerobes, results are read after 24 hr of incubation	
An-IDENT	Used for the identification of anaerobes, results are read after 4 hr of incubation	
Rapid NFT	Used primarily for the identification of non-fermenters	
API 20S	For the identification of *Streptococcus* spp.	
Rapid Strep	For the identification of *Streptococcus* spp., results are read after 4 hr of incubation	
Staph-IDENT	For the identification of staphylococcus aureus and coagulase-negative staphylococci	
API 20C	For the identification of yeasts	
Oxi/Ferm	For non-fermenters and oxidase positive fermenters	Roche Diagnostics, Nutley, New Jersey
Enterotube II	For Enterobacteriaceae	
R/B Enteric	For Enterobacteriaceae	
Flow N/F	For non-fermenters and oxidase positive fermenters	Flow Laboratories Inc., Maclean, Virginia
Enteric-Tek	For Enterobacteriaceae	
Anaerobe-Tek	For anaerobes	
Uni-yeast-Tek	For yeasts	
Micro-ID	For Enterobacteriaceae	Organon Teknika, Durham, North Carolina
Quad Eneri Panel	For Enterobacteriaceae	Micromedia systems
Minitek	Not used as widely as the API systems, has very high percentage of agreement with conventional procedures. These systems are used for the identification of Enterobacteriaceae, non-fermenters, oxidase positive fermenters, anaerobes and yeasts	BBL Microbiology Systems
IDS Rapid ANA	For anaerobes, results are read after 4 hr of incubation	Innovative Diagnostic Systems, Inc.
Quantum II	For the identification of Enterobacteriaceae and Oxidase-positive fermenters	Abbott Laboratories, Irving, Texas
Mini-ID	For Enterobacteriaceae	Scarborough Microbiologicals, Decatur, Georgia

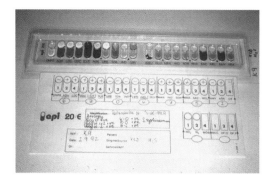

Figure 4.11d API strip card used to illustrate the typical biochemical reactions of 'type' cultures representative of bacterial species obtained from the American Type Culture Collection. See also Plate 10.

bacteriology laboratory if there is ready access to true type cultures and access to required typing sera. The latter can be done by generating type-specific antibodies in laboratory rabbits but this requires appropriate training and facilities. Commercial antisera are also available. In most cases strains of bacteria that require serotyping are sent to specialist reference centres where strains can be compared with archived type cultures. Strain identification is sometimes necessary because some bacteria such as *Escherichia coli* have a range of serotypes and only some are pathogenic. Strain characterization can also be important for survey work and for epidemiological studies. Sero-typing tests are usually performed by mixing a specific antiserum (containing antibody) with a pure sample of the isolate under test (antigen) on a microscope slide. If the antigen and antibody are specific to one another 'agglutination' (clumping) will take place. A more detailed description of the agglutination test is given in Chapter 6.

The serological typing capability in a larger regional laboratory may include confirmation of specific serotypes of the following organisms.

- Specific *E. coli* (pig, calf pathogens) various serotypes. Demonstration of significant fim-

brial antigens (K88, K99, F41, 987Por F165) is usually required and/or the enterotoxin of enterotoxigenic strains.
- Specific Salmonellae (calf, poultry pathogens). Various subgroups (I, II, IIIa, IIIb, IV, V and VI).
- Specific *Clostridium* sp. Numerous strains with different toxins identified, that is, *Clostridium perfringens* types A, B, C, D and E (rare).
- Streptococci (Lance field grouping of haemolytic streptococci). Streptococci of groups A, B and C are the most regularly encountered in veterinary medicine.

Some panels of antisera are available commercially but larger laboratories may prepare their own. It is quite common practice for veterinary laboratories to send Salmonellae and *E. coli* isolates to human health laboratories to further characterization and to check whether or not animal strains are also found in humans.

Quantitative tests –counting bacteria

In some situations, it is necessary to assess the number as well as the type of bacteria present in a sample; this is particularly important in water and milk samples as part of quality control programmes. It is accepted that in some water samples there will be some bacterial contaminants but in most cases there will be guidelines for what is acceptable, for example, coliform count in town water supplies. There are a range of methods and some kit tests available to estimate the number of bacteria present in a given amount of sample, one simple method is outlined in Figure 4.12. In this method, serial dilutions are made of a sample. Samples are plated out using a plate flooding technique. A plate with 30–100 colonies is usually chosen for counting and the number of bacteria in the initial samples can then be calculated by multiplying by the dilution used for the plate counted. Other

A 1/10 serial dilution

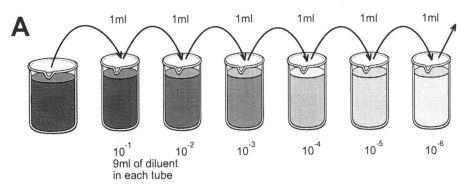

Colony counting with the surface spread technique (1 ml volume) to assess the number of *Escherichia coli* colonies grown on MacConkey agar. In this case the 10^{-6} dilution is suitable for counting.

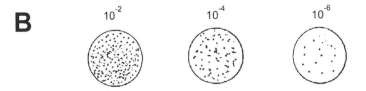

Calculation: 35 × (*E. coli* colonies counted) in a 10^{-6} dilution = 35 × 10^{6} per ml of broth (or sample)

Figure 4.12 The dilution technique can be used to count bacteria (see text). Serial dilutions are made from the original sample or a broth culture. Aliquots of selected dilutions (that is, 10^{-2}, 10^{-4}, 10^{-6}) are plated out using a spread technique and cultured for 24 h at 37°C. (A) Preparing a serial dilution (ten-fold). (B) Incubated culture plates after surface spread technique.

methods use plating out of specific aliquots of a sample or estimation of the turbidity of a sample or an inoculated broth in a spectrophotometer. In the latter test, there needs to be a reference standard against which the test sample is measured and/or a standard curve prepared to allow an accurate estimate of the number of bacteria represented by a specific reading. This method is simple to perform but has the disadvantage that it measures both live and dead bacteria whereas the colony counting techniques measure only the viable bacteria (colony forming units). Other methods used in quality control procedures for public health services use enzyme linked immunosorbent assay (ELISA) technology and antigen capture kit tests (see Chapter 6).

4.5 Antimicrobial sensitivity testing and antimicrobial use

Karen Liljebjelke and Susan C. Cork

The principle of the antibiotic sensitivity test is to determine whether or not bacterial growth occurs when cultured colonies are exposed to a selection of antibiotics. This does not precisely mimic the conditions found in the animal host but the results can still be useful when choosing a specific antibiotic treatment (Figure 4.13a).

For many clinical purposes, it is not necessary to accurately measure the minimum inhibitory and minimum bactericidal concentrations (MIC and MBC) of selected antibiotics for an isolate. It is usually sufficient to use a simple qualitative assay, which distinguishes isolates as 'sensitive' or 'resistant' to a particular antibiotic in line with the probable outcome of therapy.

There are circumstances when quantitative information about antibiotic sensitivity will benefit the treatment outcome, such as when a pathogenic strain is multiple-drug resistant, or when there are limited drugs that can be used with that patient due to multiple-drug resistance, physiological limitations of the patient, or withdrawal time considerations. In these circumstances, measuring the MIC (mg/ml) will allow the consulting veterinarian to choose antibiotics for which the bacterial strain may have a MIC which falls in the 'intermediate' range of sensitivity. Bactericidal antibiotic compounds may still be clinically effective when dosed in such a manner that the antibiotic reaches a blood or tissue concentration in the 'intermediate' sensitivity range.

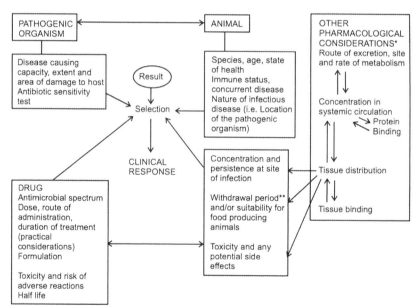

* Pharmacology is the study of how medicines work and requires detailed evaluation of the formulation of the medicine and its main active ingredients as well as the distribution, metabolism and excretion of the drug once it has been administered. ** Withdrawal period is the period of time specified, for a particular product, between the last date of treatment and the time at which human consumption of meat or other products from the animals is allowed. The duration of time specified depends on the active ingredients of the drug and the time required for the clearance of the drug from the animal's body. Note that some drugs must not be used in animals providing milk for human consumption. Always read the information leaflets supplied before any medication is given to an animal.

Figure 4.13a Factors affecting the choice of an antibacterial drug.

Of the three commonly used antimicrobial sensitivity assay types: disc diffusion, E-strip and micro-broth dilution, the E-strip and micro-broth dilution methods are quantitative, and yield MIC values in mg/ml, and the disc diffusion method is qualitative, and yields a dichotomous result of sensitive/resistant. The ability to use any of the three test types depends on a couple of factors: commercial availability of discs or E-strips for a particular antibiotic, availability of established MICs and breakpoints for specific antibiotics for specific bacteria, and the physical properties of the antibiotic compound itself. Some antibiotics require specific conditions for solubility such as addition of methanol or specific pH of the solution which may interfere with the assay.

Many variables may affect the quality of test including the amount of the inoculum, age of the bacterial culture, the nature and thickness of the culture medium, the length of inoculation and the composition of the applied antibiotic(s). These variables should be kept to a minimum and a standard protocol developed. A basic protocol for the most commonly used antimicrobial sensitivity testing method is outlined below.

Disc diffusion antibiotic sensitivity testing

The Kirby-Bauer disc diffusion assay is simple, relatively inexpensive, reliable, and suitable for routine sensitivity testing. The test uses a selection of small discs of a standard filter paper containing pre-determined amounts of chosen antibiotics. These discs are placed on large size (140 mm) Petri dishes of culture medium. The choice of culture medium used in the assay will determine interpretation criteria used for determining sensitivity. If Muller-Hinton agar (MH) +/– 5% sheep blood, is used, the interpretation criteria of the Clinical and Laboratory Standards Institute (CLSI) can be used. If Luria-Bertani (LB) or Brain-Heart Infusion agar (BHI) are used, interpretation criteria will be determined

by comparing inhibition zone measurements between the sample strain and control strains.

Petri dishes are spread uniformly with an inoculum of the bacterial isolate to be tested, and after incubation at 37°C for 18–24 h, determination of sensitivity or resistance is made by measuring the visibly clear area around the discs, which is the zone of inhibition of bacterial growth. The 'zone of inhibition' depends on the diffusion of the antibiotic from the discs into the surrounding agar, and the size depends upon characteristics of the growth medium and of the antibiotic compound that influence diffusion of the antibiotic and does not directly relate to the degree of sensitivity or resistance of the bacteria to the antibiotic (that is, a large zone of inhibition does not by itself indicate resistance). Sensitivity disc diffusion assays can be performed on significant primary cultures (that is, predominant or pure growth of suspect pathogen) from clinical specimens. The advantage is that results can be available as early as the day following the receipt of the specimen. However, the disadvantage of primary culture tests is that the number of bacteria in the inoculum cannot be standardized and sometimes is so small that results cannot be read or properly assessed; a further test must then be done on a pure subculture of the pathogen. Heavily mixed cultures are not suitable for the assay. It is recommended that the bacterial inoculum for disc diffusion susceptibility assays be cultured on blood agar, LB, or BHI media, as some ingredients in more selective media such as MacConkey media may interfere with the assay.

To perform the test, it is more convenient to use dry antibiotic discs prepared and supplied commercially. The discs are 6 mm in diameter and consist of absorbent (filter) paper impregnated with a known amount of antibiotic. Each disc is marked with a letter to show which antibiotic is present. There are protocols available that describe how to make standardized antibiotic filter discs in the laboratory, which may be a cost-saving measure. For example: prepare a solution of ampicillin with final concentration of 0.5 mg/ml (0.5 μg/μl), add 20 μl to each disc, and let dry. This will result in 10 μg of ampicillin per disc.

Dry discs can be stored at –20° C for at least a year without loss of potency and for a shorter period at 4°C. The manufacturer's expiry date should be observed for purchased discs, and the antibiotic expiry date must be observed for in-house prepared discs. Wet discs are also commercially available, these retain their potency for several months at 4°C.

Figure 4.13a outlines the wide range of factors that should be taken into consideration before an antibiotic is prescribed for an animal. It can be seen that determining the antibiotic sensitivity of bacteria isolated from a clinical sample will assist the veterinary or other animal health professional to make a decision with regard to treatment, but that other factors may be important. In some cases, a bacterial isolate which shows limited sensitivity to an antibiotic in the laboratory may still be effective when given to the animal. Generally, only qualified veterinary professionals are permitted to prescribe medicines and to treat animals so recommendations should be made following discussion with the veterinary officer in charge.

Choice of drug for disc test

When large numbers of sensitivity tests have to be done on isolates from clinical specimens, it is convenient to restrict the routine 'first-line' tests to the number of antibiotics that can be accommodated on a single culture place. It is standard practice to put six to eight antibiotic discs on a 140 mm plate. Using more than this may cause the zones of inhibition to overlap. In special cases requiring tests against additional antibiotics, a second plate may be set up with a supplementary set of discs. In some cases, it may

be more efficient to only assay antibiotics that are available for subsequent treatment.

Examples of suitable 'first-line' test antibiotics

1 For cultures from specimens other than urine: (for example, blood, exudates, pus, sputum, swabs from wounds and so on) test ampicillin, cotrimoxazole, erythromycin, penicillin (benzyl), tetracycline and gentamicin (or cephaloridine or streptomycin).

2 For cultures from urine test ampicillin, cotrimoxazole, nalidixic acid, nitrofurantoin, sulfonamide and kanamycin (or cephaloridine or tetracycline).

3 For *Staphylococcus* spp. test erythromycin, fucidin, penicillin (benzyl), tetracycline, lincomycin (or clindamycin), cotrimoxazole (or gentamicin or movabiocin).

4 For other Gram +ve cocci test ampicillin, cephaloridine cotrimoxazole, erythromycin, penicillin (benzyl).

5 For most Gram –ve bacteria test ampicillin, cephaloridine, cotrimoxazole (or sulphonamide), streptomycin (or kanamycin) and tetracycline, for *Salmonella typhi* include chloramphenicol.

6 For *Haemophilus* sp. test ampicillin, cephaloridine, chloramphenicol, cotrimoxazole, sulfonamide and tetracycline.

7 For *Pseudomonas aeruginosa* test carbenicillin, colistin (or polymyxin B) and gentamicin.

8 For Clostridia and other anaerobic Gram +ve bacilli test ampicillin, clindamycin, fucidin, penicillin (benzyl) and tetracycline.

Culture medium

A nutrient medium should be used which is as free as possible from substances inhibitory to the action of the antibiotics (and sulphonamides). Muller-Hinton agar is the most commonly used growth medium for disc diffusion assays. The pH should be 7.2–7.5. Glucose and reducing substances such as thioglycollate should not be added to media. Blood should be added when testing *Streptococcus pyogenes* or *Pneumococcus* spp. and heated ('chocolated') blood should be used when testing *Haemophilus* spp. A volume of 20 ml agar in a standard sized flat-bottomed Petri dish 100 mm in diameter gives a uniform layer of agar 3–4 mm deep. Plastic Petri dishes have a fill line. The thickness of the agar should be standardized for quality control.

Test procedure

Dry the culture plate in the incubator with the lid ajar until its surface is free from visible moisture. Do not allow the plate to become too dry as this will interfere with the diffusion of the antibiotic in the medium. Inoculate the bacteria to be tested by one of the procedures given below. Apply the chosen antibiotic discs at adequate spacing (2 cm or more apart) to the surface of the plate with sterile fine-pointed forceps and press gently to ensure full contact with the medium and a resultant moistening of the disc. It is important to ensure full contact of the disc. Incubate immediately for the minimum time needed for normal bacterial growth, usually for 18–24 h at 37°C.

Inoculation of primary cultures direct from specimen

- Swab: Rub a moist swab, well soaked in sample material, evenly over the plate.
- Milk or Urine: A loop full (0.95 ml) is rubbed evenly over the plate.

From pure sub-culture

Flooding: this is done by using 5 ml of saline containing a small amount of bacterial culture such as an overnight broth culture, or a suspension of a few colonies in broth. With a sterile Pasteur pipette transfer 1 or 2 ml of the dilute suspension to the plate, tip the plate in different directions to wet the whole of its surface. Remove all the excess fluid with the pipette, partially dry the plate with its lid ajar in the incubator for up to 30 min or on the bench for an hour and finally apply the discs.

Spread method: The day of the test, dilute your liquid cultures back and grow to mid-log phase, around an OD of 0.5 at 600 nm. Alternatively, pick four to five colonies from the sub-culture plate, and suspend them in sterile broth to an OD 0.5 at 600 nm. Spread 150 μl of the culture evenly throughout the plate using a sterile glass or plastic spreader, or sterile swab. Allow any extra liquid to dry on the plate.

Inoculating the plate evenly is an important quality control measure for the assay.

Control cultures

It is occasionally necessary to evaluate the assay by checking the result of the test with a standard control organism which has a known sensitivity to specified antibiotics. For example, the Oxford strain of *Staphylococcus aureus* is a suitable control for tests against most kinds of antibiotics (except polymyxin). For a 'control' for urinary tract cultures an 'antibiotic-sensitive' strain of *E. coli* is recommended.

Reading and interpretation of results (see Figure 4.13b)

The diameters of the zones of inhibition of growth (including the 6 mm diameter of the disc itself) are measured with callipers or by viewing the plate against a ruler or ruled screen. The measurement of the zones of inhibition can be compared with those published by CSLI.[15] Alternately where 'control strain' bacteria are available, determination of sensitivity can be made by comparing the zones produced by each drug for the test bacteria with the zones produced

Figure 4.13b Antibiotic sensitivity testing. *Staphylococcus aureus* growing on test agar incubated with discs containing gentamicin (GN), enrofloxacin (ENO), chloramphenicol (C) and tetracycline (TT). This isolate appears to be 'resistant' to tetracycline. (a) zone of inhibition. If a > 5 mm it is likely that the bacterial strain tested is sensitive to the antibiotic(s) tested.

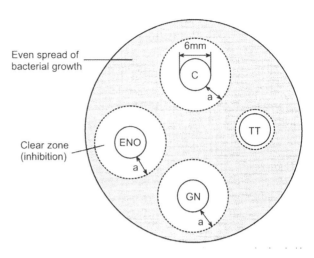

with the control strain bacterium. It is important to assay the control strain at the same time as the sample strain to reduce variability. The method for determination of sensitivity using this method is explained below.

Sensitive

An organism is considered to be 'sensitive' if the zone of inhibition around the antibiotic disc is greater than, equal to, or not more than 4 mm less than that of a 'sensitive' control strain (see figures 4.13c and d).

Moderately sensitive

An organism is considered to be moderately 'sensitive' if the zone diameter is at least 12 mm but is reduced by more than 4 mm compared to a 'sensitive' control culture.

Resistant

An organism is considered to be 'resistant' if the zone of growth (or the diameter of the zone) is not more than 10 mm (that is, 2 mm on each side of a 6 mm disc).

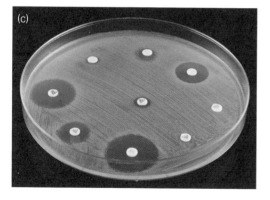

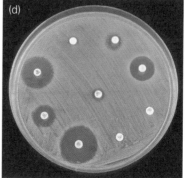

Figure 4.13c and d Antibiotic sensitivity testing plates. This is an example of the Kirby-Bauer disc diffusion assay for antibiotic sensitivity. The clear zones in the solid 'lawn' of bacteria around discs indicate bacterial growth has been inhibited by the antibiotic. Where bacteria have grown to the disc edge indicates resistance to that antibiotic.

A NOTE ON ANTIMICROBIAL RESISTANCE

Over the past few decades a wide range of highly effective antimicrobial drugs have been developed to control bacterial diseases in humans and animals. Much of this success has been due to our greater understanding of how bacteria cause disease (Table 4.4a) and the development of antimicrobial drugs targeted against specific bacterial virulence factors. Many of the newer classes of drugs target a narrow range of bacteria thereby reducing the side effects associated with depletion of normal gut flora. However, antibiotic use, especially when antibiotics are not used correctly (that is, when used to enhance production in healthy animals or where underdosing of clinical cases occurs), may lead to the selection of resistant forms of microorganisms. A variety of antimicrobial resistance mechanisms have been identified in bacteria and some of these are outlined in Table 4.4b. Microorganisms are constantly evolving. It is known that resistance to antibiotics existed long before antibiotics were widely used. However, there is now a demonstrated link between antimicrobial use in humans and animals and the development of resistance in environmental, commensal and some potentially pathogenic bacteria. The degree to which the use of antibiotics in veterinary medicine contributes to antimicrobial resistance in bacteria isolated from humans is not currently known but it is evident that improper use of antibiotics in agriculture and aquaculture can lead to resistance in bacteria isolated from farmed livestock and fish. See also Chapter 13.

ACQUIRED RESISTANCE

Antibiotic resistance can occur when genetic material is transferred between bacteria, this can occur in several ways; for example, via plasmids, which are small extrachromosal DNA molecules or via integrons and transposons, which are short DNA sequences, can be transmitted both vertically (that is, from bacteria to bacteria) and horizontally (during replication) and can code for multi-resistance. It is thought that a lot of the acquired resistance observed in bacteria is plasmid-mediated. In comparison, mutational resistance develops as a result of

Table 4.4a Bacterial mechanisms to enhance survival in the host.

Mechanism	Example
Capsule production	Many bacteria use this to prevent or limit phagocytosis
Capsular antigen	Some Gram –ve bacteria incorporate sialic acid which has a –ve effect on complement fixation
M protein production	*Streptococcus equi* uses this to prevent phagocytosis
Antigenic variation of surface antigens	Facilitates evasion of the host immune response, seen with *Mycoplasma* sp. and *Borrelia* sp. infections
Antigenic mimicry of host antigens	Facilitates evasion of the host immune response, seen with *Mycoplasma* sp. (also seen with some parasitic infections)
Production of leukotoxins	*Mannheimia haemolytica*, *Actinobacillus* sp. and others produce toxins that can lyse phagocytes
Escape from phagosomes	*Listeria monocytogenes* and rickettsiae
Interference with phagosome-lysosome fusion	*Mycobacteria* use this to survive within phagocytes
Coagulase production	*Staphylococcus aureus* may convert fibrinogen to fibrin to isolate the site of infection from the active immune response

Table 4.4b Mode of Action of some commonly used antibacterial drugs.

Antibacterial drug	Mode of Action	Comments
BACTERIOCIDAL		
ß – Lactam antibiotics e.g. Penicillins, Cephalosporins	Inhibition of cell wall synthesis	Low toxicity
Aminoglycosides, e.g. Steptomycin, Neomycin	Inhibition of protein synthesis	May be ototoxic and nephrotoxic
Trimethoprim	Inhibition of nucleic acid synthesis	Usually administered with sulphamethaxazole as a 'potentiated sulphonamide
Vancomycin	Inhibition of cell wall synthesis	May be used in cases of methicillin resistant *Staphylococcus aureus* (MRSA)
Polypeptides, e.g. Polymixin	Inhibition of membrane function	May be nephrotoxic and neurotoxic
BACTERIOSTATIC		
Nitrofurans	Inhibition of protein synthesis	Relatively toxic. Broad spectrum of activity
Tetracyclines	Inhibition of protein synthesis	Development of resistance common, has been widely used for prophylactic treatment in livestock
Chloramphenicol	Inhibition of protein synthesis	Potentially toxic. Use prohibited in food producing animals in many countries
Macrolides, e.g. erythromycin, Tylosin	Inhibition of protein synthesis	May be effective against *Mycoplasma* sp.
Quinilones, e.g. Enrofoxican, Nalidixic acid	Inhibition of nucleic acid synthesis	May be used in combination with other drugs for treatment of mastitis
Sulphonamides	Inhibition of nucleic acid synthesis	Effective against rapidly growing bacteria
Nitroimidazoles, e.g. Metronidazole	Disruption of DNA structure and inhibition of repair	Effective against anaerobic bacteria and some protozoa, e.g. *Giardia*

spontaneous mutations in the loci on the microbial chromosome that determine susceptibility to specific antibiotics. It is thought that the presence of the antibiotic serves as a selecting mechanism to suppress susceptible microorganisms and promote the growth of resistant strains. Spontaneous mutations can also occur and these are transmissible vertically, that is, they are passed on during bacterial replication.

MULTIPLE RESISTANCE

The development of multiple resistance in bacteria depends on several different mechanisms. More than one mechanism may operate for the same antibiotic. Microorganisms resistant to a certain antibiotic (see Table 4.4b) may also be resistant to other antibiotics that share a mechanism of action or attachment. This type of 'cross-resistance' is most common with antibiotics that are closely related chemically (for example, polymyxin B and colistin, neomycin and kanamycin), but may also exist between unrelated chemicals (for example, erythromycin-lincomycin).

EPIDEMIOLOGY OF RESISTANCE

The type of antibiotic resistance patterns seen in animals, especially farmed livestock, are largely affected by antibiotic exposure but will also vary according to:

- the size of the population of microorganisms
- pre-exposure prevalence of resistance genes
- the fitness of the selected population of microorganisms in competition with other microorganisms present in the environment which have not been exposed to antibiotics.

How to select an antimicrobial drug

The selection of the appropriate antibiotic treatment should take a number of factors into account. The clinical experience of the clinician or veterinary officer attending the case will influence the choice of antibiotic as will the availability of a suitable formulation of the drug for the animal(s) to be medicated. In many cases the attending veterinarian will prescribe a course of treatment before receiving microbiology culture and sensitivity results from the laboratory. In some cases, samples will not be collected because testing may not be considered necessary. However, if samples are submitted to the laboratory there will be an expectation that laboratory staff can provide advice to submitting veterinarians and extension staff on the selection and appropriate use of antibiotics.

The choice of the right antibiotic to treat an individual animal or a population should be based on the following:

- accurate diagnosis
- antibiotic sensitivity testing
- treatment efficacy
- duration and feasibility of treatment
- drug withholding period.

ACCURATE DIAGNOSIS

Any recommendation to prescribe antibiotics should be based on the clinical evaluation of the animals under the care of the prescribing veterinary surgeon and, where possible, on the results of laboratory tests. When it is not possible to make a direct clinical evaluation, the diagnosis should be based on knowledge of the farm epidemiological status and on ongoing sensitivity testing. Antibiotic therapy should only be used if it will help to shorten duration and severity of the infection or reduce risks of systemic complications.

ANTIBIOTIC SENSITIVITY TESTING

Antibiotics should only be used when it is known that the bacteria involved is likely to be susceptible to the therapy and that the correct dose of the antibiotic can be given over a sufficient duration to be effective. However, in certain circumstances, for example, a disease outbreak involving high morbidity and/or mortality or where the disease is spreading rapidly among contact animals, precautionary treatment might be started on the basis of clinical diagnosis prior to determining the likely efficacy based on laboratory tests. The effect of the drug *in vivo* (that is, in the animal) depends on its ability to reach the site of infection in a high enough concentration, the nature of the pathological process and the capability of the host's immune response to limit the infection. The response *in vivo* cannot always be predicted from the laboratory results *in vitro* due to the wide range of factors that can influence the host–pathogen interaction.

TREATMENT SELECTION AND EFFICACY

The choice of the right antibiotic for a given species or disease is dependent on host and pathogen involved as well as the pharmacokinetic parameters of the available antibiotics such as bioavailability, tissue kinetics and distribution and the drug half-life to ensure that the selected antibiotic reaches the site of infection at a sufficient dose.

A product with a narrow spectrum of activity is often the best choice where the organism has already been cultured and a diagnosis confirmed. However, broader spectrum antibiotics are favoured when the diagnosis is pending. In these cases, it is important to select an antibiotic that is likely to be effective against the suspect pathogen but has a minimal effect on other (beneficial) microorganisms.

DURATION AND FEASIBILITY

The selected antibiotic will need to be available in a formulation that is easily administered to the animal(s) undergoing treatment. The route of administration of the available antimicrobial products must be considered carefully, that is, prolonged oral use should be avoided. This is because the development of resistance is often associated with the selection and transfer of genetically resistant bacteria that inhabit the gut and also because prolonged oral treatment can destroy the beneficial microbial flora. Insufficient duration of treatment, or the inability to administer the required dose, can lead to resurgence of the infection.

DRUG WITHHOLDING PERIOD

Duration of drug withdrawal periods may also be a factor in choosing suitable products, that is, the milk from dairy animals, eggs from hens and the meat from meat animals should not be used for human consumption within the withdrawal period as the antibiotic concentration may remain above acceptable limits in the product.

ALTERNATIVES TO ANTIBIOTICS

In conclusion, antibiotics can provide a valuable treatment option for a range of bacterial diseases. However, their use should be part of, and not a replacement for, integrated disease control and prevention programmes. These programmes usually include recommendations for appropriate hygiene and disinfection procedures, effective biosecurity measures, improved husbandry, that is, reduction of stocking rates, better nutrition and preventative interventions such as vaccination.

4.6 Mycology

Mycology is the study of fungi. Fungi may be parasitic or saprophytic (feeding off dead organic matter) and are abundant in nature. Fungi are a common cause of damage to crops, food stuffs, fabric and building materials. In animals, fungi are usually secondary invaders, rather than primary pathogens, and therefore fungal infections often imply poor immune function or disruption to normal flora. For example, *Candida* sp. yeast infections are more common in animals that are debilitated due to concurrent disease or nutritional deficiencies; they are also more frequent following antibiotic use. However, healthy animals can be severely affected by mycotoxins produced by fungal contaminants in forage and feed supplies (for example, aflatoxins produced by *Aspergillus* sp.). Screening feed for aflatoxins is outlined in Chapter 7. Many of the most common fungal diseases in animals are cutaneous and spread by direct contact (for example, ringworm) but airborne and other routes of transmission are possible and can result in systemic disease (for example, blastomycosis, histoplasmosis). Some fungal diseases (mycoses) of veterinary importance are discussed below.

Moulds and yeast

The term mould is generally used to describe a fungus that produces branching tubular structures. These structures are normally divided into sections (septa) by a cell wall. Unlike most bacterial colonies, fungal colonies tend to spread radially on a culture plate due to their branching growth. The peripheral filaments that make up a fungus are called hyphae (see Figure 4.14a).

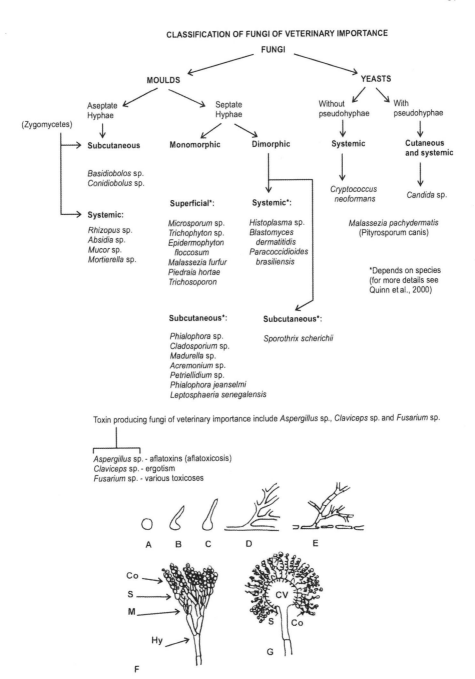

CLASSIFICATION OF FUNGI OF VETERINARY IMPORTANCE

Figure 4.14a The morphology of yeasts and fungi; this figure demonstrates some of the morphological features of yeasts and the species of fungi (*Penicillium* spp. and *Aspergillus* sp.). Germ tubes and hyphae: (A) yeast cell forming germ tubes (B+C) and alter, non-septate (D) and septate (E) hyphae. Hyphae and Conidia: (F) *Penicillium* sp. branching conidiophore bearing metulae (M) sterigmata (S) and conidia (Co). (G) *Aspergillus* sp. Cross section. Conidiophore with conidiophore vesicle (CV) bearing two rows of sterigmata (S) and conidia (Co). Diagrammatic view, as seen under the microscope (20×).

Figure 4.14b Diagrammatic representation of some fungal species.

Dermatophytes (ringworm)	Culture characteristics on Sabouraud's dextrose agar	Microscopic characteristics
Microsporum sp.	Variable depending on the species involved	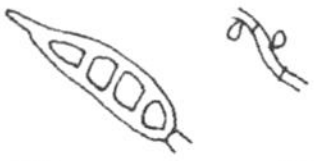
Trichophyton sp.		
Contaminants (feed spoilage and/or disease related complication) *Aspergillus* sp.	Often blue-green with velvet texture but can be yellow and black depending on the species	
Pencillium sp.	Blue-green with velvet texture but shades vary with species	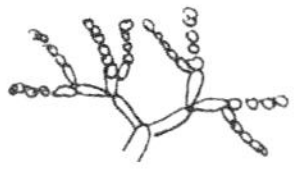
Mucor sp.	Low but rapid flat growth, white grey or yellow	
Absidia sp.	Rapid growth, woolly and white to grey in colour	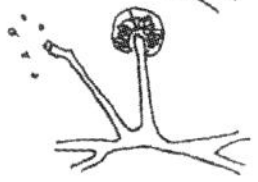
Yeast *Candida albicans*	Mucoid appearance on standard media, characteristic smell	

The rate of growth, the colour and the form of the fungal colony is distinct for each species and these characteristics are important in identification.

Single cell fungi that reproduce by budding are generally referred to as yeasts. In the yeasts hyphae are absent or represented only by pseudohyphae which are elongated budding cells often linked in branching chains. Some opportunistic fungal pathogens are dimorphic, that is, they occur in two different forms. Dimorphic fungi are filamentous (mycelial) in culture and yeast-like in infected tissues. Most fungi reproduce by forming spores, which may be produced in many different ways. The size, shape and colour of spores and their manner of production are also of value in identifying fungal species (Figure 4.14b).

Clinical importance

Fungi are common in the environment but rarely cause significant mortality in healthy animals. Mycoses (and mycotoxins) can nevertheless be a cause of morbidity in animals. The range of diseases that can be caused by fungi are numerous and potentially complex. Fungal infections are especially common in tropical countries where humidity is high. Some fungal diseases are caused by primary pathogens but most are opportunistic, that is, they are ordinarily harmless but may infect an animal when predisposing factors, such as immune-suppression or concurrent disease, are present.

A confirmatory diagnosis of fungal infection is usually achieved by direct microscopy of samples (usually skin scrapings or tissue smears) but culture may be required to definitively identify the pathogen involved. For systemic infections, the appearance of stained fungal hyphae in histological sections may help in the identification of fungal pathogens but this is more likely to be at necropsy vs. ante-mortem.

Laboratory diagnosis

Fungal diseases are diagnosed in a similar way to bacterial diseases, that is, starting with aseptic collection of specimens from affected tissues followed by culture on specialized media (for example, Sabouraud dextrose agar). The specific approach will depend on the location and type of lesions seen. For necropsy cases the approach followed is outlined in Chapter 8.

Fungal infections of the skin

One of the most common fungal diseases seen in temperate regions is ringworm. Although this skin condition is caused by a fungus it is often discussed along with parasitic skin diseases. Confirmatory diagnosis of ringworm and other cutaneous fungal infections involves clinical examination, collection of skin scrapings and hair, microscopic examination of samples and (potentially) culture. In cases where an unusual fungal pathogen is involved samples may need to be sent to a mycology reference centre for identification.

The collection of skin specimens is described below and in Chapter 3 (section 3.9) along with the diagnosis of diseases caused by ectoparasites.

Specimens to collect

1 Cleanse the affected area with 70% alcohol.
2 Collect skin scales, crusts, and pieces of nail or hair on a clean piece of paper. If available use dark coloured paper because the specimen will be easier to see against a dark background.

- Skin scales: collect by scraping the surface of the margin of the lesion using a sterile scalpel blade.
- Crusts: collect by removing part of the crust nearest to healthy skin using sterile scissors and tweezers.
- Nail, horn or hoof piece: collect by taking a small section of the infected part of the nail or hoof/horn using sterile tweezers and scissors.
- Hair: pluck complete hairs with the surrounding skin segment if possible.

After collecting representative specimens place them in a labelled envelope or plastic container. The label should indicate the name of the submitter, the identity and species of the infected animal, the source of the sample material and the date.

Examples of specific fungal diseases

Ringworm

Ringworm is a skin disease caused by a group of fungal species known as dermatophytes. The most common dermatophytes in temperate countries include *Microsporum* sp. and *Trichophyton* sp. Depending on the severity of the infection and the presence or absence of concurrent disease and predisposing factors, clinical signs may or may not be present. In mild cases, there may only be evidence of minor hair loss over specific regions of the body. In severe cases the animal may suffer significant hair loss due to secondary bacterial infections and resultant irritation and rubbing. A percentage of *Microsporum* sp. infected hairs fluoresce under a Wood's lamp (UV light). Although this is a popular diagnostic test for domestic pets a negative result does not necessarily mean that no fungus is present because many other dermatophyte fungi do not fluoresce.

The direct microscopic examination of hair specimens, with or without culture, is the recommended method for detecting the presence of dermatophytes.

Direct microscopy

Fungi are usually larger than bacteria and can readily be seen by direct microscopy provided the material from skin, hair or nails is first softened and cleared with a strong alkali such as 200 g/1 (20% w/v) potassium hydroxide (KOH). The purpose of the alkali is to digest the keratin surrounding the fungi in samples so that the hyphae and spores can be seen. The method of examining a KOH preparation is as follows.

Place a drop of 10–20% potassium hydroxide (KOH) solution and sample on a slide. Proceed with caution as the KOH solution is corrosive. Mix with a drop of blue-black fountain pen ink or lactophenol blue and put a coverslip over the preparation. The dye will highlight morphological features but these are not necessarily diagnostic for specific fungi. Place the slide in a Petri dish or other container with a lid, together with a damp piece of filter paper or cotton wool to prevent the preparation from drying out.

Note: To assist clearing, hairs should not be more than 5 mm long and skin scales, crusts and nail (horn/hoof) snips should be not be more than 2 mm in diameter.

Hairs will clear within 5–10 min. Skin scales and crusts usually take 20–30 min. Pieces of nail, however, may take several hours to clear. Clearing can be hastened by gently heating the preparation over the flame of a spirit lamp or pilot flame of a Bunsen burner taking care to prevent drying or splashing of the corrosive KOH solution (wear protective glasses when performing this procedure).

As soon as the specimen has cleared, examine it microscopically using 10× and 40× objectives with the condenser iris diaphragm closed sufficiently to give a good contrast. If too intense a light source is used the contrast will not be adequate and the fungi will not be seen.

Appearance of skin scales, nails or crusts in KOH preparation

Look for branching hyphae, chains of angular or rounded arthrospores (spore resulting from hyphal fragmentation) or a mixture of both. All species of ringworm fungi have a similar appearance. Ringworm fungi must be distinguished from epidermal cell outlines, elastic fibres and artefacts such as intracellular cholesterol and strands of cotton or vegetable fibres. Fungal hyphae can be differentiated from these structures by their branching growth, uniform width and cross-walls (septa) which can be seen when using the 40× objective.

Culture

Identification of fungi is a specialist job. It should also be noted that 10–20% of specimens collected from ringworm lesions, where fungi are seen microscopically, are negative on culture. This is usually due to the material collected being non-viable. In addition, contaminating bacteria or non-pathogenic fungal spores in a specimen may overgrow pathogenic species if steps are not taken to limit or remove them before attempting fungal culture.

Isolation of ringworm and other fungal pathogens can be achieved by:

1 adding inhibitory substances to the medium (see below)
2 leaving the specimen to dry out, so contaminating bacteria, but not fungi, will die off
3 adding acetone or alcohol to the specimen and culturing after a few minutes.

Choice of media

1 Blood agar with added antibiotics to inhibit bacterial growth.

2 Sabouraud dextrose agar alone and/or Sabouraud dextrose agar with cyclohexamide 0.5 mg per ml, penicillin 20 units per ml, and streptomycin 40 mg per ml.

The specimen should be pressed into the centre of the agar plate (or slope), using sterile forceps, and incubated in the dark at a temperature of 24–26°C. The culture should be examined at intervals of a few days over a period of 4 weeks. Identification of the fungus involved is based on microscopic examination of stained slides of the cultured colonies and the recorded macroscopic characteristics. In most cases lactophenol cotton blue (LPCB) (see box) stain is used.

Lactophenol cotton blue stain (for staining fungal elements in wet preparation)

Phenol crystals 20 g

Glycerine 40 ml

Lactic acid 20 ml

Distilled water 20 ml

The ingredients are dissolved by gently heating over a steam bath. When in solution, 0.05 g of cotton blue dye is added. The solution should be mixed thoroughly.

Fungi that are commonly isolated from cases of ringworm and other skin conditions in animals

- Cattle: *Trichophyton verrucosum, Trichophyton mentagrophytes*
- Horses: *Trichophyton equinium, Trichophyton verrucosum, Microsporum equinum*
- Dogs and cats: *Microsporum canis* (some infected hairs may fluoresce green under a Wood's UV lamp). *Trichophyton mentagrophytes.*

All the above fungi form arthrospores outside the hair shaft.

Other fungal pathogens

A number of fungal species have the potential to cause systemic disease in animals although this is more common when fungal infections occur secondary to other debilitating disease(s). For example, *Candida albicans*, a yeast, can cause a range of diseases in debilitated animals, for example, 'canker' in the horse's foot and inflammation of the crop in birds ('sour crop'). Thrush (or candidiasis) is especially common in people and animals after prolonged treatment with broad spectrum antibiotics, which destroy the normal flora of the gastrointestinal and reproductive tracts, allowing an overgrowth of yeast. *Candida* sp. yeasts can be readily seen in smears prepared from lesions, identification can be confirmed by culture at 37°C for up to 4 days.

Aspergillus fumigatus can be readily seen in smears and histological sections prepared from lung and air sac lesions collected from birds with aspergillosis. Culture of material from lesions should be done at 37°C with samples incubated for 1–4 days. *Aspergillus* species fungi can also cause guttural pouch infections in horses, nasal granulomas in a range of species and Mycotic abortion in cattle.

Dimorphic fungi include *Blastomyces dermatitidis* (blastomycosis), *Histoplasma capsulatum* (histoplasmosis) and *Coccidioides immitis* (coccidioidomycosis). The spores of these fungi, which can cause systemic disease, usually enter the body via the respiratory route. Culture requires incubation at 37°C over 1–4 weeks for the yeast phase and at 25°C for 1–4 weeks for the mould phase.

There are a range of antifungal drugs available to treat fungal diseases but it is usually just as important to treat the underlying problem. A full evaluation of the animal's health and accurate identification of the fungal species involved is essential. Each case of fungal infection, other than ringworm, should be discussed with the nearest veterinary officer so that its significance can be determined.

Mycotoxins

Mycotoxin production in crops (for example, groundnuts, maize and oil seed) can pose a significant health hazard. The presence of fungi, and subsequent toxin production, is dependent on the environmental conditions before and after harvesting and can also occur in feed stuffs after processing. Although mycotoxins are produced by fungal organisms, the absence of fungi in animal feeds does not imply the absence of mycotoxins. Aflatoxin, produced by *Aspergillus* sp. fungi is particularly common in humid conditions but also occurs in temperate climates especially where feed becomes damp during storage. There are several forms of naturally occurring aflatoxins (including B1, B2, G1 and G2). Aflatoxins M1 and M2 are hydroxylated metabolites of B1 and B2 and can be excreted in the milk of animals that have consumed the toxins. Aflatoxin type B1 is the most common and one of the most toxic mycotoxins. Even low levels of aflatoxins can cause cumulative liver damage where ingestion of spoiled feed is prolonged.

Other toxins that can be found in feed include ergot (*Claviceps purpurea*), Zearalenone (*Fusarium* spp.), Ochratoxin (*Penicillium* spp. and *Aspergillus* spp.). For the purposes of diagnosis, the fungal toxin levels that should be considered significant depend on the amount of the contaminated diet which will be fed and the livestock species involved. For example, in poultry $< 100 \mu g$ aflatoxin/kg feed can be fatal and in pigs $140 \mu g$ aflatoxin/kg may produce liver damage. These species are very susceptible to aflatoxicosis but ruminants and horses are more resistant. Mycotoxin contamination can be avoided by making sure that the raw ingredients of animal feeds are of good quality and that concentrate feeds are stored in dry conditions.

4.7 Virology

M. Faizal Abdul-Careem, Susan C. Cork and M. Sarjoon Abdul-Cader

Viruses are widely distributed in nature and cause a broad range of diseases in plants and animals (including bacteria). Many hundreds of viruses have now been described.

Credit for the first discovery of a virus is given to Dmitri Iwanowski, a Russian botanist, who in 1892, presented a paper to the St Petersburg Academy of Science demonstrating that the agent that produced the mosaic disease of tobacco plants could be freely passed through bacteria-retaining filters. In 1898, Friedrich Loeffler, working in conjunction with Paul Frosch in Germany, identified a filterable agent, smaller than any bacteria previously discovered, that caused foot-and-mouth disease. With the advent of the electron microscope in 1939, by Ruska and Knoll, our understanding of the morphological structure of viruses rapidly progressed. In 1931, Ernest Goodpasture, an American pathologist, discovered that viruses could be propagated in embryonating chicken eggs. Later, in 1954, John Franklin Ender was able to culture poliovirus *in vitro*. These developments began an era of research where virus–cell interactions could be studied in the laboratory at the cellular level. Subsequent to this, the use of molecular tools has helped us to better understand the genetic structure and phylogeny of viruses. Some DNA and RNA viruses of veterinary importance are listed in the next section (see Tables 4.7a and b).

In this section, we will cover the traditional approach to diagnosing diseases caused by viruses. The molecular tools used in the areas of bacteriology and virology will be considered in the supplementary section at the end of this chapter.

Properties of viruses

Viruses are obligatory parasites requiring living cells for replication. Similar to other microorganisms they are infectious, many are host specific and not all are pathogenic. Mature viruses, also called virions, vary considerably in size. For example, porcine circoviruses are the smallest animal virus with the size of approximately 17 nm ($\times 10^{-9}$m) in diameter, foot-and-mouth disease (aphthovirus) virions are approximately 27 nm in diameter, while the poxviruses (largest animal virus family) are nearly 300 nm in diameter. Recently, Pandora viruses with the size of 1000 nm have been described inhabiting protozoa rebutting the fact that the viruses are not visible under a light microscope.

Methods used to diagnose viral infections

In common with the diagnosis of other diseases, the identification of a viral disease depends on assessment of clinical signs and consideration of the epidemiology of the disease condition or outbreak. As indicated in Figure 4.15, a thorough clinical evaluation allows the selection of suitable samples for laboratory diagnosis.

As shown in Figure 4.16, various components of the virus structure can be detected or quantified using modern laboratory assays. This is often based on specific antigen detection using serological techniques (see Chapter 6). Appropriate sample collection is a critical step that will impact the ability of the laboratory

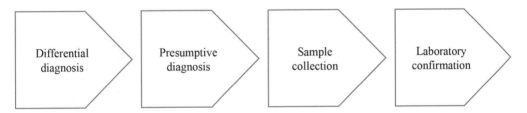

Figure 4.15 Steps in viral disease diagnosis starting from the evaluation of clinical signs leading to laboratory confirmation.

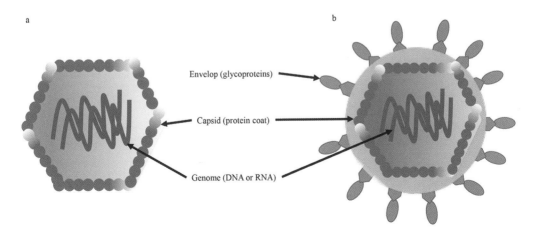

Figure 4.16 Modern laboratory assays are available targeting various structural components of the enveloped and naked viral particles.

to provide a confirmatory diagnosis. The samples required for laboratory confirmation of a presumptive diagnosis depends on the type of the laboratory assay. For example, diagnosis based on virus isolation requires sterile samples maintained in ice or dry ice to prevent bacterial contamination whereas for assays based on detection of viral nucleic acid or antigens, the samples may not be required to be sterile. It is important to contact your regional or reference laboratory for instructions on collecting suitable samples and maintaining samples during shipping.

Laboratory methods used to diagnose viral infections

Knowledge of the morphology and genetics of viruses has increased significantly as new tools and techniques to study them have developed. This has also resulted in re-classification and re-naming of some viruses.

Figure 4.17 summarizes a few standard methods used to isolate and identify some viruses of veterinary importance. This information is given to enable technical and extension staff to appreciate what is involved in virological examination although most district or even regional veterinary laboratories will not usually have

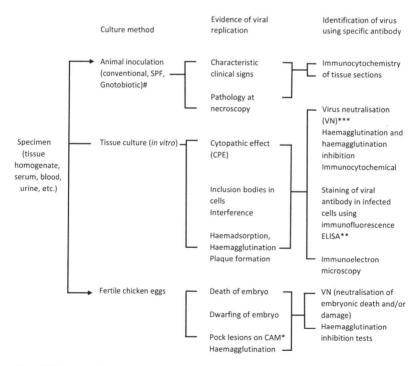

Notes: *CAM = chorioallantoic membrane, **ELISA = enzyme linked immunosorbant assay, ***VN = neutralization of virus seen by reduction of CPE, plaques etc.

#Source of animals: SPF = specific pathogen free (free of specified pathogenic organisms, and usually supplied by specialist animal supply companies), gnotobiotic = free of pathogens, born and raised in aseptic conditions (expensive and technically difficult, supplied by specialist facilities, used in research and not commonly used in diagnostic work). Conventional = 'normal' laboratory supplied experimental animal (usually mice, rats, rabbits or designated species for specified viral agent).

Figure 4.17 Identification and isolation of viruses from clinical samples.

the facilities required to perform these tests. In most countries, samples for virus isolation, especially in cases where a new or emerging disease is suspected, are sent to specialist facilities for a definitive diagnosis. National Reference Laboratories, Research Institutes and some veterinary schools may also offer a set range of virology services for a fee. In most regional or district laboratories, especially if molecular tools are not available, the use of serological screening or antigen capture technology (for example,

ELISA kits[16]) is the most common approach for the initial diagnosis of viral diseases (see also Chapter 6 and Table 4.5).

Light microscope

It is possible to see the elementary bodies of some viruses, in stained histological sections, viewed with the ordinary light microscope. In infected tissue sections, or prepared cell culture, stained with haematoxylin and eosin there may

Table 4.5 Isolation and identification of some viruses of veterinary importance.

Virus	Specimen(s)	Host species	Evidence of viral replication	Identification
BVD (Bovine viral diarrhoea)/ Mucosal disease complex)	Aborted foetal tissues, intestine, blood (white cells in the buffy coat*)	Cell culture** (bovine origin, usually calf kidney or foetal lung cell lines)	Cytopathic effect (CPE), FA (Fluorescent antibody for non-cytopathic isolates)	Virus neutralization (VN), FA (also fluorescent antibody test on a fecal antigen
IBR (Infectious bovine rhinotracheitis virus and/ or infectious vulvovaginitis)	Nasal and ocular swabs, tracheal scraping, foetal liver	Cell culture (as above)	CPE Intranuclear inclusions	VN, FA
PI3 (Bovine parainfluenza 3)	Nasal swabs, nasopharyngeal scrape, lymph nodes	Cell culture (as above)	CPE, HA (Guinea pig red blood cells)	NV, FA, heamaglutination inhibition
Swine fever (Hog cholera)	Spleen, tonsilar material, lymph nodes	Cell culture (porcine)	FA	FA
Equine viral arteritis	Nasal swabs, blood	Cell culture (equine)	CPE	VN
Rabies	Brain tissue	Cell culture (neuroblastoma cell line), Laboratory mice (intracerebral inoculation)	Central nervous system signs and death	FA (Negri bodies), VN, FA

Notes: *Buffy coat – the white line seen when blood is centrifuged in a haematocrit separating the red cells from the plasma. In many infectious diseases, and in some blood disorders, the layer of white cells can be significantly increased making the buffy coat very easy to see (see Chapter 5). **Cell culture requires specialized facilities with a designated clean section where aseptic techniques are applied and skilled technical staff have experience to grow and maintain cell lines. The latter are usually purchased from specialist suppliers. FA = fluorescent antibody; HA = haemagglutination.

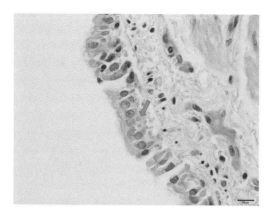

Figure 4.18a Canine airway epithelium with intra-cytoplasmic inclusions of canine distemper virus (the arrow indicates eosinophilic intracytoplasmic inclusion bodies). See also Plate 11. Photo: Dr Jennifer Davies, University of Calgary, Canada.

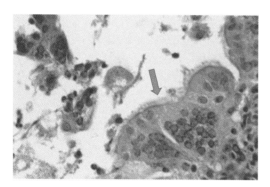

Figure 4.18b Trachea of a chicken infected with infectious laryngotracheitis virus with multinucle-ated syncytial cell formation (the arrow indicates a multinucleated syncytial cell). See also Plate 12. Photo: Dr M. Faizal Abdul-Careem, University of Calgary, Canada.

be evidence of viral inclusion bodies in host cells (Figure 4.18a) and other virus induced cellular changes such as syncytial cell formation (Figure 4.18b). The appearance of these will depend on the stage of infection as well as the type of virus present. The morphology of inclusion bodies and other changes can be characteristic and help to identify the virus.

Techniques such as immunohistochemistry (Figure 4.19a and b) and the use of special stains have also been devised to highlight the presence of viral antigens in cell or tissue preparations. As with bacteria, differential staining makes it easier to distinguish the viral antigens from other small particles seen under the microscope. However, although it is possible to resolve

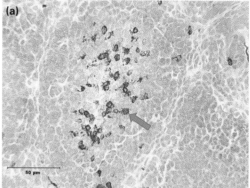

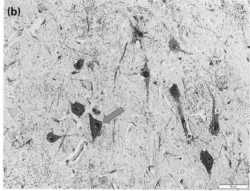

Figure 4.19 Immunohistochemistry staining can be performed in frozen sections or formalized sections to visualize viral antigens. (a) A frozen section of Bursa of Fabricius of a chicken infected with Marek's disease virus (the arrow indicates the brown colour stained viral antigens). Photo: Dr Shayan Sharif, University of Guelph, Canada. (b) A formalized brain section of a dog infected with canine distemper virus (the arrow indicates the brown colour stained viral antigens). See also Plate 13a & b. Photo: Dr Cameron Knight, University of Calgary, Canada.

particles as small as 250 nm in diameter, it is very difficult to identify them with certainty.

Electron microscope

The electron microscope is a highly specialized instrument. Sample preparation and the use of the electron microscope requires considerable technical training and skill. The basic principles of magnification using the electron microscope are similar to those used in the light microscope except that electrons are used in place of light rays and the electron rays are focused by magnetic fields.

For preparations to be viewed using the electron microscope, phosphotungstate and uranyl sulphate or gold are often used to provide good contrast for the viral particles.

The preparation method used depends on whether scanning or transmission electron microscopes are used. Transmission microscopes allow views of tissues section by section whereas the scanning microscope allows three-dimensional surface views of cells and viral particles. Figure 4.20 illustrates a corona viral particle visualized using an electron microscope.

Immunofluorescent microscope

Fluorescent dyes bound to antibody raised against specific viral antigens can be added to cell suspensions or tissue sections. Binding, and the associated fluorescence, indicates the location of the virus within cells. The presence or absence of fluorescent labelled antibody-antigen complex can be seen using a microscope with an ultraviolet light source (see Figure 4.21). Acridine orange can also be used to visualize aggregates of viruses.

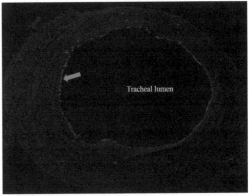

Figure 4.21 Image captured under fluorescent microscope following staining of trachea infected with infectious bronchitis virus demonstrating nuclear antigen of the virus. Arrow points at the epithelial lining facing the tracheal lumen. The section was counterstained with fluorescent dye staining nuclei. See also Plate 14 Photo: Dr M. Faizal Abdul-Careem, University of Calgary, Canada.

Viral culture methods

In the laboratory, viruses may be cultivated using animal inoculation, tissue cultures and inoculation of embryonated chicken eggs (see Figure 4.26).

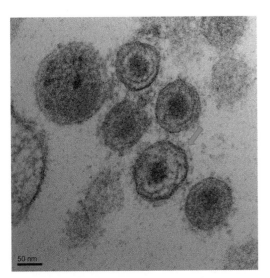

Figure 4.20 Transmission electron microscopy (TEM) imaging of macrophages infected with infectious bronchitis virus (corona virus). Photo: Dr M. Faizal Abdul-Careem, University of Calgary, Canada.

Laboratory animals

Animal inoculation was the first method used for virus cultivation and for many years this was the only means of virus propagation. However, except for the use of embryonated chicken eggs, concerns about animal welfare have led to the development of alternative virus isolation methods such as cell culture. Animals are now rarely used for the purposes of diagnostic testing although they are still used to determine the virulence of some strains of viruses (for example, highly pathogenic avian influenza and the Paramyxovirus (APMV-1) responsible for Newcastle disease). Laboratory animals are also used in scientific research and for the production of some biological materials for laboratory tests, that is, antisera, compliment and so on. The use and maintenance of laboratory animal colonies is beyond the scope of this book and will not be discussed further here.

Cell culture

There are many techniques for the cultivation of viruses in cell cultures in the laboratory (*in vitro*) and the choice of media and conditions will depend on the cell type and cell line selected. This in turn is determined by the nature of the virus to be cultivated. The labware used for cultivating tissue cells must be absolutely clean and free from contaminants and toxic substances and cell culture specific disposable labware are readily available. The chemicals and distilled water used for cleaning labware should be deionized. The various manipulations required for handling cells and media should be carried out in a room with restricted access that is free from dust and air currents. The equipment required for cell culture is expensive and includes level two biosafety cabinets, cell culture incubators, centrifuges with temperature control, water baths and inverted microscopes (Figure 4.22). All instruments and other equipment must be sterile and handled in such a manner as to avoid contamination.

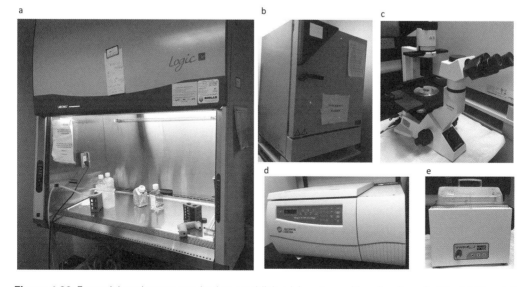

Figure 4.22 Essential equipment required to establish a laboratory with cell culture facility. (a) Biosafety cabinet, (b) cell culture incubator, (c) inverted microscope, (d) centrifuge, and (e) water bath.

Initial growth of cells is done in various types of cell culture flasks, disposable glass flasks and polyethylene flasks of various sizes are commonly used. The main applications of cell culture are to grow viruses to high titres, to identify viruses based on the cytopathic effects induced in cell culture, and for the quantification of replicating viruses in samples using plaque assays (Figure 4.23). However, some viruses are more difficult to culture than others. The correct cell line and the ideal environmental conditions required can only be selected when the sample submitter has provided a clear list of diagnostic differentials in advance. For these procedures, establishment of monolayers of cells in six-, twelve- and twenty-four-well plates is generally required.

Once a good monolayer of cells has been produced, the media used to grow the cells can be replaced with nutrient media containing the virus of interest (this can be extracted from samples using filtration to remove bacteria and other microorganisms). After a short period of incubation ($\approx$30–60 min) to allow attachment of the virus to the cells, the fluid is poured off and replaced. Cell culture media usually contains a red dye that changes colour (due to the change of pH) when the media needs replacing. Following a period of incubation ($\approx$2–5 days) with the virus, necrotic cells may be observed due to virus growth within the cells (Figure 4.24). The extent of necrosis (cell death demonstrated by shrunken nuclei and withering cell margins or plaque formation) can be noted and recorded to determine the relative number of viral particles present in the virus inoculum. In general, it is assumed that each area of necrosis or plaque formation results from infection by one viral particle.

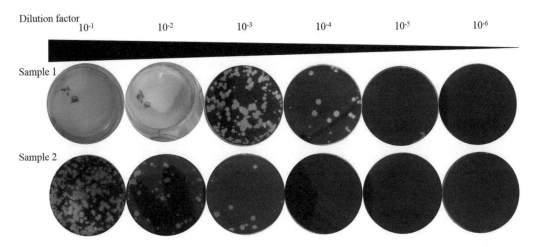

Figure 4.23 In plaque assay the virus inoculum is ten-fold serially diluted in phosphate buffered saline and inoculated into the monolayer of cells which are permissive to the inoculating virus. In this figure, avian influenza virus titration has been done in Madin-Darby Canine Kidney (MDCK) cells. After two days of inoculation of the MDCK cell monolayer, the cells have been stained with crystal violet to see the extent of cell damage due to the virus replication. The clear areas represent the loss of cells due to viral replication. The number of infectious virus particles in sample 1 is higher than the sample 2. See also Plate 15. Source: M. Sarjoon Abdul-Cader, University of Calgary, Canada.

a b

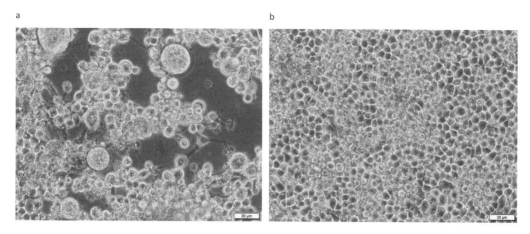

Figure 4.24 Infectious laryngotracheitis virus (herpesvirus) replicates and produces cytopathic effects on monolayer of leghorn male hepatoma (LMH) cells. The cells were captured under inverted microscope two days following the infection. The infected (a) and uninfected cells (b) are shown. Photo: M. Sarjoon Abdul-Cader, University of Calgary, Canada.

a b c

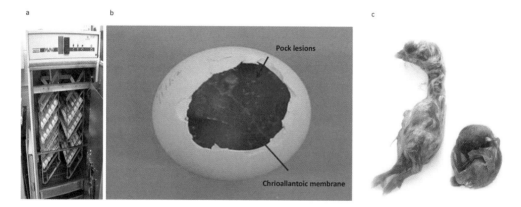

Figure 4.25 (a) Egg incubator is an essential component of a virology laboratory. For virus isolation and for egg inoculation for detecting lesions in 9–11 days old embryos SPF eggs are necessary. (b) Pock lesions on the CAM of embryonated chicken egg two days of inoculation. Nine to twelve days old embryonated chicken eggs are inoculated with the test inoculum prepared from tracheal swabs. The shell over the broader end of the egg has been removed to visualize the CAM. (c) The embryo on the left is the uninfected control. The embryo on the right, inoculated with infectious bronchitis virus shows stunting and curling. Photos: Dr Davor Ojkic, Animal Health Laboratory, University of Guelph, the Netherlands.

Virus propagation in embryonated chicken eggs

The chicken embryo is susceptible to a great many viruses and has become useful in the diagnosis of viral diseases, to grow viruses to high titres and also in vaccine production (Figure 4.25).

Chicken eggs to be used for the purpose of embryo inoculation should be procured from specific pathogen free (SPF) flocks since in the field setting, some infectious agents (including viruses) may pass from the infected hen into the egg. Maternal antibodies may also be present in eggs laid by recovered infected birds, which will interfere with the growth of certain avian viruses.

For virus inoculation, 9–11-day-old embryonated eggs are used. Inoculation can occur *via* a number of routes (Figure 4.26). The most common routes are the chorioallantoic membrane (CAM), allantoic cavity and yolk sac. The less common routes of inoculation are intraembryonic, intracerebral and intravenous.

Serological methods used to diagnose viral diseases

Viruses encode antigens in their capsids and envelopes which readily stimulate the production of antibodies in infected hosts. There are a wide range of serological techniques used to detect specific antibodies to viral antigen(s).

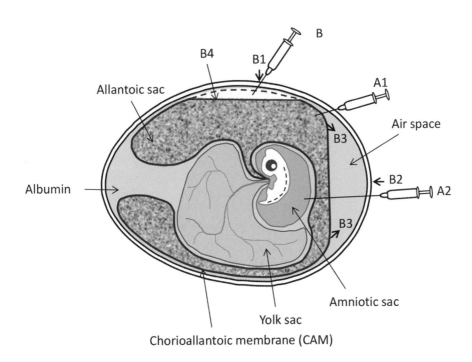

Figure 4.26 Egg inoculation routes using an embryonated chicken egg (at 9–11 days of incubation). (A1) Allantoic sac route (the air space remains intact). (A2) Amnion. (B) CAM route (this a bit more complicated). Holes are drilled through the shell at (B1) and (B2) and air is sucked out from the air space (B3) to allow separation of the layers of the CAM. Inoculation is made through (B1) onto the dropped CAM (B4), which extends as air is removed.

These tests are particularly useful for the diagnosis of viral infections (see also Chapter 6). Antibodies can also be used for the identification of an unknown virus in the laboratory (see Table 4.6). Some viruses will agglutinate erythrocytes, this is a characteristic that can also be used for virus characterization (for example, haemagglutination is used to identify sub-types of avian influenza virus).

The serological methods commonly employed in viral disease diagnosis include the following:

- agglutination and precipitation
- complement fixation (CFT)
- serum neutralization (SN) and virus neutralization (VN)
- inhibition of cytopathic effects (CPE) in tissue culture
- haemagglutination and haemagglutination-inhibition (HA and HI)
- enzyme-linked immunosorbent assay (ELISA)

Agglutination and precipitation

Serological reactions such as agglutination and precipitation are *in vitro* reactions that remain popular methods for the diagnosis of diseases and for identification of specific antigens and antibodies. The main difference between these two serological reactions pertains to the size of the antigens. In the case of precipitation, antigens are soluble molecules while in the case of agglutination; antigens are large, insoluble molecules (see also Chapter 6). Another difference between precipitation and agglutination is that the agglutination reaction is often more sensitive than the precipitation reaction because a lot of soluble antigens and antibody molecules are required to form visible precipitation. However, it is possible to make a precipitation reaction more sensitive by converting it into agglutination reaction. This can be achieved by attaching soluble antigens to large, inert carriers such as erythrocytes or latex beads.

A NOTE ON THE AGAR GEL IMMUNODIFFUSION TEST (AGID)

The basis for the AGID test (see Figures 6.8a–c) is the concurrent migration of antigen and antibodies towards each other through an agar gel matrix. The test is commonly used for the detection of antibody to avian influenza viruses, paramyxoviruses, haemorrhagic enteritis virus and equine infectious anaemia virus. When the antigen and the specific antibodies come in contact, they combine to form a precipitate in the gel matrix resulting in a visible line. The precipitin line forms where the concentration of antigen and antibodies is optimum. Differences in the relative concentration of the antigen or antibodies will shift the location of the line towards the well with the lowest concentration or result in the absence of a precipitin line. Electrolyte concentration, pH, temperature, and other variables also affect precipitate formation.

Complement fixation

The complement fixation test (CFT) is an immunological test that can be used to detect the presence of either specific antibody or specific antigen in serum. It was widely used to diagnose viral infections that are not easily detected by culture methods (see also Chapter 6). However, in clinical virology the CFT has been largely superseded by other serological methods such as the ELISA and by molecular methods of pathogen detection such as the polymerase chain reaction (PCR).

Neutralization tests

Neutralization tests are still frequently used in virology. There are two neutralization techniques commonly employed: SN and VN and these are often considered together. SN tests employ the use of a known virus to which unknown (test) serum is added (Figure 4.27). If antibodies for the virus are present in the serum, the virus will

Dilution factor	Replicate 1	Replicate 2	Replicate 3

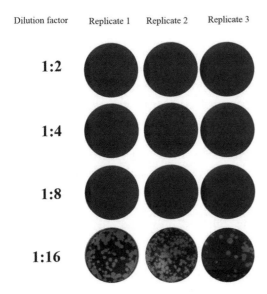

Figure 4.27 In SN assay, the unknown serum sample is two-fold serially diluted and titrated against a known quantity of virus. The serum blocks virus infection at the 1 : 2, 1 : 4 and 1 : 8 dilutions, but not at all at 1 : 16. Each serum dilution has been tested in triplicate, which allows for more accuracy. In this sample, the SN titre would be 8, the reciprocal of the last dilution at which infection was completely blocked. See also Plate 16. Source: M. Sarjoon Abdul-Cader, University of Calgary, Canada.

be neutralized or rendered non-infective when added to a 'test system'. The amount of virus, the amount of serum and the test system (that is, cytopathic effect in cell culture) must be modified to suit the conditions necessary for virus growth and to obtain satisfactory endpoints. In VN tests the principle is similar but a given range of sera are used with an unknown (test) virus. VN tests are highly specific and may also be used to serotype viral isolates.

A note on CPE inhibition

The neutralization test performed in cell culture is one of the most sensitive and accurate methods for detecting specific antibody or for identifying a virus isolate. For identification of a cytopathic virus, a known antibody (for example, anti-bovine viral diarrhoea virus (BVDV) antibody) is added to tissue culture media to see if it blocks the CPE of the subject virus (BVD). This test system can also be used to detect the presence and titre of specific antibodies, for example, a cytopathic virus (BVDV), is added at a known concentration and antibody is titrated into tissue culture wells to see if it can inhibit the CPE. If sufficient BVD antibodies are present the CPE will be inhibited. Details of how to perform the test are given below.

Neutralization test (Inhibition of CPE):

1 Determine the virus infectivity (titre) prior to the test (For a rapid identification one may select a dilution based on the rapidity with which it induces CPE).

2 Add equal volumes of a constant virus dilution containing approximately 100 $TCID_{50}$ per 0.1 ml to a known type-specific antiserum at a concentration of 20 units and mix well. (Note: one unit equals the highest serum dilution that neutralizes 100 $TCID_{50}$ of virus infectivity).

3 Allow the virus-serum mixture to remain at room temperature for 1 h; inoculate 0.2 ml of the virus-serum mixture into each of two to four culture tubes or wells if using a multi-well plate.

4 For the virus control, inoculate 0.2 ml of each serial 10-fold dilution into a set of culture tubes or wells, two to four tubes or wells per dilution.

5 Check all tubes or wells for CPE daily for 5–7 days.

6 Complete inhibition of CPE at a challenge dose of 100 $TCID_{50}$ by a known antiserum type is considered a positive SN test and indicates the identity of the virus. (It is always advisable to use serum pools, if available, for preliminary identification of an isolate. Final identification can then be confirmed by using type-specific antiserum).

Virus haemagglutination (HA) test

In 1941, Hirst observed that agglutination of chicken embryo erythrocytes occurred when they were mixed with amniotic and allantoic fluids, which contained high levels of influenza virus. Subsequent research has shown that many viruses will agglutinate chicken and some mammalian erythrocytes. Haemagglutination tests are available to test for the presence of a number of viruses (for example, influenza A and some paramyxoviruses) and can also be used to show the quantity of virus present in a given fluid. The minimum quantity of virus that will produce haemaggluntination can be quite accurately titrated by the HA test. In the performance of the HA titration, doubling dilutions of the virus-containing material are made in saline. After a suspension of washed erythrocytes is added, the tubes are shaken and then allowed to stand at room temperature until sedimentation of the cells occurs and the results can be read. If agglutination is present, a granular mat, often with curling at the edges, will be observed in the bottom of the tube. Absence of agglutination is shown by the cells settling to the very bottom of the tube as a rather compact round 'button' (see Figure 6.7a).

Haemagglutination-inhibition (HI) test

The discovery that certain viruses would agglutinate erythrocytes was promptly followed by the observation that antibodies to the virus could inhibit the reaction. HI tests have been a convenient method of detecting the presence of specific antibody in the serum of infected, or convalescent, individuals. Furthermore, by using serial dilutions of the serum, the comparative amount of antibody can be determined. The viral antigen used in this test must be titrated accurately by using the HA test previously described. More details are provided in Chapter 6.

ELISA

Enzyme-linked immunosorbent assay (ELISA), also known as an enzyme immunoassay (EIA), is a biochemical technique used to detect the presence of an antibody or an antigen in a sample. The ELISA is widely used as a diagnostic tool in veterinary and human medicine. ELISAs can be developed to detect either antigen or antibody. In simple terms, for a direct antibody ELISA, a known amount of antigen is affixed to a surface, and then serum is added over the surface so that any antibody present can bind to the antigen. A specific anti-species antibody linked to an enzyme is then added, and in the final step a substrate is added so that the enzyme can convert the substrate to some detectable signal, most commonly a colour change in a chemical substrate (see Figures 6.13a and b).

Performing an antigen ELISA involves at least one antibody with specificity for a particular antigen of interest. The sample with an unknown amount of antigen is immobilized on a solid support (usually a polystyrene microtitre plate *via* adsorption to the surface or *via* capture by another antibody specific to the same antigen as in a 'sandwich' ELISA). After the antigen is immobilized, the detection antibody is added, forming a complex with the antigen. The detection antibody can be covalently linked to an enzyme or can itself be detected by a secondary antibody that is linked to an enzyme. Between each step, the microtitre plate (or other surface) is typically washed with a mild detergent solution to remove any proteins or antibodies that are not specifically bound. After the final wash step, the plate is developed by adding an enzymatic substrate to produce a visible colour change, which can be measured using a spectrophotometer to determine the quantity of antigen in the sample (see also Chapter 6).

A note on prions

Unlike viruses, bacteria and parasites, prions are entirely proteins. The misfolding of cellular prion proteins (PrPc) into pathogenic prion proteins (PrPsc) and subsequent accumulation of PrPsc leads to spongiform diseases such as scrapie, in sheep, bovine spongiform encephalopathy (BSE) in cattle, and chronic wasting disease (CWD) in deer.

Anti-mortem diagnosis based on combination of neurological manifestations and real-time quaking-induced conversion (RT-QuIC) assay using various samples are being investigated, currently, the diagnosis of prion diseases is based on combination of neuropathology in the central nervous system and western blot assay targeting the detection of proteinase K resistant PrPsc (Figure 4.28) based on brain samples, particularly obex samples.

Benefits and limitations of rapid 'pen side' tests

Point of care or 'pen side' diagnostic tests need to be quick, simple to use and easy to interpret with little training. These tests must also be completely self-contained with no maintenance or calibration required. Rapid diagnostic tests which can detect pathogens in as little as five minutes are considered to be a good 'pen side' screening option to support more comprehensive diagnostic laboratory services. However, despite success in the laboratory, in the hot, humid areas where the test kits are likely to be used, questions remain as to how reliable these tests are. Few commercially available kits are validated for field use in the populations of interest and they are rarely developed to handle harsh climatic conditions. However, the benefit of using pen side tests (for example, DirectigenTM flu A+B, Figure 4.29) is that a rapid result is available and that the test can be performed with minimal equipment and with limited training. The limitations of these tests relate largely to the difficulty in determining the diagnostic sensitivity (DSe) and specificity (DSp) in the population of interest. Although most of the pen side kits available do provide data to support claims of high DSe and DSp, test performance should really be established for the specific population of interest. This can be done by comparing the results obtained with the kit test and those obtained using a 'gold standard' test such as classical culture or a traditional serological screening test.

In the laboratory setting there may be scope to develop and validate new 'in-house' diagnostic tests but this is rarely justified when there are a wide range of commercial kits readily available for the detection of antibodies (that is, serological test) or microorganisms (antigen detection test) in clinical samples (Table 4.6). Although

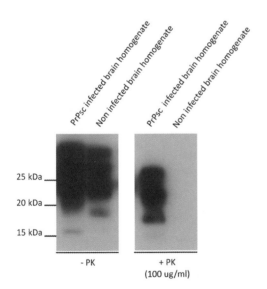

Figure 4.28 Western blot assay demonstrating pathogenic prion proteins, PrPsc, in brain homogenates. The brain homogenates (10%) originated from pathogenic prion infected and non-infected C57BL6 mice were either non-treated with proteinase K (-PK) or treated with proteinase K (+PK) for 1 h at 37°C and assayed. Photo: Sabine Gilch, University of Calgary, Canada.

Table 4.6 Examples of commercial 'kit tests' available for the diagnosis of viral infections in livestock (table compiled by Dr Regula Waeckerlin, Faculty of Veterinary Medicine, University of Calgary, Canada).

Ruminants

Disease	Species	Detection method	Component detected	Sample	Supplier
Bluetongue	Ruminants	AGID (BTV and EHD)	Antibodies	Serum	VMRD Inc. Pullman, WA, USA
	Ruminants	Competitive ELISA	Antibodies	Serum	
	Sheep, goats, cattle	Competitive ELISA, VP7 antigen	Antibodies	Serum, plasma	IDEXX Corp Ludwigsburg, Germany
	Cattle, sheep, goats	Double recognition ELISA, VP7 antigen	Antibodies	Serum	Prionics AG Schlieren – Zurich, Switzerland
	Cattle	ELISA VP7 antigen	Antibodies	Milk	Ingenasa SA Madrid, Spain
	All species	Multiplex real-time PCR All BTV serotypes, differentiation of BTV-8	RNA	Whole blood	Labordiagnostik Leipzig Leipzig, Germany
Bovine leukaemia	Cattle	ELISA	Antibodies	Serum	VMRD Inc. Pullman, WA, USA
	Cattle	AGID	Antibodies	Serum	IDEXX Corp Ludwigsburg, Germany
	Cattle	ELISA, screening Ultrapurified virus lysate antigen ELISA, X2 Antigen from improved cell line	Antibodies	Serum, plasma	
	Cattle	ELISA, Ultrapurified virus lysate antigen	Antibodies	Milk	
	Cattle	Blocking ELISA, gp51 antigen	Antibodies	Serum, single or pools	
	Cattle	Blocking ELISA, gp51 antigen	Antibodies	Serum, single or pools Milk	Ingenasa SA Madrid, Spain
BSE	Cattle	Immunohistochemistry	Prion Protein	Brain stem samples	VMRD Inc. Pullman, WA, USA

Ruminants

Disease	Species	Detection method	Component detected	Sample	Supplier
BSE / TSE / CWD	Cattle, small ruminants, cervids	Western Blot	Prion Protein	Brain stem samples	Prionics AG Schlieren – Zurich, Switzerland
	Cattle, small ruminants, cervids	ELISA	Prion Protein	Brain stem samples	
	Cattle, small ruminants, cervids	Strip test	Prion Protein	Brain stem samples	
Caprine arthritis-encephalitis	Goats	cELISA	Antibodies	Serum	VMRD Inc. Pullman, WA, USA
	Sheep	Indirect ELISA	Antibodies	Serum	ID.VET Montpellier, France
	Goat	AGID, p28 transmembrane protein antigen	Antibodies	Serum	IDEXX Corp Ludwigsburg, Germany
	Goat, sheep	ELISA Virus lysate antigen	Antibodies	Serum, plasma, milk	
	Goat	ELISA, p28 transmembrane protein antigen	Antibodies	Serum	
Bovine viral diarrhoea	Cattle	SNAP test	Antigen	Ear notch, serum	IDEXX Corp Ludwigsburg, Germany
	Cattle, cell culture	ELISA	Antigen	Leukocytes, nasal swab, cell culture	
	Cattle	ELISA	Antigen	Serum, plasma, milk, ear notch	
	Cattle	ELISA	Antigen	Ear notch	IDEXX, Portland, OR, USA
BVD and border disease	Cattle	Competitive ELISA	Antibodies	Serum, plasma, milk	IDEXX Corp Ludwigsburg, Germany
	Sheep			Serum	
BHV-1 / IBR	Cattle	ELISA	Antibodies	Serum, plasma, milk	Labordiagnostik Leipzig Leipzig, Germany

Table 4.6 *continued*

Ruminants

Disease	Species	Detection method	Component detected	Sample	Supplier
Foot-and mouth disease	Cattle, sheep	ELISA, 3ABC antigen (suitable for DIVA diagnostics)	Antibodies	Serum, plasma	IDEXX Corp Ludwigsburg, Germany
	Cattle, sheep, goats, pigs	Competitive ELISA Serotype 0	Antibodies	Serum	Prionics AG Schlieren – Zurich, Switzerland
	All species	Competitive ELISA NS protein	Antibodies	Serum	
Calf diarrhoea (rotavirus VP7, coronavirus S, *E. coli* K99)	Cattle	ELISA	Antigen	Faeces	IDEXX Corp Ludwigsburg, Germany
Rotavirus	Cattle, pigs, dogs	SNAP rapid antigen test Direct sandwich	Antigen	Diarrhea faeces	Labordiagnostik Leipzig Leipzig, Germany
Bovine respiratory complex (adenovirus, parainfluenza virus – 3, respiratory syncytial virus)	Cattle	ELISA	Antibodies	Serum	IDEXX Corp Ludwigsburg, Germany

Equine

Disease	What is being detected?	Detection method	Sample	Supplier
Equine infectious anaemia	Antibodies	AGID	Serum	IDEXX Corp Ludwigsburg, Germany
	Antibodies	Competitive ELISA	Serum	
	Antibodies	Double antigen ELISA Anti p26	Serum	ID.VET Montpellier, France

Equine

Disease	What is being detected?	Detection method	Sample	Supplier
Influenza A	Antibodies	Blocking ELISA	Serum	IDEXX Corp Ludwigsburg, Germany
				ID.VET Montpellier, France
	Antigen	ELISA	Broncho-pulmonary lavage fluid	ID.VET Montpellier, France
West Nile virus	IgM Antibodies	ELISA	Serum, cerebrospinal fluid	IDEXX Corp Ludwigsburg, Germany
	Antibodies	Competitive ELISA E-glycoporotein antigen	Serum	ID.VET Montpellier, France
				Ingenasa SA Madrid, Spain
African horse sickness	Antibodies	ELISA VP7 antigen	Serum	Ingenasa SA Madrid, Spain
	Antigen	ELISA	Spleen	
Equine arteritis	Antibodies	ELISA	Serum	Ingenasa SA Madrid, Spain
				ID.VET Montpellier, France

Pigs

Disease	Component detected	Detection method	Sample	Supplier
Classical swine fever	Antibodies Vaccine and field strains	Blocking ELISA Envelope protein E2 (A domain)	Serum	Prionics AG Schlieren – Zurich, Switzerland
	Antigen	ELISA	Whole blood, plasma, serum, tissue	IDEXX Corp Ludwigsburg, Germany
	Antibodies DIVA diagnostics for E2 subunit vaccine	ELISA E^{rns} antigen Double mAb assay	Serum	

Table 4.6 *continued*

Pigs				
Disease	**Component detected**	**Detection method**	**Sample**	**Supplier**
African swine fever	Antibodies	Blocking ELISA VP73 protein antigen	Serum	Ingenasa SA Madrid, Spain
	Antigen	ELISA Double antibody sandwich for VP73 antigen	Spleen samples	
Aujeszky / pseudorabies (suid herpesvirus – 1)	Antibodies, vaccine and field strains	Blocking ELISA Glycoprotein B	Serum	Prionics AG Schlieren – Zurich, Switzerland
	Antibodies, For DIVA diagnostics	Blocking ELISA Glycoprotein E	Serum	
				IDEXX Corp Ludwigsburg. Germany
				ID.VET Montpellier, France
Influenza	Antibodies Influenza A	Blocking ELISA	Serum	IDEXX Corp Ludwigsburg. Germany
	Antibodies H1N1	ELISA	Serum	
	Antibodies H3N2	ELISA	Serum	
	Antibodies Influenza A	Blocking ELISA	Serum	Ingenasa SA Madrid, Spain
	Antibodies H1N1, H1N2, H3N2	ELISA	Serum	
	Antibodies	Competitive ELISA Also for birds and horses	Serum	ID.VET Montpellier, France
	Antigen	Capture ELISA	Broncho-pulmonary lavage fluid	
Hepatitis E	Antibodies	ELISA Multi-species conjugate	Serum	ID.VET Montpellier, France
Porcine parvovirus	Antibodies	Blocking ELISA	Serum	Prionics AG Schlieren – Zurich, Switzerland

Pigs

Disease	Component detected	Detection method	Sample	Supplier
Transmissible gastroenteritis / porcine respiratory coronavirus	Antibodies	Blocking ELISA Coronavirus S protein	Serum	Ingenasa SA Madrid, Spain
	Antigen	SNAP rapid antigen test Direct sandwich	Diarrhoea faeces	BioNote Gyeonggi-do, Korea

Poultry

Disease	Species	What is being detected?	Detection method	Sample	Supplier
Avian encephalomyelitis	Chicken	Antibodies	ELISA	Serum	IDEXX Corp Ludwigsburg, Germany
Avian influenza	Multiple species	Antibodies	Blocking ELISA	Serum	IDEXX Corp Ludwigsburg, Germany

ID.VET Montpellier, France |
	Birds	Antibodies	Competitive ELISA N1 antigen	Serum	ID.VET Montpellier, France
	Birds	Antibodies	Competitive ELISA H5 antigen	Serum	
	Birds	Antibodies	Competitive ELISA H7 antigen	Serum	
	Birds	Antigen	Capture ELISA	Bronchopulmonary lavage fluid Cloacal / tracheal swab, faeces	
	Birds	Antigen	SNAP rapid antigen test Direct sandwich All Influenza A viruses	Cloacal swab, faeces	BioNote Gyeonggi-do, Korea
	Birds	Antigen	SNAP rapid antigen test Direct sandwich AIV H5	Cloacal swab, faeces	

Table 4.6 *continued*

Poultry

Disease	Species	What is being detected?	Detection method	Sample	Supplier
Avian leukosis	Chicken	Antibodies	ELISA Subtypes A, B and J	Serum	IDEXX Corp Ludwigsburg, Germany
	Birds	Antigen	Capture ELISA p27 antigen All subtypes	Albumin, cloacal swabs	
Avian pneumovirus	Chicken, turkey	Antibodies	ELISA Serotypes A, B and C	Serum	
Avian reovirus	Chicken	Antibodies	ELISA	Serum	
Chicken anaemia virus	Chicken	Antibodies	Blocking ELISA	Serum	
Infectious bronchitis	Chicken	Antibodies	ELISA	Serum	
	Chicken	Antibodies	ELISA Recombinant Nucleocapsid-protein antigen	Serum	Labordiagnostik Leipzig Leipzig, Germany
Infectious bursitis (Gumboro)	Chicken	Antibodies	ELISA	Serum	IDEXX Corp Ludwigsburg, Germany
	Chicken	Antibodies	ELISA Recombinant structural protein	Serum	Labordiagnostik Leipzig Leipzig, Germany
	Birds	Antigen	SNAP repid antigen test Direct sandwich	Cloacal / bursal swab	BioNote Gyeonggi-do, Korea
Newcastle disease	Chicken, turkey	Antibodies	ELISA	Serum	IDEXX Corp Ludwigsburg, Germany
	Chicken, turkey	Antibodies	ELISA Recombinant Nucleocapsid protein antigen	Serum	Labordiagnostik Leipzig Leipzig, Germany
	Birds	Antigen	SNAP rapid antigen test Direct sandwich All strains	Cloacal / tracheal swab	BioNote Gyeonggi-do, Korea
Avian reticuloendotheliosis virus	Chicken	Antibodies	ELISA	Serum	IDEXX Corp Ludwigsburg, Germany

Poultry

Disease	Species	What is being detected?	Detection method	Sample	Supplier
West Nile virus	Birds	Antibodies	Blocking ELISA E-protein antigen	Serum	ID.VET Montpellier, France

Domestic carnivores

Disease	Species	What is being detected?	Detection method	Sample	Supplier
Canine parvovirus	Dog	Antigen	SNAP rapid antigen test Direct sandwich	Blood, faeces	IDEXX Corp Ludwigsburg, Germany BioNote Gyeonggi-do, Korea
Canine distemper	Dog	Antigen	SNAP rapid antigen test Direct sandwich	Urine, blood, serum	BioNote Gyeonggi-do, Korea
Canine distemper, canine parvovirus	Dog	Antibodies	SNAP rapid antigen test Direct sandwich	Urine, serum, blood	BioNote Gyeonggi-do, Korea
Rabies	Dog, cattle, raccoon	Antigen	SNAP rapid antigen test Direct sandwich	Saliva, brain homogenate	BioNote Gyeonggi-do, Korea

Note: For the most up to date guidelines on testing for specific diseases in livestock species check the online edition of the OIE Manual of Diagnostic Tests and Vaccines for Terrestrial Animals http://www.oie.int/standard-setting/terrestrial-manual/access-online/.

technology and testing platforms continue to evolve there is likely to remain a need for traditional microbiology methods especially with respect to identifying and characterizing new and re-emerging pathogens.

The OIE Manual of Diagnostic Tests and Vaccines for Terrestrial Animals is a compilation of diagnostic procedures and a ready source of information for any veterinary diagnostic laboratory. The Manual has been designed for practical use in the laboratory setting and contains over 110 chapters on infectious and parasitic diseases of mammals, birds and bees. http://www.oie.int/en/standard-setting/terrestrial-manual/access-online/.

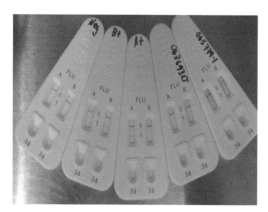

Figure 4.29 Influenza A and B viruses can be differentiated based on rapid antigen ELISA test conducted on a membrane. The figure shows five Directigen™ Flu A+B test membranes. Nasopharyngeal swabs can be screened using this antigen detection test. This is a rapid (takes about 15 min) and qualitative test. Negative results in suspect clinical cases should be confirmed using other means. In positive samples (that is, those containing antigen), the antigen-antibody reaction is determined by visual colour development, that is, samples 1, 4 and 5 are negative for both influenza A and B antigens. Samples 2 and 3 are positive for influenza B virus and influenza A virus respectively. Photo: Davor Ojkic, Animal Health Laboratory, University of Guelph, Canada.

4.8 Molecular microbiology and its application as a diagnostic tool

Julie Collins-Emerson and M. Faizal Abdul-Careem

In the following section, some of the current molecular techniques used in diagnostic microbiology are discussed. Many of these techniques are applicable for the detection of a range of potential pathogens including bacteria, viruses, fungi and multi-cellular parasites, however, the main focus here is on bacteria and viruses.

Introduction

Traditional approaches to clinical microbiology revolve around the detection and isolation of microorganisms and the analysis of phenotypic characteristics such as, bacterial colony morphology, viral cytopathic effects, serology, biochemical and drug resistance markers. These characteristics are used to identify species and strains and to inform treatment and epidemiological investigations. These methods however do have some drawbacks. Many organisms are either non-culturable, low in number or too difficult or slow to grow. In addition, cultures are subject to contamination and the sample type submitted to the laboratory may carry components that are inhibitory in culture media. Preserving the quality (including viability) of field-derived specimens, particularly when transport over long distances is required, can also pose problems that impact the successful isolation and culture of microorganisms for further characterization. In addition, some microorganisms are highly dangerous pathogens and culture requires specialized facilities with high levels of biosecurity and biosafety. These factors can make conventional microbiology both time-consuming and expensive and may require considerable technical expertise.

In more recent times, the rapidly evolving field of molecular biology (Figure 4.30) has significantly broadened the repertoire of diagnostic tools available to laboratories. In particular, the speed, sensitivity and specificity of many of these techniques have helped circumvent the difficulties and limitations presented by some of the traditional phenotypic methods. While the initial costs for establishing some molecular techniques in the laboratories can be considerable, the routine running costs and the high throughput possibilities for some processes can make the techniques economical. The market place changes rapidly with increasing numbers of commercially produced, molecular based

Figure 4.30 PCR work stations are used for the prevention of product carry over or cross contamination. It is vital to physically separate pre- and post-PCR steps and that could be done using dedicated PCR work stations and pipettes and frequent change of gloves preventing potential cross contamination. Photo: Dr Davor Ojkic, Animal Health Laboratory, University of Guelph, Canada.

diagnostic kits available. Although frequently expensive, many kits have the advantage of being produced under quality assured conditions offering the user a degree of confidence around repeatability and validation. Also, many kits have been designed to be 'user-friendly' and therefore often do not require highly specialized technical training for use. In addition, it should be noted that diagnostic labs may also provide 'in-house' molecular tests for organisms of interest in their region.

Techniques

Molecular techniques fall into the following general categories: hybridization, DNA restriction, amplification and sequencing. A brief overview of these techniques and some of their applications are discussed below.

Hybridization

DNA : DNA/DNA : RNA

Hybridization relies on the direct detection of nucleic acids. When using whole genomic DNA, the degree of relatedness between organisms can be established. DNA from one organism is annealed to a solid support (for example, nitrocellulose), denatured and mixed with labelled, single-stranded DNA from another organism and allowed to form hybrid, double stranded DNA. Closely related organisms will have a high degree of sequence similarity and consequently will re-anneal more tightly, requiring a higher energy state (melting point) to separate than that of DNA from less closely related organisms. Alternatively, labelled DNA can be used to hybridize to bound RNA. While this method for whole genome comparison has been used to identify or separate different strains of organisms, it requires culturing the organisms in order to obtain sufficient volumes of DNA for the method. It was useful for the identification of microorganisms and for epidemiological studies in the past, but it is a technique that has generally been superseded by amplification and sequencing approaches.

PROBES AND MICROARRAYS

Probes (RNA, DNA or peptide nucleic acids) are used to hybridize to specific sequences of DNA or RNA. Previously, probes tended to be derived and selected from cloned libraries of the organism's genome though now, with large amounts of sequence data available for numerous organisms on internet databases (for example, GenBank), probes are more commonly synthetically manufactured in a similar manner to primers. Probes have been developed for application on a variety of tissue specimens, for example, blood or blood culture, body fluid swabs, cultures, urine, CSF fluid and chromosomal preparations. Detection can be achieved a number of ways: radioactive probes, non-radioactive probes (for example, biotin labelled), fluorescent (for example,

Fluorescence *In Situ* Hybridization [FISH]) and, branched DNA technology (bDNA) that uses an enzyme-labelled probe plus a chemiluminescence-based reaction for signalling).

In virology, the technique can be employed to detect DNA or RNA viruses (Table 4.7a and 4.7b) and their transcripts directly in the affected tissue using DNA or RNA probes. Both frozen and formalized tissue sections can be used and, depending on the technique utilized, the laboratory visualizes the probes using either autoradiography, immunohistochemistry or fluorescent microscopy. The relative insensitivity of the direct *in situ* hybridization assay can be improved by prior amplification of the target viral sequences in the tissue employing *in situ* polymerase chain reaction (PCR) or *in situ* real-time-PCR. This technique provides many useful features for viral disease diagnosis; the main advantage being the ability of the assay to demonstrate the viral pathogens within the lesion. Additionally, the ability of the assay to detect DNA or RNA viruses and their replication stage can be used to demonstrate active infection.

Technological advances now make it possible to assemble thousands of spot probes onto solid surfaces (glass or silicon chip) called microarrays using covalent bonding. This technology has many uses and some of those applicable to the clinical and epidemiological settings include; assessing gene expression, overall genetic relatedness between organisms, single nucleotide polymorphisms (SNPs), specific gene detection to identify particular organisms and antibiotic resistance genes.

For viral disease diagnosis, the microarray probes used may be selected to represent the nucleotide sequences of all the viruses, or a group of viruses, that result in similar clinical and pathological manifestations in an animal species. The probes should be designed from conserved regions of the viral genome such that the probes detect all the viruses in a given genus or species and do not hybridize with sequences of other viral genera or species. These fixed probes are hybridized by target complementary viral genome sequences in the nucleic acid samples (that are labelled *in vitro*) extracted from clinical material. A microarray reader (Figure 4.31a and b) is used to record the resultant fluorescent intensity which is dependent on the amount of hybridized target genome sequence on the

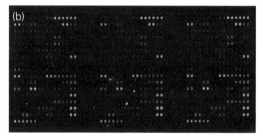

Figure 4.31 (a) A microarray reader is necessary to scan the amount of fluorescent labelled probes hybridize with the target nucleic acid in the sample. The fluorescent labelled spots that binds the target nucleic acid due to hybridization are excited by a laser and scanned at suitable wavelengths to detect the fluorescent dye. The read fluorescence intensity corresponds to the amount of bound nucleic acid. (b) Scanned array image showing positive (yellow) and negative results (black). Each spot shown in the array corresponds to a specific fluorescent value, determined by the strength of hybridization, in a semi-quantitative manner. See also Plate 17. Photos: Dr Shayan Sharif and Dr Jennifer Brisbin, University of Guelph, Canada.

Table 4.7a The classification of some DNA viruses of veterinary importance.

Virus family	Cattle	Sheep/goats	Pigs	Horses	Dogs	Cats	Avian species
Parvoviridae			Porcine parvovirus (SMEDI)		Canine parvovirus	Feline parvovirus (panleuko paenia)	
Papovaviridoae	Bovine papilloma virus		Equine papilloma virus		Canine oral papilloma virus		
Adenoviridae				Equine adenovirus	Canine adenovirus – (infectious canine hepatitis, tracheobronchitis)		Egg drop syndrome
Hepadnaviridae							Duck hepatitis B virus
Herpesviridae	Bovine herpes virus; Malignant catarrhal fever virus; Aujeszky's disease (pseudorabies virus)	Malignant catarrhal fever (bovine herpes virus 2); Aujeszky's disease (pseudorabies virus)	Porcine herpes virus 1 (Aujeszky's disease); Porcine herpes virus 2 (inclusion body rhinitis)	Equine herpes viruses (i.e. rhinopneumonitis, coital exanthema)	Canine herpes virus; Aujeszky's disease (pseudorabies virus)	Feline herpes virus (rhino tracheitis); Aujeszky's disease (pseudorabies virus)	Infectious laryngo tracheitis; Marek's disease virus; Duck herpes virus
Poxviridae	Bovine popular stomatitis virus; Pseudocowpox virus; Lumpy skin disease virus	Orf virus; Sheep pox virus; Goat pox virus	Swine pox virus				Fowl pox virus; Pigeon pox virus; Canary pox virus

Table 4.7b The classification of some RNA viruses of veterinary importance.

Virus family	Cattle	Sheep/goats	Horses	Pigs	Dogs/cats	Avian species
Picornaviridae	Foot & mouth disease virus, Bovine rhinovirus, Bovine enterovirus	Foot & mouth disease virus	Equine rhinovirus	Foot & mouth disease virus, Swine vesicular disease virus, Porcine enteroviruses (e.g. SMEDI)		Avian encephalomyelitis virus (Epidemic tremor)
Caliciviridae				Vesicular exanthema virus	Feline calicivirus	
Reoviridae	Bluetongue virus, Bovine rotavirus	Bluetongue virus, Ovine rotavirus	African horse sickness virus, Equine rotavirus		Canine/feline rotavirus	Avian reovirus
Birnaviridae						Infectious bursal disease virus (Gumboro)
Flaviviridae	Bovine viral diarrhoea virus	Louping ill virus, Border disease virus	West Nile virus	Japanese encephalitis virus, Swine fever virus		
Togoviridae			Equine arteritis virus, Equine encephalomyelitis virus (Eastern, Western and Venezuelan)			
Orthomyxoviridae			Influenza A virus (some subtypes can infect many species)	Swine influenza virus	Influenza A virus (some subtypes can infect many species)	Avian influenza/influenza A viruses – some strains highly pathogenic (i.e. H5 and H7 subtypes) but not to all species

Paramyxoviridae	Bovine parainfluenza virus, Bovine resp. Syncytial virus, Rinderpest (now eradicated)	Ovine parainfluenza virus, Bovine resp. Syncytial virus, Peste-des-petits ruminants virus			Canine distemper virus, Canine parainfluenza virus	Avian paramyxoviruses (some pathogens e.g Newcastle disease, other strains less pathogenic)
Coronaviridae	Bovine coronavirus		Transmissible gastroenteritis virus		Canine coronavirus, Feline infectious peritonitis virus	Infectious bronchitis virus
Rhaboviridae	Rabies virus, Vesicular stomatitis	Rabies virus	Rabies virus	Rabies virus, Vesicular stomatitis virus	Rabies virus	
Bunyaviridae	Rift valley fever, Akabane virus	Rift valley fever, Akabane virus, Nairobi sheep disease virus				
Retroviridae	Bovine leukaemia virus	Maedi-visna virus, Caprine arthritis encephalitis		Equine infectious anaemia virus	Feline leukaemia virus, Feline immunodeficiency virus	Avian leukosis/ sarcoma viruses

known spot. The obtained data are semi-quantitative and may be validated using the real-time PCR technique. Commercial diagnostic microarrays is a developing field. In the near future, it is expected that this technique may become more popular in diagnostic laboratories due to its ability to screen for multiple microbial agents in a single assay.

DNA restriction

PULSED FIELD GEL ELECTROPHORESIS (PFGE)
Pulsed field gel electrophoresis is a method that uses enzymes that cut at infrequently found sites (rare-cutting restriction enzymes) to cleave whole genomic DNA into large fragments that are then separated by size on an electrophoresis gel. SNPs in any one of these rare restriction sites will lead to an altered banding pattern on the electrophoresis gel and have been used to identify strain types for some organisms, for example, *Escherichia coli* O157:H7, *Salmonella*, *Shigella*, *Listeria*, or *Campylobacter*. Differences between operator handling and in the equipment used can lead to inter-laboratory variations. In order to facilitate comparisons between strains from laboratories around the world, protocol standardization and laboratory certification processes have been established (for example, PulseNet). The patterns obtained are then photographed, standardized and the data stored digitally which enables ready comparison to strains in other laboratories.

AMPLIFICATION
DNA for PCR must be free of the enzymatic inhibitors that can be present in growth media or in specimen samples, for example, stool specimens in particular. There are many DNA preparation methods, some as simple as a rapid boil preparation in water, to sophisticated commercially available DNA kits, some of which are for use on direct tissue samples. However, some methods are more successful than others, depending on the sample type. Any DNA extraction and purifying process is likely to experience some loss in total DNA from the original preparation and this needs to be taken into consideration.

Amplification relies on the use of primers (short segments of site-specific oligonucleotides) to anneal to opposite strands of the target DNA and by utilizing PCR for synthesizing and amplifying the targeted sequence many fold. Primers need to be specific to prevent cross-reactivity with non-targeted sequences. Primer design programmes are available on the internet and custom-made primers are manufactured commercially to order. The PCR process basically involves heat-mediated denaturation of DNA to a single-stranded state, followed by annealing of the primers to the target sequence. Annealing temperature should be such that it is stringent enough to prevent non-specific binding. This step is followed by extension (DNA synthesis step) at a higher temperature. The process is repeated around 35 times, theoretically doubling the amplified fragment (amplicon) at every cycle resulting in over 10^9 copies. Target based amplification is therefore generally much more sensitive than probe based methods for detecting organisms in clinical specimens.

There are a number of factors that need to be considered when using PCR based detection. Mutations in the primer binding site resulting in reduced sensitivity or non-amplification and possible amplification of closely related sequences resulting in false positives are such examples. Given the underlying principles of PCR, one of the greatest problems to manage is preventing amplification of non-specific DNA, either in the target organism or through contamination of the reaction. It is therefore very important that PCR protocols be thoroughly validated in order to estimate the specificity and sensitivity of any test and that positive and negative controls be included in each experimental run.

Following PCR, the resulting amplicon can be checked a number of ways to determine whether it is the expected targeted sequence and the method will depend on what type of amplification has been used. However, results need to be assessed in context along with other diagnostic evidence as, just because DNA from an organism has been successfully amplified, it does not automatically follow that this organism is the cause of the disease, for example, it could be a case of mixed infection.

CONVENTIONAL PCR

Conventional PCR uses a pair of primers designed to anneal to the opposite strands of DNA at either end of the targeted sequence and the generation of multiple copies of this region by repeated synthesis cycles in a thermocycler (Figure 4.32). The products of all these reactions are visualized and the fragment size estimated by running on electrophoresis gels. The amounts of DNA product can be estimated against standards on gels, but it is, at best, only semi-quantitative. Alternatively, the amplicons can be analysed by sequencing them which then allows comparison/matching with sequences in databases. There are a number of variations on this basic technique.

- Nested PCR, where specificity is increased by using a second set of internally positioned primers (or partially internal as is the case with hemi-nested PCR) to amplify a smaller section of the original amplicon.
- Multiplex PCR, where multiple primer sets are included in a reaction to simultaneously identify a number of targeted regions, for example, multiple antibiotic gene markers in an isolate.
- Random amplification of polymorphic DNA (RAPD). PCR reactions are conducted with whole genome DNA and short random primers. The resulting pattern is used to characterize strains and can be useful for

organisms for which there is currently not much sequence data.
- Arbitrarily primed PCR (AP-PCR) is a related technique.
- Asymmetric PCR, which is used primarily to produce multiple copies of a single DNA strand.
- Reverse transcriptase PCR, where reverse transcriptase is used to produce copy DNA (cDNA) from RNA (and hence, can also be used for RNA virus detection). The cDNA then acts as a template for conventional PCR.
- Combinations of PCR plus enzyme restriction are also used, for example, amplified fragment length polymorphism (AFLP).

There are some limitations to conventional PCR when compared to the real-time PCR including; (1) it is not quantitative and only a detection method, (2) it is less sensitive than real-time PCR, (3) time consuming and, (4) prone to contamination and brings health hazards because of post-PCR processing. In fact, non-quantitative conventional PCR techniques are not recommended for viral disease diagnosis in situations where viruses that are found in healthy animals can cause clinical disease. For these reasons, many laboratories are increasingly applying real-time PCR for diagnostic purposes.

REAL-TIME PCR (qPCR)

Real-time PCR assays use either fluorescent reporter probes (of which there are different types) that bind to the target sequence or, DNA dyes that intercalate with double stranded DNA. Both detection methods rely on changes in the fluorescent signal to measure double stranded DNA. Therefore, one of the main benefits of real-time PCR is the direct quantification of PCR product (hence the abbreviation, qPCR) by fluorescence and post-PCR processing is not required (Figure 4.33). This technique can also be employed for quantification of target RNA by using reverse transcriptase to produce cDNA and

Figure 4.32 (a) Thermocycler used in conventional PCR technique. This equipment provides a cycling temperature for amplification of nucleic acid and the end product detection is done by gel electrophoresis. (b) Loading of PCR products onto a agarose gel for electrophoresis. (c) Gel electrophoresis of conventional PCR products. Template was amplified with three pairs of porcine circovirus capsid gene specific primers and the samples were run in a 2% agarose gel. Lane 1: DNA ladder; lanes 2–4 pig number 1 three different primers, lane 5–7 pig number 2 three different primers, lane 8 positive control only one primer and lane 9 no template control. Photos: (a) Dr Davor Ojkic, Animal Health Laboratory, University of Guelph, Canada; (b) Dr Markus Czub, University of Calgary, Canada.

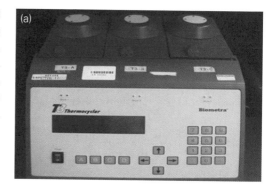

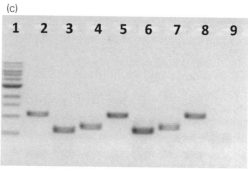

therefore, can be used for RNA viruses too. The change in fluorescent signal as double stranded DNA dissociates with increased temperature, may also be coupled with real-time technology in a process called high resolution melting analysis (HRM). This is very sensitive and can be utilized for analysis of SNPs in the targeted sequences. Real-time PCR therefore has applications in detecting small quantities of DNA or RNA in specimen samples, can be used for identification of biomarkers such as antibiotic resistance genes or SNPs for strain identification or epidemiological studies and, because of its good quantitative capabilities, be used to monitor disease progress or treatment, for example, tuberculosis.

In virology, real-time PCR methods based on different detection systems have been developed for the quantification of the viral genome in different tissues. These methods are dependent on TaqMan/hybridization probes or using SYBR green chemistry and determine the viral genome loads based on either relative increase of viral genomes and gene transcription levels, but not virion replication *in vitro*, or absolute quantification. Both real-time PCR methods are equally efficient in quantifying viruses. The

SYBR green-based, real-time PCR assay does not require probes and, therefore, is economical and easily adoptable from the conventional PCR system. Further, the SYBR green-based, real-time PCR assay detects the target PCR product accumulation independent of the sequence; as such, it allows the quantification of the viral genome with minor variations in the sequence. This may not be possible with probes as they operate in a highly target sequence-specific manner and small changes in target sequence may abrogate probe binding leading to the generation of false

(a) A modern real-time PCR machine

(b) Real-time PCR amplification curves

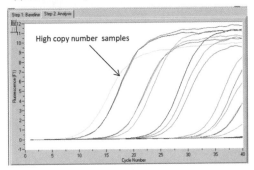

(c) Real-time PCR melting peaks

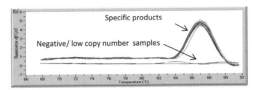

Figure 4.33 Real-time PCR amplification curves and the melting peaks. (A) A modern PCR machine capable of running 384 PCR reaction in a single plate. (B) Representative amplification curves for the quantification of viral DNA by real-time PCR based on SYBR Green chemistry. The DNA sample was serially 10-fold diluted and each dilution has assayed in triplicate. High DNA copy number samples (lower dilution show amplification curves early during the amplification and the low copy number or negative and no template control samples (higher dilution) show late or no amplification. (C) Melting peak analysis shows the specific amplification. Depending on the sensitivity of the assay low DNA copy number samples yield no peaks as there is no template control. See also Plate 18. Photos: (a) Dr Davor Ojkic, Animal Health Laborartory, University of Guelph, Canada; (b and c) Dr M. Faizal Abdul-Careem, University of Calgary, Canada.

negative results. However, sequence-independent detection in SYBR green-based assay is expected to have low specificity. The specificity of SYBR green-based assays can be improved to a level comparable to probes by optimizing the real-time PCR conditions, coupled with melting peak analysis (Figure 4.33c). Additionally, in the real-time PCR method, specific binding of the dye is controlled using a heat activated fast start Taq DNA polymerase which prevents non-specific binding of the dye before the specific reaction has started. Fast start Taq DNA polymerase is inactive at room temperature and activated at the pre-incubation step of the reaction (95°C for 60 s).

BINARY TYPING USING PCR

With PCR binary typing (P-BIT), a panel of genes is tested by PCR amplification in a group of isolates resulting in a simple yes, present/no, not present result that can be expressed in a binary fashion as a string of digits that identifies the isolate type. This digitalization of data can be stored on databases for ready comparison of isolates between laboratories. This can be a relatively cheap and discriminatory process for use in sub-typing.

ISOTHERMAL AMPLIFICATION

Isothermal amplification is a non-PCR based molecular technique that requires inexpensive laboratory equipment such as a heating block to maintain constant temperature (Figure 4.34a). Two main assay types of isothermal amplification have been developed; loop-mediated isothermal amplification (LAMP) and, nucleic acid sequence-based amplification (NASBA). Unlike PCR, the strand displacement step of genome amplification in both formats is accomplished

(a) Heating block

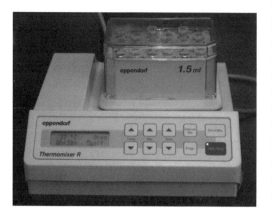

(b) LAMP isothermal amplification products

Visual detection of magnesium pyrophosphate

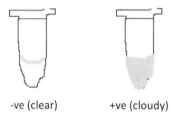

-ve (clear)   +ve (cloudy)

Visual detection using SYBR green

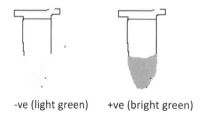

-ve (light green)   +ve (bright green)

Figure 4.34 (A) Isothermal amplification requires only a simple heating block to maintain constant temperature up to an hour. Photo: Dr Regula Waeckerlin, University of Calgary, Canada. (B) LAMP isothermal amplification products can be visualized using naked eye due to the accumulation of PCR by product magnesium pyrophospahte (cloudy) or colour change using SYBR green. See also Plate 19.

by enzymes so that complex temperature cycling is not required; only a constant temperature for nucleic acid amplification is required. The LAMP technique was originally developed by Eiken Chemical Co. (Tokyo, Japan) and can be employed for the amplification of the genome of DNA viruses using four primers recognizing six sequences of the DNA template at 61°C. The NASBA technique relies on two primers and three enzymes employed for the amplification of target nucleic acid of RNA viruses at a constant temperature of 41°C. Both versions of isothermal amplification result in approximately 10^9–10^{10} copies of the target genome within 1 h of assay time.

There are number of endpoint and real-time monitoring systems available for the detection of amplicon produced by LAMP technique. The optimized endpoint detection systems for the detection of LAMP products include visualization of the amplification product using the naked eye (Figure 4.34b), turbidometry or fluorimetry. The more sensitive real-time monitoring system using fluorochrome or turbidity also has been developed. Detection of amplicons in the NASBA technique is based on endpoint estimation of products *via* hybridization analysis using an electrochemiluminescent detection system or real-time monitoring using fluorophore-tagged oligonucleotide probes.

This molecular technique of viral genome amplification is increasingly being recognized as a diagnostic tool particularly in resource-poor situations or field situations due to (1) simplicity, (2) speed, (3) specificity and sensitivity and (4) cost effectiveness. Once total nucleic acids are extracted from the sample, either assay requires only simple equipment which is, a heating block to maintain constant temperature at 61°C or 41°C, primers and/or enzymes. The main disadvantage of the technique is the extra effort required for the design of the four to six primers needed for developing one assay. Multiple primers are necessary for increasing sensitivity and a web tool is available for designing the primers for LAMP.

Sequencing

Nucleotide sequencing reveals the sequential arrangement of four nucleotides (adenine, guanine, cytosine and thiamine) in a stretch of genome (Figure 4.35). Sequencing is based on identifying genetic variations in the nucleotide sequence of variable genome regions and is therefore a highly discriminatory technique. For example, it is routinely used in diagnostic laboratories for differentiating highly pathogenic avian influenza viral pathotypes from low pathogenic avian influenza pathotypes. In this particular case, sequencing targets the region of cleavage site of the haemagglutinin (HA)

(a) Nucleotide sequencing equipment

(b) Sequencing output showing arrangement of nucleotides

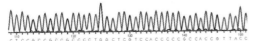

Figure 4.35 (a) Modern sequencing equipment used for nucleotide sequencing. Photo: Dr Davor Ojkic, Animal Health Laboratory, University of Guelph, Canada. (b) A representative of sequencing output. The blue, red, black and green peaks represent cytosine, thiamine, adenine, and guanine, respectively. See also Plate 20. Photo: Dr Rikia Dardari, University of Calgary, Canada.

gene of the influenza viral genome. In addition, nucleotide sequencing can be used for tracking the spread of infections based on constructing a phylogenetic tree and this is an important application since finding the source of infection complements the control measures preventing future disease outbreaks. Identifying genetic variation in immunodominant epitopes using sequencing technique also may explain vaccine failures and facilitate use of appropriate vaccines for the control measures (Figure 4.36).

Methodology can vary from the analysis of sequences from small regions of targeted genes to comparing large segments or entire genomes. Some examples of the different approaches are given below.

MULTI-LOCUS SEQUENCE TYPING (MLST)

MLST is a technique introduced by Maiden et al. (1998) that uses sequence data from a number of genes to identify isolates by their sequence type (ST). There are genes that are considered essential for the survival of the organism and therefore are present in all isolates and are strongly conserved. These are referred to as housekeeping genes. MLST uses a selection of these housekeeping genes that are distributed around the genome for identification purposes. For example, in *Campylobacter jejuni*, conserved sequences within seven housekeeping genes are amplified by PCR, sequenced and the sequence data analysed. Each different ST for a given gene (that is, allelic type) in the scheme is given a unique number to identify it. The seven genes used in this scheme are then concatenated to form an allelic profile comprising seven numbers. This combination of seven allelic numbers denotes a unique strain type and has proven to be quite robust in that allelic profile strongly reflects the genetic relatedness in the case of *Campylobacter* isolates and STs that differ by one allele can be grouped in clonal complexes (CCs) and are useful groupings for phylogenetic analysis of populations.

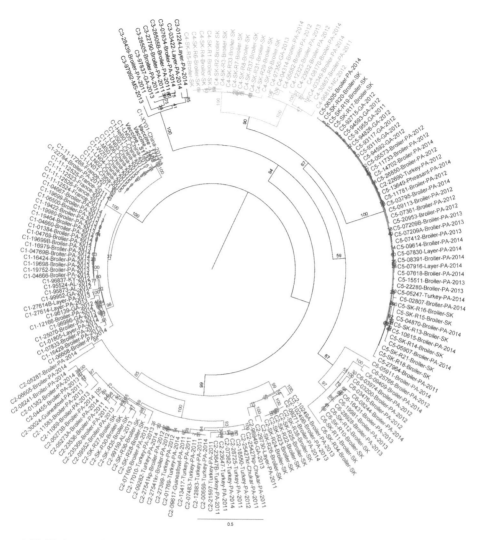

Figure 4.36 Phylogenetic Tree of a total of over 180 partial σC gene sequences of avian reovirus. The analysis included sequences obtained from the GenBank. Vaccine strains are shown in a bold red colour. The tree shows several clusters of avian reovirus and most of the sequences are far from the sequences of the avian reovirus vaccines. This raises the question whether vaccine induced immunity is protective against the disease induced by these diverse avian reo viral strains. See also Plate 21. Source: Victor Palomino-Tapia, University of Calgary, Canada.

The power of this type of analysis is that it allows the digitalization of strain type. This can be stored readily on databases and is hence highly transportable allowing easy comparison of strains from around the world. The other advantage is that these sequence data are defini-tive in that they are not open to interpretation in the same way as banding patterns on gels or biochemical tests can be. A number of databases for an increasing number of organisms have been established allowing comparison of strain data from around the world.

DIGITALIZED ISOLATE INFORMATION, BIOINFORMATICS AND EPIDEMIOLOGY

The use of digitalized data for the identification of isolates, as described in the preceding sections, combined with computer-based modelling that draws on mathematics, statistical principles (bioinformatics) and geographic information systems, has enabled the development of powerful tools for use in the field of epidemiology. Strain type information can be used along with such programmes to identify the likely infection source, overall risk assessment from various sources as well as monitoring disease outbreaks and the change in population composition and distribution over time (Figure 4.36), and can be used to inform public health policies and disease management strategies (French, 2009).

NEXT-GENERATION SEQUENCING AND METAGENOMICS

In recent years, different platforms have been developed that enable the rapid sequencing of whole genomes or metagenomes very quickly and have the advantage of being highly automated (Figure 4.37). There are a number of different sequencing techniques used, but the chain termination process (Sanger sequencing) and pyrosequencing (sequencing by synthesis) are common formats. Of the two techniques, pyrosequencing is a better method for sequencing a short stretch of DNA (up to 30 bases) due to (1) cost effectiveness, (2) less time consuming and (3) real-time monitoring of data whereas the Sanger method of sequencing is more optimized for sequencing long stretches of DNA (up to 600 bases). High throughput formats (formerly referred to as 'next-generation sequencing') is a quickly evolving area. Where once the struggle lay with the acquiring sequence data, the challenge now is how to store and manage the vast quantities of sequence information that are generated and the interpretation of these data. Sequencing equipment is developing rapidly and will soon be within reach of many laboratories

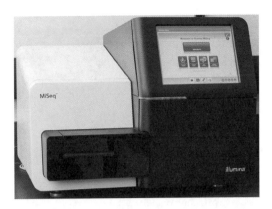

Figure 4.37 Next-generation sequencing equipment used for rapid sequencing of whole genomes or metagenomes. Photo: Paul Gajda, University of Calgary, Canada.

with portable units that can be used in the field (for example, MinIon® from Oxford Nanopore Technologies). Such technology has been used to randomly sequence DNA from a specific environment (metagenomics), for example, soil or from the gut, and by matching sequencing data against large databanks of DNA sequences, can be utilized to determine what organisms are present. As the technology matures, the possibilities for different uses in the molecular bacteriology field will expand accordingly.

MALDI-TOF

Matrix-assisted laser desorption/ionization – time-of-flight (MALDI-TOF) is a form of mass spectrometry that is emerging as a useful technology for the identification of microbial identification and disease diagnosis. Briefly, the bacterium or fungus of interest (the analyte in this case) is dried onto the surface of a metal plate (a chip), and over-laid with a matrix material. A pulsed laser is then fired at the sample. The energy from the laser causes vaporization and ionization of proteins from the prepared sample. A time-of-flight mass spectrometer measures the mass-to-charge ratio of these ionized proteins by determining the length of time

Figure 4.38 MinIon sequencing device (Oxford Nanopore Technologies).

it takes for the molecule to travel the length of the flight tube. The spectrum created from the sample is characteristic of the organism and is compared to profiles in a database of known organisms to predict its likely identity (Singhal et al., 2015) (Figure 4.38).

Endnotes

1 To prepare a wet smear: emulsify the specimen on a microscope slide with a small drop of saline, iodine or a specific stain, (India ink, or methylene blue, for example). Carefully place a coverslip over the suspension taking care that fluid does not go beyond the edges of the coverslip. This can then be examined under the microscope.

2 The GasPak system is available commercially and a new packet of chemicals must be used each time. Typical constituents of GasPak sachets are as follows:

1 sodium borohydride – $NaBH4$
2 sodium bicarbonate – $NaHCO3$
3 citric acid – $C3H5O$ $(COOH)3$
4 cobalt chloride – $CoCl2$ (catalyst).

3 See Table 4.2b. There are a range of haemolysins present in bacteria. Some bacterial colonies have a clear area around them on blood agar although this varies in appearance and size depending on the type of haemolysin present.

4 Available from suppliers of reagents (see Appendix 5) usually in 25 or 100 mg vials.

5 Hugh-Liefson medium and Bromothymol blue (3′,3′Dibromothymolsulfonephthalein) are all available from various suppliers (Appendix 4).

6 Koser citrate medium is available from various suppliers (see Appendix 4) in 100 and 500 g units.

7 Rabbit plasma can be bought from standard commercial suppliers of immunological reagents.

8 TSI is available from reagent suppliers (Appendix 4). Instructions for the production and handling of TSI are supplied with the medium. It is useful to keep a range of product catalogues as these often contain methods, additional information and references which may be useful.

9 Methyl red (4-Dimethylaminoazobenzene-2′-carboxylic acid) pH 4.2 (pink)–pH 6.2 (yellow) available from suppliers (Appendix 4).

10 α-naphthol and creatine are available from suppliers (Appendix 4).

11 Sulfanilic acid (4-aminobenzene sulfonic acid) is available from suppliers (Appendix 10.4).

12 Oxidase reagent (NN-dimethyl-P-phenylenediamine oxalate) darkens readily when exposed to air and is available from suppliers (Appendix 10.4).

13 Urea Agar Base (Christensen) available from suppliers as a base media in 500g units (Appendix 10.4)

14 Motility test medium (Tryptose, agar and sodium chloride) available from suppliers (Appendix 10.4)

15 http://clsi.org/.

16 See the OIE Register of diagnostic kits for veterinary use http://www.oie.int/scientific-expertise/registration-of-diagnostic-kits/the-register-of-diagnostic-kits/.

17 $TCID_{50}$ = Total cell infective dose (50) or the amount of a pathogenic agent that will produce pathological change in 50% of cell cultures inoculated. Expressed as $TCID_{50}$/ml.

18 Test development and validation is well explained in the OIE Manual of Diagnostic Tests and Vaccines for Terrestrial Animals at http://www.oie.int/standard-setting/terrestrial-manual/access-online/.

19 http://www.ncbi.nlm.nih.gov/genbank/.

20 https://www.luminexcorp.com/clinical/infectious-disease/respiratory-disease/respiratory-pathogens-flex-test/.

21 http://www.cdc.gov/pulsenet/.

22 http://www.cdc.gov/pulsenet/.

23 http://primerexplorer.jp/e/.

24 http://pubmlst.org/.

25 GenBank: http://www.ncbi.nlm.nih.gov/genbank/.

Bibliography

Cann, A.J. (2005) Principles of Molecular Virology, 4th edn. Elsevier Academic Press, London.

Cheng Y.C., Hannaoui, S., John, T.R., Dudas, S., Czub, S., Gilch, S. (2017) Real-time quaking-induced conversion assay for detection of CWD prions in fecal material. Journal of Visualized Experiments Sep 29; (127).

French, N.P. and the Molecular Epidemiology and Veterinary Public Health Group (2009) Enhancing surveillance of potentially foodborne enteric diseases in New Zealand: human campylobacteriosis in the manawatu: project extension incorporating additional poultry sources. http://www.foodsafety.govt.nz/elibrary/industry/enhancing-surveillance-potentially-research-projects/finalreportducketc2009.pdf.

Hugh-Jones, M.E., Hubbert, W.T., Hagstad, H.V. (2000) Zoonses: Recognition, Control and Prevention. Iowa State Press, Ames, IA.

Mahy, B.W.J., van Regermortel, M.H.V. (2010) Desk Encyclopaedia of Animal and Bacterial Virology. Elsevier Academic Press, London.

Maiden, M.C.J., Bygraves, J.A., Feil, E., Morelli, G., Russell, J.E., Urwin, R., Zhang, Q., Zhou, J., Zurth, K., Caugant, D.A., Feavers, I.M., Achtman, M., Sprat, B.G. (1998) Multilocus sequence typing: a portable approach to the identification of clones within populations of pathogenic microorganisms. Proceedings of the National Academy of Sciences of the United States of America 95: 3140–3145.

OIE (2017) Manual of Diagnostic Tests and Vaccines for Terrestrial Animals, 6th edn. Volumes 1 and 2. OIE, Paris.

OIE Manual of Diagnostic Tests and Vaccines for Terrestrial Animals. Chapter 1.1.2 Collection, submission and storage of diagnostic specimens. http://www.oie.int/fileadmin/Home/eng/Health_standards/tahm/1.01.02_COLLECTION_DIAG_SPECIMENS.pdf.

OIE Manual of Diagnostic Tests and Vaccines for Terrestrial Animals. Chapter 1.1.3 Transport of biological materials. http://www.oie.int/fileadmin/Home/eng/Health_standards/tahm/1.01.03_TRANSPORT.pdf.

OIE Manual of Diagnostic Tests and Vaccines for Terrestrial Animals. Chapter 1.1.6 Principles and methods of validation of diagnostic assays for infectious diseases. http://www.oie.int/standard-setting/terrestrial-manual/access-online/.

OIE Manual of Diagnostic Tests and Vaccines for Terrestrial Animals. Chapter 1.1.7 Standards for high throughput sequencing, bioinformatics and computational genomics. http://www.oie.int/fileadmin/Home/eng/Health_standards/tahm/1.01.07_HTS_BGC.pdf.

Quinn, P.J., Markey, B.K. (2003) Concise Review of Veterinary Microbiology. Blackwell Publishing, Oxford.

Quinn, P.G., Carter, M.E., Markey, B., Carter, G.R. (2000) Clinical Veterinary Microbiology. Mosby – Wolfe, London.

Singhal, N., Kumar, M., Kanaujia, P.K., Jugsharan S., Virdi, J.S. (2015) MALDI-TOF mass spectrometry: an emerging technology for microbial identification and diagnosis. Frontiers in Microbiology, https://doi.org/10.3389/fmicb.2015.00791.

Songer, J.G., Post, K.W. (2004) Veterinary Microbiology: Bacterial and Fungal Agents of Animal Disease. Elsevier Saunders, Philadelphia, PA.

Wagner, E.K., Hewlett, M.J. (2004) Basic Virology, 2nd edn. Blackwell Publishing, Oxford.

chapter 5

Haematology

Susan C. Cork and Roy Halliwell

5.1 Introduction

Haematology is the study of blood. A number of organs are involved in the production and recycling of red and white blood cells in the body, these organs form the haematopoietic system. In this chapter, we will outline the principles of haematology and explain some of the terminology used. The collection of blood samples and the techniques used to assess them are also outlined in this chapter. A more detailed description of white cell function is given in Chapter 6.

The haematopoietic system: terminology

Haematopoiesis (haemopoiesis) is the formation of red and white blood cells. The haematopoietic system is widely distributed in the body and includes organs, such as the liver, which have additional functions. The main role(s) of the organs involved in blood cell production and recycling are summarized in Table 5.1. In cases of anaemia or in disorders of the white blood cells it is important to evaluate the function of the whole haematopoietic system in order to identify the source of the problem.

Blood volume and the appearance and characteristics of blood cells

Red blood cells are also known as erythrocytes. In most domestic animals, erythrocytes account for about 30–40% of the total blood volume. The remaining components of the blood include the white blood cells, plasma proteins and water. The total blood volume is generally considered to be 10% of the body weight of the animal, for example a 300 kg cow has about 30 l of blood whereas a 25 kg dog has about 2.5 l of blood.

A healthy animal will develop signs of shock if it loses > 25% of its blood volume but it is usually safe to take up to 10% of the blood volume from a donor animal for the purposes of blood transfusion, that is, 3 l from a 300 kg cow, 5 l from a 500 kg horse and so on.

Mature red blood cells in mammals do not have a nucleus and are shaped like biconcave discs in many species. However, in deer and camelids (camels, lamas and so on) the red cells are elliptical or sickle shaped. In birds, reptiles and fish mature red blood cells are nucleated and are oval to round in shape.

White blood cells, also known as leukocytes, are an important component of the immune system in animals, although they only occupy a very small percentage of the blood volume. The number of white blood cells and their composition will change in response to immune triggers, such as infection and inflammation. The relative

Table 5.1 Components of the haematopoietic system and their function.

Tissue or organ	Function in haematopoiesis
Bone marrow	Produces red and white cells, stores iron and is a source of precursor cells.
Lymph nodes and follicles	Produce lymphocytes and plasma cells, which are important in the immune response.
Liver	Stores vitamin B_{12}, folic acid and iron. Produces clotting factors (prothrombin), fibrinogen and proteins (albumin and gamma globulins). Coverts haemoglobin breakdown products (free bilirubin) into products for excretion (bound bilirubin) in bile. Produces the precursor for erythropoietin (renal hormone). Retains potential for erythropoiesis.
Spleen	Produces lymphocytes and plasma cells and participates in the production of an immune response. Stores red blood cells and iron. Destroys abnormal red blood cells, removes foreign material.
Stomach and intestinal muscosa	HCl is produced in the gastric mucosa to release iron from complex organic molecules. Produces intrinsic factor that assists in the absorption of vitamin B_{12} across the intestinal epithelium.
Reticuloendothelial system	Destroys abnormal and aged red blood cells in the circulation, converts haemoglobin to iron, free bilirubin and other components. Stores iron.
Kidney	Involved in the production of erythropoietin, which stimulates the production of new red blood cells.
Thymus	Involved in the differentiation of precursor lymphocytes (in bone marrow) into immunocompetent lymphocytes (mammals). In birds the bursa of Fabricius has a similar role (B- and T- lymphocytes have a different function but look similar in blood smears, see text).

proportion of each type of white blood cell and their morphology can provide a lot of information about the nature of a disease process. However, because the normal white cell range for each species is quite different it is important to be familiar with the normal range as presented in Tables 5.3 and 5.4, below. Some of the different cell types seen in the study of haematology are illustrated in Figure 5.1 and in the photographs in Figures 5.6 to 5.10.

The general function and characteristics of each cell type are outlined below.

Lymphocytes

The majority of these cells have no granules. There is one sub-group known as large granular lymphocytes, which contain a few magenta granules in the cytoplasm. Lymphocytes are derived from the bone marrow, spleen and lymph nodes. There are two main types, B- and T-lymphocytes, but their appearance is similar. The granular lymphocytes are a subgroup thought to represent natural killer cells. Lymphocytes increase in number (lymphocytosis) in some acute viral infections. Generally, the B-lymphocytes are involved in antibody production (plasma cells) and the T-lymphocytes are involved in cell mediated immunity, for example, in tuberculous lesions and in other chronic diseases. In some viral diseases, for example, canine parvovirus or feline panleukopaenia, the total white cell count falls (leukopaenia) but the relative percentage of lymphocytes will often remain high.

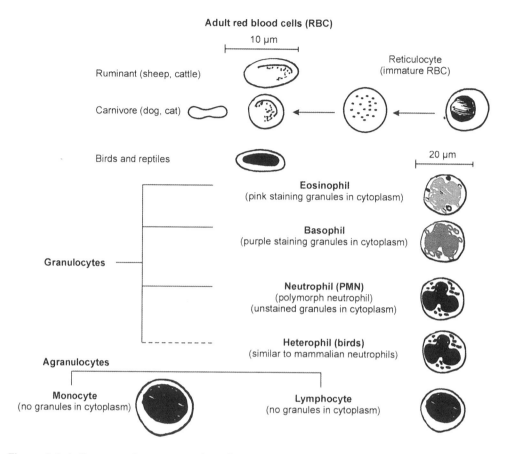

Figure 5.1 A diagrammatic representation of the appearance of blood cells in stained blood films (not to scale). The illustrations represent a stylized version of different blood cell types common in domestic species but there are quite a few important few species differences. For more detail see the bibliography listed at the end of the chapter. Note that red cells are generally much smaller than the white cells (see Figures 5.7–5.9).

Monocytes

In most species, these cells become more common in chronic disease states but they are the predominant cell in normal elephant blood smears. Monocytes contain few or no granules. In the inflammatory response (particularly tissue necrosis and abscess formation) these cells migrate into tissues and are known then as tissue macrophages. They belong to the family of histiocytic cells. Reptilian azurophils also belong to this group of leukocytes and are the exception with respect to cytoplasmic granulations.

Platelets

Platelets are also known as thrombocytes. These are not actually cells but are pieces of cytoplasm from bone marrow derived megakaryocytes. Thrombocytes are important for the normal clotting of blood. In haemorrhagic diseases, there may be a reduction in the number of thrombocytes (thrombocytopaenia) although in some bleeding disorders and myeloproliferative diseases (leukaemia) there may be an increase (thrombocytosis). Physiological stress can also lead to a thrombocytosis which is not pathologic in nature.

Granulocytes

Granulocytes include the polymorph neutrophils (PMN), eosinophils and basophils and avian/reptilian heterophils.

Neutrophils

Neutrophils may contain granules but they do not stain with Romanowsky stains. The nucleus usually has three or more lobes but in times of excess production (for example, due to infection) band forms of nuclei are seen as well as segmented forms. When many 'band' forms are present this is known as a 'shift to the left'.[1] In acute bacterial diseases, the proportion of neutrophils may increase (neutrophilia) and the total white cell count may also increase (leukocytosis). In chronic diseases, this may not occur because white cells can become sequested in tissues and the production of new white cells can be interrupted if essential resources become scarce, for example, in severely debilitated and malnourished animals. Neutrophils form a large component of 'pus', which is a thick protein rich material that also contains necrotic material and cellular debris and is commonly seen in tissues infected with bacteria. The number of circulating neutrophils may also increase when animals are exposed to physiological stress, for example, crowding or handling so it is important to minimize stress when blood samples are collected.

Eosinophils

Eosinophils contain eosinophilic (acid/pink) staining granules. They are more numerous in some allergic reactions and are attracted by histamines which are released from damaged cells. An increase in eosinophils (eosinophilia) can indicate that the animal has a parasitic infection (for example, lungworm).

Basophils

Basophils contain basophilic (blue/basic) staining granules. Depending on the species these range from a very pale grey (cats) to deep blue (horses). They are not common.

5.2 Collection of specimens

For haematology, blood is usually collected from the neck (jugular vein) in cattle (see Figures 1.18b–d), horses, sheep and goats and from the front leg (brachiocephalic vein) in the dog and cat. If haemoparasites are suspected then blood smears should also be made from blood collected from peripheral vessels such as the ear vein (in large animals). Blood for haematology should be collected into anti-coagulant (for example, ethylene-diaminetetra-acetic acid (EDTA[2]), heparin or sodium citrate) to prevent clotting. For most livestock, a 10 ml sample of blood collected into one or more vacutainers is ideal for laboratory testing. If commercial vacutainers are not available, or vacutainers are re-used, they may be prepared in the laboratory but it is important to make sure that there is sufficient vacuum present. Vacutainers containing EDTA have purple tops; those with heparin have green tops (see also Chapter 7). It is important to collect at least 5 ml of blood for full haematology although for some tests, especially for small animals and birds, smaller volumes (2 ml) may be acceptable. Smaller volumes can also be used if automated microtitre systems are available, these systems use the principles of flow cytometry to measure the size and number of cells but they do need to be appropriately calibrated for the species of interest (see Chapter 2).

Unclotted whole blood samples are used in the laboratory to determine the packed cell volume (PCV), the total red and white cell counts, the differential white cell count and the haemoglobin content of the sample. These parameters can then be used to examine a number of other characteristics of the red and white blood cell population. It is recommended that a total white blood cell count is performed along with a differential white cell count as this provides the required context to help interpret the test results. If it is difficult to collect a large volume of blood then it is acceptable to collect a small amount in two capillary tubes for estimation of PCV along with two blood smears to allow a rough assessment of the differential white cell count and the red and white blood cell morphology (see figures 5.6–5.9).

Always ensure that the animal(s) to be sampled are appropriately restrained. In some species it may be necessary to clip the hair in the area around the vein to allow easy access. The vein may also be more easily visualized if the skin is swabbed with alcohol or methylated spirit. The usual collection site and the size of needle recommended for collecting blood from common species of domestic animals is outlined in Table 5.2.

Preparation and staining of blood and bone marrow smears

The preparation of a good blood film requires the use of clean grease free slides (immersion of slides in 70% alcohol will remove grease but ensure that the slide is dry before use). A blood film can be made from a single drop of peripheral blood or a drop from an EDTA sample submitted in a vacutainer. The prepared film should have a random distribution of white blood cells throughout the film (Figure 5.2). Erythrocytes should be distributed in a single layer on part of the blood slide. A spreader is used for preparation of smears, this should have a smooth, even edge with the corners cut to give a side slightly narrower than the width of the microscope slide. For examination of haemoparasites, it is preferable to make thick and thin smears from peripheral blood (see also the parasitology section – Figure 3.36). Note that the haematocrit will affect the time it takes for the smear to spread, that is, in anaemic animals it will spread quickly and haemoconcentrated specimens will take longer.

Table 5.2 Blood collection sites in different species of animal.

Species	Site	Needle size*
Horse	Jugular vein	16–19 gauge/1.5–2 inch** length
Cow	Jugular vein, tail vein	16–19 gauge/1.5–2 inch length
Sheep/goat	Jugular vein	18–20 gauge/ 1.5–2 inch length
Pig	Anterior thoracic vena cava, ear veins	20 gauge/1.5–4 inch
Cat	Brachiocephalic, jugular or saphenous vein	20–25 gauge/1.5 inch
Dog	Brachiocephalic, jugular or saphenous vein	20–25 gauge/1.5 inch
Small primate	Femoral vein	20–26 gauge/1.5 inch
Mouse	Tail vein	Microhaematocrit tube

Note: *The larger the needle diameter, the smaller the gauge size, that is, a 16 gauge needle has a wider bore than a 25 gauge needle. In general, you should try to use the largest gauge needle suitable as this reduces the risk of haemolysis when the sample is drawn. **1 inch is equal to 25.4 mm.

Method

Place a small drop of blood 1.5–2.0 cm from one end of a clean slide. Hold the spreader at an angle of 30° in front of the drop of blood and bring it back to touch the blood (see Figures 3.36 and 5.2a) allowing it to spread along the edge. Complete the film by pushing the spreader

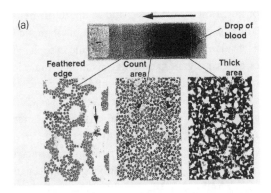

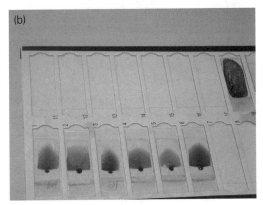

Figure 5.2 (a) Blood smear modified from Pratt (1997). Gross and microscopic views of different areas on a blood film. The three major areas of the blood film (feathered edge, count area, thick area) are indicated by the lines connected to the macroscopic view of the slide. There are often many artefacts in the feathered edge. In the count area the red blood cells are present in a monolayer and white cells are flattened to show intracellular detail. Avoid counting cells in the thick layer. (b) Samples 1–6 are blood smears, sample 18 is a tissue smear. Photo: Dr Nicole Fernandez, WCVM.

forward in a direct and even movement. The film should be about 3–4 cm long. Air dry rapidly by waving the film in the air (this avoids crenation of the erythrocytes). Label the film either by writing with a pencil in the film itself or mark the slide using a diamond tipped pen. If staining of the smear will be delayed by more than 24 h fix the film in methanol for 2 min, this should be done as soon as the film is dry. If the films will be stained within 24 h then air dry and place out of the light and protect from dust. Air drying without fixation prior to staining is preferred but not always possible in the field. If staining (which includes a fixation step) will be delayed by more than 24 h the film can deteriorate rapidly and so it is advisable to carry a small bottle of methanol on field trips for this purpose. In the laboratory, the slides can be stained with Giemsa stain. This stain, along with Leishman stain and Wright's stain, is known as a Romanowsky stain. There are a number of staining methods recommended but the following has been found satisfactory in our experience (see also Chapter 3). Several quick kit test stains are also available for example, DiffQuick™.

Stain preparation

Prepare a stock Giemsa stain (Giemsa powder 1 g, glycerol 66 ml; mix thoroughly and heat to 56°C for 90 min, add 66 ml methanol. Mix thoroughly and leave to stand for 7 days; filter and store in labelled glass-stoppered bottles).

The buffer should be prepared and held at pH 7.2 but some laboratories have found that the stain will work in the range of pH 6.8–7.2. Each new batch of stain should be tested to determine the optimum pH and staining time.

1 Fix the blood smear in methanol for 2 min and air dry.
2 Stain in 10% Giemsa (stock solution diluted in pH 7.2 buffer) for 20 min (15 min may be sufficient).

3 Wash off excess stain with buffer.
4 Flood with excess buffer and leave until differentiation is complete (usually for 1–2 min).
5 Dry carefully with blotting paper or leave upright to dry in a rack.

5.3   Cell counts and white cell indices

The white cells (or leukocytes) in the blood are especially important for the body's defence against infectious disease. There are different types of white cells which can be identified morphologically in a blood smear (see Figures 5.8–5.10). The total number of white cells, along with the relative proportion of each type of cell, can be used to identify the nature of the disease process in an animal. Leukocytes, which include granulocytes (neutrophils, eosinophils and basophils) and mononuclear cells (monocytes and lymphocytes), are produced in the bone marrow and lymphoid tissue, along with red blood cells (erythrocytes) and platelets, and are released into the circulation in greater numbers in response to some diseases and physiological states. A high number of white cells in the circulation (leukocytosis) most commonly indicates that an inflammatory process is occurring such as in response to infectious diseases. In viral diseases, the proportion of lymphocytes may rise (lymphocytosis) or fall (lymphopaenia) whereas in bacterial diseases, the proportion of neutrophils usually increases (neutrophilia). In allergic and some parasitic diseases, the number of eosinophils increases (eosinophilia). However, this is not always the case and in very debilitated animals the total number of white cells may fall (leukopaenia). In animals with chronic intra-cellular bacterial diseases, such as tuberculosis and brucellosis, the white cell count may be normal although an increase in monocytes (monocytosis) may be evident. Evaluation of the white cell population involves performing a total white cell count (TWCC) and a differential count.

The blood smear should also be examined for the presence of abnormal white cells, for example, unusual cell nuclei or staining characteristics and for haemoparasites (for example, *Babesia* spp. *Theileria* spp. and so on) and inclusion bodies (dense staining material in the nucleus or cytoplasm). Inclusion bodies in the nucleus or cytoplasm of lymphocytes or monocytes may indicate viral or toxic insult or nutritional deficiencies and so on. In the neutrophils the presence of cytoplasmic change such as a foamy appearance, a basophilic cytoplasm or excessive granulation may also indicate toxic change. The size and shape of the leukocytes, as well as their granules, varies a little between species but some of the common characteristics are outlined in section 5.1. In Giemsa stained smears the red cells appear pink and the nuclei of white cells appear purple or blue. The nuclei of the granulocytes (neutrophils, basophils and eosinophils) are usually multi-lobular with three to five segments (variable). Granulocytes are important in the acute inflammatory response and their granules contain digestive enzymes to kill bacteria and to remove cellular debris. Precursor PMN have a band shaped nucleus and are referred to as band cells. If these are present in large numbers (that is, a proliferative response) it is interpreted as being evidence of a strong immune response to an infectious disease, for example, a neutrophilia with a left shift.

In Romanowsky stained smears, the cytoplasm of eosinophils, in most species, contains pink granules and that of basophils contains blue granules. The nuclei of monocytes may be lobulated and in lymphocytes are usually round in shape.

Monocytes are generally larger than lymphocytes but in blood or tissue smears it is not always easy to tell these two cell types apart. Lymphocytes are present in the spleen, lymph nodes and in other reticulo-endothelial tissues of the body, such as the liver. The lymphoid system acts as the defence system for the blood by filtering out foreign material. There are two populations of lymphocytes, 'T' and 'B' types named after the thymus and bursa derived populations in chickens. It is not possible to distinguish the two types in blood or tissue smears. In simple terms, the B-lymphocytes produce antibodies to foreign material (antigens) and the T-lymphocytes are important in cell mediated immunity (see Chapter 6 for more details). Monocytes are often present in large numbers in chronic inflammation. The peroxidase test (below) can be used to help distinguish between monocytes and lymphocytes.

Peroxidase reaction

This staining system may assist in the identification of cell types in blood smears from some species.

1 Fix air dried blood films in 9 parts ethanol (95%) and 1 part formalin for 5 s.
2 Wash the slide in tap water for 1 min.
3 Stain the slide in 10% Giemsa stain for 30 min.
4 Pour off the stain and flood with 0.5% aqueous copper sulphate ($CuSO_4.5H_2O$) for 5 s.
5 Prepare a 0.1% aqueous solution of benzidine and add 2 drops of 3% hydrogen peroxide (H_2O_2) per 100 ml.
6 Pour the copper sulphate off the slide and place the slide in the benzidine solution for 2 min.
7 Wash the slide in tap water and air dry.
8 Examine under oil immersion (1000×). Blue granules are present in developing and mature neutrophils, a few granules are usually seen in monocytes and no granules are seen in lymphocytes.

Differential white blood cell count (diff. WBC)

To perform a differential white cell count a thin, even blood film should be prepared and either air dried and placed in a covered slide container or, if there will be a delay of greater than 24 h, fixed in methanol. The slide can then be stained with Giemsa (or an alternative Romanowsky stain) as outlined previously. In a well-prepared blood smear the white cells should be distributed evenly. However, even in well-made smears there is potential for a large degree of error. If smears are too thick it is not possible to accurately differentiate cells, especially monocytes and lymphocytes. If smears are too thin there is an unrepresentative distribution of cells towards the edges and tail of the smear. Equally, an uneven edged spreader will give a misleading representation of different cells throughout the smear. For cell counts, the blood film should initially be scanned under low power both 200× (20× objective) and 400× (40× objective), then high power (oil immersion 1000×) to evaluate the shape, size and staining characteristics of the red and white cells in the film. The presence of any blood parasites (for example, *Trypanosoma* sp., *Babesia* sp., *Theileria* sp. and *Rickettsia*, see Chapter 3, section 3.6) should also be noted. Depending on the parasite you will find them either on scanning on the lower magnification, for example, *Dirofilaria*, or on high magnification, for example, *Babesia*, *Ehrlichia* and so on. One of the most popular counting methods is the Battlement method (Figure 5.3).

The Battlement method involves examining the slide field by field along a horizontal line for three fields, then for two fields up, three fields horizontally, then a further two fields down to the edge of the film. Counting is continued until 100 white cells have been counted on each section of the film. For routine purposes, at least 200 white cells should be counted and then each cell type can be expressed as a percentage of the total.

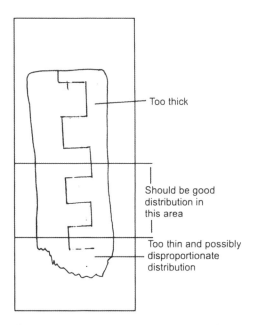

Too thick

Should be good distribution in this area

Too thin and possibly disproportionate distribution

Figure 5.3 Battlement counting method. Correctly prepared blood smear with a think zone at one end and a thin 'feather' at the other end (see also Figure 5.2).

TOTAL CELL COUNTS

For red and white cell counts a haemocytometer is required, preferably one with the Neubauer ruling (see Figures 5.4a and 5.4b). Use whole blood collected in EDTA anti-coagulant. Use calibrated pipettes with disposable tips or a haemocytometer pipette for making the required dilutions. If a haemocytometer pipette is used it should be well washed and rinsed in distilled water after each use. Mouth pipetting is not recommended. As with most manual methods the following techniques are subject to a high degree of error especially when carried out by inexperienced personnel.

Total white cell count

PREPARE THE DILUTING FLUID

Diluting fluid: 2% acetic acid in distilled water coloured pale blue with Gentian violet.[3]

Prepare a 1/20 dilution of EDTA blood by adding 50 μl (0.05 ml) of blood to 950 μl (0.95 ml) of diluting fluid in a mixing tube. Seal the tube and mix by inversion and rotation. If a haemocytometer pipette is used fill with EDTA blood to the 0.5 mark and then add diluent to the 101 mark. Mix by rotating the pipette. Note that the pipette has three markings on it, 0.5 mark in the middle of the stem, 1 mark at the junction between stem and bulb and 101 mark above the bulb. The total volume of the pipette is 101 parts, of which one part is in the stem and 100 parts in the bulb.

Before loading the haemocytometer ensure that the diluted blood is well mixed and if using the haemocytometer pipette, discard the first few drops before loading the haemocytometer. Leave the counting chamber for a few minutes to allow the white blood cells to 'settle' before counting. Count all the cells in the four large outer squares (see Figure 5.4b) (1 mm × 1 mm) under the 10× or 20× objective. Include cells on the left and upper lines but not cells on the lower and right lines. If nucleated (that is, immature) red cells (nRBC) are encountered in mammalian blood smears, they should be recorded separately from the leukocytes and reported as the number of nRBC/100 WBC. This will be noted while performing the differential count as they will not be able to be differentiated on the haemocytometer.

(a)

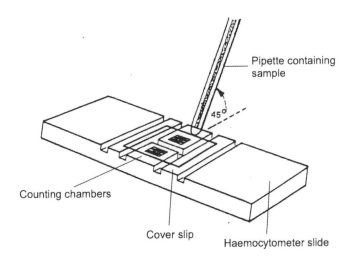

Pipette containing sample

45°

Counting chambers

Cover slip

Haemocytometer slide

(b)

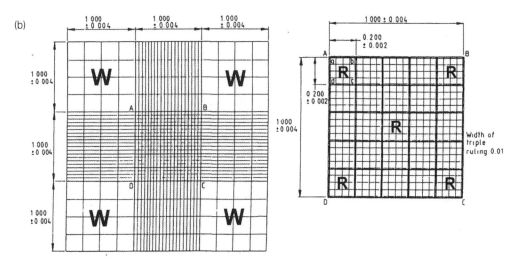

Figure 5.4 (a) Filling the counting chamber using a Pasteur pipette. (b) Improved Neubauer ruling for a blood counting chamber. All dimensions are in millimetres. The central portion of the square is also illustrated.

Calculation (depends on the type of haemocytometer used)

Total number of cells counted = N (for example, 240); volume = 4 mm² (area counted) × 0.1 mm (depth of chamber) = 0.4 mm; dilution = 1/20

White cells per microlitre = N × 1/0.4 × 20 = N × 50

If N = 240 the TWCC = $12,000/\mu l$ = 12.0 × $10^3/\mu l$ = 12.0 × 10^6/ml = 12.0 × 10^9/l

Total red cell count

PREPARE THE DILUTING FLUID

Dilute fluid: formol-citrate 1% in 3% aqueous trisodium citrate.[4]

Physiological salt solution (0.85%): 8.5 g sodium chloride (NaCl) in 1 l of distilled water.

Prepare a 1/200 dilution of blood in formol-citrate solution by adding 20 μl of blood to 4 ml of diluent. Wipe the end of the pipette. If a haemocytometer pipette is used, it is necessary to draw up blood to the 0.5 mark on the unit, wipe the tip and then draw up diluent to the 101 mark to make a 1/200 dilution. Allow 2 min for mixing to occur while rotating the pipette. Discard half of the contents and then fill the chamber. Flush the pipette with the blood/diluent mixture to ensure that all the blood has been added to the mixing tube. Seal the tube and invert two or three times to mix well. Fill the counting chamber of the haemocytometer using a fine Pasteur pipette. Make sure that the counting chamber is just filled and that no fluid flows into the surrounding channels. Allow the cells to settle for a few minutes and count using the 40× objective. Count the red cells (erythrocytes) in the four corners and the middle square in the central area, that is, 5 × 16 small squares, including those on the right and lower lines and excluding those on the left and upper lines (see Figure 5.4). In normal practice a minimum of 500 cells should be counted (that is, 100 per square).

Calculation (depends on the type of haemocytometer used)

Total cell counted = N (for example, 500); volume = 0.2 mm² (area counted) × 0.1 mm (depth of chamber) = 0.02 mm³

Dilution = 1/200 red cell per microlitre = N (500) × 1/0.02 × 200 = 500 × 10,000 = 5 × 10⁶/μl or 5 × 10¹²/l

5.4 Red cell indices and anaemia

Red cell indices include the measurement of PCV, total red cell count (TRCC) and the haemoglobin (Hb) concentration (g/dl). In the clinical assessment of cases of anaemia it is of value to use these indices to calculate the mean cell volume (MCV), mean cell haemoglobin (MCH) and the mean cell haemoglobin concentration (MCHC).

The MCV indicates the volume of the 'average' red cell in a sample. It is expressed in femtolitres (fl; 10^{-15} l). Traditionally, MCV was a calculated parameter, derived by using the following formula:

$$MCV = (PCV/RBC) \times 10$$

where RBC is red blood cells. Therefore, if an animal has a PCV of 42% and a RBC of 6 × 10⁶/μl = 70 fl.

Red cell populations with the MCV below the reference interval are termed *microcytic* and those with the MCV above the reference interval are termed *macrocytic*. Automated systems calculate this parameter directly.

MCH is the mean cell haemoglobin. This represents the absolute amount of haemoglobin in the average red cell in a sample. Its units are picograms (pg) per cell. The MCH is calculated from the haemoglobin concentration and the red cell count using the following equation:

$$MCH \text{ (pg)} = [Hb \text{ (g/dl)}/RBC (\times 10^9 \, \mu l)] \times 10$$

Mean cell haemoglobin concentration (MCHC)

MCHC is the mean cell haemoglobin (Hb) concentration, expressed in g/dL. It can be calculated from the Hb and the PCV using the following formula:

MCHC = [Hb (g/dl)/PCV %] × 100

The normal MCHC value for many common domestic species is about 33%. If red cell populations have values below this they are referred to as 'hypochromic'. This can be seen in cases of regenerative anaemia, where there is an increased population of young red cells (reticulocytes) with low Hb content. Low MCHC can also occur in iron deficiency, where small (micro-cytic) hypochromic red cells are produced as a result of the lack of iron for haemoglobin synthesis (that is, microcytic anaemia). Occasionally there may be an increase in MCHC, this is often observed if lipaemic samples are examined and is considered to be an artefact.

Examination of blood smears allows assessment of the degree of reticulocytosis (red cell production) and the morphology of the red cells as well as the detection of haemoparasites (for

Table 5.3 Characteristics of blood cells in different species of animal.

Species	Red cell characteristics		Mean diameter (µm)/special features	Reticulocytes in peripheral blood (%)	Approx. neutrophil/ lymphocyte ratio	White cells characteristics/ special features
	Rouleaux formation	Central palour				
Dog	+	++	7.0/uniform size. Crenation not common	+/– 1.0	70 : 20	Basophils rare, eosinophil granules vary in size. Cytoplasm pale blue
Cat	++	+	5.8/crenation. Slight aniocytosis	+/– 0.5	60 : 30	Basophils rare. Eosinophil granules rod like. Few band cells (PMN)
Cow	-	+	5.5/aniocytosis common	0	28 : 58	Eosinophil granules small, round and densely staining. Vacuoles common in monocytes
Sheep	+/–	+	4.5/regular size and shape.	0	30 : 60	Neutrophil nucleus multibolutated. Granules in lymphocytes. Monocyte nucleus amoeboid in clumps
Horse	+++	+/–	5.7/marked rouleaux formation	0	55 : 35 HB 50 : 45 CB	Eosinophils contain large granules and fill cell. Lymphocytes small. Monocytes have a kidney shaped nucleus

Table 5.3 *continued*

Species	Red cell characteristics		Mean diameter (µm)/special features	Reticulocytes in peripheral blood (%)	Approx. neutrophil/ lymphocyte ratio	White cells characteristics/ special features
	Rouleaux formation	Central palour				
Pig	++	+/–	6.0/crenation. Slight aniocytosis	+/– 1.0	35 : 50	Eosinophil granules ovoid. Band PMN present in healthy pigs, Lymphocytes may be small or large
Camelid			Elliptical RBC; 7.7 x 4.4		50 : 40	
Deer			Sickle shaped cell		57 : 38	
Elephant	+++		Large RBC, rouleaux formation common		18 : 29	Many monocytes, these are the predominant cells (52%)

Note: HB = hot blooded; CB = cold blooded; PMN = polymorph neutrophil.

Table 5.4 Normal haematological parameters for livestock and companion animals.

Species	Ruminants			Pig	Horse*	Carnivores	
	Cow	Sheep	Goat			Dog	Cat
RBC (x10^{12}/l)	5–10	5–15	8–17	5–8	6–13	5–9	5–10
Mean value	**7**	**12**	**13**	**6.5**	**9.5**	**6.8**	**7.5**
Hb (g/dl)	8–15	8–16	8–14	10–16	11–19	12–18	8–15
Mean value	**11**	**12**	**11**	**13**	**15**	**15**	**12**
PCV (%)	24–46	24–50	20–40	30–50	30–52	36–55	24–45
Mean value	**35**	**38**	**28**	**42**	**42**	**45**	**37**
WBC x 10^9/l	4–12	4–12	4–13	11–22	5–13	6–17	6–20
Mean value	**8**	**8**	**9**	**16**	**9**	**11.5**	**12.5**
	Differential leukocyte count						
PMN %	15–45	10–50	30–50	28–50	30–65	60–77	35–75
Band (mean)	0.5	Rare	Rare	0.4	0.5	0.8	0.5
Lymphocytes (%)	45–75	40–75	50–70	40–62	25–70	12–30	20–55
Monocytes (%)	2–7	0–6	0–4	2–10	2–7	3–10	2–4
Eosinophils (%)	2–20	0–10	2–8	2–11	0–11	2–10	2–12
Basophils (%)	0–2	0–3	0–1	0–2	0–3	Rare	Rare

Note: Carnivores tend to have a higher proportion of neutrophils to lymphocytes whereas ruminants have a higher proportion of lymphocytes to neutrophils. *The normal red cell parameters vary from breed to breed with the 'hot blooded' breeds of horse such as the Arab and Thoroughbred having a higher PCV than pony breeds. There are some variations depending on age, breed and physiological state in all species and this should be taken into consideration when interpreting the results of blood tests. RBC= red blood cell count, Hb = haemoglobin, WBC = white blood cell count, PMN = neutrophil.

example, *Babesia* sp. *Theileria* sp. and *Rickettsia*). Normal values for the red cell indices of common domestic species are shown in Tables 5.3 and 5.4 along with the normal white cell parameters.

Anaemia

Red cells (erythrocytes) contain haemoglobin, which is responsible for oxygenating the blood. Anaemia is a shortage of either red cells or of haemoglobin. As anaemia progresses the animal may become very weak and the mucous membranes become visibly pale. Oxygen is essential for most metabolic functions in the body. Red cells circulate in the blood stream providing an oxygen supply to tissues and when they are aged they are removed from the circulation by the spleen and the breakdown products are recycled by the liver. Parasitized and abnormal red cells are also removed from the circulation. Anaemia can occur when there is rapid removal or destruction (for example, intravascular haemolysis) of circulating red cells (as in anaplasmosis, *Babesia* infections, immune mediated haemolytic anaemias and in some toxic insults) or when the production of new red cells is delayed (that is, in iron deficiency anaemia and in malnutrition).

In haemolytic diseases such as anaplasmosis, where many red cells are broken down over a short period of time, the liver may become overwhelmed and waste products enter the circulation. If waste products such as bile, which is formed from degradation of haemoglobin, accumulate in the body tissues these tissues can turn yellow. This is known as jaundice and is usually evident on examination of the mucous membranes. The iron from the haemoglobin is re-used but other components are normally excreted from the biliary system into the intestine giving faeces a typical brownish colour (more information is given in the biochemistry section of Chapter 7).

Deficiency of iron, or other components of haemoglobin, or deficiency of proteins and other nutrients may deplete bone marrow of the precursors required to produce more red cells, especially if demand is high. Healthy animals respond to blood loss by releasing immature red cells (reticulocytes) into the blood and these cells may still have their nucleus (nRBC) or have a basophilic stippling (purple spots over the surface as seen in Giemsa stained preparations). In many types of anaemia, the red cells may vary in size (aniocytosis) and shape (poikilocytosis). In most mammals, the red cells in the circulation are of a uniform size, circular (like a disc) in shape and have no nucleus. As noted earlier, in birds and reptiles the red cells are elliptical in shape and have a prominent nucleus.

Packed cell volume (PCV)

Packed cell volume is the volume of red cells per 100 ml of blood. The PCV provides a useful preliminary assessment of a case of suspected anaemia (indicated by a low PCV). Whole blood samples collected in EDTA anticoagulant should be used for the determination of PCV.

There are two commonly used methods, the Wintrobe method and the microhaematocrit method. If a microhaematocrit centrifuge is available the latter is the preferred method. However, the Wintrobe method has the advantage of providing additional information as outlined below.

WINTROBE METHOD
Wintrobe tubes should have a graduated surface divided into 1.0 mm intervals to 100 mm. These should be kept clean and dry and washed immediately after use. The tubes are filled to the top with EDTA blood using a Pasteur pipette and centrifuged at 4000 rpm for 30 min. In order to avoid breakages make special tube holders to allow the tubes to fit securely into the centrifuge cups. When reading the result, the height of the

column of red cells is expressed as a percentage of the total volume of the blood in the tube.

After the tubes have been spun there will be three layers:

1 a bottom red layer of compacted red blood cells
2 a middle whitish layer (the buffy coat) of white blood cells and platelets
3 a top clear to pale yellow layer containing the plasma (serum + clotting proteins).

From the PCV result using this method a rough estimate of other parameters can be made as follows:

PCV = 45 %

Normal Haemoglobin (Hb) = 1/3 PCV (45%) = 15 g/dl

Normal erythrocyte count (TRCC) = 1/6 PCV × 10^6 = 7.5 × $10^6/\mu l$

Leukocytosis (high TWCC) = 1.5 mm or greater in the Wintrobe tube

Leukopaenia (low TWCC) = 0.5 mm or less in the Wintrobe tube

Cloudy plasma = lipaemic (fatty blood)

Dark yellow plasma = jaundice (bile pigments in the blood)

Note that a high PCV may indicate dehydration. Dehydration can therefore mask anaemia.

MICROHAEMATOCRIT METHOD

For this method capillary tubes, sealant and a microhaematocrit centrifuge (at 12,000 g) will be required along with a calibrated reader (see Figure 5.5).

Fill a pair of capillary tubes with the blood sample (use EDTA blood) using capillary attraction until the tube is filled to two-thirds of its length. Paired samples are prepared in case of breakage. Wipe the outside of the tube and seal the ends using sealant (or seal the free end in a flame). Make sure that the tubes are well sealed to prevent leakage. Place the tubes in a micro-haematocrit centrifuge opposite each other and (where several samples are handled together) take note of the position of the haematocrit tube(s). Most divisions in the haematocrit rotor will be numbered. If the tubes are not well positioned or the rubber insulating material around the inside edges of the haematocrit rotor is worn the tubes may break so check this before securing the lid of the unit. Centrifuge the samples at 12,000 g for 4 min. After centrifugation note that there are two obvious layers, the buffy coat is less readily detected in this method than with the Wintrobe method. Make sure that there is no haemolysis or clotting of the sample. Measure the height of the red cell column using a haematocrit reader (often provided on the lid of a haematocrit centrifuge).

Determination of haemoglobin content

Haemoglobin is the red pigment contained in red cells. This pigment is important in the transfer of oxygen to body tissues. The measurement of haemoglobin is usually recorded as grams per 100 ml of blood. There are various methods used to measure the haemoglobin level in blood samples and most depend on measuring the intensity of colour produced by haemoglobin. One of the simplest methods is Sahli's acid haematin method as it requires little equipment and is easy to perform.

This method is, however, subjective and has a high degree of error unless performed regularly by the same person. If a colorimeter is available along with a haemoglobin standard the methods outlined in the biochemistry section (see Chapter 7) are recommended. The advantage of using a colorimeter is that the results are less

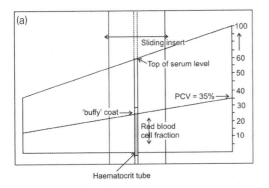

Figure 5.5 (a) Microhaematocrit reader. The haematocrit tube can be used to provide useful information about the health status of an animal. The tube should be filled with fresh blood and the ends sealed before it is placed in a micro-haematocrit centrifuge. Usually, several samples are examined at one time so make sure that the position of each sample is recorded in the centrifuge. After the tubes have been spun, examine the resultant three fractions. The upper clear layer is the serum (or plasma if EDTA blood is used), the small white band separating the red and clear layers is the 'buffy' coat and is composed of white cells. The red layer is the compacted red cells. In some cases a large band of white cells can indicate that a disease process is present but a complete blood count will give more accurate results. In any case, examine a blood smear from each case to allow differentiation of white cells present. Place the prepared haematocrit tube in the sliding device and move it along the top line until the top of the serum level matches the line. Read the level of the red cells at this point. The PCV (or haematocrit) of the sample illustrated is 35%. Some microhaematocrit centrifuges have a microhaematocrit reader etched on the lid. (b) Microhaematocrit centrifuge. Photo: S. Cork, University of Calgary, Canada.

subjective and more readily repeatable but the machine must be calibrated correctly and a standard preparation should be run with each batch of test samples.

Sahli's acid haematin method

This method relies on matching a colour change in the sample with a coloured glass standard block. Note that the colour standard glass will deteriorate over time and that the measured solution (acid haematin) is not stable. A com-plete unit can be purchased in a kit form which consists of a haemoglobin standard and comparator, a graduated tube, a 0.02 ml pipette, a dropper bottle containing 0.1 M HCl and a glass mixer.

METHOD

Fill the graduated tube to the 20 mark with 0.1 M HCl (hydrochloric acid). Add 0.02 ml of blood to the tube using a pipette. Rinse the pipette by drawing up the acid mixture several times. Leave for about 5 min (the time varies with the kit so check first). Mix until the colour of the acid-blood mixture matches that of the glass or graduated Perspex standard (yellow/

brown). Compare the colour to that produced by a titrated preparation of the standard.

CYANMETHAEMOGLOBIN METHOD

For this method a colorimeter is required along with a standard haemoglobin solution and Drabkin's reagent.

Drabkin's reagent:

Potassium cyanide (KCN) 0.05 g

Potassium ferricyanide ($K_2Fe(CN)_6$) 0.20 g

Distilled water 1000 ml (1 l).

This reagent is stable for 1 month. Store in a stoppered bottle and label. Note that cyanide compounds are toxic so handle with care.

Add 0.02 ml of blood to 4 ml of Drabkin's reagent in a stoppered bottle. Mix by inversion two or three times and allow to stand for 10 min. Compare against a commercially available standard in a colorimeter. Most haemoglobin standards have a concentration of 12.0 g or 16.0 g/100 ml (12 g/dl or 12%). If available, modern colorimetric techniques can also be used, for example the Diaspect Tm handheld haemogobinometer described in Chapter 2 (Figure 2.45).

Erythrocyte sedimentation rate

The rate at which RBC settle in a set time is known as the erythrocyte sedimentation rate (ESR). The ESR is influenced by a number of factors, for example, the number of immature red cells present, rouleaux formation, the presence of abnormal cells and so on. In some species it is used as a guide to the effectiveness of long-term therapy, for example, in cases of inflammatory disease. In order to fully evaluate the significance of the ESR it is important to consider it in reference to the PCV and the species (for example, horses tend to have a higher ESR).

Generally, because of mechanical factors, a high PCV is associated with a low ESR. Equally a low PCV may be associated with an elevated ESR.

Method

Fill a Wintrobe haematocrit tube with blood to the top of the 100 mm scale and place in a vertical position at room temperature. The level to which the top of the RBC column falls in exactly one hour is recorded as the observed ESR.

Interpretation](simplified)

The rate of settling is influenced by: (1) species; (2) the number of red cells (that is, PCV); (3) the concentration and composition of plasma proteins, that is, the ESR is often high in chronic disease due to the increased amount of fibrinogen and gamma globulins; (4) the maturity of the erythrocytes, that is, reticulocytes and immature red cells exhibit a greater tendency to clump or form chains (rouleaux formation) than mature red cells; and (5) physical factors such as cleanliness of pipettes, the position of the ESR tubes, vibrations, ambient temperature and so on.

5.5 Collection and examination of smears from bone marrow

The primary function of bone marrow is the production of red blood cells, granulocytes (neutrophils, basophils and eosinophils), agranulocytes and the precursors of thrombocytes (platelets). Active bone marrow is red and resting marrow is yellow in colour. In newborn animals, most of the bone marrow tissue is red marrow and is actively producing blood cells. In adults, yellow marrow fills much of the bone cavity space in the long bones (those of the limbs) but this may convert to red marrow if the demand for red cell production increases as in blood loss and chronic anaemia. The flat bones (ribs, pelvis and skull) and the short bones of the vertebrae contain red marrow throughout life.

Bone marrow samples collected from freshly dead animals can provide useful and relatively easily collected diagnostic material but only appropriately qualified and experienced veterinary clinicians should collect bone marrow from a live animal.

The most popular puncture site chosen in live animals is the iliac crest or the head of the femur (dog, cat) but the ribs and sternum may also be used (cow, horse). Once the animal is appropriately restrained (or under anaesthesia), local anaesthetic (for example, lignocaine) is injected around the region and a skin incision made over the selected bone. A sterile wide bore needle is required to aspirate the marrow material using a 10 or 20 ml syringe. It is important to collect as much cellular material as possible and to minimize the amount of fluid in the sample. However, if too much fluid is aspirated the aspirate can be centrifuged and the sediment examined on a microscope slide. The prepared bone marrow smear can be fixed in methanol and stained with Giemsa stain as for a routine blood smear. A differential count is usually made after counting 500 nucleated cells. The myeloid/erythroid (white cell/red cell) ratio can also be determined from the smear and can be used to assess the bone marrow response to a blood loss anaemia. Interpretation of the results requires experience and depends on the species, this will not be considered further here.

5.6 Determination of normal values

Every diagnostic laboratory needs to establish the normal range of haematological profiles for the species likely to be tested in a given region. Typical haematological ranges for common domestic species are available in text books but because each laboratory uses different techniques it is necessary to develop specific standard values and value ranges for each veterinary laboratory. It is standard practice to list the normal range of values alongside the test values to allow the submitting veterinary or livestock extension officer to assess the results against the normal range for the laboratory. The normal

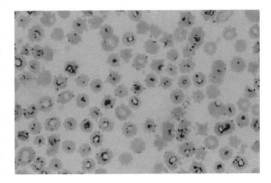

Figure 5.6 Blood smear from a cow which later died following fever, haematuria and weight loss over a period of several days. The animal had concurrent *Babesia bovis* and *Theileria* sp. infection following a spring peak in tick populations, Pemagatshel, Eastern Bhutan. Giemsa 20x magnification. The darker pigmented areas in the red blood cells indicate stages of the life cycles of the two protozoal species. See also Chapter 3 and Plate 22a.

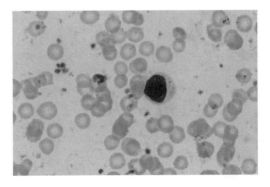

Figure 5.7 Poor quality bovine blood smear stained with Giemsa 100x oil immersion. Note the presence of a large mononuclear cell. The red cells are irregular in shape and there is a lot of artefact. The presence of granules suggests that new stain should be prepared. See also Plate 22b.

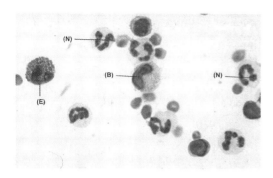

Figure 5.8 Equine blood smear viewed under oil immersion (Diff Quick 1000×) illustrating numerous polymorph neutrophils (N), an eosinophil (E) and a basophil (B). Diff Quick™ stain. RBC are also present. See also Plate 23.

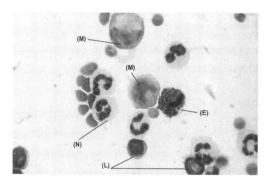

Figure 5.9 Equine blood smear viewed under oil immersion 1000× illustrating granulocytes (E and N) and agranulocytes (M and L). (E) Eosinophil, (N) polymorph neutrophil. It is not always possible to distinguish between monocytes and lymphocytes in blood smears. In this smear, the monocytes (M) are activated and are larger than dormant monocytes or lymphocytes (L). See text. Diff Quick™ stain. RBC are also present. See also Plate 24.

published ranges for common domestic species are given above in Table 5.4.

5.7 Interpretation of results

Interpretation of results requires a good understanding of the haematopoietic system, the immunological response and the nature of disease. Before evaluating haematology results, consider the following questions and points.

1 What was the purpose of the test, that is, is the animal unwell or has the sample been sent for the purposes of general health screening? What is the presenting clinical problem, that is, does the animal have anaemia or is the submitter interested in the white cell response to infection?

2 Check the submission form. Has the submitter only requested a general haematology screen or are specialized tests also requested? Some more specialized tests, for example, assessment for blood clotting disorders, may need to be done at a referral laboratory.

3 Test results for each case should be compared with the laboratory's list of normal values.

4 Do the results fit with the submitting veterinarian's list of differential diagnoses for the case? If not, it may be important to discuss the case further with the submitter to decide whether additional tests are required or if tests should be repeated on a fresh sample. Where there are unusual or unexpected results it is important to check that the test results are accurate, that is, ensure that quality control measures have been complied with.

Haematology can be used to provide useful information about the health of an animal and about the animal's response to disease. However, the interpretation of results can be challenging. In many cases sequential samples may be needed to confirm a diagnosis. Sequential samples can also be used to help the submitter assess the likely prognosis for a case, that is, is the animal's immune system responding effectively? Some common, non-specific, changes seen in the white cell fraction of the haematology profile are outlined in Table 5.5.

Haematology can provide a great deal of valuable information about a case but special-

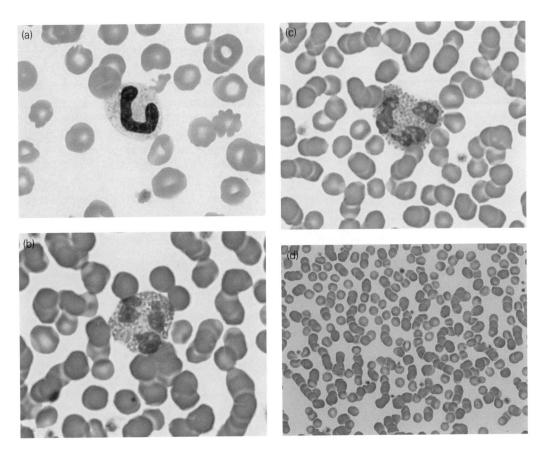

Figure 5.10 (a) Canine band neutrophil (immature PMN). Note the band shaped nucleus with smooth sides and lack of constrictions. This cell also shows signs of toxic change with a foamy appearance due to the presence of vacuoles. Modified Wright's stain 100×. (b) Feline eosinophil. Note reddish granules in the cytoplasm and segmented nucleus. The nuclear segments are connected by a fine strand. Wright's stain 100×. (c) Feline basophil. Note pale lavender granules in the cytoplasm and segmented nucleus. The colour of the cytoplasmic granules varies from pale grey to dark purple depending on the species. Wright's stain 100×. (d) Feline blood smear showing rouleaux formation. This is commonly seen in equine and feline blood smears. Wright's stain 20×. See also Plates 25, 26, 27 and 28. Photos: Dr Nicole Fernandez, WCVM.

ist knowledge and experience are required to assess more subtle changes in the blood picture. In order to learn more about the topic it is advisable to speak to the laboratory manager or the regional veterinary officer to see if there are any suitable training courses available. Additional information can be found in the texts listed in the bibliography and on websites, such as the Cornell University Hematology Atlas.[5]

Table 5.5 Simplified explanation of various changes to white cell populations in blood samples. For more information consult the bibliography at the end of the chapter.

Test	Cell type	Result and interpretation
Total white cell count*	All white cells	High (leukocytosis) – Inflammatory response, may indicate infection or severe tissue damage anywhere in the body. It may also indicate abnormal white cell production. Further interpretation requires a differential white cell count. Mild elevation may be seen as part of the normal 'stress' response.
		Low (leukopaenia) – May indicate immunosuppresion due to a disease process (especially viral disease) or abnormal white cell production as seen in leukaemia and many neoplastic diseases. Further interpretation requires a differential white cell count and assessment of cell morphology.
Differential white cell count	Neutrophils	High (neutrophilia) – Usually indicates the presence of an infection (including bacterial). Immature neutrophils are present in the blood when demand is greater than supply, this is known as a 'left shift'. This can indicate a guarded prognosis if the number of mature neutrophils starts to fall significantly.
		Low (neutropaenia) – May occur if circulating neutrophils are concentrated in tissues at the site of inflammatory response.
		A lower magnitude increase in neutrophils can also be a result of physiologic stress (cortisol induced).
	Eosinophils	High (eosinophilia) – May indicate an allergic reaction or parasitism. Small rises occur in stress but numbers tend to be low in healthy animals.
	Basophils	Numbers are low in normal animals.
	Lymphocytes	High (lymphocytosis) – May indicate an active immune response to a viral or other infectious agent or a neoplastic change.
		Low (lymphopaenia) – May be associated with overwhelming viral attack, stress or shock. If persistent, it can indicate a very guarded prognosis.
	Monocytes	High (monocytosis) – Levels are elevated in chronic disease as numbers increase to remove built up tissue debris. A mild monocytosis can occur with physiological stress.
Platelet count	Platelets	Low (thrombocytopaenia) – Often associated with chronic disease and may indicate bone marrow damage due to neoplastic, viral or auto-immune disease. It can also occur transiently after acute haemorrhage and is a feature of DIC. Thrombocytopaenia is often associated with blood clotting defects but further tests will be required to determine the cause of the problem.

Notes: *The total white cell count may be high, low or normal during a disease process and the differential white cell count may alter during the course of the disease. DIC = disseminated intravascular coagulation.

5.8 A note on automated systems

There are currently a large range of automated systems available for haematology (for example, Coulter counter[6], Cell Dyn[7]) but they tend to be expensive and may require constant replenishment of reagent kits. These systems are fairly simple to use once they are set up but regular quality control and system maintenance is essential. Most of the high throughput automated systems can use small volumes of blood and can be calibrated for a range of species but the 'normals' for each species must be regularly calibrated to ensure accurate results. As outlined in Chapter 2, most of the automated systems rely on the use of flow cytometry.

In most district laboratories staff will usually need to rely on manual methods but these methods need to be performed regularly if the haematology results are to be reliable. For the purpose of quality control, it is a good idea to periodically send duplicate blood samples to a laboratory which has an automated haematology analyser to check that the results from manual techniques are acceptable.

Endnotes

1 Note that if the proportion of band cells exceeds the number of mature neutrophils or if the neutrophil count is low but bands are present this could be a sign of per-acute inflammation or sepsis and the animal should be closely monitored.
2 The dipotassium salt of EDTA is the preferred formulation although the disodium salt can also be used.

3 Gentian violet (hexamethyl pararosanilinie chloride) and acetic acid are available from suppliers of laboratory reagents (see Appendix 4).
4 Formalin and trisodium citrate are available from most suppliers of reagents (see Appendix 4). The RBC diluting fluid must be isotonic with blood so that haemolysis does not occur. Normal saline can be used but it can cause crenation of the cells. Formalin acts as a preservative and sodium citrate prevents coagulation.
5 http://ahdc.vet.cornell.edu/clinpath/modules/heme1/intro.htm.
6 https://www.labcompare.com/10-Featured-Articles/162042-Hematology-Analysers-From-Complete-Blood-Counts-to-Cell-Morphology/.
7 https://www.corelaboratory.abbott/us/en/offerings/category/hematology.

Bibliography

Feldman, B.F., Zinkl, J.G., Jain, N.C. (eds) (2000) Schalm's Veterinary Haematology, 5th edn. Lippincott, Williams and Wilkins, Pennsylvania.

Harvey, J.W. (2001) Atlas of Veterinary Haematology: Blood and Bone Marrow of Domestic Animals. Saunders, Philadelphia, PA.

Hendrix, C.M., Sirois, M. (2007) Laboratory Procedures for Veterinary Technicians, 5th edn. Mosby, St Louis, Missouri. MO.

Pratt, P.W. (1997) Laboratory Procedures for Veterinary Technicians. 3rd Edition. Mosby. St Louis, Missouri

Reagan, W.J., Irizarry Rivera, A.R., DeNicola, D.B. (2008) Veterinary Haematology Atlas of Common Domestic and Non-domestic Species. Wiley-Blackwell, Ames, IA.

Schalm, O.W., Jain, N.C., Carroll, E.J. (1986) Veterinary Haematology, 4th edn. Lea & Febiger, Philadelphia, PA.

Thrall, M.A. (2006) Veterinary Haematology and Clinical Chemistry. Blackwell Publishing, Ames, IA.

chapter 6

Serology and immunology

Susan C. Cork, M. Faizal Abdul Careem and M. Sarjoon Abdul-Cader

6.1 Introduction

Serology is the branch of laboratory medicine that studies blood serum for evidence of infection, particularly antibody-mediated immune response by evaluating antigen–antibody reactions *in vitro*. It includes the study of antibodies (or antigens) present in serum, although antibodies can also be detected in other body fluids such as cerebrospinal fluid, saliva, milk and so on. A serology technician typically carries out techniques involving antigen–antibody reactions that are highlighted and measured in different ways (for example, agar gel immunodiffusion, precipitation or agglutination tests, complement fixation and so on). Serology, using paired sera taken over a 1–2 week period (during acute phase and convalescent period) can be used to detect current infections (that is, if a rising antibody titre can be demonstrated) and can also be used to indicate historical exposure to specific antigens.

The broader study of immunology includes serology as well as the evaluation of the cell-mediated immune response (for example, as detected in delayed-type hypersensitivity skin tests such as the tuberculin test). Research into the cause and treatment of cancer and immune-related diseases in humans and animals has improved our understanding of the cell-mediated immune response. Laboratory technicians and veterinary staff should have a basic understanding

of the principles of immunology, especially in relation to the tests that are carried out in the diagnostic laboratory. In this chapter, we will present an overview of the immune response along with the laboratory tests most commonly used at the district and regional level.

The immune response

In simple terms, the immune response in animals can be divided into 'specific' and 'non-specific' components (Figure 6.1a) and these two arms of the immune system are linked (Figure 6.1b). The 'non-specific' response refers to a general defence against all invaders, which is also called the 'natural' or 'innate' immune response and is potent, quick to respond (hours to days), broadly effective and lacks memory. The components involved in the non-specific immune system include the intact skin surface, intact mucous membranes and surface mucus, which often contains immunoglobulin (IgA). Non-specific macrophage engulfment of foreign material is also an important component along with the process of inflammation, which involves heat, and the release of chemical mediators to attract phagocytic cells (see Figures 6.2a and 6.2b). Recent discoveries show that natural killer (NK) cells, type 1 interferon (IFN) and pathogen recognizing receptors such as toll-like receptors (TLRs) are also essential contributors

(a)

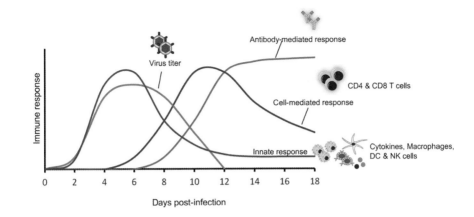

(b)

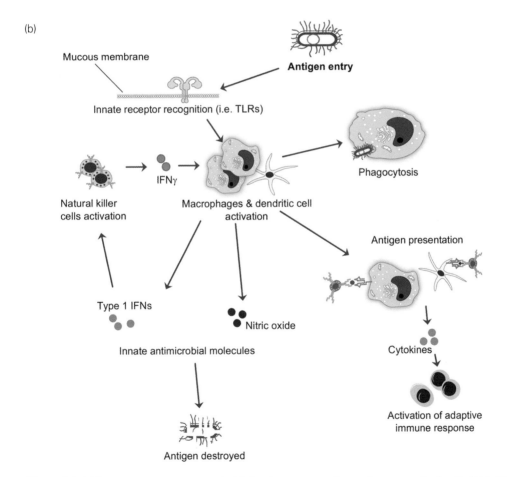

Figure 6.1 (a) Immune response generated following exposure to a pathogen can be innate (non-specific) response and adaptive (specific) response. See also Plate 29. (b) Innate immune response leads to antigen presentation and subsequent development of adaptive response. Source: M. Sarjoon Abdul-Cader, University of Calgary, Canada.

Figure 6.2(a) Diagrammatic representation of some aspects of phagocytosis, which is part of the innate immune response. Source: M. Sarjoon Abdul-Cader, University of Calgary, Canada.

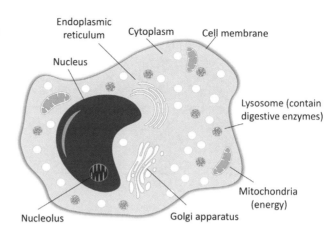

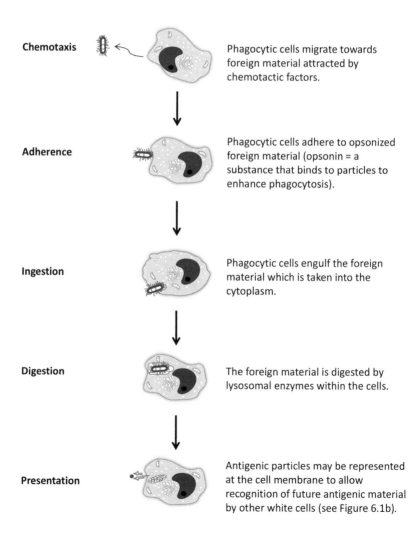

Chemotaxis — Phagocytic cells migrate towards foreign material attracted by chemotactic factors.

Adherence — Phagocytic cells adhere to opsonized foreign material (opsonin = a substance that binds to particles to enhance phagocytosis).

Ingestion — Phagocytic cells engulf the foreign material which is taken into the cytoplasm.

Digestion — The foreign material is digested by lysosomal enzymes within the cells.

Presentation — Antigenic particles may be represented at the cell membrane to allow recognition of future antigenic material by other white cells (see Figure 6.1b).

of innate host responses (Figure 6.1b). Factors that influence this first line of defence in the immune response include genetic factors, the level of stress such as the presence of concurrent diseases and the level of challenge. The second line of defence and the associated immune response is often referred to as 'acquired' or 'adaptive'. For the purposes of this chapter we will use the terms 'non-specific' and 'specific' immunity.

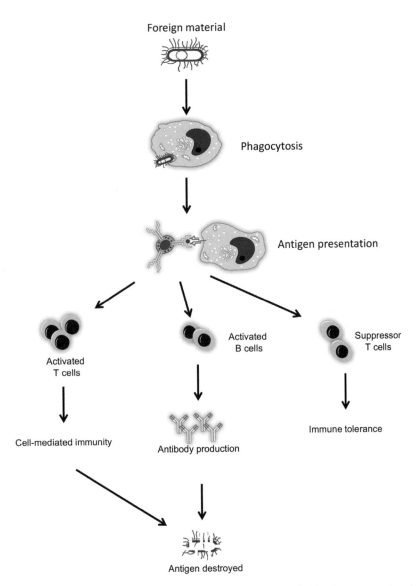

Figure 6.2(b) Essential features of the immune response. Foreign material is phagocytosed and presented to other cells of the immune system. This can stimulate a cascade of other responses resulting in (1) tolerance, (2) antibody production and/or (3) cell-mediated immune response (see also Figure 6.2a). Source: M. Sarjoon Abdul-Cader, University of Calgary, Canada.

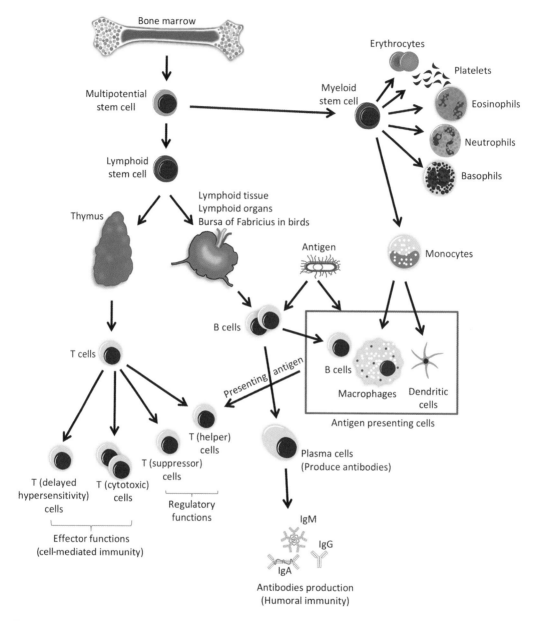

Figure 6.3 A simple chart outlining the cells involved in the specific immune response which consists of (1) humoral response and (2) cell-mediated response. In some disease conditions one will predominate, for example, cell-mediated immunity in tuberculosis and antibody-mediated immune response in acute viral infections, but in most situations both responses can be detected. Source: M. Sarjoon Abdul-Cader, University of Calgary, Canada.

Specific immunity is more 'specialized' than non-specific immunity in that it is a response to a specific 'invader', that is, a specific bacteria, virus, parasite or other foreign protein. Cells involved in the specific response recognize and remember individual invaders following antigenic presentation (Figure 6.2b) and a cascade of events occurs to deal with these invaders and any other subsequent invader with similar antigenic components. In this context, an 'antigen' is a substance that when introduced into the body stimulates the production of cell- and or antibody-mediated immune responses. Some bacteria share a similar antigenic 'coat' so cross reactions may occur which can result in false positive reactions to a given antigen, for example, *Yersinia enterocolitica* and *Brucella abortus* in laboratory serology tests. Cross reactions such as this may also confer a degree of immune protection if an animal is exposed to another pathogen with a similar antigenic profile.

The specific defensive response may broadly be split into two types, although they are interlinked with each other and with the non-specific response (Figure 6.2b and Figure 6.3). One is called the 'humoral' or antibody-mediated and the other the 'cell-mediated' response. The humoral response involves the production of B-lymphocytes from bone marrow, lymph nodes, liver and spleen. B-lymphocytes produce antibody specific to an antigen. The cell-mediated response involves the production of T-lymphocytes from the thymus. T-lymphocytes produce and respond to cell-mediators to facilitate phagocytosis and cell destruction. Tests to measure the cell-mediated response include the tuberculin skin fold test which is used to detect cattle exposed to *Mycobacterium bovis*. However, although new assays are being developed to measure the cell-mediated immune response (for example, lymphocyte proliferation assays, IFNγ assay) most of the commonly used diagnostic assays are still based on serological screening

and either measure immunoglobulins (Ig) (predominantly IgG and IgM – see Figure 6.4a) or are designed to detect antigens.

6.2 Antigens and antibodies

Antigen

An antigen is an immunogenic substance, usually of large molecular weight, which has a chemical grouping on its surface rendering it capable of stimulating an immune response in the animal host. Proteins tend to be more immunogenic than most carbohydrates although lipopolysaccharides may stimulate a strong immunogenic response and are frequently used in the preparation of 'killed' vaccines.

If purified antigen is injected into an animal's body it causes an immune response to take place resulting in the production of specific antibodies that are detectable in the serum a few days later, the time frame can be variable depending on the antigen type, route of administration and dose as well as the host's previous immune status (that is, a more rapid and higher antibody titre would be seen if prior exposure to the same antigen had occurred). This process is commonly used for the production of specific antisera (usually using laboratory rabbits) for the purposes of sero-typing and the production of test reagents and controls. However, not all antigens stimulate a strong serological response, for example, intracellular bacteria such as *Mycobacterium bovis*. For these organisms, the cell-mediated immune response is evaluated (that is, the tuberculin skin fold test).

Antibody/immunoglobulins

There are five classes of antibodies, or (Ig) in mammals (Figure 6.4a). The serum antibody found in birds is referred to as IgY rather than

Figure 6.4(a) The structures of classes of antibodies. The IgG molecule is made of two identical 'heavy' chains (H) and two identical 'light' chains (L). The main body of the heavy chains is variable in each antibody class but the light chains tend to be antigenically similar in each class. The ends of both light and heavy chains are composed of a variable sequence of amino acids (A–C and B–D), this is the antibody-antigen binding site. Source: M. Sarjoon Abdul-Cader, University of Calgary, Canada.

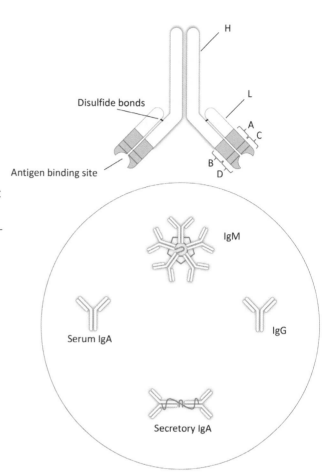

IgG but it is the functional equivalent to mammalian IgG.

In mammals, IgA, IgD, IgE, IgG and IgM are recognized. In simple terms, IgG provides long lasting immunity; IgM is part of the primary antibody response and is a short-lasting antibody with a large molecular weight; IgA is present in secretions (respiratory tract, tears, saliva, milk and so on) and IgE is produced in allergic reactions and is commonly associated with immunity to helminth parasites. The function of IgD is not fully understood but it has been shown to be active on B-lymphocytes as an antigen receptor. IgD has also been shown to activate basophils and mast cells to produce antimicrobial factors.

Antibodies are proteins produced by plasma cells which themselves are derived from 'B' cells and secreted into body fluids in response to antigen exposure. Immunoglobulins circulating in the blood make up approximately 15% of serum protein.

Antibodies combine specifically with the antigen that stimulated their production. The high degree of specificity of antibody for its respective antigen is its most important characteristic and forms the basis for all antigen – antibody reactions whether they occur *in vivo*, as a defence mechanism in the host, or *in vitro* in a serological test performed in the laboratory.

Specific antibodies are named by placing the prefix 'anti' before the antigen with which the antibody reacts. For example, antibody specific for the rotavirus antigen is called 'anti-rotavirus'. The species of animal in which the antibody is raised will determine the nature and extent of the antibody reaction. If rotavirus is inoculated into a pig and the pig produces antibody in response, the antibody would be called 'pig anti-rotavirus' antibody. A wide range of prepared antisera are now available commercially.

Structure and function of antibodies

A typical antibody molecule consists of four protein chains bound together so that they form a shape similar to the letter 'Y'. Each arm of the 'Y' contains a binding site which reacts with the specific site (that is, antigenic determinant or epitope) on an antigen (see Figure 6.4a).

The specific antigen–antibody reaction is a physical binding (Figure 6.4b). It can be compared to a lock and key fit, with the antibody being the lock and the antigen the key. In this way, the antibodies are able to inactivate antigens. This can also make antigens more readily phagocytized by white blood cells. The specific immune system, in a healthy host previously exposed to a range of antigens, is able to recognize and immediately respond to attack from many potential pathogens.

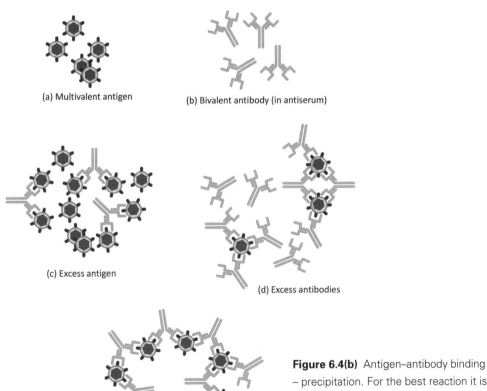

(a) Multivalent antigen

(b) Bivalent antibody (in antiserum)

(c) Excess antigen

(d) Excess antibodies

(e) Optimal proportion of antigen and antibodies

Figure 6.4(b) Antigen–antibody binding – precipitation. For the best reaction it is important to add antibody and antigen at the correct concentration. In many cases this must be determined by using serial dilutions. Source: M. Sarjoon Abdul-Cader, University of Calgary, Canada.

Primary antibody response

The primary antibody response is the first anti-body response seen following exposure to a new antigenic challenge (Figure 6.5). The first type of antibody detectable after initial exposure to the antigen is predominantly IgM, which usually appears 3 to 4 days after exposure. IgG is detectable one to 2 weeks after antigen exposure and decreases over a period of months or years (depending on the antigen involved). In contrast, the IgM titre (concentration) quickly peaks and then declines rapidly over a few weeks. Detection and measurement of IgM can provide information about the time frame in which an individual animal was exposed to an antigen

(or organism), that is, because IgM is produced early and declines rapidly, detection of antigen specific IgM indicates recent exposure (acute disease). A rising titre of IgG in serum samples collected 2 to 3 weeks apart also indicates recent exposure and possibly current active infection. Sero-conversion is the term used when antibody becomes detectable in the serum of an animal which previously tested negative.

The secondary antibody response or anam-nestic 'booster' response is the response seen after re-exposure to an antigen (Figure 6.5). This is what provides protective immunity in a vaccination regimen. In this case, because immune cells 'remember' the antigen, the IgM and IgG levels rise immediately (within 2 or 3 days)

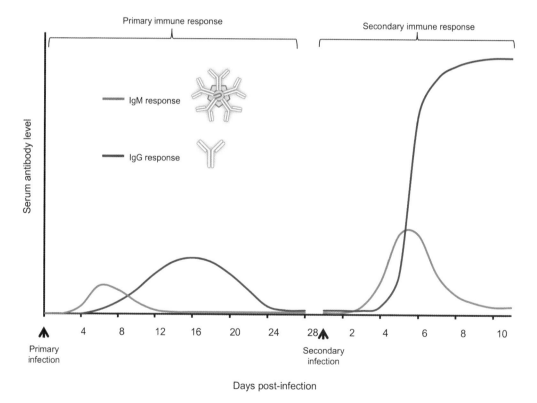

Figure 6.5 Serum antibody concentrations following primary and secondary infections. The initial IgM response follows the IgG response. The magnitude of the secondary response is greater and the time taken for the response to occur is shorter. See also Plate 30. Source: M. Sarjoon Abdul-Cader, University of Calgary, Canada.

following antigen re-exposure. In the secondary response IgG reaches higher levels than in the primary response and usually remains detectable in serum for several years. In mammals, the pregnant female is able to pass on some of this 'immunity' to the newborn in the first milk (colostrum). Intake of colostrum containing antibodies provides 'passive immunity' to the newborn but this protection does not last long and immunity needs to be boosted to ensure ongoing protection. Colostrum is crucial for newborn mammals because most species receive no passive transfer of immunity *via* the placenta before birth, so any antibodies that they need have to be ingested. This oral transfer of immunity can occur because the newborn's gut is porous. This means that large proteins (such as antibodies) can pass through the gut wall. The newborn animal must receive colostrum within 6 h of being born for maximal transfer of antibodies to occur. In birds, the IgY is passed to the chick *via* the egg yolk.

Before we move on to consider specific tests it is important to define some commonly used terms.

Definition and use of terms

Sensitivity

Sensitivity is the ability of a test to detect known positive samples (positive reactions). This should be considered in the context of analytical sensitivity as well as diagnostic sensitivity. The latter relates to the performance of a test in the field whereas the former relates to the detection limits of the test in the laboratory.

Specificity

Specificity is the ability of a test to distinguish between true positive samples and 'cross' reactions which may give false positive results. Ideally a test should be both highly sensitive and specific.

Agglutination

Agglutination is the clumping together of biological material, such as red blood cells (RBC) or bacteria, suspended in liquid, usually in response to a particular antibody. The reaction between an antigen and specific antibody can be visualized more clearly by attaching a coloured dye to antigen or antibody to highlight the reaction. RBC and latex beads can also be used.

Complement fixation

Complement fixation is the binding of active serum complement to an antigen–antibody pair. This is the basis for various diagnostic tests to detect the presence of a specific antigen or antibody. The complement cascade is an important chemical mediator of the inflammatory response. Activation of guinea pig complement in an *in vitro* system which contains sensitized red blood cells (complement fixation test [CFT]) is often used to test for the presence of antibody in a test sample.

Enzyme-linked-immunosorbent assay (ELISA)

ELISA tests are available in various formats but most involve specific antibody or antigen in the test system that is fixed to a plastic plate or dish. Test serum or other material is added. Antibody or antigen in the test material binds to bound antigen or antibody. The reaction is visualized (coloured) by addition of an enzyme linked to antibody or antigen and its specific substrate. Results can be measured by eye or using a colorimeter. There are now a lot of pen side ELISA kit tests available for preliminary serological screening. These have been widely used for influenza A screening as well as in the screening of a number of small and large animal viral infections and are available in both antibody or antigen capture formats (see Figure 4.29).

Fluorescent antibody test

Fluorescent dye is bound to antibody, which attaches to test antigen or antibody and is visualized using a fluorescent microscope.

Haemagglutination

Haemagglutination refers to agglutination of RBC. Some antigens are able to form a link between test RBC and facilitation, or inhibition, of this response can be used to highlight whether a test antigen–antibody is present (see Figure 6.7a).

Precipitation

Precipitation is the interaction of soluble antigens with IgG or IgM antibodies which leads to precipitation reactions. Precipitation reactions depend on the formation of lattices and occur best when antigen and antibody are present in optimal proportions. Excesses of either component decrease lattice formation and subsequent precipitation. An agar gel can be used to support the formation of a precipitate by an antigen and homologous antiserum. This is shown as an opaque line which is readily visible (see Figure 6.8b).

6.3 Tests used to measure antibody/ antigen reactions

All the serological tests that are used in district and most regional laboratories are likely to be based on a simple antigen–antibody reaction. These reactions take place at the microscopic level, which is generally not visible to the naked eye. Different variations have been developed to highlight or visualize the antigen–antibody reaction at the macroscopic level so that it can be seen and measured. Measurement may be visual (that is, using the naked eye) or by using instruments such as a spectrophotometer or

fluorescent microscope. Some examples are provided in the following section.

Suppliers of reagents and kits used in serological tests will state clearly how their product(s) should be used (see Appendix 4 for potential suppliers). It is important to follow instructions precisely and to use glassware, plastic ware and water which are free from contaminants. Most serological tests will be time and temperature dependent.

Agglutination and precipitation tests

At optimal proportions of antibody and antigen very large insoluble complexes can form, which can readily be visualized by naked eye. However, in cases of antibody excess and antigen excess only very small insoluble immune complexes are produced. Bivalent antibodies linking soluble antigens to form precipitates may also cross link particular antigens resulting in clumping or 'agglutination'. Agglutination can occur when a specific antibody is mixed with bacterial antigen as seen in field tests such as the pullorum test for *Salmonella pullorum* or the Rose Bengal test for *Brucella abortus*. In the pullorum test a drop of dyed antigen, prepared from cultures of *S. pullorum*, is mixed with a drop of whole chicken blood. In the presence of antibodies to *S. pullorum* clumping of the stained antigen occurs and is readily seen by naked eye. This is a simple test which can easily be performed in the field (Figure 6.6).

In some cases, the preferred serological test depends on the type of antibody present and this may change during the course of an infectious disease. For example, serum levels of IgG tend to peak after those of IgM. This is illustrated in the Table 6.1, which outlines the extent of reaction for IgG compared to that of IgM.

Simple agglutination tests are not always very specific nor are they particularly sensitive but they are easy to carry out in the field

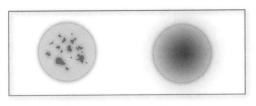

Positive   Negative

Figure 6.6 Diagrammatic representation of what can be seen in the slide agglutination test. This is a very basic test which is simple to perform and useful for initial screening. A drop of antigen is added to a drop of serum or blood and mixed using a tooth pick. If there is specific antibody present, there will be 'clumping' of the antigen (agglutination response). It is often standard practice to use a colored antigen for field based tests (for example, Rose Bengal antigen for *Brucella abortus*) so that it is easy to read the result. If the test is positive, it is usually followed up with further tests to provide a quantitative analysis of the level of antibody present in the serum sample. Source: M. Sarjoon Abdul-Cader, University of Calgary, Canada.

Table 6.1 Extent of reaction for IgG compared to that of IgM.

Property	IgG	IgM
Agglutination	+	+++
Precipitation	+++	+
Complement fixation	+	+++
Neutralising	+	++
Time after exposure to antigen	3–7 days	2–5 days
Time to reach peak titre	7–21 days	5–14 days

and are inexpensive. Owing to the potential for cross reactions to occur there should be a careful assessment of the test results in view of the clinical presentation and herd/flock history before any action is taken. Follow up testing, using more specific tests, of any preliminary positive samples is recommended. This can be

done using an ELISA test (if available) or a tube agglutination test. The latter is best performed using paired serum samples collected 10–14 days apart to check for a rising antibody titre.

Slide agglutination

One drop of a suspension of bacteria mixed on a microscope slide with serum containing the specific antibody will produce clumping that can usually be seen with the naked eye. As mentioned above, the reaction can be seen more clearly by binding a coloured dye to the antigen as in the tests for Brucellosis in cattle (Rose Bengal antigen – see Figure 6.6) and for some Salmonellae in poultry, for example, pullorum test antigen.

Where type specific antisera are available, a similar procedure can be used to identify the different groups of bacteria such as streptococci.

Materials

- glass slides (clean, grease free)
- micro- or Eppendorf pipette (0.03 ml) or a Pasteur pipette (for 1 drop)
- a box of tooth picks or match sticks
- stained antigen
- control and test sera.

Procedure

- Allow the antigen to reach room temperature and mix well before use.
- Place one drop of the test serum on a glass slide (or 0.03 ml).
- Mix one drop of serum with one drop of the antigen using a match stick or tooth pick.
- Rock the slide gently from side to side for about 4 min.
- Read the result (where border line reactions occur experience is required to assess the result, in cases of doubt the test should be

repeated and/or an alternative test should be used).

As with most diagnostic tests, positive and negative controls should be run with every test, or batch of tests, performed. In this case, the positive control will be a known positive antiserum. A negative control will be serum from a known sero-negative animal.

Recording the results

+++: rough agglutination (large quantities) with clear fluid

++: smooth agglutination with obvious border and the fluid a little turbid

Negative: no agglutination, comparable to the negative control.

If samples are positive on the slide agglutination test it is often advisable to determine the titre of the response using the tube agglutination test (for example, for Brucellosis).

Tube agglutination

Agglutination or clumping can be visualized in test tubes by using relatively large volumes of suspension. Agglutination is seen as a clearing of the suspension due to the aggregated clumps falling to the bottom of the tube resulting in cleaner supernatant. In veterinary laboratories this test is used to titrate serum antibody levels for *Brucella abortus* against a standard strain and strength of *B. abortus* antigen. The principles of the tube agglutination tests are outlined in Figures 6.7a and b.

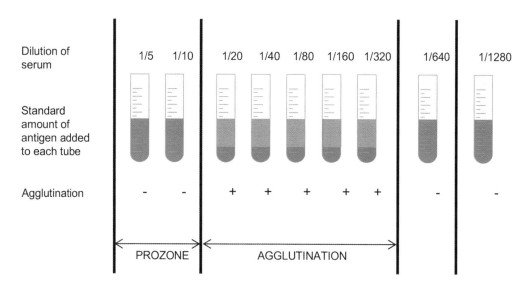

Figure 6.7(a) Tube agglutination test with doubling dilutions beginning at 1 in 5. The prozone is the 'zone' in which the amount of antibody present is very high and does not form a complex with the antigen present. Agglutination appears where levels of antibody and antigen form complexes which can be readily visualized when the concentrations are within an optimal range (see also Figure 6.7b). Source: M. Sarjoon Abdul-Cader, University of Calgary, Canada.

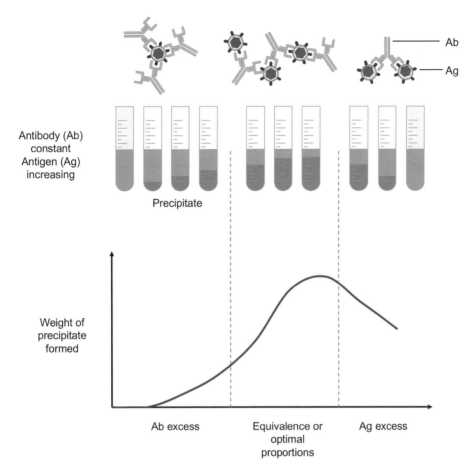

Figure 6.7(b) The quantitative precipitin test. When an increasing amount of soluble antigen is added to a constant amount of antibody, antigen–antibody complexes of different ratios are formed. At equivalence, all of the antigen–antibody is precipitated yielding the greatest amount of insoluble precipitin. Source: M. Sarjoon Abdul-Cader, University of Calgary, Canada.

Example of the tube agglutination test

Materials

- 37°C water bath
- microtitre pipettes (200–1000 μl)
- tubes and racks.

Method

- Place tubes in a rack and number (1–10).
- In tube No. 1 put 0.8 ml of phenol saline solution (0.85% NaCl and 0.5% phenol) and in tubes No. 2 to No. 8, add 0.5 ml of phenol saline solution.
- Deliver 0.2 ml serum to be tested in tube No. 1 and mix thoroughly. Next transfer 0.5 ml of suspension from tube No. 1 to tube No. 2 and mix. Twofold serial dilution is continued to tube No. 8 from which 0.5 ml is discarded. Hence, tubes No. 1 to No. 8 contain diluted serum from 1 : 5 to 1 : 640.
- Into each tube deliver 0.5 ml of the antigen so that the final dilution of sera is now 1 : 10 to 1 : 1280.

- For the negative control, add 0.5 ml phenol saline and 0.5 ml antigen into tube No. 9.
- For the positive control, add antibody (known positive) in 0.5 ml phenol saline and 0.5 ml antigen into tube No. 10.
- Incubate the tubes at 37°C in a water bath for 24 h.
- Read the results.

Results

++++: 100% agglutination with clear upper layer and sedimented aggregate

+++: 75% agglutination with 25% turbid upper layer

++: 50% agglutination with 50% turbid upper layer

+: 25% agglutination with 75% turbid upper layer

-: no agglutination.

The highest dilution showing 50% agglutination is usually considered to be the 'titre'.

Interpretation

This test allows determination of the titre (or degree) of the antibody response. A second blood sample collected a few weeks later may show a higher titre which would indicate a 'rising titre' and therefore is suggestive of current infection rather than previous exposure to the antigen. Paired sera should preferably be tested at the same time in the same test.

To facilitate the interpretation of 'in house' semi-quantitative and quantitative test results it is necessary to calibrate the test system to determine a 'cut off' point. This can be achieved by using known strong and weakly positive samples from known infected animals, and known negative samples, and titrating each of these out to assess level of the reaction. The 'cut off' titre

may be higher in tests where there is likely to be a lot of cross reaction, that is, in tests with low specificity. However, raising the 'cut off' titre can reduce the sensitivity of the test. For tests which used commercially prepared reagents published guidelines on 'cut off' titres are available. For tests using 'in house' reagents, the principles of test development and proper test validation are outlined in the OIE Manual for Vaccines Diagnostic Tests and for Terrestrial Animals (see reference list).

Immunodiffusion tests

When soluble antigen is mixed with specific antibody in solution in optimum proportions an immune complex (precipitate) is formed and the reactions can be visualized in an agar gel. This is the basis for the agar gel immunodiffusion test (Figures 6.8a and b) and is used frequently in regional laboratories. The test is simple to carry out and is fairly sensitive (and specific) as long as the reagent and the control serum are not the same. The immunodiffusion test can be used to measure reactants (that is, it is semi-quantitative) but it is usually used in regional laboratories to detect an unknown reactant (antigen or antibody) by comparison with a known reactant (that is, a qualitative test).

In the double diffusion Ochtelony test, antigen and antibody are placed in wells cut in agar (Figures 6.8a–c) and allowed to diffuse towards one another. At the point where they meet in the correct proportions a precipitate is formed which is seen as an opaque line in the gel. Care must be taken in the interpretation of the precipitation lines and only when there is a line of confluence can a positive result be determined with confidence. Manufacturers of various reagents will prescribe the optimum conditions required for the test. Immunodiffusion tests are usually carried out in petri-dishes. The gel used is usually made with a purified agar (for clarity) which

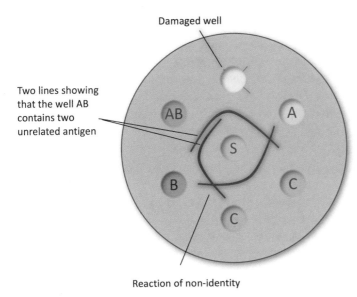

Figure 6.8(a) Typical layout used for an 'Ochtelony' diffusion test (see also Figure 6.8b). The wells are usually made using a metal template to cut the agar in the pattern shown. Damaged wells should not be used. S = antiserum containing precipitating antibodies to all three antigens (A, B and C). A, B, C = solutions of antibodies A, B and C (and a combination of A and B). Source: M. Sarjoon Abdul-Cader, University of Calgary, Canada.

Figure 6.8(b) Agar gel immunodiffusion test. When antigen solution is mixed with specific antibody in optimum proportions a precipitate is formed and this reaction can be visualized in gels. This is the principle of the agar gel immunodiffusion test which is widely used in veterinary laboratories as a simple and reliable diagnostic screening test. In the illustrations (A–C) we can see the method of testing an unknown antigen (a) and a known antigen (b) against the antibody raised (c) against the known antigen, this is based on the double diffusion or 'Ochtelony test'. In (A), the test and control antigen are the same resulting in a smooth and continuous line of precipitation forming with the antibody. In (B) there is 'partial identity' in which there is some recognition of the test (a) and control antigen (b) but they are not identical and in (C) there is 'non-identity' indicating that the reaction observed between the two antigens (a_2, b) and the control serum is not the same. Source: M. Sarjoon Abdul-Cader, University of Calgary, Canada.

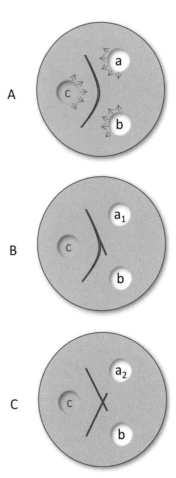

is buffered to optimum pH and may contain a preservative such as phenol to prevent microbial contamination. The layer of gel that is poured into Petri dishes should be level so that the wells cut will contain equal volumes of reactants. Incubation times and temperatures will vary depending on the test reactants. It may be necessary to take precautions to prevent the gel from drying out if incubation times are lengthy. An

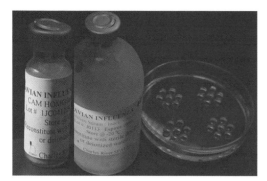
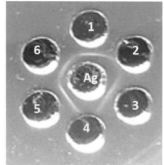

Figure 6.8(c) Agar gel immunodiffusion is a qualitative test used for the detection of antibodies against influenza A virus routinely. A semi-solid agar gel base is used to cut the wells as indicated in the Figure 6.8(a). The central well is allocated for placing the antigen of interest and the well numbers 1, 3 and 5 are used for placing the positive serum to the antigen of interest (a). The well numbers 2, 4 and 6 have been used for the serum samples to be evaluated. Within 48 hours of placing the sera and antigen precipitin lines are identifiable. The left image shows the standard avian influenza antigen, serum and a prepared agar gel base. The right image shows the antigen–antibody reactivity as visualized by precipitin lines. Well number 2 = weak positive, well number 4 = negative and well number 6 = strongly positive. Photos: Dr Davor Ojkic, Animal Health Laboratory, University of Guelph, Canada.

oblique electric light may assist with the reading of opaque lines. Positive and negative controls should be included in each test. Occasionally a haze forms around the well due to lipids or other material in which case the test may need to be repeated with a fresh sample.

Haemagglutination tests

HA is the agglutination of RBC, which can be caused by chemical agents or viruses. Haemagglutination inhibition (HI) is the inhibition of HA, for example by using hyperimmune serum specific to a haemagglutinating virus to prevent haemagglutination occurring. The property of some viruses to agglutinate the RBC of poultry and certain other animals is used in a range of systems to detect viral antigen and also to titrate specific antibody. The agglutination of RBCs is accomplished by the binding of proteins on the outer coat (envelope) of the virus with receptor sites on the red cells. Presumptive tests to detect haemagglutinating virus in cell cultures or in faecal emulsions (for example, parvovirus in dog faeces) can be carried out by demonstrating haemagglutination with appropriate RBC. Haemagglutinating properties of viruses are most often used in diagnostic laboratories to detect and titrate antibodies by using the HI. However, before the HI test can be carried out, the haemagglutinating strength of the virus must be determined. This is done in order to standardize the test so that the HI test is carried out in different laboratories using a uniform strength of antigen. Doubling dilutions of antigen are mixed with washed RBC and allowed time to agglutinate, the dilution at which agglutination stops is considered to contain 1 haemagglutinating unit (1 HAU). By counting back two doubling dilutions the dilution containing 4 HAU is determined and this is the strength of antigen usually used in the HI test. Note that there are a number of methods published for HI tests, here we will outline a method that we have used.

Haemagglutination inhibition test

This test is usually performed using a 96 well microtitre plate (Figure 6.9a and b). In this test, each well used will contain equal volumes and strengths of antigen, an equal volume of 1% RBCs and the same volume of doubling dilutions of test serum (antibody). Haemagglutinating properties of the virus are blocked by antibody binding to the virus. Blocking takes place up to the dilution of serum (antibody) where there is no longer sufficient antibody to combine with, and eliminate the virus. At this dilution virus will be free to agglutinate RBCs.

Protocols for HA and HI tests for Newcastle disease (NCD) in poultry follow.

HA TEST

Preparation of chicken RBCs (1.0%)

1 Collect chicken RBCs in an equal volume of Alselver's solution at least 3 days before the test.
2 Wash RBCs three times in PBS (pH 7.2 to 7.4 isotonic saline).
3 Mix 1.0 ml of packed RBCs with 99.0 ml of PBS (1% VV).

Twofold dilutions of a 0.025 ml (25 μl) volume of antigen are made in sterile PBS (pH 7.0 to 7.4). To do this, place 0.025 ml of PBS in each well, 0.025 ml of virus solution is added to the first well, mixed, and 0.025 ml is taken to the

Figure 6.9 (a) Photograph of a plate used to perform the HI test (note that this is not the same test plate as that shown in Figure 6.9b). Where there is a 'button' of red cells, this indicates that agglutination has not taken place. The HI test is used for the detection of antibodies against, for example, influenza virus. The test sera are treated to eliminate any non-specific haemagglutinins and inhibitors and are two-fold diluted with PBS in 96 well micro titre plate. Antigen (in this case influenza virus) is then added. Column 1–10 HI titres = 64, 128, 128, 32, < 4, 16, 32, 16, 16, 32 respectively. See also Plate 31. Photo: Dr Davor Ojkic, Animal Health Laboratory, University of Guelph, Canada. (b) Diagrammatic representation of the haemagglutination inhibition test. Rows F, G and H contain reagent controls. The button of RBCs indicates no agglutination. See also Plate 32. Source: M. Sarjoon Abdul-Cader, University of Calgary, Canada.

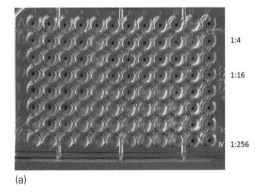

(a)

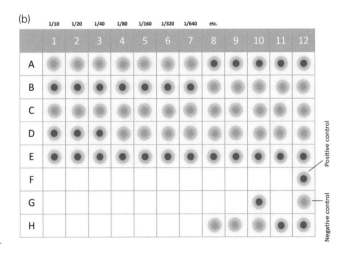

second well and double diluting is continued to the last well from which 0.025 ml is discarded. Then 0.025 ml of 1% v/v of washed chicken RBC is added to each well and gently mixed.

The test is carried out in a 96 well (U-shaped bottom) microtitre plate and the results are recorded after 60 min. Agglutinated cells will be prevented from settling. The test should be read at the highest dilution giving complete aggluti-nation. This is assessed by tilting the plate and the highest dilution where no 'streaming' occurs is the endpoint and represents one HA unit.

Some technicians have difficulty understand-ing this titration and its conclusions. Imagine that neat virus solution has 100 agglutinating units, then a 1 : 2 dilution will have 50, and a 1 : 4 will have 25 and so on. A point will be reached when there is less than 1 unit in the diluted virus (in this case 1 : 128) and at this point the RBC will no longer agglutinate. In other words, it takes at least 1 unit to agglu-tinate RBC. In this example, a 1 : 64 dilution was the last dilution to have at least 1 unit. If we take this dilution as 1 unit then 1 : 32 must be 2 and 1 : 16 must be 4 HA units. 1 : 16 is the dilution of antigen (virus) which must be used in subsequent HI tests. The reason why this step is necessary is that different preparations of viral antigen may have different levels of reactivity, that is, the titre is variable and must be stan-dardized so that test results can be compared between test batches and between laboratories.

HI TEST

Test and control sera are diluted to 1 : 2 in PBS and inactivated at 56°C for 30 min (to remove complement). Twofold dilutions of 25 μl amounts of serum to be tested are made in PBS. Then 25 μl of diluted virus (4 HAU) is added to each well. After a reaction time of 15 to 30 min, 25 μl of 1% v/v washed chicken RBC are added to each well. After gentle mixing the plate is left for 45 min at room temperature (22–24°C). If the ambient temperature is high, put the plate

in the fridge and leave for 60 min at 4°C. The HI titre is the highest dilution of serum caus-ing complete inhibition of agglutination. This is assessed by tilting the plate and observing the highest dilution of serum at which the button of red cells 'streams' at the same rate as the positive control RBC.

Both the HA and HI tests can usually be done at room temperature unless this is extremely high, in which case the reagents should be used directly from a refrigerator and the plates placed at 4°C during the test. Titres may be expressed as a dilution, for example, 1 : 8 or 1 : 256; as the reciprocal of the dilution, that is, 8 or 256, or as log$_2$ of the reciprocal, for example, 3 or 8. Alternatively, as mentioned previously, the titre can be represented as the highest dilution that inhibits haemagglutination multiplied by the number of HA units used.

REAGENTS

Diluent: Sterile isotonic saline (0.85%) preferably use PBS to ensure a pH of 7.2 to 7.4.

RBC (2–3 ml blood); ideally, these should be collected into an equal volume of Alselver's solution from not less than four (2–4-week old) specific pathogen free (SPF) chickens. The cells should be washed not less than three times in sterile isotonic saline followed by centrifugation at 1000 g for 5 min. The washed cells should be diluted to a 1.0% suspension which can be determined photometrically or volumetrically. The working suspension can be used for up to 36 hours if stored at 4°C prior to use. Blood cells can be collected from birds known to be free of NCD but it may be difficult to interpret the results if they have been vaccinated.

Complement fixation test

The fact that antibody, once it combines with antigen, activates complement can be used for diagnostic purposes. The standard CFT uses sheep RBCs sensitized with rabbit anti-sheep RBC antibody as an indicator system. In the body, the complement cascade is a chain of chemical reactions which is involved in the control of inflammation and the immune response. The CFT has the disadvantage that it is difficult to perform. This is largely because it is difficult to obtain and standardize the reagents used.

Guinea pig serum is commonly used as the source of complement for the haemolytic CFT as it has a strong ability to lyse RBCs. Activated (antibody coated) sheep RBCs are usually used as the 'indicator system' in this test. In the CFT, if specific antibody is present in the test serum sample, the sheep RBCs do not lyse as the antibody interferes with the action of the added guinea pig complement. If antibody is not present then the guinea pig complement is available to cause cell lysis. The basis for this test is illustrated in Figure 6.10. The advantage of the test is that it is usually very easy to read the

Positive samples
Test system fixes complement (C)

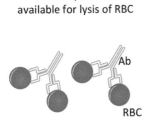

Negative samples
Complement is not fixed in the absence of antibody (Ab)

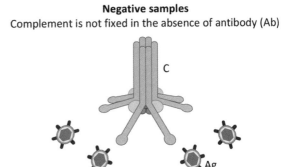

No complement
available for lysis of RBC

Complement is available
to lyse RBC

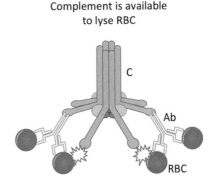

Figure 6.10 Principle of the CFT. In a positive sample the antigen–antibody reaction binds with the complement and inhibits the lysis of RBCs in the test sample. Source: M. Sarjoon Abdul-Cader, University of Calgary, Canada.

endpoint of the reaction. It may be necessary to titrate the antigen prior to the test to determine that the optimum concentration of antibody is used.

Fluorescent antibody tests

Fluorescent dyes such as fluorescein isothiocyanate (FITC) and rhodamine can be conjugated (coupled) with antibodies without interfering with the ability of the antibody to combine with a specific antigen. This antigen–antibody-fluorescent dye combination can then be viewed using a fluorescent microscope (Figure 6.11). This principle is employed in veterinary laboratories to detect microorganisms.

Two ways of carrying out this test are detailed below.

Direct method

In this test, the fluorescent dye is conjugated with a specific antibody to the antigen that is to be demonstrated. For example, if it is rabies virus (negri bodies in brain tissue) under test, then the brain smears are incubated with anti-rabies (IgG) coupled with fluorescein.

Indirect method

In this test, an unconjugated antibody is applied directly to the test sample. An antigen–antibody reaction (coupling) at this stage may take place but cannot be viewed under a fluorescent microscope since no fluorescent label is present. The tissue is then washed to remove the unattached antibody. A fluorescent conjugated anti-immunoglobulin to the unconjugated antibody is now applied. If antibodies are still present (which will only be the case if they are bound to specific antigen in the test material) conjugation of the immunoglobulin/anti-immunoglobulin/fluores-

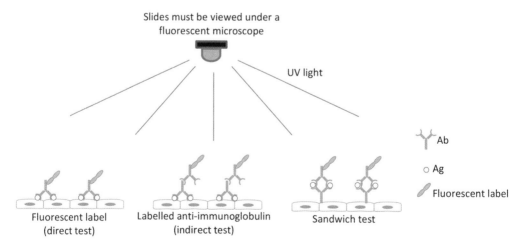

Figure 6.11 Fluorescent antibody technique/immunofluorescence. The test can be done using a direct, an indirect or 'sandwich' technique. Direct technique: a specific antibody, raised against the antigen of interest, is labelled with fluorescent dye and incubated with the test tissue. Indirect technique: a fluorescein labelled antibody is used that attaches to specific antibody raised against the antigen of interest in the test tissue. Sandwich technique: labelled antibody reacts with antigen bound by antibody present (produced by plasma cells) in the tissue section. Source: M. Sarjoon Abdul-Cader, University of Calgary, Canada.

cein will take place and this can be visualized under a fluorescent microscope. This test is more sensitive than the direct test and fluorescence is often brighter because there are more combining sites for the fluorescent anti-immunoglobulin.

These tests should be carried out in accordance with instructions provided with the reagents. To ensure that the test performs well the following should be considered.

- Make thin smears of material under test.
- Stick rigidly to pH and incubation recommendations.
- Do not use the fluorescence microscope if the voltage is fluctuating.
- Do not use too much mounting fluid and watch for air bubbles.
- Run positive and negative controls with each test.
- Do not allow preparations to dry out.

The test can also be performed using non-fluorescent labels, such as horseradish peroxidase and other enzymes, that react to form a colour with specific substrates. The latter forms the basis of immunohistochemistry (immunocytochemistry, Figures 6.12a and b).

Enzyme linked immunosorbent assay (ELISA)

As outlined earlier, the technology associated with the ELISA has made the use of diagnostic serology and antigen capture tests much more accessible for smaller veterinary laboratories. The ELISA is used to detect and quantify antigens and antibodies for a wide range of diseases. The majority of these tests are now available in kit form and have the advantage that some results can be read by the naked eye. Many of the modern kits use ELISA technology for single sample screening at the pen side (see Figure 4.29). However, for high throughput testing

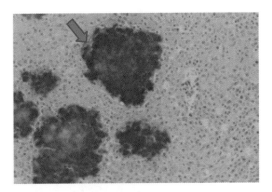

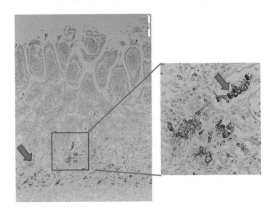

Figure 6.12 (a) Immunocytochemical staining of a section of liver (20×) from a bird that died following infection with *Yersinia pseudotuberculosis*. The slide was incubated with antibody against *Y. pseudotuberculosis* followed by incubation with a second antibody bound to a peroxidase conjugate. Diaminobezidine substrate was then added to illustrate the presence of Yersinia bacteria in the liver lesions. This is depicted by the darker staining areas in the slide. Photo: Dr Susan Cork, University of Calgary, Canada. (b) Immunocytochemical staining of a section of intestine from a dog which died following infection with parvo viral infection. The parvo viral antigens are demonstrated in the crypt area and not in the mature enterocytes of the tip of the intestinal villi. Photo: Dr Cameron Knight, University of Calgary, Canada.

that requires accurate and precise readings an ELISA plate reader, which determines the specific absorbance level for the colour change in the test, should be used. The ELISA plate reader can be calibrated to give an assessment of 'titres' for test samples by using titrations of known positive samples along with negative controls and preparing a standard curve or assessing values against a predetermined 'cut off' point. The plate reader will record optical density readings and calculate which samples have readings above the 'cut off' point, that is, those that are likely to be positive. In single sample test kits the result is usually clear cut with a colour change observed in the positive control and a similar reaction seen for positive samples. Negative samples will usually show no colour change or a different colour change to that of the positive control. The instructions provided with the kit tests will outline what to expect.

ELISA remains one of the most sensitive and specific of the serological tests carried out at the level of regional laboratories although the relative degree of sensitivity or specificity varies with the test used. In this context, the term 'sensitivity' is the ability of the test to detect true positive samples and the term 'specificity' is the ability of the test to distinguish between true positive and negative samples. In some tests, the sensitivity is increased at the expense of specificity, that is, all positive samples are detected but some false positives also occur. These tests are useful when the aim is to detect and eliminate a disease, but in the later stages of a disease control programme it is often necessary to re-test suspected 'false positive' cases using other more specific methods of diagnosis. The ELISA is becoming one of the most widely used serological tests for rapid diagnosis of animal diseases especially in surveillance programmes, but the following disadvantages must be noted.

- The cost of the kits can be high and purchase may require foreign currency.

- For high throughput screening, some systems are not sufficiently robust to give good repeatable results under less than perfect conditions (for example, water quality may have to be high and consistent).
- Local conditions may affect results (voltage fluctuations, low voltage, power cuts, poor refrigeration capacity and unreliable storage of reagents, availability of a generator).
- Ambient temperatures inside working laboratory rooms may fluctuate and can be variable. This may be a factor to consider.
- Tests done in small numbers will create waste (except where single sample kits are used).
- Kits do not always contain all the required items and replacement consumables and reagents may not be available locally.
- Many kits have not been fully validated for use on field samples from animals of unknown disease status, that is, cut off levels and potential cross reactivity with antigen–antibody used in the kits. Kit performance may need to be further examined by undertaking base line screening on samples from the population(s) of interest and comparing the ELISA result with those of a gold standard test, that is, AGD, CFT, HI and so on.

Nonetheless, ELISA test technology is a huge step forward in the diagnostic capacity of laboratories in developing countries.

ELISA test kits and systems now come in a variety of formats but essentially; in the ELISA enzymes are coupled to antibodies to mark or highlight immune complexes. The subsequent action of a substrate specific for the enzyme will then produce a colour reaction indicating the presence and strength of the immune complex.

A solid phase system is usually used for the test which involves binding protein (antigen or antibody) onto a polyvinyl or polystyrene surface as a first step. This can take place in the well of a microtitre (96 well) plate (8 × 12 well strips)

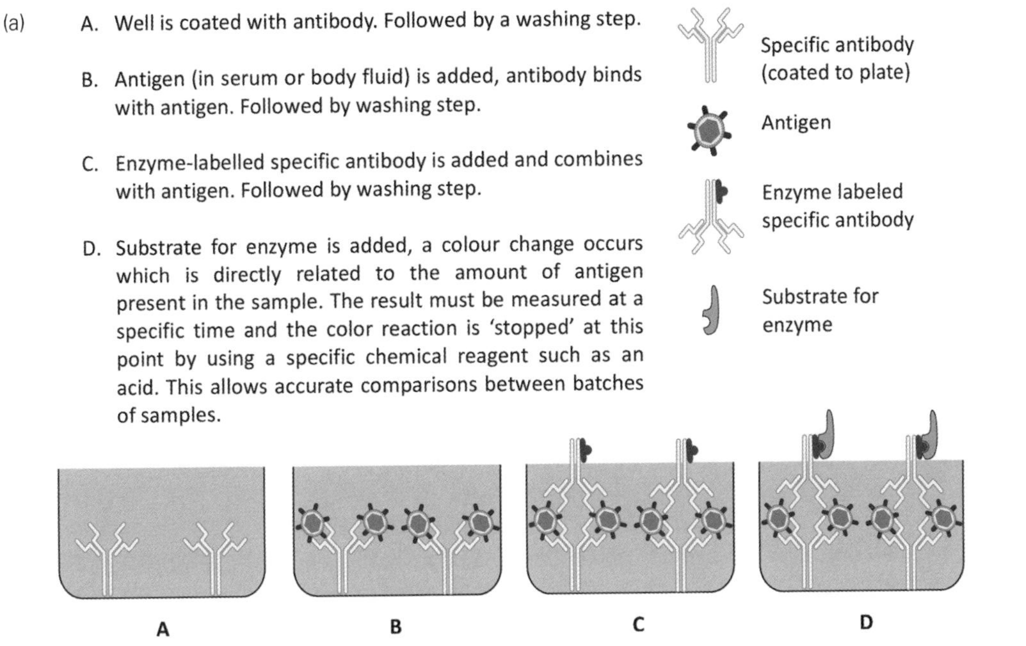

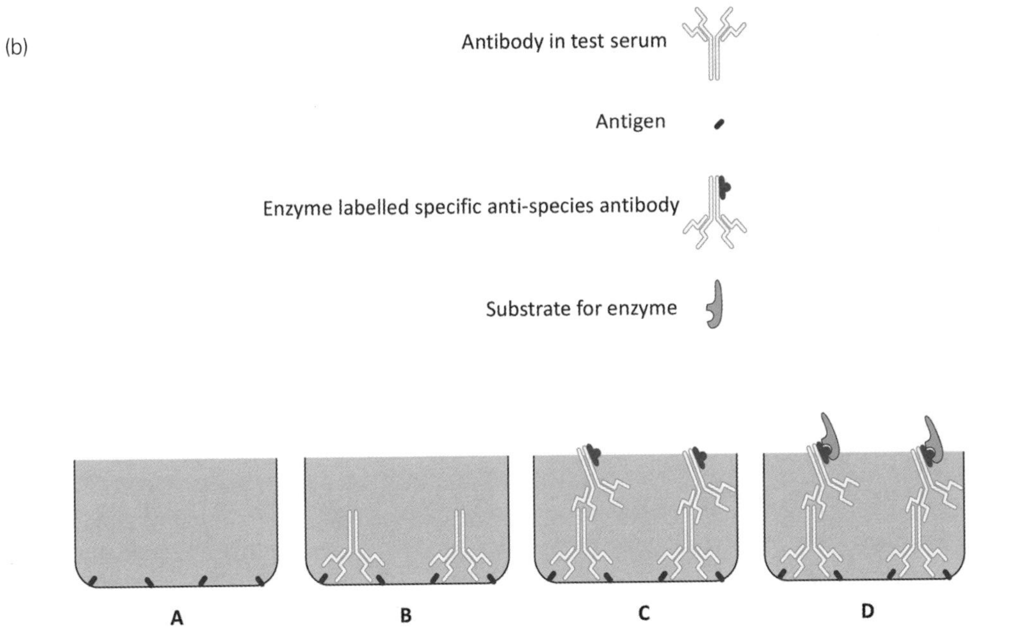

(a)

A. Well is coated with antibody. Followed by a washing step.

B. Antigen (in serum or body fluid) is added, antibody binds with antigen. Followed by washing step.

C. Enzyme-labelled specific antibody is added and combines with antigen. Followed by washing step.

D. Substrate for enzyme is added, a colour change occurs which is directly related to the amount of antigen present in the sample. The result must be measured at a specific time and the color reaction is 'stopped' at this point by using a specific chemical reagent such as an acid. This allows accurate comparisons between batches of samples.

Specific antibody (coated to plate)

Antigen

Enzyme labeled specific antibody

Substrate for enzyme

A   B   C   D

(b)

Antibody in test serum

Antigen

Enzyme labelled specific anti-species antibody

Substrate for enzyme

A   B   C   D

Figure 6.13 (a) ELISA technique for measuring the amount of antigen in a test sample (indirect method). (b) ELISA technique for measuring the amount of antibody in a test serum (indirect method). (A) Antigen coated to plate; (B) antibody binds to antigen, the plate is washed; (C) anti-species antibody bound to enzyme is added; (D) after washing the plate, substrate is added which reacts with bound antibody-enzyme complex. A colour reaction can then be measured. Source: M. Sarjoon Abdul-Cader, University of Calgary, Canada.

or in a single test format. Thereafter testing involves the following crucial steps.

1 Washing to remove unbound liquid phase reagents.
2 Addition of a buffer containing blocking agent to fill in un-utilized binding site spaces.
3 Addition of conjugate which is an antibody or antigen linked with enzyme.
4 Addition of chromogenic substrate which is specific for the enzyme used and will produce a colour as a result of enzyme/substrate/chromogen interaction.
5 Stopping solution which stops the enzyme reaction after the optimal level of colour development is attained.

There are several ELISA procedures which will highlight reactants in different ways, some of these are illustrated in Figure 6.13. ELISA tests can be developed which measure antigen (Figure 6.13a) or antibody (Figure 6.13b) in samples, some kit tests currently available for the diagnosis of viral diseases are outlined in Table 4.6. The following ELISA tests are those likely to be carried out in a regional laboratory.

The direct ELISA

• Antigen capture: Solid phase-antibody coated microtitre plate (or another format).
• Add enzyme labelled antigen.
• Antibody detection: Solid phase-antigen coated microtitre plate (or another format).
• Add enzyme labelled antibody.

The indirect ELISA:

Indirect antigen capture (that is, sandwich ELISA) (Figure 6.13a):

• Solid phase-antibody coated microtitre plate (or another format).
• Add test antigen material.

• Add detecting antibody to antigen.
• Add enzyme labelled anti-species antibody.
• Add substrate.

Indirect antibody capture (Figure 6.13b):

• Solid phase-antigen coated microtitre plate.
• Add test antibody (serum).
• Add enzyme labelled anti-species antibody.
• Add substrate.

Points to remember.

1 Follow the kit test instructions carefully.
2 pH is very important, so ensure that it is correct.
3 Care should be taken when pipetting to ensure accuracy and to avoid contamination of reagents between wells.
4 Use carefully washed glassware (several changes of distilled water).
5 Ensure adequate washing between steps.
6 Mix reagents well, check shaker settings before use.
7 Do not contaminate wells from droplets on plastic covers.
8 Watch for uneven coating of plates (or other format) and distribution of reagents in wells through poor mixing or bubbles.

6.4 Quality control and interpretation of results

As with any laboratory test, quality control is very important. See Chapters 1 and 7 for specific details about quality control in test systems. Most serological kit tests provide controls (positive and negative), a standard and often a reference sample. It is advisable to ensure that the laboratory has its own supply of positive (that is, hyperimmune) and negative sera for the tests that are performed most frequently. These can often be obtained commercially but known positive sera from vaccinated animals or from animals that tested positive in gold standard tests can also be used.

The results for serological tests are usually expected to fall within a set range but occasionally there may be an unexpected result (very high or very low), in these cases it is necessary to re-test the sample. For the diagnosis of infectious diseases most laboratories will request two samples collected over a 10–14 day period. This is to check for a rising antibody titre suggestive of active infection vs. previous exposure or vaccination. Interpretation of diagnostic and survey results often requires a lot of experience so it is necessary to discuss the results with the senior laboratory technician or veterinary officer in charge before they are passed to the submitting officer and/or owner of the animal.

6.5 Epidemiology and sampling plans

Serology is a valuable tool for diagnostic work and for general disease surveillance and screening programmes. Serological surveys require careful planning with a sampling plan based on a good understanding of the pathogenic agent involved, the disease process and the epidemiology of the disease(s) to be tested for. More information about the principles of epidemiology are outlined in Chapter 9 and are well described in the reference texts given at the end of this section. In most cases the design of a serological survey will involve input from a range of staff with complementary expertise. The degree of test sensitivity and specificity required will depend on the nature of the disease and the purpose of the screening. It is important that all associated staff understand how serological tests are chosen and that they are able to understand the procedures involved in assessing the reliability of the results.

Endnotes

1 See also the OIE Standards for diagnosing NCD available on line, http://www.oie.int/fileadmin/Home/eng/Health_standards/tahm/2.03.14_NEWCASTLE_DIS.pdf.

Bibliography

Hendrix, C.M., Sirois, M. (2002) Laboratory Procedures for Veterinary Technicians, 5th edn. Mosby, Maryland Heights, MO.

Kememy, D.M. (1991) A Practical Guide to ELISA. Pergammon Press, Oxford.

OIE (2008) Quality Standard and Guidelines for Veterinary Laboratories: Infectious Diseases. Various Authors. World Organisation for Animal Health (OIE), Paris, France.

OIE Manual of Diagnostic Tests and Vaccines for Terrestrial Animals. http://www.oie.int/standard-setting/terrestrial-manual/access-online/.

OIE Manual of Diagnostic Tests and Vaccines for Terrestrial Animals. Chapter 1.1.2 Collection, submission and storage of diagnostic specimens. http://www.oie.int/fileadmin/Home/eng/Health_standards/tahm/1.01.02_COLLECTION_DIAG_SPECIMENS.pdf.

OIE Manual of Diagnostic Tests and Vaccines for Terrestrial Animals. Chapter 1.1.3 Transport of biological materials. http://www.oie.int/fileadmin/Home/eng/Health_standards/tahm/1.01.03_TRANSPORT.pdf.

OIE Manual of Diagnostic Tests and Vaccines for Terrestrial Animals. Chapter 1.1.6 Principles and methods of validation of diagnostic assays for infectious diseases.

OIE Manual of Diagnostic Tests and Vaccines for Terrestrial Animals. Chapter 1.1.7 Standards for high throughput sequencing, bioinformatics and computational genomics. http://www.oie.int/fileadmin/Home/eng/Health_standards/tahm/1.01.07_HTS_BGC.pdf.

Quinn, P.J., Carter, M.E., Markey, B.K. and Carter, G.R. (1993) Clinical Veterinary Microbiology. Mosby, Maryland Heights, MO.

Tizard, I. (2008) Veterinary Immunology, 7th edn. W.B. Saunders Co., Philadelphia, PA.

chapter 7

Clinical chemistry

Susan C. Cork, Willy Schauwers and Roy Halliwell

7.1 Introduction

The discipline of clinical chemistry includes the biochemical analysis of serum and/or plasma, urine and other body fluids (for example, cerebrospinal fluid, transudates and so on). In this chapter, we will outline some of the common tests used for disease diagnosis with brief reference to toxicological tests and endocrine studies. Most district veterinary laboratories are only able to perform basic biochemical tests on serum/plasma, and body fluids and urinalysis. In larger laboratories, automated blood biochemistry systems may be available that can be used to perform a range of biochemical profiles on whole blood or serum plasma from common domestic species, but these systems can be expensive to run. Commercial diagnostic laboratories and some private veterinary clinics commonly use automated blood chemistry systems but cost recovery requires a minimal turnover of samples. These will not be considered in detail here. The samples required for biochemical tests depend on what tests are to be performed.

District laboratories are not usually equipped for comprehensive quantitative clinical pathology although semi-quantitative and simple qualitative tests may be possible Fouchet's test to detect bilirubin in urine or the 'dip-stick' kit tests which rely on simple chemical reactions which change the colours of reagent(s) in the strips (that is, as seen in reagent strips for blood

urea nitrogen levels, urine protein, urine or blood glucose, pH and so on). More specialized tests for toxicology and for endocrine assays are often only done in specialized referral centres. If you are likely to send samples to these reference centres you will need to be familiar with specific postal regulations for transporting clinical/pathological specimens in the standard post (see Chapters 1 and 8).

7.2 Blood sample collection and handling

Most biochemical assays performed require serum from clotted blood (that is, collect samples using a red topped vacutainer) but for some tests, plasma from heparinized blood (that is, a green topped vacutainer) or blood collected in calcium oxalate (that is, a grey topped vacutainer) is preferable, that is, for glucose assessment. Blood collected in EDTA tubes (purple topped) is generally used for haematology (see Chapter 5). Plasma may be collected from EDTA samples but these are not the first choice for biochemistry tests because the anticoagulant may interfere with the test. In general, 2 ml of serum or plasma is the minimum volume required (that is, from > 5 ml blood collected). This volume allows re-testing if necessary and also makes it possible to perform additional tests where indicated. Where possible, the submitting animal health professional should identify which tests are needed to confirm the suspected diagnosis and should contact the laboratory in advance to check the sample requirements, that is, before the samples are collected. Blood samples should be handled carefully to prevent haemolysis. A delay in centrifugation or poor collection technique may result in a poor-quality sample and false results due to transfer of ions across the red blood cell membranes and leakage into plasma or serum. Heparinized (and fluoride/oxalate) samples may be spun down (4000 rpm for 2–3 min or 2500 rpm for 10 min) and the plasma separated directly. Whole blood samples should be collected in glass containers and left to stand overnight at room temperature

(in temperate climates) or left in the refrigerator or cold room to clot. Refrigeration helps to contract the clot; the following day the serum can be removed carefully using a pipette and stored in labelled serum vials. Note that plastic syringes are not suitable for the collection of samples for some tests and that plastic tubes may promote haemolysis of whole blood during the clotting process. A summary of the recommended collection and dispatch requirements for some standard biochemical tests on blood samples is provided in Table 7.1. If haemolysis occurs during sample collection, or while separating the serum/plasma, the test(s) may give invalid results and will need to be repeated.

Table 7.1 Summary chart for blood sample collection and recommendations for specific biochemical tests.

Tests/indications	Specimens	Container*	Stability and comments**
ALT (alanine amino transferase)/enzyme released in tissue damage	3–5 ml clotted blood (serum)	Dry glass container, i.e. red topped vacutainer	Haemolysis interferes with the test. Stable in blood at 4°C for 12 h and in serum for up to 36 h
AST (aspartamine animotransferase)/ enzyme released in tissue damage	3–5 ml clotted blood (serum)	Dry glass container, i.e. red topped vacutainer	Haemolysis interferes with the test. Stable in blood at 4°C for 12 h and in serum for up to 36 h
Albumin/protein levels low in liver damage and starvation	2–3 ml clotted blood (1 ml serum)	Dry glass container, i.e. red topped vacutainer	Stable in blood at 4°C for 8 h and in serum for 4 days
Protein/usually determine fractions using electrophoresis	3–5 ml clotted blood (serum)	Dry glass container, i.e. red topped vacutainer	Stable in blood at 4°C for 8 h and in serum for 4 days
Amylase/elevated levels in pancreatic and neoplastic disease	3–5 ml clotted blood (serum)	Dry glass container, i.e. red topped vacutainer	Poor stability in whole blood, stable in serum sample for up to 12 h at 4°C. Can store longer if frozen
Bilirubin/elevated levels in obstructive liver disease and some haemolytic conditions	3–5 ml clotted blood or 3 ml with anticoagulant (serum or plasma)	Dry glass container, i.e. plain red topped vacutainer or with EDTA (purple topped)	Protect sample from light, table in whole blood at 4°C for 12 h and in serum for 48 h
Creatinine and urea/ elevated levels in some renal diseases, dehydration	5–7 ml clotted blood (serum)	Dry glass container, i.e. vacutainer	Stable in whole blood at 4°C for 8 h and in serum for 4 days

Table 7.1 *continued*

Tests/indications	Specimens	Container*	Stability and comments**
Electrolytes (Na, K, Cl), may alter in metabolic and other physiological disorders	5–7 ml clotted blood or 5 ml with anti-coagulant (serum or plasma)	Dry glass container, i.e. red topped vacutainer or Lithium heparin (green topped)	
Glucose	3 ml blood in grey-topped tube (calcium oxalate)	Dry glass container	Also use for samples collected for glucose tolerance studies
Calcium/low in milk fever (see section 7.7)	1–2 ml clotted blood/serum (not plasma)	Use dry glass container, acid washed glass, well rinsed	Stable in blood at room temperature or 4°C for 3 h/stable in serum for up to 72 h

Notes: *Plastic syringes may be used to collect samples but some types of plastic storage containers may contaminate the samples. Check with the supplier. **For referral, try to send at least 1 ml of serum in a chilled container to reach the specialist facility within 24 h. Make sure that the serum samples are correctly labelled and, where appropriate, send a case history with the submission form. It is important to add the collection date. Referral laboratories may batch samples together for some tests. Information collated from various sources.

Sample collection

Serum versus plasma

The required sample for most biochemical tests performed on blood is serum or heparinized plasma. As mentioned before, submitting animal health professionals should check what type of samples the laboratory recommends *before* collecting the sample.

Definitions: Serum is the fluid that separates from clotted whole blood (or from blood plasma) that is allowed to stand. Plasma is serum plus the proteins associated with clotting. Plasma is required for tests used to determine clotting times.

Make sure that the animal(s) to be sampled are appropriately restrained. This is to avoid injury to the animal and to the handlers but will also ensure a better-quality sample. Animal health staff and technicians should gain experience in the best method of blood sample collection from the species commonly kept in their region. When collecting blood, it is generally preferable to use the widest gauge needle suitable for the size of the blood vessel to be sampled because this will reduce the risk of haemolysis during sample collection (see also Chapter 5).

Transport and processing

If there is to be a delay in processing samples the serum (or plasma) should be separated from the red blood cells as soon as possible after collection. This will reduce the risk of haemolysis and deterioration of the sample, especially if samples are going to be transported for long distances in the heat. The samples will keep best when stored in glass containers in a cool place. If possible, it is good to set up a 'cold chain' especially if many samples need to be collected (that is, for disease surveillance, health monitoring and trace element survey work). For small numbers of samples a cool box will be adequate as long as fresh cold packs are available when needed.

Storage and preservation

In many cases serum samples will also be used for concurrent serological tests or will be stored in a serum bank for future reference. Samples that need to be kept for more than 2–3 days should be frozen at –20°C. Repeated freezing or thawing may damage stored samples and lead to false results in future tests so, since serum banks are valuable, there should be a reliable alternative power supply available in case of electricity cuts.

7.3 General health assessment

Serum biochemistry may be used for routine assessment of an animal's health or to assist in the process of confirming a clinical diagnosis. It is also useful to assess the health of a population, for example, samples can be collected from a representative percentage of animals in an area to assess the levels of serum trace elements (for example, copper) or mineral levels (for example, magnesium) especially in areas where nutrient deficiencies have occurred in the past or in high producing stock that may be susceptible to metabolic diseases.

Advances in technology have resulted in the development of semi-automated systems (see also Chapter 2) that allow full serum biochemistry

Table 7.2 Normal* serum values for biochemical parameters in common domestic species. Conversion units for common biochemical tests are provided in Table 7.3.

Test parameter	Cattle	Sheep	Pig	Horse	Dog	Cat
ALT (alanine aminotransferase) or SGPT** (iu/l)	11-40	22-38	31-38	3-23	6-24	5-15
Alkaline phosphatase (AP)** (iu/l)	35-350	68-387	120-400	95-233	0-170	0-150
Creatine phosphokinase (CPK)** (iu/l)	65	65	65	65	20-56	20-135
AST (aspratamine aminotransferase)/SGOT** (iu/l)	60-150	260-350	25-57	200-400	<50	<50
Albumin (g/l)	21-36	24-30	19-24	29-38	25-40	20-38
Globulin (g/l)	31-46	34-55	14-38	29-40	18-35	25-49
Total protein (g/l)	55-80	60-80	35-60	60-77	52-73	55-75
Total bilirubin (μ mol/l)***	0-32	0-6	0-4	4-102	0.3-7.6	0-1.3
Direct creatinine (μ mol/l)	90-240	110-170	90-240	110-170	90-150	90-150
Urea (m mol/l)	2-10	3-7	3-9	3-7	2.8-8.3	5.0-10.8
Glucose (m mol/l)	2-3	1.7-3.6	3.6-5.3	3.3-5.6	2-5.5	3.3-5.5
pH	7.35-7.4	7.32-7.5		7.32-7.55		
Electrolytes (m mol/l)						
Sodium (Na)	132-152	145-160	140-150	132-150	137-155	140-155
Potassium (K)	3.8-5.8	4.8-6.0	4.7-7.0	3-5	3.7-5.8	4.0-5.0
Chloride (Cl)	95-110	98-110	100-105	98-110	100-115	108-120

Source: Adapted from various sources.

Notes: *In each species there is a wide range of what may be considered 'normal' and this will depend on age, sex, physiological status and so on. Values will also vary from laboratory to laboratory and depend on the test methods used. Each laboratory will publish a list of 'normal' values alongside the test values for the sample to allow comparisons to be made. Note that the units used for reporting results may also vary from laboratory to laboratory. Some simple conversion factors are provided in Table 7.2. Interpretation of results is discussed in section 7.5. **Enzyme levels are usually measured in international units (iu/litre) which usually refer to the measured activity of the substance (catabolic reaction) in a set time period at 25°C. ***Bilirubin (mg/dl) normal values range from 0–0.19 in cattle and 0–0.4 in sheep. Conversion factors are provided in Table 7.3.

Table 7.3 Conversion units for biochemical parameters.

Biochemical parameter	Metric unit	SI unit	Conversion factor
Albumin	g/100 ml	g/l	10.0
Bilirubin	mg/100 ml	μ mol/l	17.1
Creatinine	mg/100 ml	μ mol/l	88.4
Glucose	mg/100 ml	m mol/l	0.055
Protein	g/100 ml	g/l	10.0
Urea	mg/100 ml	m mol/l	0.17
Calcium	mg/100 ml	m mol/l	0.25
Magnesium	mg/100 ml	m mol/l	0.41

Notes: mg/100 ml = milligrams per 100 ml, g/l = grams per litre; m mol/l = 10^{-3} moles per litre, μ mol/l – 10^{-6} moles per litre.

profiles to be assessed using only a few drops of blood. However, these are expensive and reagents may not be easily obtained without foreign currency. Examples include the BC Reflotron and the Kodak Ektachem systems that provide a test profile (1–15 parameters) on a single sample. An example of a profile would be serum creatinine/urea and protein for renal function and estimates of bilirubin, enzymes (alanine aminotransferase [ALT], aspartamine aminotransferase [AST]) and serum albumin levels for liver function and so on. Common reference ranges for some biochemical parameters in serum are given in Table 7.2 for common domestic species.

The value of biochemistry results depends on (1) the appropriate choice of samples, (2) the quality of the samples and (3) the reliability of the tests used. Serum biochemistry is usually used to confirm a diagnosis or to help the animal health professional to make a prognosis for the case (that is, it can allow an assessment of the degree of organ damage and therefore the likelihood of recovery). However, if the results do not support the clinical assessment of an animal it may be worthwhile to repeat the test or to try an alternative test, because single test results can be misleading. In cases where an animal is assessed for response to treatment a series of samples may be collected to assess recovery.

In larger veterinary laboratories, trained clinical pathologists are usually able to assist in the interpretation of results and will give advice on sample submission, selection of tests, diagnosis and treatment recommendations. In smaller facilities, the technicians performing the tests may confer directly with the veterinary staff on location to determine what information should be provided to the submitting animal health professional.

Simple dip-stick tests

There are a wide range of simple dip-stick type tests available commercially. These include products that measure urea (for example, Azostix™) and glucose (for example, Dextrostix™, Glucostix™ and B-M test glycaemie) in whole blood. Many of these were originally developed for human medicine but can be used for veterinary purposes and give reasonably accurate results when used according to instructions. Azostix™, Dextrostix™ and Glucostix™ are manufactured by Ames, and can be procured through veterinary and medical wholesalers but this may require foreign currency. There may be local retailers who can provide local equivalents or it may be possible to request supplies through a local hospital. Most of these dip-stick type tests require collection of a very small drop of blood that remains in contact with a colour reagent strip for a specified period of time. The colour strip is then washed and the colour change read alongside a colour chart which relates to a known amount of urea or glucose in the blood. Some systems, for example, B-M test glycaemie, produced by BCL utilize double colour blocks which may give a more accurate result, and these systems may be read using a special reference system. Similar tests are also available for testing urine and other samples (Figure 7.1).

When using commercial reagent strips the instruction sheets should be read carefully to ensure that the correct test parameters are used. Before use it is advisable to check that the reagent strips are (1) not out of date and (2) that they are stored correctly. Most dip-stick tests come with detailed information about the optimal test method and how to interpret the colour change. Interpretation of the test result will also depend on the presumptive diagnosis based on clinical history and examination of the animal. If test results do not fall within expected parameters it may be necessary to reconsider the diagnosis but it is possible that the test result is incorrect, particularly if the sample has not been collected or handled correctly. If results are doubtful it is best to repeat the test on a fresh sample.

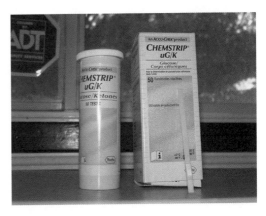

Figure 7.1 Chemstrip® uG/K are test strips used in human medicine for the semi-quantitative determination of glucose in urine and for the detection of ketone bodies (for example, acetone) in urine. Similar products are available for veterinary use. In human medicine these strips can be easily used by patients to monitor diabetes control. The test is based on dipping the reaction area of the test strip into freshly voided urine. There are two test reaction pads on these strips that change colour in proportion to the amount of glucose or ketones respectively, present in the sample. In normal healthy individuals neither of the reaction pads on the test strip will change colour. Any colour change noted after the recommended test reaction time is read against a colour coded key provided on the container. Similar test strips are available for testing blood and serum samples for the presence of glucose, creatinine and blood urea nitrogen. When used correctly these strips are fairly accurate and provide a good supplement to other diagnostic tests. Photo: Dr Susan Cork, University of Calgary, Canada.

Photometer methods

Most, if not all, biochemistry test systems are based on colorimetry in one form or another. The principles and methodology of quantitative biochemical tests are explained in section 7.4. There are a range of automated high throughput systems available (see Chapter 2) some of which require a large initial expenditure and have fairly high maintenance costs with ongoing requirements for reagent kits and servicing. In laboratories where large numbers of biochemical tests are performed regularly these systems would be worth considering but they may not be justified where the budget is restricted or where technical support for servicing repair is unreliable. There are also some newer and less expensive kit test systems which could be considered for smaller veterinary laboratories where cost recovery is an option.

There are two main types of colorimetry system, these can be classified as 'open' or 'closed'. Open systems may be adapted for different types of test procedure and are useful in a veterinary laboratory specializing in health screening and survey work. Closed systems are designed to be sold as compact units with specified reagent kits provided with each system. The advantage of the closed systems is that they are very easy to use. The disadvantage is the expense and the restriction of having a predetermined set of test profiles offered.

7.4 Colorimetry explained

The following section has been added to provide some background to the subject of colorimetry with the aim of allowing the laboratory worker to recognize the potential limitations of colorimetric tests and to be able to see the importance of calibration, quality assurance and monitoring to ensure that test results are valid.

The scientific principles

Most quantitative or semi-quantitative methods used in clinical biochemistry are colorimetric.

Many substances are either coloured in solution or can be altered to produce a coloured derivative. In most cases the intensity of the colour will be related to the amount of substance in the solution. In colorimetry, the colour intensity of a solution, containing an unknown amount of the test substance, is compared with a reference or standard solution which represents or contains a known amount of the substance of interest. Colorimetry is based on the principle that a coloured solution can absorb light at a given wavelength in the visible spectrum (Figure 7.2) and that the extent to which this occurs is dependent on the colour intensity. To appreciate the concept of colorimetry it is necessary to have an understanding of light energy. The source of light energy is electron activity. Light energy is emitted in waves of varying lengths and these are measured in nanometres (nm = 10^{-9} m) depending on the distance between peaks and troughs (Figure 7.3).

Colour and light absorption

Light may be absorbed, reflected or transmitted when it falls on a coloured solution in a tube. A coloured liquid seems to appear coloured because it transmits a particular wavelength of light from the visible spectrum (Figure 7.4).

The number of vibrations of wave motion per second is known as wave frequency and is measured in Hertz (Hz)

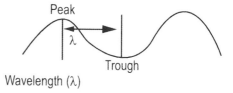

Wavelength (λ)

A wavelength (λ) can be expressed in the formula:
(λ) = V (velocity of light) / frequency (Hz)

V = velocity is measured as distance travelled by light per second

Typically, the lower the wave frequency, the longer the wavelength. The range of frequencies over which electromagnetic radiation is transmitted is referred to as the electromagnetic spectrum.

Figure 7.2 The principle of wavelength.

380-780 nm (visible light); 10-400 nm - UV; 0.75-1000 μm - IR; 10-0.01nm - x-rays

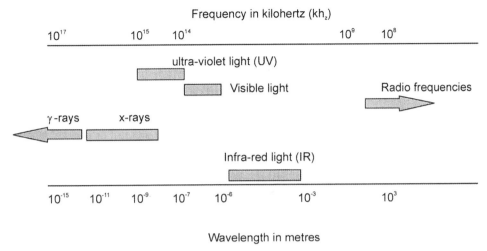

Figure 7.3 The electromagnetic spectrum.

Wavelength in nanometres (nm)

400 424   490   575 585   650 700

Violet Blue   Green   Yellow Orange   Red   Infra-red(>700nm)

UV(<400nm)

Visible spectrum

Figure 7.4 The visible part of the electromagnetic spectrum.

It has been shown that, under appropriate conditions, when illuminated with light of a suitable wavelength the amount of light absorbed by a coloured solution is directly proportional to the concentration of the coloured solution and the length of the light path through the solution.

The amount of light absorbed by a substance in solution equates to absorbance (or optical density [OD]) and this depends on (1) the light path and (2) the wavelength(s) used. The light path is usually kept constant for a particular piece of equipment by choosing good quality optically matched cuvettes (Figure 7.5).

Essentially, the concentration of the reference solution determines the amount of light absorbed by the reference solution (= absorbance). Using this principle, and a standard reference solution, the concentration of an unknown test solution can be determined as follows.

> Concentration of an unknown solution = absorbance of test solution × conc. of reference divided by the absorbance of the reference solution

For example, if the reference solution has a concentration of 20 mg/ml and an absorbance value of 250 (that is, the OD reading), an unknown test solution with an absorbance value of 240 has a concentration of 19.2 mg/ml.

> 20 mg/ml is to 250 what Z mg/ml is to 240

> So $Z = 20 \times 240/250 = 19.2$ mg/ml

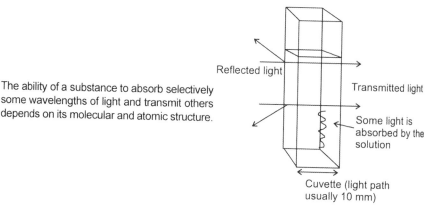

The ability of a substance to absorb selectively some wavelengths of light and transmit others depends on its molecular and atomic structure.

Reflected light

Transmitted light

Some light is absorbed by the solution

Cuvette (light path usually 10 mm)

Figure 7.5 Light reflected, absorbed and transmitted when it falls on a coloured solution.

However, this is only true if the relationship between absorbance and concentration is linear (see next section, Figure 7.6).

For biochemical tests that come in kit form, the wavelength required to read the test will be recommended in the test protocol provided. A standard reference reagent will also be provided to allow calibration. Even with kit tests it is necessary to standardize each test by running a 'blank' sample which contains all the reagents for the test except the test substance.

A standard (that is, reference) solution of known concentration, along with the 'blank' allow appropriate calibration of the colorimetry equipment. This should be done before each batch of samples is processed and is an essential part of the quality control process.

Practical aspects

The reliability of colorimetry based tests depends on the uniformity of the solution to be tested and the suitability of the equipment used. The cuvettes used must be of good quality and free of smears or dust. Plastic cuvettes should not be re-used. Some glass cuvettes may be re-used if properly cleaned.

In most cases it is necessary to prepare a standard curve to calibrate equipment for specific colorimetric tests. The procedure is outlined below.

- Prepare serial dilutions of a known amount of the substance to be measured and add the test kit reagents.

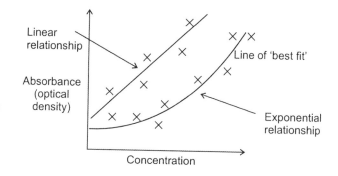

Figure 7.6 Calibration graph showing the potential linear or exponential relationship between absorbance readings and concentration.

Linear relationship

Absorbance (optical density)

Line of 'best fit'

Exponential relationship

Concentration

- The required graph is prepared by plotting the absorbance (OD) values obtained against the known concentration of the standard reagent.

If there is a linear relationship between the concentration of a test reagent and colour absorption (OD) then the concentration of the test substance is directly proportional to the colour intensity of the solution. However, in most cases the relationship will be exponential (Figure 7.6) and there will only be a specific area on the linear part of the 'standard curve' where accurate readings can be taken. In these cases, very low or high values may not be valid and concentrated solutions or samples should be diluted and re-tested.

Summary

- The colorimetry equipment used must be set at the correct wavelength to read the specific colour range absorption recommended in the protocol selected for a particular test.
- Equipment should be calibrated using a control or standard reagent and a 'blank' sample prior to running the test.
- As part of the quality control process, every batch of samples for a biochemistry test must be run along with standard controls, that is, a known positive and a known negative control. In the case of quantitative tests, a dilution series of the known standard reagent should also be run.

Quality assessment

It is standard procedure to periodically send duplicate serum/plasma samples to other laboratories for biochemical tests to check that laboratory results are reliable. It is also important to run an internal quality control scheme by keeping a record of all test results and preparing

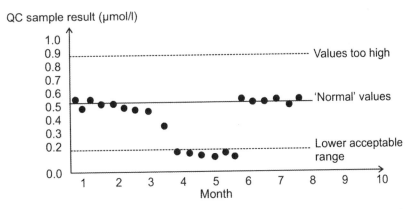

Figure 7.7 Plotting values for quality control (QC). The method of recording QC results illustrated above allows detection of abnormal results which occur due to laboratory error. The QC samples are samples which contain a known amount of test substance and are run along with each batch of test samples. The example above shows that samples tested in months 4–5 gave unacceptably low values but that the problem appears to have been sorted out by month 6. It is important to detect potential problems as early as possible and therefore monitoring of QC samples should be done on a daily basis in a very busy laboratory. The acceptable and 'normal' values for each set of QC samples should be determined for each test performed in the laboratory and a portion of the QC sample used should be sent to another diagnostic laboratory to make sure that values are correct.

a graph of 'standard' and/or 'control' values to make sure that the test procedure gives accurate (representative) and precise (repeatable) results each time. Limits of acceptability are predetermined and if the values of 'normal samples' lie outside set limits (see Figure 7.7) it is important to check the methods used. In some cases, the tests may need to be repeated. Quality Management Systems are discussed in more general terms in Chapter 1, section 1.6.

7.5 Interpretation of clinical chemistry results

There are a wide range of 'reference' tables that list the normal or expected range of biochemical values in different species of animal. It is important for a laboratory to validate its own list of 'normal values' for the tests performed by using blood, serum/plasma and other samples obtained from a selection of healthy animals in the local area. To check that the laboratory

values are representative is it also useful to run through a few test samples, with values assessed by other laboratories, to make sure that the results fall within acceptable parameters. There may be some variation in results between laboratories but if the samples tested are of good quality (that is, collected and stored correctly) the variation should be small if the same test procedures are followed.

The range of values considered to be 'normal' for a given series may be quite wide for some tests such that it is not always clear where the 'cut off' point for 'abnormal' values should begin. An example of the distribution of results for a typical biochemical assay are illustrated in Figures 7.8 and 7.9.

The interpretation of the values for a given test will depend, to a large extent, on the clinical examination and history of the case involved and the experience and expertise of the submitting animal health professional. In addition, the laboratory staff will be able to comment on the reliability of the test results and whether

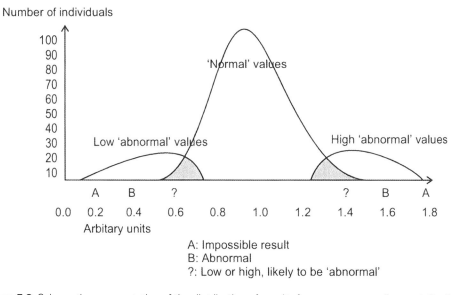

Figure 7.8 Schematic representation of the distribution of results for a serum assay ('normal distribution'). Note that 'normal' and 'abnormal' ranges may overlap.

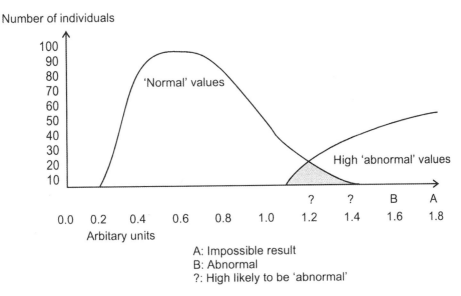

Number of individuals

A: Impossible result
B: Abnormal
?: High likely to be 'abnormal'

Figure 7.9 Schematic representation of the distribution of results for a laboratory test measuring a substance which has a 'skewed' range of values. Note that 'normal' and 'abnormal' ranges may overlap significantly with this type of distribution. Some tests for example, for levels of some toxic substances or metabolites associated with a disease process, will usually have little or none of the substance normally present therefore there are no 'low' values.

or not additional tests are required. In larger laboratories, there is likely to be a specialist in charge of the clinical pathology section who can provide advice on the significance of any abnormal biochemical test results and provide recommendation for follow up testing. Some general guidelines for interpreting common test results are provided below. Normal values for some common domestic species are provided in Table 7.3.

Nitrogenous substances

Urea, also referred to as blood urea nitrogen (BUN), is often elevated in kidney disease, circulatory failure and dehydration. Mild elevation can occur where weight loss has occurred with resultant muscle breakdown or when dietary nitrogen sources are increased. Low BUN is less common but can be seen in the early stages of some dietary deficiencies and some metabolic abnormalities.

Creatinine is another nitrogen component that may be elevated in circulatory failure and dehydration, and is a more specific indicator of kidney damage than BUN. The ratio of creatinine to urea is often used to determine the extent of renal impairment. Interpretation of urea and creatinine levels in a case is preferably done after consideration of urine specific gravity, which provides an indication of the animal's state of hydration. Ammonia may be measured in blood. Elevated levels are associated with abnormality of liver function. Samples for ammonia estimation must be tested shortly after collection and stored on ice.

Plasma proteins

Total plasma protein is often measured to assess the health and the level of hydration of an animal, raised levels often imply dehydration. Low plasma protein may indicate poor nutritional status and is seen in generalized debility. Assessment of the relative albumin and globulin levels can give a better indication of nutritional and health status of the animal. Albumin levels are low in starvation and/or liver disease and elevated in dehydration. Globulin levels are elevated in some immunological disorders and also in chronic disease.

Bilirubin and bile acids

In cases of severe liver damage or in haemolytic anaemia the plasma and mucous membranes may be yellow in colour; this is often due to elevation of the levels of bilirubin in the blood and tissues (jaundice). Bilirubin is a metabolite of haemoglobin which is usually metabolized in the liver. Levels rise if the breakdown of blood is too rapid (pre-hepatic jaundice) or in liver disease (post-hepatic or hepatic jaundice). If it is possible to determine whether or not the bilirubin is conjugated (that is, has been processed through the liver) or not, this will help distinguish pre-from post-hepatic jaundice. In haemolytic anaemia and other causes of pre-hepatic jaundice there will usually be other clinical signs such as elevated temperature and red urine (haemoglobinuria). In cases of pre-hepatic jaundice, it is helpful to assess the packed cell volume (PCV) of the animal to check for concurrent anaemia. Biliary obstruction secondary to liver fluke infection, neoplasia and other causes will result in elevated levels of conjugated bilirubin. Bile acids may also be measured in the assessment of jaundice and often give some indication of the severity of liver damage.

Cholesterol

Cholesterol levels may be measured to evaluate metabolic processes; levels are raised in some endocrine disorders such as hypothyroidism but can also be elevated after a fatty meal in monogastric species.

Enzymes

A range of enzymes can be measured in blood samples and are present in increased amounts following cell damage. Cells of different tissue types contain varying levels of different enzymes such that elevation of certain enzymes implies damage to specific tissues, however, the interpretation of enzyme assays must be based on clinical history and examination of the animal as a range of variations are possible and may lead to misleading conclusions. Creatinine kinase (CK or CPK) is fairly specific for striated muscle and is used as an indication of the degree of muscle damage, that is, in 'downer' cows. Alkaline phosphatase (AP) occurs in bone (osteoblasts), liver, the intestinal wall and so on. Levels are elevated in biliary stasis and fatty liver syndrome especially when associated with diabetes and metabolic disorders. Lipase and amylase are present in the pancreas and are elevated in pancreatitis and some other acute abdominal conditions. AST (formerly SGOT) and ALT (formerly SGPT) are associated with both cardiac and skeletal muscle and liver tissue. Gamma glutamyl transferase (SGGT) tends to mimic changes in ALT in some species. If there is elevation of AST/ALT and bilirubin with low albumin levels this would indicate that liver damage had occurred. Non-specific elevation in enzyme levels can occur with a range of abdominal or systemic diseases.

Glucose

Slight elevations in blood glucose may occur after a meal in monogastrics. Marked elevations (hyperglycaemia) are indicative of diabetes mellitus (sugar diabetes due to pancreatic damage or primary endocrine disorders) but a mild rise can be seen in 'physiological stress' and a range of metabolic disorders. Low blood glucose (hypoglycaemia) may result in collapse and coma and can be associated with a range of disorders. Marked hypoglycaemia may occur after administration of insulin in diabetic animals and moderate hypoglycaemia occurs in starvation and some other endocrine or neoplastic disorders. Metabolic diseases such as 'twin lamb' disease and bovine ketosis occur when the diet is inadequate in calories or too low in glycogenic precursors (that is, propionic acid) with resultant depletion of liver glycogen stores and low blood glucose. The level of ketones ('ketone bodies') is elevated in most cases of hypoglycaemia and these may be detected in the urine. Diabetic animals may develop ketosis due to inability to use the glucose in the blood.

Electrolytes

Electrolyte levels may be measured to determine the nature and degree of metabolic disturbances especially in monogastric species following vomiting (depleted Cl– and H+ leading to alkalosis), diarrhoea (depleted K+ and HCO– leading to acidosis) or blood loss. Dehydrated animals may require fluid therapy and this must be selected according to the specific needs of the animal. Imbalance of sodium and potassium may occur in some endocrine disorders, drug therapy for heart disease, and in renal disease.

The correct balance of electrolytes in the blood is usually maintained by the respiratory and renal system and is important for the maintenance of fluid balance.

Minerals

Normal serum calcium and magnesium levels must be maintained to ensure correct function of nerve and muscle tissues. Levels of calcium depend to some extent on the relative ratio of calcium and phosphorus in the diet and also on the correct function of the parathyroid glands, gut and kidneys which regulate the levels of Ca/P. Disorders such as hypocalcaemia (low blood calcium) are often the result of rapid calcium demand associated with late gestation or early lactation. If the parathyroid gland is unable to produce parathyroid hormone to promote release of calcium from the bones and enhance uptake from the kidney and across the intestinal tract, the animal's calcium levels fall with resultant staggering and ultimate collapse. The latter is common in high yielding dairy cows and can be prevented by provision of adequate calcium supplements, vitamin D in late gestation and early lactation and not too much during early gestation (this would make the parathyroid gland unresponsive later in pregnancy). Calcium imbalance may also occur in renal disease or when the diet is too high in phosphorus with resultant changes in bone density and secondary bone damage; this is more common in dogs and cats. Low levels of magnesium (hypomagnesaemia) occur when livestock graze pasture low in magnesium; this can result in staggering and behavioural changes. Levels of magnesium are also low in ruminants during early lactation so the appropriate treatment for animals that 'go down' during lactation may depend on the extent of hypocalcaemia and/or hypomagnesaemia. Laboratory tests can help animal health professionals in making decisions about treatment options although in many cases treatment may need to be initiated before test results are available.

Plasma pepsinogen

Pepsinogen is an enzyme released when the abomasal mucosa is damaged during helminth infestation, especially *Ostertagia* sp. in cattle (see

Chapter 3). Plasma levels of pepsinogen in ruminants can also be used to assess the severity of abomasal damage.

7.6 General principles and methodology of basic biochemical tests

Most automated biochemistry systems use quantitative methodology to determine the exact amount of a given substance in a sample. Manual tests may also give quantitative results but to ensure reliability it is very important to follow the protocols exactly. In both systems, clean glassware and good quality water are required. For accurate chemical analytical work or when using new glassware it is advisable to soak it in 2–5% (20 ml/l) hydrochloric acid (HCl) followed by washing and rinsing in two changes of tap water and three changes of de-ionized water (see Chapter 2) before use.

The following section outlines the principles behind common biochemical tests and the practical aspects which need to be considered. Quantitative and semi-quantitative analysis is commonly done using either of the following methods:

1 volumetric analysis, in which a substance in solution is measured by titration
2 colorimetric techniques, by which a substance reacts with reagents to give a coloured product which is measured by an absorption filter colorimeter/ spectrophotometer.

The majority of modern routine biochemical assays are based on chemical reactions that produce coloured substances. Other methods, including flame emission and liquid chromatography will not be considered further here.

Principles of chemical reactions

To understand the principles behind biochemical tests it is necessary to have a basic knowledge and understanding of chemical reactions. There are a wide range of text books on the subject of biochemistry and it is a good idea to become familiar with the periodic table of the elements and the atomic weight of commonly used elements and compounds. Some useful texts are listed in the bibliography at end of this chapter.

Basically, compounds and their constituents (elements) are made up of atoms. Each atom is composed of neutral (neutrons), positive (protons) and negative (electrons) subunits that determine the structure and reactivity of an element. An element is defined by its atomic number, mass and position in the periodic table (see Table 7.4), for example, carbon has a nucleus containing six neutrons and six protons (Figure 7.10), an atomic number of six and an atomic mass of $(6 + 6) = 12$. The number of electrons depends on the 'charged state' of the element, carbon atoms may gain or lose electrons depending on the compound that is formed. These characteristics determine how carbon atoms will behave in chemical reactions. The way in which elements combine to form compounds, by the transfer or sharing of electrons, is known as valency, for example:

$$Na^+ + C + H^+ + 3\ O^{2-} = NaHCO_3\ \text{(sodium bicarbonate)}$$

$$2H^+ + O^{2-} = H_2O\ \text{(water)}$$

$$Na^+ + Cl^- = NaCl\ \text{(salt)}$$

In these examples carbon (C) has a valence of $+4$, hydrogen (H) and sodium (Na) have a valence of $+1$, chlorine (Cl) has a valence of -1 and oxygen (O) has a valance of -2.

Some elements have more than one valance, for example iron (Fe^{2+} and Fe^{3+}). Elements are composed of one type of atom but may combine

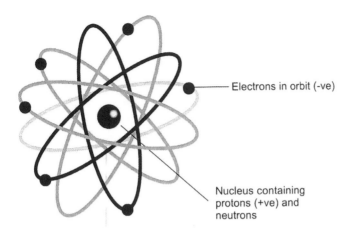

Figure 7.10 Atomic structure of carbon.

Electrons in orbit (-ve)

Nucleus containing protons (+ve) and neutrons

with other elements to form a compound composed of two or more types of atom. A molecule is the smallest particle of an element or compound which can exist independently; the relative mass of a molecule is known as its molecular weight.

Oxidation and reduction

Oxidation is defined as a reaction in which an atom loses electrons (usually the combination of oxygen with a substance or the removal of hydrogen from it). Reduction is a reaction in which an atom gains electrons (usually the removal of oxygen from a substance or the addition of hydrogen to it).

Electrolytes and electrolysis

Electrolytes are acids, bases or salts that conduct electric current when dissolved in water. Positively charged particles are known as cations and negatively charged particles are known as anions. In an electrolytic system (for example, a battery), positively charged particles (for example, Na^+) are attracted to the –ve pole (cathode) and negatively charged particles (for example, Cl^-) are attracted to the +ve pole (anode).

Table 7.4 International atomic weights of selected elements.

Name	Symbol	Atomic weight*
Aluminium	Al	26.98
Arsenic	As	74.91
Bromine	Br	79.916
Calcium	Ca	40.08
Carbon	C	12.011
Chlorine	Cl	35.457
Copper	Cu	63.54
Hydrogen	H	1.008
Iron	Fe	55.85
Magnesium	Mg	24.32
Nitrogen	N	14.01
Oxygen	O	16.00
Phosphorous	P	30.98
Potassium	K	39.10
Selenium	Se	78.96
Sodium	Na	22.99

Notes: *Atomic weight = relative atomic mass. For natural elements with more than one isotope, it is an average for a mixture of isotopes. This is different from the atomic number which relates to the position of the elements in the periodic table.

Acids and bases

A base is a substance that liberates hydroxide ions (OH^-) in solution and accepts a proton. The common bases include oxides and hydroxides of metals, such as sodium hydroxide ($NaOH$), potassium hydroxide (KOH) and so on. Alkaline solutions turn litmus blue and react with acids to form a salt and water. An acid is a substance that liberates hydrogen ions (H^+) in a solution and donates a proton. When reacted with a base, an acid produces a salt and water. Acids turn litmus red and react with carbonate (CO_3^{2-}) to produce carbon dioxide (CO_2).

Most acids are corrosive, for example, sulphuric acid (H_2SO_4), hydrochloric acid (HCl) and so on. Example:

HCl (acid) + NaOH (base) = NaCl (salt) + H_2O (water)

The strength of an acid or base is described by its pK (pK = $-\log_{10} K$) where K = ionization constant. A strong acid or base has a low pK and a weak acid or base has a high pK. Concentrated and low pK acids and bases can be corrosive and very reactive and must be handled with care. The pH of a solution will indicate whether or not it is acidic or basic, measuring pH is outlined in Chapter 2, in the section on buffers.

Buffers

Buffer solutions contain a mixture of a weak acid and a salt of a strong base, or a weak base and its salt with a strong acid. Owing to their composition, buffers are able to resist changes in pH. For example, if a small amount of hydrochloric acid is added to a buffer solution the hydrogen ion content does not increase very much because it combines with the base in the buffer resulting in only a slight decrease in pH. Buffers are used in clinical chemistry and in other disciplines when the pH needs to be carefully controlled, for example, when measuring enzyme activity.

Indicators

Indicators are substances that change colour or shades of colour at different pH values. For example, phenol red changes from yellow at pH 6.8 to a deep red at pH 8.4. Indicators are used to determine the pH of liquids and the 'endpoint' of acid-base titrations.

Neutral solutions have an equal concentration of hydrogen ions and hydroxide ions, that is, pure water:

$$[OH^-] + [H^+] = H_2O$$

Whether a solution is acidic or alkaline, there are always both hydrogen ions and hydroxide ions present, in most cases the solution is described by its $[H^+]$ where (H^+) is the hydrogen ion concentration.

$$pH = \log_{10} [1]/[H^+] \text{ In pure water } [H^+] = 100 \text{ mmol/l}$$

$$pH \text{ (pure water)} = \log_{10} /10^{-7} = 7$$

As outlined in Chapter 2 the pH scale is from 0–14 with values > 7 indicating an alkaline solution and < 7, an acid solution at 25°C.

Techniques for preparing solutions

A solution is composed of a solvent and a solute. The solvent is the 'dissolving medium' and the solute (that is, a chemical) is the substance dissolved. In a well-prepared solution there should be an even distribution of the solute throughout the solvent. When preparing a solution decide whether the solution requires an accurate volumetric preparation (for example, to prepare a standard), or a less accurate method of preparation (for example, to prepare a stain). For the latter, chemicals must be weighed and added to the appropriate volume of solvent but the need for precision is less than when preparing solutions such as standards for quantitative tests. Some guidelines for the preparation of accurate solutions are provided below.

Guidelines for preparing accurate solutions

- Use a sensitive balance.
- Select analytical grade chemicals.
- Hygroscopic (deliquescent) chemicals need to be weighed rapidly to avoid errors.
- Use accurately calibrated clean glassware.
- Use a funnel to transfer the chemical(s) from the weighing container into a volumetric flask.
- Wash any chemical(s) remaining in the container into the flask with a little of the solvent.
- Make the solution up to its final volume only when it has cooled to the temperature used to graduate the flask (this temperature is written on the flask).
- To avoid over-shooting the graduation mark use a Pasteur pipette or wash bottle to add the final volume of solvent to the flask.
- Make sure the bottom of the meniscus of the fluid is on the graduation mark when viewed at eye level (Figure 7.11).

Expressing the concentration of solutions

As mentioned above, in a well-prepared solution there should be an even distribution of the molecules or ions of the solute throughout the solvent. Concentrations of solutions can be expressed as the percentage solution (weight for volume [w/v] or volume for volume [v/v]), or as a molar (M) solution. In considering the ways of expressing the concentration of solutions, it should be remembered that in chemical reactions it is 1 mole of a substance that reacts

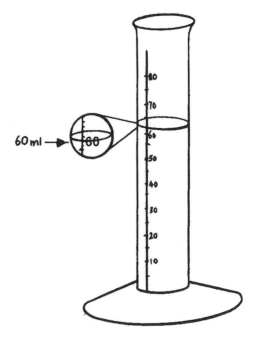

Figure 7.11 Reading the level of a fluid column (meniscus). When using a calibrated flask, beaker or measuring cylinder to measure a specified volume of a liquid read the level at the base of the 'meniscus'. This is the bottom of the concave surface of the fluid seen when 'eye level' is the same as that of the liquid. See also Figure 2.58.

with 1 mole of another substance. Owing to the fact that chemicals interact in relation to their molecular masses it is recommended that the concentration of solutions be expressed in terms of the number of moles of solute per litre of solution. However, if the relative molecular mass of a substance is not known, the concentration of such a substance in solution be expressed in terms of mass (weight) concentration, that is, grams or milligrams per litre (per 100 ml is less commonly used).

MOLE PER LITRE SOLUTIONS
A mole per litre (mol/1) solution contains one mole of solute dissolved in and made up to 1 l with solvent.

MOLE

A mole is defined as the amount of substance that contains as many elementary units (atoms, molecules, or ions) as there are carbon atoms in 12 g of the pure carbon C. One millimole is one millionth of a mole, that is, 1×10^{-6} mole.

MOLECULE

A molecule is an atom or a finite group of atoms that is capable of independent existence and has properties characteristic of the substance of which it is the unit.

MOLECULAR WEIGHT

Molecular weight is the relative molecular mass of a molecule of a substance relative to that of an atom of ^{12}C taken as 12.000.

Preparation of molar (mol/l) solutions and dilutions

To dilute a fluid is to reduce the concentration of the dissolved solute in the solvent. The solute is the solid substance that has been dissolved, for example, salt (NaCl), and the solvent is usually water but may be other fluid substances such as alcohol. The following examples outline how to prepare solutions and how to dilute them.

To prepare a molar (mol/1) solution, use the following formula:

Required mol/1 solution × molecular mass of substance = number of grams to be dissolved in litre of solution

EXAMPLES

1 To make 1 l of sodium chloride (NaCl), 1 mol/l:

required mol/l concentration = 1

molecular mass of NaCl = 58.44.

Therefore, 1 l of 1 mol/l NaCl contains: $1 \times 58.44 = 58.44$ g of the chemical dissolved in 1 l of solvent.

For recording purposes, the concentration is usually written after the name of the substance, that is, NaCl 1.0 mol/l.

2 To make 1 l of physiological saline, that is, sodium chloride (NaCl), 0.15 mol/l:

required mol/l concentration = 0.15

molecular mass of NaCl = 58.44.

Therefore, 1 l NaCl, 0.15 mol/l contains: $0.15 \times 58.44 = 8.77$ g of NaCl (solute) dissolved in 1 l of water (solvent).

3 To make 50 ml of physiological saline (NaCl) 0.15 mol/l:

required mol/l concentration = 0.15 molecular mass of NaCl = 58.44.

Therefore, 50 ml NaCl, 0.15 mol/l contains: $0.15 \times 58.44 \times 50/1000 = 0.438$ g of NaCl (solute) dissolved in 50 ml of water (solvent).

PREPARED DILUTIONS

1 To prepare 500 ml sodium hydroxide (NaOH, 0.25 mol/1) from a 0.4 mol/1 solution:

$$X = \frac{C \times V}{S}$$

where: C is the concentration of solution required (mol/l); V is the volume of solution required (ml); S is the strength of the stronger solution (mol/l); X is the amount of stronger solution required (ml).

C = 0.25 mol/1 V = 500 ml S = 0.4 mol/1
$$\frac{0.25 \times 500}{0.4} = X = 312.5 \text{ ml}$$

Therefore, measure 312.5 ml NaOH, 0.4 mol/1, and make up to 500 ml with distilled water.

2 To make 1 l hydrochloric acid (HC1), 0.01 mol/1 from a 1 mol/1 solution:

$$C = 0.01 \text{ mol/1}$$
$$V = 1 \text{ l}$$
$$S = 1 \text{ mol/l}$$

Volume of stronger solution required (X):

$$\underline{0.01 \times 1000/1} = X = 10 \text{ ml}$$

Therefore, measure 10 ml of HC1 (1 mol/1), and make up to 1 l with distilled water.

Saturated solutions

Saturated solutions are solutions which contain the maximum amount of solute which can be dissolved in a given solvent at a given temperature and pressure. Supersaturated solutions are those in which more solute is added than can be dissolved. Salt crystals from saturated solutions may form around the neck of glass stoppered bottles causing the lid to stick, so it is advisable to use screw capped Borex bottles instead.

DILUTING BODY FLUIDS AND CALCULATING DILUTIONS

Samples such as body fluids may need to be diluted in specific solvents for biochemical tests. These need to be prepared carefully and handled according to the relevant laboratory protocol.

EXAMPLES

1 To make 10 ml of a 1 : 20 dilution of blood:

volume of blood required: 10/20 = 0.5 ml.

Therefore, to prepare 10 ml of a 1 : 20 dilution, add 0.5 ml of blood to 9.5 ml of diluting fluid.

2 To make 5 ml of a 1 : 2 dilution of serum in physiological saline:

volume of serum required: 5/2 = 2.5 ml.

Therefore, to prepare 5 ml of a 1 : 2 dilution add 2.5 ml of serum to 2.5 ml of physiological saline.

3 Calculate the dilution of blood when mixing 0.05 ml of blood and 0.95 ml of diluting fluid:

total volume of body fluid and diluting fluid: 0.05 + 0.95 = 1.0 ml.

Therefore, dilution of blood: 1/0.05 = 20, that is, 1 : 20 dilution.

4 Calculate the dilution of urine using 1 ml of urine and 9 ml of diluting fluid (that is, physiological saline):

total volume of urine and diluting fluid: 1 + 9 = 10 ml.

Therefore, dilution of urine 10/1 = 10, that is, 1 : 10 dilution.

Diluting techniques

Diluting techniques are simple but must be done with due care and with thorough mixing between steps. The easiest method is to perform a stepwise dilution from the original solution into a series of containers containing diluent, see Figure 7.12.

Examples of simple qualitative tests for district laboratories

Although most of the common biochemical screening tests are now done using automated systems or simple dip-stick kits there can still be occasions where these are not available. The following are five simple manual tests that can be performed in small district veterinary laboratories, and which we have found useful, for

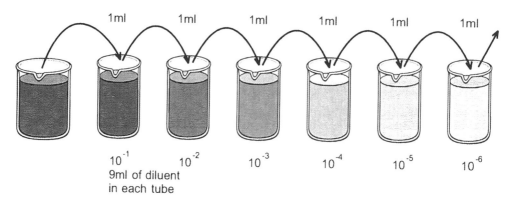

A one in ten or 'ten fold' dilution

10^{-1} 10^{-2} 10^{-3} 10^{-4} 10^{-5} 10^{-6}
9ml of diluent
in each tube

A one in two or 'two fold' serial dilution

Original sample   mix at each step   Discard
(mix well)

1:2   1:4   1:8   1:16   1:32   1:64
1ml of diluent
in each tube

Figure 7.12 Standard dilution technique for a 1 : 10 serial dilution and a 1 : 2 serial dilution.

preliminary assessments. Other manual tests are described in the reference texts listed at the end of the chapter. The reagents used in the following tests are easily obtained from most suppliers of reagents (see Appendix 4).

Iodine flocculation test for liver function

PRINCIPLE

Many tests are available for the assessment of liver function and all have their advantages and disadvantages. The iodine flocculation test is reasonably easy to carry out and has been found to be a fairly reliable indirect indicator of liver function. A healthy liver produces albumin, a protein required by the body, and gamma globulin which is required for immune function.

Iodine forms a precipitate with gamma globulins when the level of albumin in the blood is reduced, this occurs following liver damage. In the normal animal, there is sufficient albumin in the blood to prevent iodine flocculation of gamma globulins in this test but as liver damage progresses the amount of gamma globulin in a serum sample increases and is free to form complexes with iodine.

REAGENTS REQUIRED

Mix 6.7 g (52.8 mmol) of iodine (I) and 13.3 g (80.12 mmol) of potassium iodide (KI) together, add a small amount of water while stirring and continue until dissolved to make solutions up to 100 ml.

METHOD

- Add one drop of the reagent to 1, 2 and 4 drops of serum on a white porcelain tile. Allow to stand for 2 min then mix with a tooth pick. Examine after 30 min and assess the degree of precipitation.
- A positive reaction is shown by the presence of a precipitate.
- If only the 1 : 1 sample has reacted this indicates sub-clinical liver damage.
- If the 1 : 2 or 1 : 4 samples show a precipitate this indicates clinical liver damage.

Ideally serum samples should be submitted for liver enzyme tests (AST/ALT) and further analysis. If this is not available, supportive treatment for the animal should be recommended and repeat tests conducted after a few weeks to determine if the liver damage is progressive.

Rapid Spot test for blood ketones

PRINCIPLE

This test relies on the fact that serum containing ketones will produce ammonia when reacting with the test reagent and turn the reagent powder red. A positive reaction will only occur if the level of blood ketones is greater than 10 mg/100 ml.

REAGENT

Grind the following together into a fine powder using a mortar and pestle and store in a dark stoppered bottle.

1 sodium nitroprusside 3 g (10.07 mmol)
2 ammonium sulphite 100 g (756.8 mmol)
3 sodium carbonate (anhydrous) 50 g (471.7 mmol).

METHOD

1 Reagent powder is sprinkled on a white porcelain tile.
2 A drop of serum (0.04 ml) is deposited on the powder using a Pasteur pipette.
3 The reaction is read after 5 min.
4 A pink colour indicates a positive reaction.

Estimation of serum bilirubin

PRINCIPLE

Ferric chloride will oxidize bilirubin to biliverdin producing a greenish blue colour. An intense blue/green colour reaction between the test serum and Fouchet's reagent indicates the presence of excess amounts of serum bilirubin. However, this does not allow reliable distinction between bound and unbound forms of bilirubin.

FOUCHET'S REAGENT

Dissolve 25 g (153 mmol) trichloracetic acid in 100 ml of distilled water and add 10 ml of 10% (370 mmol/l) ferric chloride ($FeCl_3$).

METHOD

1 On a white porcelain tile place a few drops of serum and an equal number of drops of Fouchet's reagent.
2 Mix well with a glass rod and examine for up to 5 min for the development of a blue/green colour.
3 The interpretation of the result is outlined in Table 7.5.

Table 7.5 Interpretation of the bilirubin test.

Green or blue colour	Time (min)	Bilirubin (mg/100 ml)*
-No green/blue colour	5	<0.2
+ slight colour	5	0.2–0.5
++ moderate colour change**	5	0.5–0.8
+++ Strong colour change	5 (1)	0.8–1.4
++++ Strong***	1	1.4–2.0
+++++ Intense	1	> 2.0

Notes: *Values may also be expressed in µmol/l (x 17.1) see Tables 7.2 and 7.3. **A reaction of ++ or +++ would require further investigation. ***Results in the range of (++++/+++++) would be considered high.

Serum total protein by the Biuret method

PRINCIPLE

A violet colour complex is formed by the reaction of serum proteins and peptides with a copper sulphate (Biuret) reagent. Other nonprotein nitrogen complexes such as creatinine and urea do not react.

BIURET REAGENT

Potassium sodium tartrate 45.0 g (159.48 mmol)

Cupric sulphate 0.5 H_2O 15.0 g (60.08 mmol)

Potassium iodide 5.0 g (30.12 mmol)

Add sodium hydroxide 0.2 M (200 mmol/l) to make 1000 ml.

Dissolve the tartrate in approximately 400 ml of 0.2 M NaOH; Add, while stirring, the cupric sulphate: continue to stir until in solution. Add the potassium iodide and when dissolved make up to 1000 ml (1 l) with 0.2 M NaOH.

BLANK DILUENT REAGENT

Potassium sodium tartrate 9.0 g

Potassium iodide 5.0 g

Sodium hydroxide 0.2 M to 1000 ml.

To prepare 0.85% saline solution, dissolve 8.5 g NaCl in 1 l of distilled water. Prepare the blank reagent in a similar manner to that outlined for the Biuret reagent.

METHOD

- Pipette reagents out as outlined in Table 7.6. Mix the contents of each of the tubes and place in a water bath at 37°C for 15 min.
- Using a colorimeter set at a wavelength of 530 or 565 nm, measure the OD of the samples against a reagent blank. The blank is used to 'zero' the machine before sample ODs are read.
- Calculation – The protein concentration:

OD value of the sample × conc. of the standard used / OD value of the standard (that is, 50 g/l)

INTERPRETATION

See Table 7.3 for normal serum protein values.

Haemoglobin determination

There are a number of methods available to determine haemoglobin levels in blood samples but most are based on the principle of comparing a colour reaction in the test sample against that of a control reagent which has a known haemoglobin level (see Chapter 5). To a large extent the reliability of the methods depends on the stability of the coloured samples being measured. Cyanmethaemoglobin, used in the Drabkin method, is more stable but has the disadvantage of being a toxic chemical. There are also a number of kit tests now available for Hb determination (see also Chapter 2). The alkaline haematin method is outlined below.

Table 7.6 Preparation of reagents.

	Reagent blank	Standard	Sample	Sample blank
Test serum	-	-	200 µl	200 µl
Control serum*	-	200 µl	-	-
Saline	5 ml	5 ml	5 ml	5 ml
Biuret reagent	5 ml	5 ml	5 ml	-
Blank diluent	-	-	-	5 ml

Note: *Control serum with total protein 50 g/l.

PRINCIPLE

A dilution (1 : 10) of whole blood is made using a weak solution of sodium hydroxide (0.1 µm) to form haemoglobin, this is measured colorimetrically against a haemoglobin standard (for example, Gibson-Harrison).

REAGENTS

1 Gibson-Harrison Haemoglobin standard[1] – 16 g/dl (B.H.D)
2 NaOH 0.1M (100 mmol/l).

METHOD

Pipette 50 µl (0.05 ml) of whole blood into 4.95 ml of 100 mmol/l sodium hydroxide (NaOH), heat for 4–5 min in a boiling water bath and then cool rapidly in water.

The Gibson-Harrison haemoglobin standard should be boiled and cooled alongside the test sample. When cold, but within 30 min of heating, read the OD using a colorimeter set at 540 nm (yellow-green filter) for both samples. By comparing the OD readings for the standard haemoglobin sample and the test sample it is possible to calculate the haemoglobin content of the test sample.

Calculation: Concentration of test OD of blood sample/ OD Gibson-Harrison standard × 16 = g/dl

The standard error using this method is +/–5%.

PREPARATION OF A STANDARD GRAPH

Owing to the fact that the haemoglobin standard is expensive it is useful to prepare a stock solution of normal blood with a haemoglobin concentration of about 11% (11g/dl). This will require titration of a blood sample with known haemoglobin concentration which has been compared against a standard control such as the Gibson-Harrison standard. The procedure is as follows:

• Use any normal blood sample. Take 100 µl of blood, add to 10 ml of 0.1 M NaOH and mix. Wait for the solution to clear.
• Pipette 10 ml of the haemoglobin standard into another tube and place both tubes in a boiling water bath for 4–5 min. Remove the tubes from the water bath and allow to cool for 5 min.
• Read the OD at 540 nm.
• Calculation as before.

If the sample has a high haemoglobin concentration it can be titrated to 11 g/dl and stored until required or the OD for a range of dilutions can be recorded to produce a reference curve. Other methods of determining the haemoglobin concentration in a sample include the Drabkin's method (Chapter 5). If available, modern colorimetric techniques can also be used, for example the Diaspect Tm handheld haemogobinometer described in Chapter 2 (Figure 2.45).

See the texts in the bibliography at end of this chapter for more details. The five biochemical tests outlined above are very basic and have largely been superseded by new kit test and automated systems. However, simple biochemical tests such as these may still be useful in small district facilities where reagent supplies can only be sourced locally and equipment is limited.

7.7 Metabolic diseases and endocrine disorders

Metabolic diseases (for example, acidosis, ketosis, electrolyte or mineral imbalance) and endocrine disorders (for example, diabetes mellitus, hyper- and hypothyroidism, parathyroid disorders, adrenal gland disorders) are often linked. Tests can be done to diagnose disease in individual animals or used to monitor the health status of groups of animals, that is, metabolic profiles. Many tests for endocrine disorders (for example, adrenal disorders) require repeated sampling and/or dynamic testing (for example, ACTH stimulation test, dexamethasone suppression test) in which an animal is given an injection of a specific hormone so that the response can be measured in the next blood sample taken and compared with a pre-treatment value. Endocrine disorders will generally only affect an individual animal and it may not be practical to perform the required diagnostic tests in a small diagnostic laboratory. There are a range of texts available on the subject of endocrinology but in most cases diagnosis and treatment is complicated and would require referral to a specialist. Metabolic diseases, however, are not uncommon in livestock and samples may be submitted to district and regional laboratories for testing. Metabolic diseases can be common in high genetic merit livestock under heavy production demand and generally occur due to an imbalance between nutrient intake versus nutrient demand, for example hypocalcaemia or 'milk fever' in high yielding dairy cows, twin lamb disease and ketosis in ewes with multiple lambs. The term 'metabolic disease' has also been applied to other nutrient deficiencies (that is, copper deficiency) or functional disturbances (that is, bloat) but usually the term refers to one or more of the following conditions.

1 Acidosis/ketosis in ruminants occurs following grain overload or due to an imbalance of dietary carbohydrates resulting in a lack of glucose precursors. Ketosis also develops in starvation and diabetes.
2 Twin lamb disease in ewes occurs due to an increased demand for glucose and lack of carbohydrate intake during late pregnancy. Ketosis develops after liver glucose supplies are used up but can be avoided if ewes carrying twins are fed a low-bulk, high-quality diet.
3 Hypocalcaemia (low blood calcium), occurs due to a total or relative lack of calcium in the diet. It may also occur due to lack of ability to mobilize calcium from the bones in early lactation.
4 Hypomagnesaemia (low blood magnesium), results from an inadequate level of magnesium in the diet. This may occur in cattle grazing young fresh pasture which is low in magnesium or can be associated with hypocalcaemia in lactation tetany.

In many cases more than one metabolic condition occurs at the same time in an animal. It is useful to monitor the general metabolic status of 'at risk' animals to establish whether or not metabolic disease is likely to occur. Good management, with special attention to matching diet to physiological demand, will prevent many metabolic diseases in livestock. This topic is well covered in a number of text books on livestock nutrition (see reference list at the end of the chapter) and will not be discussed further here.

7.8 Mineral and trace element assays

For the interpretation of mineral and trace element assays it is important to also consider the general health status of the animals tested, for example, a cow with liver disease may release a lot of stored copper into the blood stream resulting in artificially elevated levels of copper in the serum. There is usually a combination of factors that contribute to dietary deficiencies (for example, poor nutrient intake, concurrent disease, debility) but the adequacy of the diet is often the first thing to consider. In cases where several animals develop specific nutrient deficiencies it is especially important to assess the diet. This may involve analysis of commercial feed (for example, for energy, protein, minerals and trace elements) or assessment of the vegetation and soil in the area the livestock graze. Due to potential interactions between nutrients and also variation in the uptake of trace element and other dietary components it can take time to determine the cause of a deficiency. For example, if pasture or soil is high in one trace element (for example, molybdenum, Mo) it may interfere with the uptake of others (for example, copper, Cu). Low values of Ca/P may indicate inadequate feed but may also be associated with parturition and heavy lactation stress (for example, milk fever), for these reasons any samples sent for trace element assessment should be submitted along with a full clinical history. Because serum levels of Ca and P are highly inter-linked physiologically, sera should be tested for both Ca and P concurrently and the results assessed together.

There are a range of commercial assays available for serum mineral and trace element assessment, most are based on colorimetry and are fairly easy to perform. Commercial kit tests (for example, QuantiChrom™, Magnesium and Copper Assay Kits, BioVision Calcium Colorimetric Assay Kit) come with detailed instructions and usually provide example calculations based on comparison with a standard sample of a known concentration. Assays for tissue trace element or mineral levels are usually more specialized and would be done on samples of liver and other tissues collected at necropsy or from a biopsy collected by a consulting veterinarian.

It should be noted that for trace elements that are co-factors, or chemical mediators, in essential biochemical processes, the chemical measured in the blood may be a specific enzyme or substance as opposed to the element itself, for example, vitamin B_{12} for cobalt (Co) and glutathione peroxidase for selenium (Se).

7.9 Urinalysis

Urinalysis is the examination of urine and can provide information about the general health of an animal as well as the urinary tract. The urinary tract consists of the kidneys (which make urine while filtering the blood) the ureters (the tubes joining the kidneys to the bladder), the bladder (the bag which holds the urine) and the urethra (the tube which joins the bladder to the outside). In most cases a fresh midstream specimen of urine is suitable but for some tests it is necessary to collect a sample aseptically by catheterization or cystocentesis (that is, collecting by syringe directly from the bladder during anaesthesia for microbiological examination). It is important to get a 'representative' specimen of urine that reflects the state of the urine as it is in the bladder, that is, without any contamination from the vagina or urethra, skin, environment or container. Although the risk of sample contamination can be reduced by inserting a catheter into the bladder, but this can be dangerous for the animal because bacteria may enter the bladder on the catheter and infect the urinary tract. Smears can be prepared (on a glass slide) from any sediment present in the sample, these are stained and examined under the microscope for the presence of any cellular material. On standing, urine samples may become cloudy due to the presence of phosphates. A cloudy fresh urine sample will often indicate a renal tract pathology.

Biochemical analysis of urine

Kidneys excrete water and many waste substances from the blood. One of the more important waste substances is urea. Substances which are not waste products sometimes get into the urine because something is wrong systemically (for example, a metabolic or toxic problem) or with the urinary tract. Urine should be examined visually before testing to determine the colour and consistency. Any abnormalities should be recorded including any abnormal odour.

The presence of sugar in the urine is called 'glycosuria' and can indicate diabetes mellitus (due to insufficient insulin). The presence of protein in the urine is referred to as 'proteinuria'. If protein is found it can indicate one of several diseases of the urinary tract including urogenital infection, inflammatory disease of the kidneys or chronic kidney (renal) or bladder damage. A sample containing protein will usually be cloudy. Urine may sometimes be a deep brownish green due to the presence of bile pigments and it may be coloured by drugs or certain dietary additives. The presence of haemoglobin or blood, results in red urine, this is abnormal, however, it is important to rule out contamination from the reproductive tract.

Microscopic examination of urine sediment

When the urinary tract is infected, protein, bacteria and/or pus cells are usually present in the urine. Finding pus cells (leucocytes especially neutrophils) in urine (pyuria) is a much more certain sign of urinary tract infection than find-

ing traces of protein. To look for leucocytes, put a drop of urine in the counting chamber of a haemocytometer slide, and count the cells, as if counting cells in a blood sample. Use the following procedure.

Method

1 Collect a clean midstream specimen of urine.
2 Mix the specimen well. If the urine is not well mixed, leucocytes may settle at the bottom of the bottle.
3 Put a cover slip on a haemocytometer counting chamber.
4 Fill the counting chamber with urine using a Pasteur pipette.
5 Count the cells in four corners of the chamber as for a white blood cell count (W). (See Chapter 5.)

Some of the elements seen in the sediments of urine samples are shown in Figure 7.13. White blood cells will appear much larger than red blood cells, although neutrophils often disintegrate. Different white blood cells (WBC) may not be readily distinguished but neutrophils are usually numerically more common. Red blood cells may appear faintly yellowish-red in urine and are often crenated (distorted). Epithelial cells, usually from the lining of the urinary tract, will appear much larger and flatter and usually have a nucleus that is easily seen. Sometimes they appear as sheets of cells.

INTERPRETATION
In excess of 2–3 WBC per high power field or 10 cells per microlitre can be considered abnormal but it is necessary to assess this alongside other test results. Note, however, that cells from the reproductive tract may contaminate urine.

LOOKING AT A CENTRIFUGED DEPOSIT
For more detailed examination of the cellular components of a urine sample it should be cen-

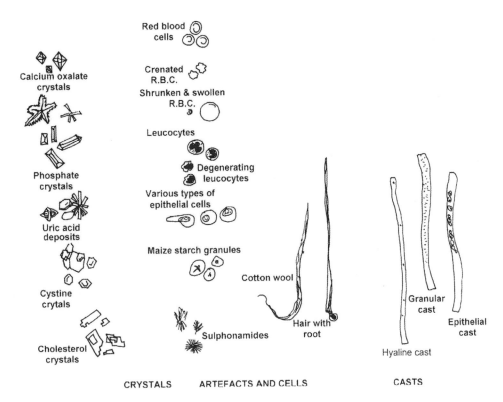

Red blood cells

Crenated R.B.C.

Shrunken & swollen R.B.C.

Calcium oxalate crystals

Leucocytes

Degenerating leucocytes

Phosphate crystals

Various types of epithelial cells

Uric acid deposits

Maize starch granules

Cotton wool

Cystine crytals

Granular cast

Epithelial cast

Sulphonamides

Hair with root

Hyaline cast

Cholesterol crystals

CRYSTALS   ARTEFACTS AND CELLS   CASTS

Figure 7.13 Microscopic examination of urine sediment (stylized and not to scale).

trifuged prior to smear preparation and staining (see Figure 7.13).

EXAMINING A CENTRIFUGED DEPOSIT

1 Mix the specimen of urine.
2 Fill a 15 ml centrifuge tube.
3 Centrifuge for about 3 min at 2000 to 3000 rpm.
4 Pour off the supernatant.
5 Using a Pasteur pipette suspend the deposit in the last drop of urine and put it on a slide under a cover slip. Search the whole area of the cover slip with a low power objective. Then look at some fields with a high-power objective or use phase contrast microscopy.
6 Fix and stain the specimen as outlined in Chapter 4, pages XX–XX.

Be careful not to pour away the deposit when the supernatant urine is discarded. If a hand centrifuge has been used, and the deposit is not well packed at the bottom of the tube, pour away only the top part of the supernatant. Remove the last few ml of supernatant with a Pasteur pipette, and mix the sediment with the last drop or two of urine and examine.

Interpretation of results

Biochemical tests

The evaluation of a urine sample can give valuable information about the health status of an animal. The characteristics of normal urine depend on the species sampled, physiological state, sex and age and are dependent on the diet. A sample of urine can be tested directly for biochemical changes using reagent strip sticks (for example, Multistix™) that change colour in response to pH, the presence of glucose, ketones, bilirubin and so on. Some of these test sticks also detect the presence of whole and haemolysed blood. The normal urinary pH for an individual depends on the animal's diet but in general, ruminants and equines have alkaline urine (pH > 7.0) whereas carnivores have acid urine (pH < 7.0). For pH tests on urine some laboratories routinely use a pH meter (see Chapter 2). Urine reagent strips are readily available and rely on a colour change on a single or multiple reagent strip; the colour reaction can be compared with the standard colour codes provided on the container or reagent strip box. The instructions that come with reagent sticks are usually quite simple and should be followed carefully. On most reagent strip containers, there will be a colour chart to allow the result to be read. However, many reagent strip testing kits are developed for human use and therefore interpretation of results for animals may not be simple. In healthy animals, there should not be any trace of protein, ketones, glucose or bilirubin in the urine. There may be some blood in urine samples collected during the oestrus period in female animals of some species, due to contamination from the reproductive tract, so this must be taken into consideration when examining free-caught samples. The presence of glucose may indicate that there is a pancreatic endocrine disorder, for example, diabetes mellitus, but mild hyperglycaemia may also be associated with 'physiological stress' or metabolic disorders in ruminants. The presence of ketones is usually a bad sign and indicates severe emaciation or the development of serious metabolic disease such as ketosis and/or toxaemia.

CASTS AND CELLS

In the kidney there are millions of tubules (small tubes) through which the urine passes on its way to the bladder. In many kinds of kidney disease these tubules become blocked with protein or the remains of dead cells. Very often the solid substance blocking the tubule becomes loose and goes down the tubule and into the urine. This material, which was blocking the tubule and has come loose is called a cast. Because it was formed inside a tubule a cast has the shape of the tubule from which it came. Casts are seen in several kidney diseases, and there are several types (see Figure 7.13). Hyaline (like glass) casts are clear or transparent, they either have no granules in them or only a few granules. Normal urine often has some of these hyaline casts, and they do not indicate that the kidney is diseased. However, in large numbers hyaline casts can indicate pathology. Granular casts are made of coarse granules. If they are found in the urine they usually indicate that the kidneys are damaged, and are therefore significant. Granular casts are found when there is protein in the urine. All urine in which protein is found (++ or more) should be centrifuged, so that the deposit can be examined for the presence of casts under a microscope. Casts are sometimes found that are partly granular and partly hyaline. Other kinds of casts are also seen – cellular, waxy and pigment casts but these are not common.

Animals that pass red blood cells in their urine are said to have haematuria (for example, as seen in cystitis, neoplasia and bracken fern poisoning). The urine will be turbid, 'smokey' or red. Red blood cells are sometimes destroyed in the circulation, and haemoglobin rather than red cells may be passed in the urine; this is known as haemoglobinuria (for example, as seen in babesiosis). It is easy to distinguish between the two conditions if the urine sample is centrifuged or is allowed to settle overnight. In cases of haemoglobinuria the sample remains red while in haematuria, red blood cells will be seen in the spun deposit or settled specimen. Sometimes there may be both haemoglobin and red blood cells in the urine (see Chapter 10).

Pus cells (white cells) may also be found in urine sediment (pyuria), these are predominantly polymorph leukocytes (neutrophils) which have migrated from the blood. They have a nucleus with several segments. As pus cells are broken down and destroyed in the urine, their nuclei may disappear and only the cell membranes remain. A few pus cells are often found in free-caught urine samples from healthy animals. Because of this, it is usual to take a number below which it is said to be normal, and above which it is said to be abnormal. Ten pus cells per microlitre is a common figure to take. Less than 10 pus cells per microlitre are thought to be normal. More than 10 pus cells per microlitre are thought to be abnormal and to indicate a urinary tract infection. This is only a guide, but it is a useful one.

Epithelial cells are large flat cells with a nucleus that can usually be seen quite easily. They come from the epithelium on the inside of the ureters, bladder and urethra. Epithelial cells are usually single but sheets of several epithelial cells are sometimes seen. Sometimes epithelial cells are found in normal urine, but if present along with large numbers of red blood cells and pus cells this is abnormal and indicates damage to the urinary tract.

When urine is formed in the kidney all the chemicals in it are in solution. After it has left the kidney, some of these chemicals may form a urinary deposit. Sometimes they form small solid granules without any special shape. Deposits of this kind are said to be amorphous (without shape). These deposits are very common. Often, however, these chemicals, as they become solid, take on the shape of crystals. When seen from above some crystals are square with a cross on them like an envelope. These are oxalate crystals. Other kinds of crystal are shown in Figure 7.13. The shape of the crystals often tells us what they are made of, but crystals in the urine are seldom significant unless they irritate the bladder lining. Crystal formation in urine (urolithiasis) can be more severe in cases of bladder damage with associated infection (cystitis) and changes in urine pH.

Spermatozoa are often found in urine and this is quite normal in male, or recently mated or inseminated female animals. *Trichomonas* protozoa can be seen in urine contaminated with preputial or vaginal discharges from infected animals. If there are any motile protozoa in the urine, they are almost certainly *Trichomonas* sp. Microorganisms are frequently seen in free-caught urine. Bacteria found in samples taken by cystocentesis or catheterization are more likely to be significant. Like all very small things, bacteria are only seen clearly in stained preparations examined microscopically using a high power or oil immersion objective. Many urinary bacteria are motile. Bacteria often grow in urine if it is left to stand in a warm room and commonly get into the sample from a dirty container or as contaminants from the skin. Finding bacteria in old urine therefore means little. When significant bacteria are seen in fresh urine there will usually also be pus cells and protein present to indicate inflammation. In some samples, there may be yeasts or the mycelia of fungi (these are longer and thicker than a bacterium, and they branch). Some objects in urine debris may not be significant but it is important to be familiar with their appearance, for example, a small bubble of air, bits of debris, pieces of cotton thread and scratches on the slides are commonly seen (Figure 7.13).

7.10 Toxicology

Judit E Smits

There are a wide range of natural and man-made compounds that are potentially toxic to animals. Numerous types of plants, toxins produced by algae, and fungi that contaminate grain are among the naturally occurring elements that can affect livestock and domestic animals. Generally, wildlife is less susceptible than domestic species, to these natural toxins because of two main factors; they have evolved to avoid eating things that make them sick, and because they can move to other food sources if they detect

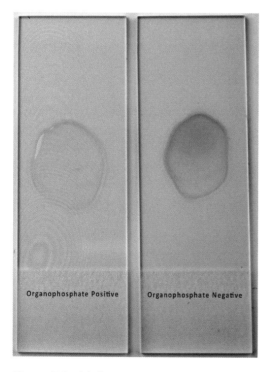

Figure 7.14 (a) Organophosphate or carbamate poisoning can be diagnosed from a serum sample of the patient using basic laboratory supplies and reagents, (b) glass slides with positive and negative serum samples are shown in the insert. See also Plate 33.

something unusual. Heavy metals, although they are naturally occurring elements in the earth's crust, generally cause toxicity only after we have concentrated them through industrial practices that contaminate water or landscapes, or, added them to products that animals can eat. Vitamin/mineral mixes are not an uncommon source of toxicity in livestock. Many pesticides that we have developed to improve food production for humans or animals, like organochlorines, organophosphates and carbamates, and more recently neonicotinoids (imidacloprid being the principal example), have a broad range of toxicity to vertebrate and non-target invertebrate species, as well as the targeted invertebrate pests. The neonicotinoids deserve special attention being the fastest growing class of pesticides worldwide (Morrissey et al., 2015). They are persistent, have high potential for run-off and leaching, and are highly toxic to a wide range of invertebrates, notably beneficial ones such as pollinators. Organophosphates are among the most widely used insecticides in the world and are responsible for much of the poisoning of wildlife, with birds being most severely affected (Smits and Naidoo, 2018). There is a diagnostic method available that can be carried out in diagnostic laboratories with modest facilities (Figure 7.14).

Medicinal products and food additives such as mineral mixture supplements, are meant to improve the health of animals and people. However, overdoses or poor mixing, or carcass residues can cause large scale toxicity problems to entire herds or wildlife populations (Jones, 2001; Kart and Bilgili, 2008; Naidoo et al., 2010; Smits and Naidoo, 2018). Plants may be toxic in small amounts, for example, yew (Burcham et al., 2012), or may require long term ingestion (for example, bracken fern, acorns or leaves from oak trees and star thistle) (Welch et al., 2012) before clinical signs will be seen. Heavy metals such as lead, mercury and cadmium may cause clinical signs in a group of animals or an individual depending on the source of poisoning. Cattle, because of their curiosity and because they tend to be less discriminating in what they taste and eat than other species such as goats, horses and swine, are the most likely to be poisoned from exposure to lead. This often occurs because of discarded batteries being left in fields. Another source of lead can develop around industrial sites and pump jacks where oil leaks out as it is being extracted from underground sources. Horses are most likely to be exposed by chewing at fences that have been painted with lead-containing paint, similar to cases of lead poisoning in puppies when they chew toys that have been painted with lead paint. Diagnosis of lead poisoning includes clinical signs such as apparent blindness and behavioural or other neurological abnormalities. At post-mortem, typical, pathognomonic brain lesions of polioencephalomalacia (laminar necrosis of the grey matter of the brain) are visible using ultraviolet light in a dark room (Figure 7.15).

Fungal contamination of animal and human food supplies which occurs because of damp weather during the crop maturation period, or inadequate storage conditions, may cause acute or chronic disease in animals ingesting the toxins (mycotoxins). In some instances, the feed may appear mouldy but this is not always the case. Aflatoxins, produced by *Aspergillus* sp. fungi, often contaminate poultry and pig feed, but at low concentrations, that do not change the appearance or smell of the feed. However, ingesting contaminated feed over a prolonged period has a cumulative effect causing poor production or overt disease in exposed livestock (Chapter 4, section 4.6 mycology). Ergot, a contaminant of rye and other grains that flourishes in damp conditions, may produce an aggressive toxin, ergotamine. Ergotamine causes constriction of peripheral blood vessels, cutting off blood supply, and resulting in dry gangrene with loss of distal limbs, ears or tails in cattle (Figure 7.15).

Major classes of insecticides that directly intoxicate cattle, sheep and wildlife also can

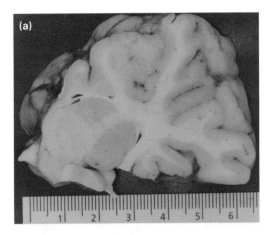

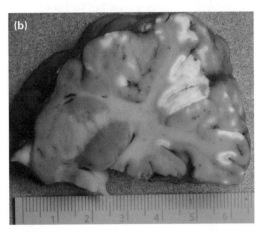

Figure 7.15 (a) A formalin fixed bovine brain with polioencephalomalacia in the grey matter, has subtle, sunken yellowish necrotic areas, (b) which show dramatic fluorescence in a dark room using UV light. See also Plates 34 and 35.

cause secondary toxicity of avian and mammalian scavengers. Among the different pesticides used in agriculture and food production, apart from toxicity caused by the insecticides discussed above (organophosphates, carbamates, neonicotinoids), rodenticides pose the greatest threat from both unintentional and malicious poisonings. Specifically, the second-generation anticoagulant rodenticides (SGARs), such as brodifacoum and bromodiolone, have higher toxicity than the original warfarin-based compounds.

Figure 7.16 Hind limbs from beef cattle suffering from severe ergotamine toxicity, show the classic lesions with the lost blood supply resulting in tissue death, with hooves and feet falling off.

Globally, these SGARs are poisoning large numbers of wildlife, heavily affecting avian and terrestrial scavengers, the species that have historically played the role of nature's clean-up crews. As well, birds of prey such as owls, falcons and hawks, feed on live rodents therefore controlling their populations. However, they are also seriously affected by rodenticide poisoning since the poisoned, but still alive rodents, are the easiest to catch. Studies have shown 80% to 100% of these raptors found on agricultural landscapes have rodenticide residues in their tissues, with 30 to 35% of those having lethal levels (Smits and Naidoo, 2018 and references therein). Eagles, vultures, wolves, coyotes, hyenas, lions and other perceived threatening or nuisance wild and domestic animals have all fallen victim to secondary poisoning, unintentionally or by design through poisoned baits.

The examination and analysis of tissues and body fluids for the presence of naturally occurring toxins and man-made toxicants, is very important to get a diagnosis (Table 7.7). In cases of sudden death, with no obvious explanation, such as trauma, it is logical and wise to consider a toxic cause and plan to collect useful samples. Most exposure to toxic compounds will be through direct contact, through skin if animals are being treated with insecticides, or through ingestion if they are feeding in fields where the chemicals have been applied. In the case of agrochemicals or pesticides, information about the product should be requested, and samples collected (Table 7.7). If a natural or manufactured poison is suspected to have been consumed, vomitus or stomach content are very important samples to collect, as well as blood and urine from affected but still alive animals. In dead animals, ocular fluid collected into a sterile syringe is useful, and often is more readily obtained than blood which will be clotted. Additional tissue samples should include pieces of liver, kidney, muscle, fat and brain, stored in plastic containers and frozen, with thin sections (< 4 mm thick) placed into formalin, if that is available.

Death due to inhalation of highly toxic gases such as hydrogen sulphide, carbon monoxide and cyanide, leaves no residues, and very little physical evidence beyond possible changes in blood colour (bright red in the case of CO poisoning). In these situations, the per-acute nature of the poisoning (animals die within minutes), and a possible source of such gases, like a sewage lagoon, or a gas pipe, should be obvious.

Toxicology facilities are often difficult to find at the district level, but samples may be submitted to a referral centre. It is not possible to test for all known toxicants in every sample, nor would it be cost effective to test unless the person submitting has a reasonably good idea as to what the poison could be. It is essential that the veterinary or animal health officer collects and submits the proper samples to the referral lab. With toxicology, it is especially important that a good case history has been taken with details of clinical signs, number of animals affected, a description of the environment where the animals were found, and note any recent changes in diets, or agrochemical treatment of crops in the area, or run-off from nearby industrial activity,

Table 7.7 Diagnostic samples to submit for various suspected toxicants.

Suspected toxicant	Source of toxic compound	Species most often affected and evidence	Samples to submit	Diagnostic feature
Metals lead (Pb), fluoride (F), mercury (Hg), cadmium (Cd)	Old or industrial paint (Pb) Mine dust, tailings (Cd, F) Released from flooding (Hg) Seed treatments (Hg)	Cattle - CNS disease Subacute mortality Birds – mortality Population declines	Hair Serum Urine Kidney	Elevated metal levels CNS cortical necrosis
Minerals	Copper sulfate (wound disinfectant)	Sheep – acute mortality Bison – subacute mortality	Blood Kidney	Haemolysis Dark discoloured kidneys
Man-made biocides, insecticides, fungicides	Consuming treated vegetation or poisoned insects	Wild birds Honey bees	Whole carcass Sample of suspected agrochemical	Population declines
Anticoagulants	Dicoumarol (mouldy sweet clover) Bracken fern Rodenticides or consuming poisoned rodents Mycotoxins?	Cattle & dogs – widespread ecchymosis or massive haemorrhage Death of healthy born calves	Vomitus/gastric content Blood Brain Liver Samples of feed	Widespread frank haemorrhage Ecchymosis or petechiae
Organophosphates and/or Carbamates	Insecticides	Cattle-ataxia, dyspnoea & acute mortality	Feed, water Stomach contents Brain Serum Liver	Chemical residues Acetylcholine-esterase activity (brain, serum)
Herbicides	Paraquat Glyphosate	Accidental exposure Aquatic wildlife	Lung Blood	History of exposure Fibrosis of lung
Pharmaceuticals (overdoses or mixing problems)	Iron (Fe) treatment Vitamin D Vitamin E/Se	Piglets Pigs Deer highly susceptible	Serum/blood Liver Liver Heart	Mineral levels Pathology
Anti-inflammatory drugs	Phenylbutazone Carprofen & deracoxib	Horses- Dogs	Gastrointestinal tract Kidney (medulla) Gastrointestinal tract	Gastrointestinal tract ulcers ulcers Renal papillary necrosis Gi tract ulcers

Table 7.7 *continued*

Suspected toxicant	Source of toxic compound	Species most often affected and evidence	Samples to submit	Diagnostic feature
Ionophores (overdoses or mixing problems) (growth promotant coccidiostats) Organic arsenicals (phenylarsonics)	Total mixed ration Salt-mineral supplement	Horses-mortality Cattle-mortality Dogs-by mistake-mortality Cattle-overdose-exercise intolerance & mortality	Serum Necropsy for heart failure Feed Stomach content	Myocardial Troponin I Residue analysis
Natural toxicants	Blue/green algae Domoic acid (red tides) Nitrates Botulinum toxin	Cattle, horses, dogs, deer Waterfowl-mortality Cattle All vertebrate species	Samples of water Aqueous or vitreous humour for nh3 Samples of plants Blood/serum	Nitrate levels
Mycotoxins on agricultural crops	Ergot Zeranalol Ht2 toxin Vomotoxin	Cattle, sheep-immunocom-promised Gangrenous necrosis of the extremities	Mycotoxological examination of feedstuffs Characterization of lesions	Dry gangrene of extremities – tail, feet
Fertilizers	Nitrate/Nitrite NH3	Cattle, sheep-dyspnoea, ataxia, mortality	Samples of water, feed Aqueous or vitreous humour for nh3 Blood (methemo-globinemia)	High nitrates Brown blood
Explosives	Dynamite Tannerite	Cattle, dogs-ataxia, mortality	Samples of water Aqueous or vitreous humour for nh3	High nitrates

and so on. This information, together with the samples, will be necessary for the specialists to work towards a diagnosis. From well described details, and the post mortem findings, it should be possible to list two or three, most likely toxicants which could be responsible and test for those specifically.

At post-mortem, the most commonly collected samples will include representative fixed tissues from each organ system (Chapter 8), stomach and/or rumen contents, heart blood, urine and fresh liver and kidney. The fresh tissues and environmental samples, which may be frozen for later analysis, are often the most valuable in cases of suspected toxicity. If possible, ante-mortem samples of ocular fluid, blood, urine and faeces should be collected. For some toxicants there may be typical cellular lesions in tissues of the kidney (for example, lead, antifreeze (ethylene glycol) or liver (for example, plant source pyrrolizidine alkaloids) but for many toxins the lesions are not specific.

Corrosive toxicants which have been ingested may leave evidence of burning, erosions or ulcerative damage in the upper digestive tract. Some clinical signs may be suggestive of a specific toxin, for example, vomiting in goats following ingestion of rhododendron plants. The clinical history often provides the most valuable information in working out toxicology cases and can give some indication of the nature of the suspect toxin, for example, sudden death in a group of animals grazing near a group of yew trees. If there are yew tree leaves in the digestive tract, leaves which are highly toxic at low doses, it is possible to confirm the cause of death at the time of post-mortem. However, most toxic plants are only toxic when ingested over a prolonged period, or are seasonal problems. Then the pathology seen at post-mortem occur weeks after the poisonous plants have been consumed.

Absence of proof is not proof of absence when it comes to toxicity. Other diagnostic strategies help to build evidence of a toxic cause in unexplained deaths. If more than one species, or different age groups within one species are affected in the same time-frame, and around the same area, it provides a strong case to look for a toxic cause. At this point, the thoroughness of the history gathered by the investigator will determine the likelihood of identifying the source of the problem.

Endnotes

1 Other standards can be used but these should preferably be within a range 16 +/−4 g/dl.

References

Anon (1978) Manual of Veterinary Investigation. Laboratory Techniques Volumes 1 and 2. Reference Books Published by The Ministry of Agriculture, Fisheries and Food. HMSO, London.

Berny P., Vilagines L., Cugnasse J.M., Mastain O., Chollet J.Y., Joncour G., Razin M. (2015) Vigilance poison: illegal poisoning and lead intoxication are the main factors affecting avian scavenger survival in the Pyrenees (France). Ecotoxicology and Environmental Safety 118: 71–82.

Booth, N.H., McDonald, L.F. (1982) Veterinary Pharmacology and Therapeutics. The Iowa State University Press, Ames, IA.

Botha C.J., Coetser H., Labuschagne L., Basson A. (2015) Confirmed organophosphorus and carbamate pesticide poisonings in South African wildlife (2009–2014). Journal of the South African Veterinary Association 86(1): 1–4.

Burcham G.N., Becker K.J., Tahara J.M., Wilson C.R., Hooser S.B. (2012) Myocardial fibrosis associated with previous ingestion of yew (*Taxus* sp.) in a Holstein heifer: evidence for chronic yew toxicity in cattle. Journal of Veterinary Diagnostic Investigation 25: 147–152.

Cheesbrough, M.C. (2005) Medical Laboratory Manual for Tropical Countries. Volumes I and II. Butterworths, London.

Christensen, D.E. (1996) Veterinary Medical Terminology. W.B. Saunders, Philadelphia, PA.

Colburn, T., and Clement, C. (eds) (1992) Chemically Induced Alterations in Sexual and Functional Development: The Wildlife Human Connection. Princton Scientific Publishing Co., Princeton, NJ.

Eason, C.T., Murphy, E.C., Wright, G.R., Spurr, E.B. (2002) Assessment of risks of brodifacoum to non-target birds and mammals in New Zealand. Ecotoxicology 11: 35–48.

Evans, G. (1996) Animal Clinical Chemistry – A Primer for Toxicologists. Taylor & Francis, London.

Fisher I.J., Pain D.J., Thomas, V.G. (2006) A review of lead poisoning from ammunition sources in terrestrial birds. Biological Conservation 131(3): 421–432.

Fraser, C.M. (1991) The Merck Veterinary Manual. A Handbook of Diagnosis, Therapy, and Disease Prevention and Control for the Veterinarian. Merck & Co Inc., New Jersey.

Henderson, K.L., Coates, J.R. (eds) (2009) Veterinary Pharmaceuticals in the Environment. American Chemical Society. Oxford University Press, Oxford.

Jones, A. (2001) Monensin toxicosis in 2 sheep flocks. The Canadian Veterinary Journal 42(2): 135–136.

Jubb, K.V.F., Kennedy, P.C., Palmer, N. (1992) Pathology of Domestic Animals: Volumes 1 and 2. Academic Press, New York.

Kart A., Bilgili, A. (2008) Ionophore antibiotics: toxicity, mode of action and neurotoxic aspect of carboxylic ionophores. Journal of Animal and Veterinary Advances 7(6): 748–751.

Mineau, .P, Fletcher, M.R., Glaser, L.C., Thomas, N.J., Brassard, C., Wilson, L.K., et al. (1999) Poisoning of raptors with organophosphorus and carbamate pesticides with emphasis on Canada, US and UK. Journal of Raptor Research 33: 1–37.

Morrissey, C.A., Mineau, P., Devries, J.H., Sanchez-Bayo, F., Liess, M., Cavallaro, M.C., Liber, K. (2015) Neonicotinoid contamination of global surface waters and associated risk to aquatic invertebrates: a review. Environment International 74: 291–303.

Naidoo, V., Wolter, K., Cromarty, D., Diekmann, M., Duncan, .N, Meharg, A.A. et al. (2010) Toxicity of non-steroidal anti-inflammatory drugs to Gyps vultures: a new threat from ketoprofen. Biology Letters 6(3): 339–341.

Ogada, D.L. (2014) The power of poison: pesticide poisoning of Africa's wildlife. Annals of the New York Academy of Sciences 1322(1): 1–20.

Radostitis, O.M., Mayhew, I.G., Housten, D. (2000) Veterinary Clinical Examination and Diagnosis. W.B. Saunders, Philadelphia, PA.

Smits, J.E.G., Naidoo, V. (2018) Toxicology of birds of prey. In: Sarasola, J.H., Grande, J.M., Negro, J.J. (eds) Birds of Prey: Biology and Conservation into the XXI Century, pages 229–250. Springer, Switzerland.

Smits, J.E.G., Fernie, K.J., Bortolotti, G.R., Marchant, T.A. (2002) Thyroid hormone suppression and cell mediated immunomodulation in American kestrels (*Falco sparverius*) exposed to PCBs. Archives of Environmental Contamination and Toxicology 43: 338–344.

Welch, K.D., Panter, K.E., Gardner, D.R., Stegelmeier, B.L. (2012) The good and the bad of poisonous plants: an introduction to the USDA-ARS poisonous plant research laboratory. Journal of Medical Toxicology 8(2): 153–159.

chapter 8

Pathology/cytology

Susan C. Cork

8.1 Introduction

The provision of pathology services should be an integral part of veterinary service development, especially in rural areas where access to a wide range of ancillary diagnostic tests is limited. In rural settings, it is of great value to have access to veterinary professionals and animal health technicians trained to recognize the gross tissue changes or lesions typical for disease conditions common to the area. In hot humid climates carcasses may putrefy quickly, making samples unsuitable for microbiology and other tests, and so gross post-mortem or 'necropsy' may be the only 'test' available. At fresh post-mortems (that is, when the animal has recently died or has been euthanized) the selection of specific samples will depend on the nature of the disease(s) suspected and on the experience of the operator (technician, veterinarian or animal health extension staff). In general, the post-mortem should be carried out according to a set protocol and performed in a methodical manner so that key findings are not missed.

More detailed examination of cellular changes in tissue samples for diagnostic purposes is known as histopathology. To ensure good tissue preservation the samples should be fixed in a fixative, such as 10% buffered formalin. This should be added to the sample container at a volume of 10 : 1 and sent to a regional or reference laboratory for further analysis. Histopathology services are usually beyond the capability of district laboratories but are discussed briefly in section 8.3. In cases where tissues are already decomposed histopathology may not be justified as any changes seen could be a result of post-mortem change as opposed to disease. Guidelines for the collection and submission of samples for other tests, for example, microbiological and parasitological examination, are outlined briefly below but are also discussed in the relevant chapters. Supplementary necropsy guidelines are available in Appendix 2 and online.

In most countries, there are strict regulations regarding the sending of pathological specimens through the post or via courier. It is the responsibility of the submitter, or the referring laboratory, to obtain a copy of the relevant transport regulations and to ensure that the correct packaging and labelling are used before sending samples on to a diagnostic facility. All specimens must be clearly addressed and marked with the words 'FRAGILE, HANDLE WITH CARE – PATHOLOGICAL SPECIMEN'. Specimens should be sent in sealed containers as in Figure 8.1 and enclosed in protective wrapping to prevent breakage and/or leakage en route to the diagnostic or referral laboratory. Guidelines for sending samples to international reference facilities are found in the OIE Manual for Diagnostic Tests and Vaccines for Terrestrial Animals (see also Chapter 1).

The OIE Manual of Diagnostic Tests and Vaccines for Terrestrial Animals is a compilation of diagnostic procedures and a ready source of information for any veterinary diagnostic laboratory. The Manual has been designed for practical use in the laboratory setting and contains over 110 chapters on infectious and parasitic diseases of mammals, birds and bees. http://www.oie.int/en/standard-setting/terrestrial-manual/access-online/

8.2 General approach for necropsy cases

Performing a post-mortem (necropsy) is an important part of disease investigation especially where there is significant mortality and morbidity. The exact procedure followed will be determined by many factors but a systematic and thorough approach should be followed for each case examined. Unless the cause of death is obvious, the initial step should include taking a full clinical history and, with due attention to

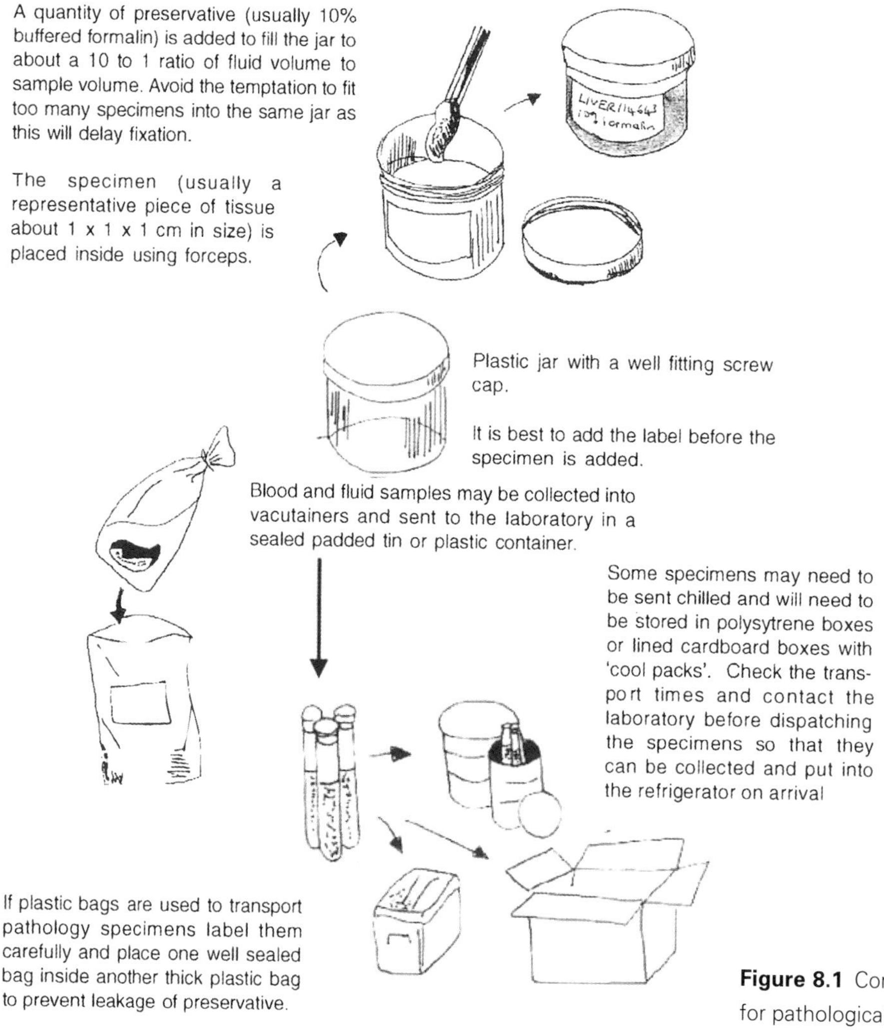

A quantity of preservative (usually 10% buffered formalin) is added to fill the jar to about a 10 to 1 ratio of fluid volume to sample volume. Avoid the temptation to fit too many specimens into the same jar as this will delay fixation.

The specimen (usually a representative piece of tissue about 1 x 1 x 1 cm in size) is placed inside using forceps.

Plastic jar with a well fitting screw cap.

It is best to add the label before the specimen is added.

Blood and fluid samples may be collected into vacutainers and sent to the laboratory in a sealed padded tin or plastic container.

Some specimens may need to be sent chilled and will need to be stored in polysytrene boxes or lined cardboard boxes with 'cool packs'. Check the transport times and contact the laboratory before dispatching the specimens so that they can be collected and put into the refrigerator on arrival

If plastic bags are used to transport pathology specimens label them carefully and place one well sealed bag inside another thick plastic bag to prevent leakage of preservative.

Figure 8.1 Containers used for pathological specimens.

biosecurity and biosafety, performing a thorough clinical examination of any live, in-contact, animals. If more than one animal has been affected, and test capabilities allow, samples from sick live animals can provide valuable supplementary information. If there is a disease outbreak it is important to determine the number of animals at risk, the morbidity rate (that is, the number of sick animals) and the mortality rate (that is, number of dead animals). The veterinarian or other animal health professional in charge of the outbreak will assess the epidemiological profile of the outbreak, which species and age groups are affected, the clinical course of the disease (that is, acute or chronic), the number and locations of properties or districts that have had cases of the disease, the presence of pre-disposing factors including changes in management, climatic factors, animal movements and so on.

Taking a good history and examining clinically affected animals will assist animal health field staff to determine which samples to collect from sick and dead animals in order to confirm a diagnosis. To assist the laboratory, for each case attended, the veterinary officer in charge should make a short list of the differential diagnoses which need to be ruled in or out. If this is not possible then it is essential to provide as much relevant information as possible on the laboratory submission form (see Appendix 2) to accompany the laboratory samples so that the appropriate tests can be selected. For submission of samples from live animals please consult the section on clinical examination, sample collection and diagnosis in Chapter 1.

Taking a case history and clinical examination

1 Determine the age, identification number, sex, species and breed of the dead animal(s).
2 Outline the duration and nature of any clinical signs observed prior to death.

3 Note any previous illness/treatment (vaccination/antimicrobial use/de-worming and so on).
4 If relevant, discuss the epidemiological features of the case with the veterinary officer/animal health professional in charge.
5 Examine the external appearance and presentation of the dead animal(s) carefully (see below).

Basic post-mortem equipment for field work – see Chapter 1 for full kit.

- note book and pen to record gross findings
- protective clothing and suitable disinfectant
- sharp knife/cleaver
- hammer and saw (to open the skull and long bones)
- scissors, scalpel blades and forceps
- tray for examination of visceral tissues
- string
- specimen jars and preservatives and so on.

Sample collection (equipment required)

- submission forms/marker pen to identify samples
- bottles and 10% buffered formalin for histological samples
- plastic bags (sealed) and sterile containers/swabs for microbiology samples
- microscope slides/ethanol for tissue impression smears
- syringes (10–20 ml)/18–20 gauge needles for taking fluid samples from body cavities, blood, urine and joint fluid.

Brief guidelines for sample collection

SAMPLES FOR HISTOPATHOLOGY

Collect samples ($1 \times 1 \times 1$ cm) representative of the normal and diseased tissue in an organ and store in a labelled bottle in ten times the sample volume of 10% buffered formalin. For intestinal sections, it may be necessary to attach the tissue to a piece of cardboard with a note outlining the area of the gut the section was taken from (for example, ileum, jejunum, caecum, colon and so on). This is because the location of lesions can be important. For brain examination, the whole brain should be fixed in a bucket of formalin. This is to allow a more comprehensive examination of different parts of the brain tissue. DO NOT FREEZE tissues for histopathology (HP).

Tissues taken from carcasses of animals which have been dead for > 12 h will show a lot of secondary changes and may therefore not be suitable for HP examination so this should be noted on the submission form along with details of any gross findings.

MICROBIOLOGY

Collect representative tissue sections, body fluids or discharges using sterile forceps, a swab, or a needle and syringe, and store in sterile well labelled containers. Collect such samples before opening the gut to avoid contamination with gut bacteria and other contaminants. Make a note of what disease is suspected in order to assist the laboratory to choose the correct growth media for culturing the causative agent(s). Impression smears taken from the freshly cut surface of organs such as the liver and spleen may be prepared and fixed in the field using ethanol or, if a flame is available, heat fix. Smears should be prepared in duplicate for staining and examination at the laboratory. Collect urine using a sterile catheter or a needle and syringe. Faecal samples for microbiology can be collected from the gut into sterile containers using a clean spatula after other samples have been collected.

PARASITOLOGY

Collect representative faecal samples (10 g) and any whole parasites and store in 70% alcohol or 10% formal saline in labelled jars or sealed plastic bags. Smears can be prepared from the mucosal lining of the intestine especially around the ileocaecal valve (for Mycobacteria) and the caecum and ileum (for *crytposporidia* and *coccidia*).

HAEMATOLOGY

Impression smears and blood smears can be prepared from the spleen, lymph nodes, bone marrow (iliac crest/rib) and heart blood to look for blood parasites and to examine the morphology of red and white blood cells.

TOXICOLOGY

Unless a specific toxin is suspected it is hard to confirm poisoning as a cause of death. Even where comprehensive toxicology facilities are available it is not always possible to confirm the presence of the toxin in tissues as the active ingredient may already have been metabolized prior to death. However, as a general rule, to investigate such deaths collect routine samples for HP (liver, spleen, heart, lung, kidney/ureter/bladder wall, and a bone marrow smear for cytology). Fresh liver and stomach contents should also be submitted along with a detailed report of the clinical signs prior to death and any gross findings. If plant toxins are suspected also send a sample of the plant. If contaminated feed is implicated as a cause of death it is important to take 1 kg from the suspect batch for analysis and to record the feed batch number (see Chapter 7).

Procedure for post-mortem examinations

The approach must be systematic to allow thorough examination of each of the main body systems (gastrointestinal, respiratory, cardiovascular, urogenital, lymphoreticular, integumentary and neurological). If neurological signs were exhibited by the animal prior to death then a detailed examination of the brain and spinal cord may be required but this is difficult and time consuming and not routinely done. If rabies is suspected care must be taken to avoid human exposure to the virus: wear gloves, a mask and, if available, use a designated kit to take the appropriate brain sections. Animal health staff likely to be exposed to rabies should be vaccinated to protect them from rabies. If in any doubt about handling the case, take/send the whole head (chilled) directly to the laboratory with a special request to test for rabies. In cases of sudden death where anthrax is common take a peripheral blood smear (usually from the ear), stain it and examine for Gram +ve bacilli before opening the carcass, however, if this is not possible bury the carcass whole (without opening it) and cover with lime. This is to prevent the release of anthrax spores which can contaminate an area for a long period of time and potentially cause an outbreak of anthrax.

During the post-mortem, it is helpful to have an assistant available to collect the samples and make a note of any gross abnormalities as the examination proceeds (see example submission form and full necropsy guidelines in Appendix 2). In some cases, the cause of death will be obvious from the clinical history or from previous examinations of the animal, however, there is often limited information available for a case. In such circumstances, it is especially important to work from basic principles going through each body system in a systematic manner. In order to recognize an abnormality it is important to be familiar with the normal anatomy of the species which are most likely to be dealt with (see Figures 8.2, 8.3, 8.4 and 8.13 for bovine and some other species).

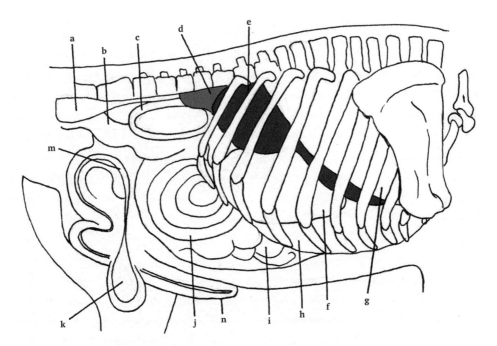

Figure 8.2 Bovine internal organs (bull), right side view. (a) rectum, (b) caecum, (c) duodenum, (d) right kidney, (e) liver, (f) omasum, (g) lung, (h) abomasum, (i) small intestine, (j) colon, (k) right testis, (m) bladder, (n) penis. Illustration: Louis Wood.

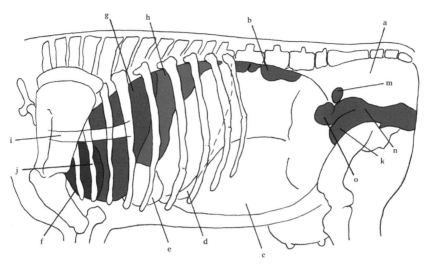

Figure 8.3 Bovine internal organs (cow), left side view. (a) rectum, (b) left kidney, (c) rumen, (d) costal line of the diaphragm, (e) reticulum, (f) heart, (g) lung, (h) spleen, (i) oesophagus, (j) ribs, (k) bladder, (m) ovary, (n) vagina, (o) uterus. Illustration: Louis Wood.

Figure 8.4 A schematic representation of the topographical anatomy of a bird. There is quite a lot of anatomical variation among avian species depending on their diet and evolutionary background. The diagram shown is based on the pigeon which is similar to domestic fowl. (No) nostril, (Tr) trachea, (Br) brain, (Sp) spinal cord (within vertebral column not shown), (Oes) oesophagus, (H) heart, (Lu) lungs, (Ki) kidney, (As) air sacs (these are located caudal and cranial to the lungs and act as 'bellows' to improve the respiratory efficiency of birds), note that most species of birds do not have a diaphragm. (Cr) crop (not all bird species

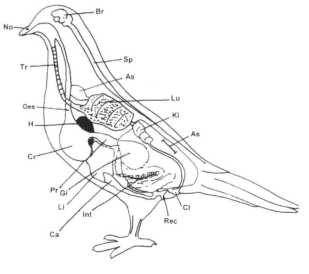

have a crop, it acts to store food and in pigeons the lining secretes a 'milk' for chicks), (Pr) proventriculus and (Gi) gizzard or ventriculus (these form the stomach of the bird, this has a thick wall and often contains stones to help the bird break down food material as birds do not have teeth to grind fibrous food, the proventriculus has a glandular lining), (Li) liver, (Ca) caecum (grain eating species have well developed paired caecae). (Int) small and large intestine. Some parasites such as coccidia can be partially identified by the region of the intestine or caecae that they prefer to invade. (Rec) rectum, (Cl) cloaca, birds do not have a separate faecal and urinary passage, both join to exit through the cloaca. Avian 'faecal' material contains the white urinary wastes (urates) and green/brown digestive wastes.

The number and type of samples required from a post-mortem will depend on the species examined and the nature of the disease(s) suspected (see Tables 10.1–10.10 in Chapter 10 for diseases associated with specific clinical signs). The following technique is suggested for a basic field necropsy. Detailed procedures for performing a more comprehensive necropsy are outlined in Appendix 2.

Performing the necropsy

1 Place the dead animal (carcass) on its back and secure it in position using blocks of wood or rocks (that is, place wedges under the pelvis/shoulders to provide balance). At the laboratory, there may be access to a post-mortem facility with specialized tables, winches and other equipment but, due to logistical challenges, most necropsies are done in the field. For poultry carcasses, it is generally preferable to remove or wet some of the feathers, especially those over the sternum, prior to beginning the necropsy.

2 Using a sharp knife, make an incision in the skin from the midline of the throat to the pelvis. In birds it is usually necessary to cut around the sternum. When cutting into the abdomen and thoracic cavity check to see if the air sacs are clear, these can become thickened and cloudy in birds with respiratory disease.

3 Make deep incisions at the axillae and the hip joints, to make the limbs lie flat. Examine the subcutaneous tissues and the superficial lymph nodes. Incise the lymph nodes (axillary, precrural, inguinal and popliteal), examine the structure and note any abnormality. When cutting through the hip joints it may be necessary to incise the round ligament (which secures the hip socket in place).

4 Note the contents of the hip joint capsule and the presence of any excess fluid.

5 Examine the head of the animal and note the colour of the mucous membranes and the position of the eyes (if sunken into the skull this is often a sign of malnutrition/dehydration as the fat pad behind the eye shrinks). Find and incise the superficial lymph nodes (these may be enlarged and have an abnormal texture in many localized or systemic infections).

6 Open the mouth and cut along the inner margins of the mandibles, free the tongue and open the pharynx to examine the teeth, tonsils and salivary glands.

7 Cut into the muscles of the neck to expose the trachea and oesophagus. Examine the glands of the neck including the thyroid glands.

8 Open the nasal cavity and look for abnormal changes in the turbinate bones (atrophic rhinitis in pigs/tumours/foreign bodies). In horses open and examine the guttural pouch and check for the presence of fungal hyphae, discharge and so on.

9 Using a sharp knife open up the rib cage by cutting posteriorly along the abdominal wall.

10 Examine the abdominal and thoracic contents and note any abnormal colouration, abnormal location of organs or the presence of excessive peritoneal fluid/ascites/peritonitis. Examine the peritoneum (colour/texture) and the mesenteric lymph nodes. In birds the colour and texture of the air sacs should be examined.

11 Examine the location, colour and texture of the lungs and heart. If there is excessive fluid around the heart (pericardial fluid) or in the thoracic cavity (pleural/mediastinal) collect a small amount (5–10 ml) using a sterile needle and syringe. This fluid can be submitted for cytological examination and microbiology. Note the volume, colour and viscosity (thickness) of any fluid collected.

12 Once the thoracic and abdominal cavities have been opened the viscera (internal organs) can be removed and examined systematically. In some cases, the viscera should be weighed and the weight recorded as part of the post-mortem.

See also Appendix 2 and the videos of avian and ruminant necropsy available on the website.

Examination of specific body systems

In the next section, we will discuss the examination of each body system in more detail. Most of the text refers to ruminants but where there are specific differences between ruminants and other species these are outlined.

Gastrointestinal system (see Figures 8.3 and 8.4)

Remove the intestines/stomach from the body cavity by tying the oesophagus and the rectum with two pieces of string 2 cm apart at each point to be detached. Make an incision between the pieces of string and remove the intestines and stomach(s) into a tray. The liver can be removed at the same time. Note the colour and size of the liver and whether or not the gall bladder is empty or full. Examine for the presence of liver fluke or tape worm cysts. These may only be an incidental finding but their presence could be important. Describe any gross lesions and remove a section of liver for histology and for microbiology (1 × 1 × 1 cm sections). Open the stomach(s) and examine the contents. Check for the presence of foreign bodies, for example, wire, plastic and so on. If poisoning is suspected keep a sample of the stomach/rumen contents and store in a labelled plastic bag. It may also be important to collect samples of suspect feed/plants for toxicology tests. In ruminants, note the presence of excessive gas in the rumen, this may indicate a primary or secondary bloat but also remember that the abdominal contents putrefy rapidly and that gas production may be a post-mortem change. Open up the intestinal tract carefully, note the presence of parasites. If possible collect a sample of any worms present for identification at the laboratory along with five to six cross-sectional pieces of intestine for histological examination. The former should be submitted in alcohol (70%) and the tissues for histopathology in 10% buffered formalin. If coccidiosis is suspected take a smear from the mucus of the jejunum or caecum and fix on a microscope slide (use a flame to heat fix or allow to dry and fix in methanol). Examine all of the lymph nodes and the ileocaecal valve. Some diseases, for example, Johne's disease (*Mycobacterium avium paratuberculosis*), cause characteristic changes at this point so this area should be submitted for histological examination.

In birds (see Figure 8.4) the anatomy of the gastrointestinal tract depends on the species examined. The chicken (*Gallus gallus*) has a storage area, the crop (at the distal end of the oesophagus), which should be examined for signs of impaction or inflammation. The caeca are paired and should be examined for lesions associated with coccidiosis (see also Chapter 3).

Cardiovascular system (see Figure 8.5)

Examine the heart and the pericardium. Look for signs of trauma such as haemorrhage or colour change (Figure 8.5). Remove the heart and open up the atria and ventricles. Examine the valves for signs of emboli and inflammatory change. Collect a section of heart muscle +/− valvular material/blood vessel for HP. Examine the major vessels in the body for signs of parasitic migration (for example, strongyles in the mesenteric vessels of horses) or evidence of inflammatory changes. In freshly dead animals, especially pigs and small animals, it is often useful to collect heart blood for microbiological examination

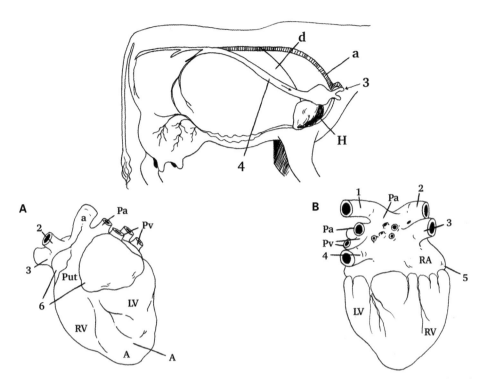

Figure 8.5 Bovine cardiovascular system. Schematic representation of a cow that illustrates the location of the heart, the aorta and the caudal vena cava. Illustrations (A) and (B) show the major anatomical features of the heart in more detail, in many cases the heart will be surrounded with some fat (pericardial fat). (A) shows the right caudal view of the heart with the pericardium removed and (B) shows the left view. (H) heart, (A) apex, (LV) left ventricle (this has a thicker layer of muscle than the right side), (RV) right ventricle, (RA) right atrium (left atrium not shown), valves (not shown) separate the ventricles from the atria to prevent regurgitation of blood during cardiac contraction (these are the mitral or bicuspid valve on the left and the tricuspid on the right), (Pa) pulmonary arteries (these vessels supply blood from the right ventricle to the lungs for oxygenation), (Pv) pulmonary veins (these vessels return oxygenated blood from the lungs to the left atrium), (a) aorta (this is the main artery in the body that receives oxygenated blood from the LV of the heart and transports it around the body via a network of branching vessels). Valves prevent back flow of blood into the heart from the aorta (aortic valve) or the Pv (pulmonary valve). (d) diaphragm, (Put) pulmonary trunk. (1) descending aorta, (2) brachiocephalic trunk, (3) cranial vena cava (brings blood from the head and neck to the right atrium of the heart), (4) caudal vena cava (brings blood from the body to the right atrium of the heart), (5) coronary vessels (these provide the heart muscle with its own blood supply), (6) left and right auricles. Note that the arteries have a thicker wall than the veins and that there is a thick layer of elastic tissue present to allow for changes in blood pressure during the cardiac contraction cycle. Illustration: Louis Wood.

especially if septicaemia is suspected. Use a sterile needle and syringe to collect 5–10 ml. Note the anatomical difference between the major arteries (for example, aorta) and the thinner walled veins (for example, jugular, vena cava). See the anatomy texts listed at the end of the chapter for more details.

Respiratory system (see Figure 8.6)

Examine the pleura which line the thoracic cavity before removing the trachea and lungs. Check that the diaphragm is still intact and look for any signs of bruising on the inside of the rib cage. Look for the presence of abnormal colouration, parasites and fluids in the airways. Note any areas of consolidation. Remove a piece of lung and see if it floats in water; if it sinks this indicates consolidation rather than post-mortem change. Note that in birds the lungs are fairly small and fixed in place and there is no obvious diaphragm. Most of the breathing cycle in birds involves the filling of the air sacs; examine the main ones which lie in the abdominal cavity and the thoracic space. In healthy birds the air sacs should be clear like the pleural lining in mammals but in animals with respiratory diseases, such as aspergillosis, the air sacs will be thickened and discoloured.

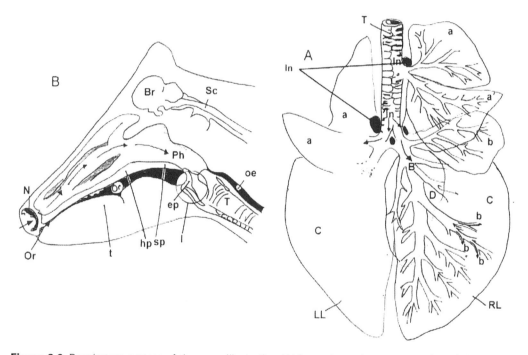

Figure 8.6 Respiratory system of the cow. Illustration (A) is a schematic representation of the trachea (T), lungs and bronchi along with some of the deep lymph nodes (In). There is some difference between the species but the general layout is similar in all mammals. In the cow the left lung has an apical (a) and a diaphragmatic (C) lobe, the apical lobe is divided into a cranial (a) and a caudal (a') section. The right lung has an apical and diaphragmatic lobe as well as the middle (b) lobe and the accessory (D) lobe. The trachea bifurcates into the left and right major bronchi (B) and these branch into the bronchioles (b). The trachea is lined with a protective layer of mucus and a ciliated epithelium to prevent access of foreign material. Respiratory disease often occurs when the protective lining is damaged. In illustration (B) you can see a sagittal section of a bovine head to illustrate the location of the upper respiratory and digestive tracts. (N) nostril, (t) tongue, (hp) hard palate, (sp) soft palate, (ep) epiglottis (this covers the entrance to the trachea during swallowing to prevent food entering the respiratory system), (oe) oesophagus, (Ph) pharynx, (Or) oral cavity, (l) larynx, (T) trachea, (Br) brain, (Sc) spinal cord, lymph nodes are coloured black (In).

Examine the throat and trachea for signs of obstruction and/or inflammatory changes. Examine the thoracic cavity for evidence of trauma.

Urogenital system (see Figure 8.7)

Examine the kidneys, ureters and bladder *in situ*. Look for the presence of cysts, polyps, haemorrhage or colour change. Remove 10–20 ml of urine using a sterile needle and syringe. If the animal has had haematuria/haemoglobinuria take a section of ureter and bladder wall for HP examination.

Cut into each of the kidneys and examine the pelvis, cortex and medullary region. If taking a sample for histopathology make sure that a cross section of cortex and medulla are included. The reproductive system should be carefully examined for the presence of infection or physical abnormality. In pregnant females, there may be lesion on the foetus or placenta that indicate the cause of death. Care should be taken when examining tissues and fluids from pregnant livestock because many of the infectious agents involved can be zoonotic. This is discussed further in Chapters 10 and Appendix 1. Examination of the reproductive system is specialized and will not be discussed further here, a range of specialist texts are available on this topic.

Lymphoreticular system (see Figure 8.8)

Note that some lymph nodes can be palpated in the live animal as illustrated in Figure 8.8, but the majority lie deep among the visceral organs and serve to drain the lymph fluid and act as a centre for white blood cell activity in the fight against disease (see Chapter 5). Examine and remove the lymph nodes, as previously described, and the spleen. A smear of bone marrow aspirated from the tuber ischium or the head of one of the long bones may also be useful especially when investigating chronic cases of anaemia or immu-

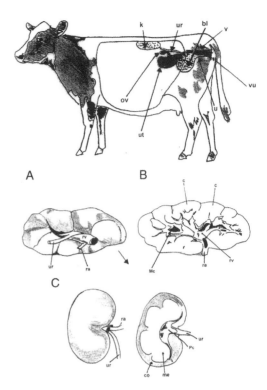

Figure 8.7 Schematic representation of the bovine abdomen showing the location of the urogenital organs of a cow. In illustrations (A and B) you can see a more detailed representation of the bovine kidney and in (C) you can see the simple anatomy of the sheep kidney (pigs, primates and most carnivores are similar). (k) kidney, (ur) ureter, (bl) bladder, (v) vagina, (vu) vulva, (ut) uterus, (ov) ovary, (u) urethra. (A) Bovine kidney, lateral view; (B) longitudinal section. (rv) renal vein, (ra) renal artery, (c) lobes of the cortex, (Mc) major calyx. (C) Sheep kidney (lateral view and longitudinal section). (co) cortex, (me) medulla, (pc) renal pelvis, (ur) ureter. At post-mortem collect a sample from the kidney which contains a section of both the cortex and medulla.

nosuppression. Also examine the sternum and lumbar vertebrae by longitudinal cross section. In young animals, active bone marrow is also present in the long bone cavity. In the chicken the Bursa of Fabricius, which is located in the

Figure 8.8 Superficial lymph flow of the bovine. This illustrates the location of the superficial lymph nodes which may become enlarged if the area drained by the lymph tissues becomes inflamed due to trauma or infection. Locate the lymph nodes on a live animal so that you can become familiar with what is 'normal'. Some lymph nodes are not palpable except when enlarged. The lymph flow keeps the tissues clear of excess fluid. Oedema, or tissue swelling, may occur when osmotic balance is disturbed, for example, damage to the liver by liver fluke may result in low serum protein and leakage of fluid from the blood stream resulting in 'bottle jaw' and fluid in other areas of the body such as the sternum. (a) mandibular lymph node, (b) parotid lymph node, (c) lateral retropharyngeal lymph node, (d) superficial cervical lymph node, (e) subiliac lymph node, (f) gluteal lymph node, (g) popliteal lymph node, (h) tuberal lymph node, (l) auxiliary and (2) inguinal lymph node, these are located on the inside of the fore and hind limb respectively. There are a large number of internal lymph glands and tissues which drain the internal organs, these cannot be palpated externally.

cloacal region, should be examined. Examine the thymus in neonates.

Neurological system (see Figure 8.9)

It is not always easy to examine the brain but sometimes it is necessary in order to confirm a diagnosis. To cut into the skull additional equipment such as a band saw is required. Where a zoonotic disease is suspected this should not be done without appropriate protective clothing.

Remove the skin from the head and make a transverse incision just behind the ears. Use a saw and/or cleaver to get through the bone. Take care not to damage the meninges. Look for signs of discolouration or increased fluid. Collect any excess fluid for cytology (use a needle and a sterile syringe). Remove the brain as a whole if at all possible. If the meninges are very congested take a sample for microbiological examination. Preserve the whole brain in formalin (10%) in a bucket. If it is necessary to examine the spinal cord hang the carcass by the hind legs and split the vertebral column with an axe. If individual vertebrae are to be examined it is possible to remove sections individually but for this it is important to have a clear idea of the clinical signs prior to death, that is, hind limb paralysis/paresis, forelimb lameness.

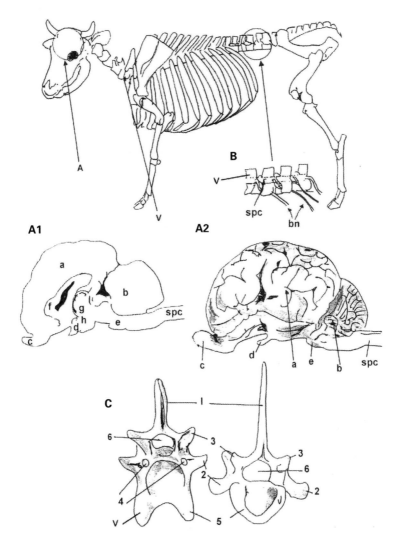

Figure 8.9 Schematic view of the bovine central nervous system. Bovine skeletal frame. The spinal cord (spc) and the brain (A) are protected by the vertebral column and the skull and are surrounded by three membranes (the meninges) and a layer of protective fluid. Illustrations (A, B and C) show some anatomical features of the brain, spinal column and vertebrae (v), respectively. (A) Lateral view (A2) and sagittal section (A1) of the brain, (a) cerebral hemisphere, cortex, (b) cerebellum, (c) olfactory bulb, (d) optic chiasma, (e) pons, spinal cord (spc), (f) corpus callosum, (g) thalamus, (h) hypothalamus. (B) View of the vertebral column illustrating the peripheral nerves (bn) which branch from the spinal cord. There is an intricate network of motor and sensory nerves which supply information to and take information from each organ in the body. If samples are collected from cases with neurological deficits it is necessary to perform a very detailed post-mortem with the guidance of an experienced pathologist who will have a detailed knowledge of the anatomy of the central and peripheral nervous system. (C) Caudal view of the 6th and 7th cervical vertebrae. (1) spinous process, (2) transverse process, (3) articular process, (4) articular process, (5) caudal cavity of vertebral body, (6) vertebral foramen (this is where the spinal cord is located).

Musculoskeletal system
(see Figures 8.10, 8.11 and 8.12)

The musculature should be examined for evidence of swelling or bruising. Note that bruises can only occur ante-mortem but that autolytic changes may occur after death. The musculoskeletal system has the same basic structure in all mammalian species but in birds the bones are filled with air spaces which may connect with internal air sacs. In some cases, broken bones may result in entrance of infectious organisms into the respiratory system of the bird and result in septicaemia and death. Figure 8.11 illustrates the general skeletal structure of the bovine. Ruminants and horses have a fixed spinal column and strong neck ligaments. The lower limbs of ruminant species and pigs have two main digits as compared with a single main digit in the horse (Figure 8.12). Note that most general lameness in the horse is usually in the foot. Examine the feet and lower limbs of a live horse to identify the main structures illustrated. The lower limbs of non-hoofed animals tend to have four or five digits. It is important to examine the limbs carefully in an animal which has had a history of lameness but if no obvious abnormalities are present in the muscles, bone, joints or associated tendons/ligaments the problem may have been in the nervous system. It is not easy to diagnose peripheral neurological damage at necropsy without histological examination of nerves, this is quite specialized and requires specialist knowledge.

Endocrine system

The endocrine system is a system of glands which secretes hormones to regulate body functions. The endocrine and neurological systems work together to maintain the body in its normal physiological state. The control centre for most endocrine functions is the hypothalamus, which is located at the base of the brain. The hypothalamus controls the production and secretion of hormones in the pituitary gland which in turn regulates other endocrine glands in a feedback

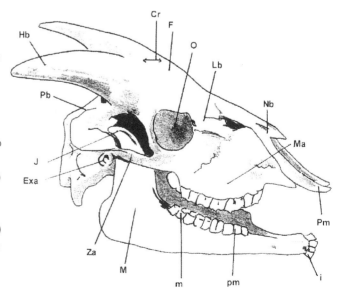

Figure 8.10 Goat skull. The following features should be easy to identify on a skull. Teeth: (i) incisor, (m) molar, (pm) premolar (the teeth can be used to age young animals due to the fact that different juvenile and adult teeth erupt at specific times, the age of adult animals can be estimated by examining the degree of wear on the incisors and molars. Note that most ruminants do not have upper incisor teeth). (Exa) external ear canal, (M) mandible, (Za) zygomatic arch, (J) coronoid process of the jaw, (Pm) premaxilla, (Pb) parietal bone, (Ma) maxilla (jaw), (Nb) nasal bone, (O) orbit (eye socket), (F) frontal bone, (Cr) cranium, (Hb) bony part of the horn, (Lb) lacrimal bone.

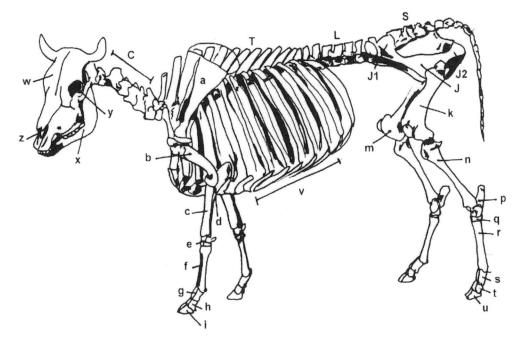

Figure 8.11 Skeleton of a cow. Vertebral column: (C) cervical (neck) vertebrae, (T) thoracic (chest) verte-brae, (L) lumbar (back) vertebrae, (S) sacral vertebrae. Fore limb (typical 'even toed ungulate', digits 4 and 3 form the hoof): (a) scapula (shoulder blade), (b) humerus (upper 'arm' or long bone), (c) radius and (d) ulna (lower 'arm' bone), (e) carpus ('wrist' bones), (f) metacarpal bone (equivalent to fused hand bones, digits 4 and 3), (g and h) proximal (upper) and middle digits (equivalent to finger bones, separate digits 3 and 4), (i) distal (lower) digits which lie within the keratin hoof. Hind limb: (J) pelvis, (J1) sacral tuberosity, (J2) ischiatic tuberosity, (k) femur (upper 'leg' or long bone). (m) stifle joint (knee), (n) tibia (fused with fibula, equivalent to lower leg bones), (p) tarsus, (q) tarsal joint or 'hock' (equivalent to the 'heel'), (r) metatarsus bone (equivalent to fused foot bones, digits 4 and 3), (s and t) proximal and middle digits (equivalent to foot bones, separate digits 4 and 3), (u) distal digits which lie within the keratin hoof. Rib cage and sternum (v): Skull, (w) frontal bone, (x) lower jaw bone, (y) orbit, (z) nasal turbinates.

loop. Occasionally these glands do not respond correctly to stimuli and metabolic changes occur in the animal. Most endocrine diseases can be diagnosed ante-mortem by clinical examination and careful evaluation of a clinical history. It is difficult to examine the brain and pituitary at necropsy unless you can saw the skull in half. The pituitary gland lies at the base of the skull in a small depression.

Examine the adrenal glands which lie just in front of the kidney. Remove the glands and cut them in cross section, examine the outer cortical layer (cortex) and the inner medullary area (medulla). The cortex produces cortisol and other steroid hormones, which are released in long-term 'physiological stress' whereas the medulla produces the catecholamines, epineph-rine and nor-epinephrine which are released as part of the 'fight or flight' response. If the adre-nal glands are enlarged this may indicate chronic stress prior to death, for example, prolonged dis-ease/malnutrition, bullying by other stock and so on, or there may be evidence of neoplastic (cancerous) growth.

Figure 8.12 Anatomy of the lower limb of the horse (typical of an 'odd toed' ungulate). The following key applies to all three illustrations. (A) lateral view of lower limb bones, (B) frontal view of lower limb bones and (C) cross section of lower limb bones *in situ*. (a) distal digit or coffin bone, (b) distal sesamoid bone (small bone which lies at flexor surface of the joint), (c) middle digit, (d) proximal digit, (e) proximal sesamoid bones, (f) lower end of metacarpal main bone (front limb, equivalent to digit three), (g) distal or lower end of small metacarpal bone (one on each side, remnant of digits 2 or 4, also known as 'splint' bones), (h) digital cushion (cushion of the underside of the foot), (i) 'frog', (j) sole of the foot, (k) hoof wall, (l) laminae (sensitive area equivalent to bed of the fingernail – laminitis is the inflammation of this area), (m) coffin joint, (n) pastern joint, (p) fetlock joint, (q) common digital extensor tendon (this keeps the foot straight), (r1) deep and (r2) superficial digital flexor tendons (allows the horse to lift the foot, that is, flexes the fetlock joint). Tendons attach muscle to bone and are well developed in the lower limbs of hoofed animals.

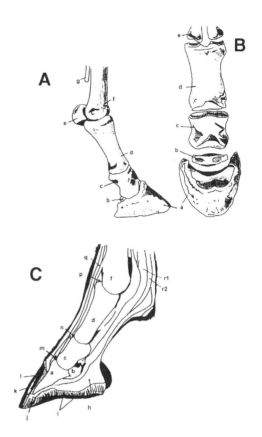

Check the thyroid gland in the midline ventral neck area, this will be enlarged in neonatal and mature animals in iodine deficient areas (goitre) or in cases of hyperthyroidism (often due to benign neoplastic changes). Diagnosis will need to be made after histological examination to assess the nature of the pathology present. The thyroid gland produces thyroid hormones which are important in the regulation of growth and metabolic rate. The parathyroid glands are usually found within the structure of the thyroid gland in birds and just outside it in most mammals. The parathyroid gland produces parathyroid hormone, which controls calcium and phosphorous balance along with calcitonin which is produced in the thyroid gland. There will rarely be evidence of gross changes in the parathyroid glands but collect these for histol-

ogy if a metabolic disease involving the bones is suspected.

The pancreas produces the sugar regulating hormones insulin and glucagon in the islets of Langerhans. It also produces exocrine enzymes and becomes inflamed in pancreatitis. The organ is located among the upper parts of the small intestine near the liver and biliary entrance to the duodenum; it is usually pale brown colour and is elongated in shape. Histological sections may be examined to assess the integrity of the secretory cells in the islets of Langerhans.

Other endocrine tissues include the reproductive glands (testes in the male and ovaries in the female) but if it is necessary to assess causes of infertility or reproductive disease in the female animal at post-mortem, try to request specialized training and/or some experience in rectal examination ante-mortem so that it is possible to assess what is normal versus abnormal at various stages of the oestrous cycle. In the male, testes can readily be examined in the scrotal sac although in some young animals one or both may not have fully descended and will be located somewhere along the inguinal canal or still in the abdominal cavity. Intra-abdominal testes are more likely to develop neoplastic changes.

A note on species differences

Although all vertebrate species have many anatomic features in common (Figure 8.13) there

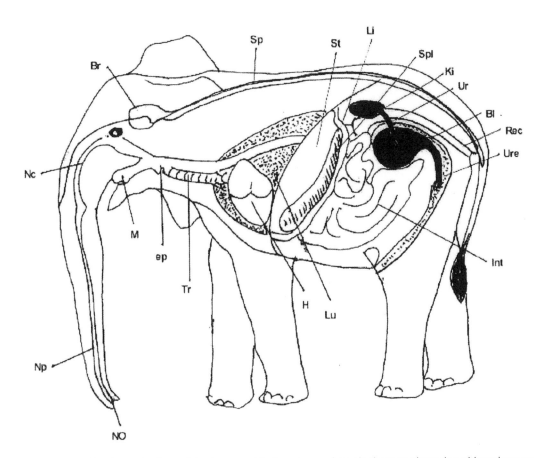

Figure 8.13 A schematic view of the topographical anatomy of an elephant to show that although many wild or exotic species may look quite different to domestic ruminants the general anatomy is similar. There will be differences in the length, size and shape of specific organs, especially the gastrointestinal tract due to different diet and habits. If it is necessary to perform a post-mortem on a species with which you are not familiar, follow the same general protocol as set out for the domestic ruminant. There will be some differences with regard to the common diseases seen in specific species but it is important to follow a standard and thorough post-mortem procedure and to seek advice from a facility with experience in the species examined. (NO) nostril, (Np) nasal passage, (Nc) nasal cavity, (M) mouth, (ep) epiglottis, (Br) brain, (Sp) spinal cord, (St) stomach, (Li) liver, (Spl) spleen, (Ki) kidney, (Tr) trachea, (Ur) ureter, (Ure) urethra, (Rec) rectum, (Bl) bladder, (Lu) lung, (H) heart, (Int) intestine.

are important species differences. Farmed species can include reptiles (for example, crocodiles), fish and a number of different avian species (for example, ducks, geese, quails) as well as fur species (for example, mustelids and other carnivores) and large mammals (such as equids, elephants and camelids) used for power and transport. Veterinary and other animal health staff should become familiar with the normal anatomy of the species that are commonly handled in their region. Practical training in conducting a necropsy should be delivered by an experienced veterinary professional and due care should be taken with regard to biosafety and biosecurity.

8.3 Histopathology

Tissues from post-mortem cases

Histology is defined as the study of tissues. Histopathology is the study of abnormal tissues. It is necessary to be familiar with the normal structure of tissues before pathological changes can be recognized. Histopathological examination can be useful to confirm a diagnosis but it may be too expensive to be justified in routine examinations. The preparation of histology slides requires skill and experience and the interpretation of slides requires specialized training. To ensure good quality sections tissues must be fixed in 10% buffered formalin (see box) although for some tissues other fixatives may be recommended (consult specialized reference texts or an expert for more detail). Once fixed, tissues are then cut, processed and blocked ready for sectioning and staining. In most cases an area of tissue which has normal and abnormal cells will be chosen for section preparation. The tissue sections are then set in paraffin wax and cut into thin slices using a microtome which is usually set at 2–5 μm. The sectioned tissues are then put onto microscope slides and stained. There are a range of automated systems available to optimize this procedure and the tissue sections produced are generally of a high quality. A range of stains can be used to highlight specific cellular structures and/or invading microorganisms. Standard stains such as Haematoxylin and Eosin (H&E)[1] are routinely used but often additional sections will be cut and stained with specialized stains such as Gram stain for bacteria, Giemsa for haemoprotozoa and Young's fungal for fungal hyphae. Stained slides can be mounted using a mounting agent to seal a cover slip in place and will remain in good condition for many years.

Neutral buffered formalin (fixation time 12–24 h)

- Formalin (40% aqueous solution of formaldehyde) 100 ml
- Sodium dihydrogen orthophosphate (monohydrate) 4 g
- Disodium hydrogen orthophosphate (anhydrous) 6.5 g
- Distilled water 900 ml

Neutral buffered formalin is suitable for most histological purposes and is preferable to formal saline as the formation of formalin pigment in tissues is avoided. The solution is isotonic so specimens can be stored in this fluid. Typically, a sample should be fixed in 10x the volume of fixative relative to the volume of the sample.

Tissues from biopsy samples

Tissue samples may be collected from live animals but these are usually limited to skin sections collected using aseptic collection methods under local or general anaesthesia. These samples are usually referred to as biopsy samples. The techniques used are outlined in standard surgical text books and should only be performed by suitably trained veterinary professionals. Liver, gut sections and other tissues (that is, lymph nodes) may also be collected under general anaesthesia. Tissues are prepared using similar techniques to those described for necropsy but in some cases quick cryostat methods are used to process tissues to speed up the process and provide rapid diagnostic results (for example, in the case of exploratory surgery for neoplastic disease). The latter is rarely performed in veterinary medicine but is not uncommon in human medicine.

8.4 Cytology

Cytology, or the study of cells, includes the examination of tissue smears and sediment analysis of urine and other body fluids such as joint fluid and transudates or exudates from body cavities. Transudates occur in association with, or secondary to, congestive heart failure, liver failure and other causes of osmotic imbalance. Exudates occur in inflammatory processes following infection or damage to tissues. Biochemical analysis of body fluids is also useful to assess the nature of the disease process. There is often a mixture of cell types in abnormal body fluids and the predominant cell types present, as well as the appearance of the cells (that is, evidence of cell death, inclusion bodies or bacteria/fungi) will give an indication of the nature and extent of the disease process. Cytology is a useful tool for ante-mortem examination of samples as well for post-mortem samples.

8.5 Examination of samples and interpretation of results

In some cases, the histological appearance of cells in stained sections will be diagnostic for a specific disease or disease process, for example, viral or toxic inclusions in specific cells, but in many cases the changes seen may be representative of a range of diseases and diagnosis will still be based on a combination of clinical examination, history and gross findings supplemented by histological findings and the results of other laboratory tests. Unstained histological sections may also be used for immune-fluorescence tests and immune-histochemistry using enzyme labelled specific antisera as a diagnostic tool (see also Chapter 6). Consult specialized texts to find specific protocols, some useful texts are listed at the end of the chapter. In most cases it will be necessary to go on a training course at a specialist centre before starting to prepare histological sections.

Interpretation

Veterinary professionals will require specialized training and years of experience before they become proficient pathologists. In order to interpret histological slides the pathologist needs to have a good knowledge of the histology of normal tissues so that abnormalities can be recognized. There are a wide range of 'artefactual' changes that may be present if (1) slides are not correctly prepared, (2) tissues were not adequately fixed or (3) tissues were autolysed prior to collection. This is why it is important to make sure that tissues selected for histopathological examination are as fresh as possible and that they are fixed in 10% buffered formalin (other fixatives may be used but check with the referral laboratory). In most cases there should be some abnormal and some normal tissue submitted in a section $1 \times 1 \times 1$ cm in proportion to ten times the volume of fixative. All specimens sent to a referral laboratory accompanied by the correct submission form (see Appendix 2 for an example), which should contain information about the case, that is, full clinical history, gross necropsy findings and so on. Some examples of clinical conditions, gross necropsy cases, histology and histopathology slides are provided in Figures 8.14 to 8.26 and additional background information

Figure 8.14 There are often a lot of deaths at lambing time but the presence of the occasional deformed lamb may not necessarily indicate that there is a disease present. Abnormalities like this one-eyed lamb ('cyclops') occur during development, abnormal embryos are usually reabsorbed or aborted early in development. Occasionally some survive to term but die shortly after birth. This lamb was one of two, the other lamb was normal. Some mycotoxins, plants or drugs may result in foetal abnormality (that is, they are teratogenic). See also Plate 36.

Figure 8.15 This old ewe was found in a bog shortly after lambing. The general appearance of the animal indicates that she is seriously ill but it is not possible to determine the cause of the problem without taking a full clinical history and examining the animal carefully. In many cases it will also be necessary to take samples for laboratory diagnosis but if the animal is terminally ill the laboratory results may be a bit misleading due to the development of secondary metabolic changes which may mask the primary cause of the problem. In this case the animal died and during the necropsy it became clear that the ewe had a significant liver fluke burden with liver damage which may have contributed to the development of ketosis associated with the heavy demands of lactation and a period of unseasonably cold weather.

Figure 8.16 This old ewe has started to lose her wool after recovering from ketosis (twin lamb disease). The ewe also had evidence of liver damage. Although it would initially appear that the ewe had a primary problem with her wool it was clear on further examination that she was slightly jaundiced and was under weight. There was no evidence of mite infection or any primary skin disease. The clinical history and laboratory tests confirmed liver damage. Wool loss often follows a period of extreme physiological stress. The animal made a partial recovery but was culled the following year.

can be found in the texts listed in the bibliography at the end of this chapter. Additional necropsy guidelines are available online and in Appendix 2.

Figure 8.17 Laboratory technician setting out to perform a post-mortem. It is often necessary to travel to villages to determine the cause of death in livestock because in many cases it is not possible for the farmer to transport the carcass to the laboratory. The technician's field kit needs to be easy to carry but should contain a good knife, a selection of sample collection materials, a note book, recording pen and marker pen, plastic gloves and a sachet of disinfectant.

Figure 8.18 If a post-mortem is carried out under field conditions it is important to explain each step of the process to the farmer. In many cases the owner of the animal may want to sell the meat from the dead animal for human consumption and may be wary of giving a full case history. It is important to develop a good relationship between laboratory and extension staff and local farmers so that the needs of all groups can be successfully met. In this case the local laboratory had only recently started performing field post-mortems at the farmer's request to comply with the requirements of a government law requiring all meat to be certified fit for consumption. The animal examined had evidence of heavy liver fluke infection and various tape worm cysts but the meat was certified suitable for consumption as long as it was well cooked. The importance of parasite control was highlighted to the famer who was subsequently more interested in accepting diagnostic help from the laboratory and anthelmintic support from the animal health extension service. Khaling village, Eastern Bhutan.

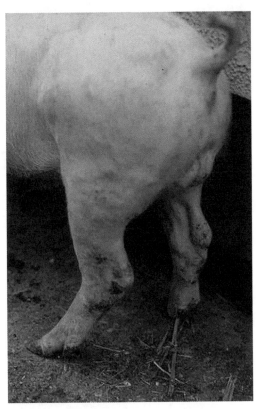

Figure 8.19 Skin rash seen in a pig with *Erysipelothrix insidiosa* sp. (now *E. rhusiopathiae*) infection. This pig was treated with antibiotics and subsequently recovered. See also Plate 37.

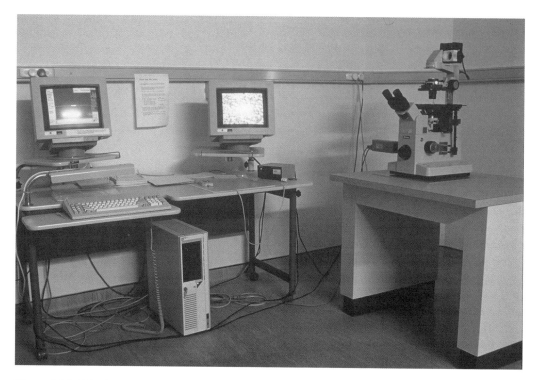

Figure 8.20 This photograph illustrates a common set up for image analysis where histology images can be viewed directly on a computer screen and specific areas can be measured and mapped based on differences in staining intensity or other differential markers. This is useful for assessing the size of lesions or the presence of abnormalities such as excessive iron storage (hemosiderosis) seen in the image on the computer screen in this photograph. The slide is a liver section from a bird stained with Perls' iron stain which indicates the presence of iron stored in tissues.

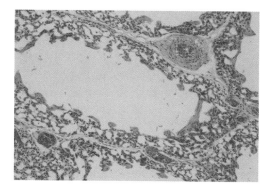

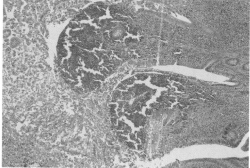

Figure 8.21 Histology section H&E 20× of a normal chicken (*Gallus gallus*) lung. H&E = Haematoxylin and Eosin stain (see main text). See also Plate 38.

Figure 8.22 Histology section H&E 20× of a bird intestine (chicken, *Gallus gallus*) illustrating haemorrhage secondary to a bacterial infection. See also Plate 39.

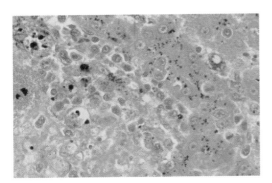

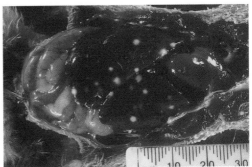

Figure 8.23 Histology section of a bird liver 40× stained with Perls' Prussian Blue iron stain to illustrate the presence of iron stored as hemosiderin in hepatocytes and Kupffer (macrophage) cells. The presence of excess hemosiderin can occur as a result of excessive intake of iron or due to excessive iron breakdown as in the case of haemolytic anaemia (that is, can be associated with avian malaria) or subsequent to sever trauma and debilitation resulting in tissue damage and/or muscle breakdown. See also Plate 40.

Figure 8.24 Illustration of gross necropsy on a freshly dead aviary bird (parakeet) illustrating an enlarged and discoloured liver with several abscesses. Cultures taken from these abscesses grew a pure culture of *Yersinia pseudotuberculosis* serotype 2. This is not an uncommon cause of death in wild and captive birds especially when predisposing factors such as immune-suppression or concurrent systemic disease are present. Iron storage diseases can also predispose to the development of pseudotuberculosis and other bacterial infections (for more information see Cork, 2000). See also Plate 41.

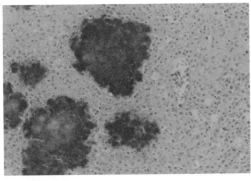

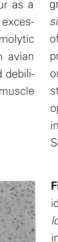

Figure 8.25 The use of immune-histochemistry to identify lesions caused by *Yersinia pseudotuberculosis* (serotype 3) in the liver of the case illustrated in Figure 8.24. Immuno-histochemistry is discussed in Chapter 6. Essentially, the cut histology sections are incubated with hyperimmune serum (in this case rabbits anti-*Yersinia pseudotuberculosis* type 2 antisera) which is bound to an enzyme or conjugate and then rinsed. A substrate is then added to the slides and if the antibody remains bound to the cells (or, in this case, bacterial lesions) on the slide this indicates the presence of antigen (for more information see Cork et al., 1999). See also Plate 42.

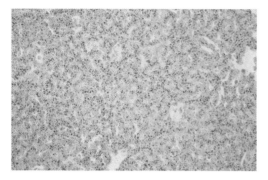

Figure 8.26 Histology section of a wild bird liver (New Zealand kokako [*Callaeas cinereal*]) 10× stained with Perls' Prussian Blue iron stain to illustrate the presence of excess stored iron. Excess iron in the liver and spleen can occur as a result of excessive uptake (usually genetic) or excessive breakdown of blood cells as seen in haemolytic diseases and also in cases of avian malaria where iron from infected and damaged cells is recycled at high rate (Cork, 2000). See also Plate 43.

Endnotes

1   H&E stain is the most commonly used and may be obtained from suppliers outlined in Appendix 4. Haematoxylin acts like a basic dye and stains acidic structures (chromatin, ribosomes) blue. Eosin is an acidic dye which stains most other components of the cytoplasm and intercellular material red or pink.

Bibliography

Bacha, W.J., Bacha, L.M. (2000) Colour Atlas of Veterinary Histology, 2nd edn. Wiley-Blackwell, Hoboken, NJ.

Cheville, N.F. (2006) Introduction to Veterinary Pathology, 3rd edn. Blackwell Publishing, Ames, IA.

Colville, T.P., Basset, J.M. (2007) Clinical Anatomy and Physiology for Veterinary Technicians, 2nd edn. Mosby, St Louis, MO.

Cork, S.C. (2000) Iron storage diseases in birds. Avian Pathology 29(1): 7–12.

Cork, S.C., Collins-Emerson, J.M., Alley, M.R., Fenwick, S. (1999) Visceral lesions caused by *Yersinia pseudotuberculosis* serotype 2 in different species of bird. Avian Pathology 28(4) : 393–399.

Dyce, K.M., Sack, W.O., Wensing, C.J. (2009) Textbook of Veterinary Anatomy, 4th edn. W.B. Saunders, Philadelphia, PA.

Getty, R. (1975) Sisson & Grossman's The Anatomy of the Domestic Animals, Volumes 1 and 2. W.B. Saunders Co, Philadelphia, PA.

King, A.S., McLelland, J. (1984) Birds: Their Structure and Function. Bailliere Tindall, London.

Maxie, G.H. (2007) Jubb, Kennedy and Palmer's Pathology of Domestic Animals, 3 volumes, 5th edn. Saunders, Philadelphia, PA.

McGavin, M.D., Carlton, W.W., Zachary, J.F. (2000) Thompson's Special Veterinary Pathology, 3rd edn. Mosby, St Louis, MO.

McGavin, M.D., Zachary, J.F. (2006) Pathological Basis of Disease, 4th edn. Mosby, St Louis, MO.

OIE Manual of Diagnostic Tests and Vaccines for Terrestrial Animals. Chapter 1.1.2 Collection, submission and storage of diagnostic specimens. http://www.oie.int/fileadmin/Home/eng/Health_standards/tahm/1.01.02_COLLECTION_DIAG_SPECIMENS.pdf.

OIE Manual of Diagnostic Tests and Vaccines for Terrestrial Animals. Chapter 1.1.3 Transport of biological materials. http://www.oie.int/fileadmin/Home/eng/Health_standards/tahm/1.01.03_TRANSPORT.pdf.

OIE (n.d.) Terrestrial Animal Health Code. Chapter 4.12 Disposal of Dead Animals. http://www.oie.int/index.php?id=169&L=0&htmfile=chapitre_disposal.htm.

Shapiro, L. (2009) Pathology and Parasitology for Veterinary Technicians, 2nd edn. Delmar, New York.

Plate 1 Modified nematode larval culture: (A) Weigh required amount of feaces. (B) In a wide mouth cup put an inch of vermiculite and layer the feces on top. (C) Mix it by moistening with water (0.1% Sodium carbonate solution will prevent mould growth during incubation). (D) Create a hole in the middle to increase aeration to culture. (E) Place a rolled-up paper towel on the rim and use a petri dish as a lid to cover the cup. Incubate at room temperature for 10-14 days. Periodically moisten the culture. (F) on final day of incubation fill the culture with warm water. Place the Petri dish on top and make sure no large air bubbles trapped under the lid. (G) carefully invert the cup and pour lukewarm tap water into Petri dish and leave at room temperature for 4 h. (H) Suck the fluid from the Petri dish, concentrate by sedimentation or centrifugation, and examine under microscope. Photos: Mani Lejeune, Holly White and Jaime Hazard. College of Veterinary Medicine, Cornell University.

Plate 2a Nematode eggs: (A) *Ancylostoma caninum* – dog; (B) *Toxocara cati* – cat; (C) *Trichuris ovis* - sheep; (D) Strongyle type – horse; (E) *Nematodirus* sp. – camel; (F) *Oxyuris equi* – horse: Photos: Mani Lejeune

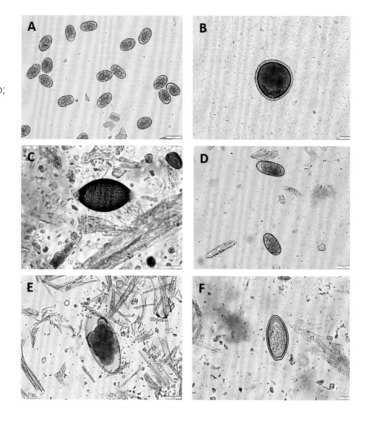

Plate 2b Nematode eggs and larvae: (A) *Parascaris equorum* – horse; (B) capillarid egg – duck; (C) Spirurid type – pig; (D) *Dictyocaulus viviparus* larvae – cattle; (E) *Aelurostrongylus abstrusus* – cat; (F) *Eucoleus aerophila* – cat. Photo: Mani Lejeune.

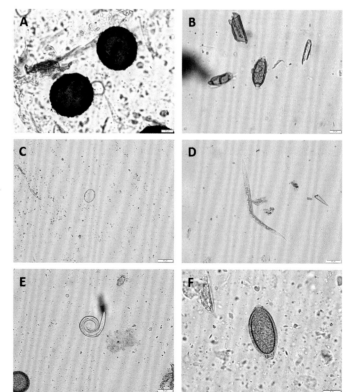

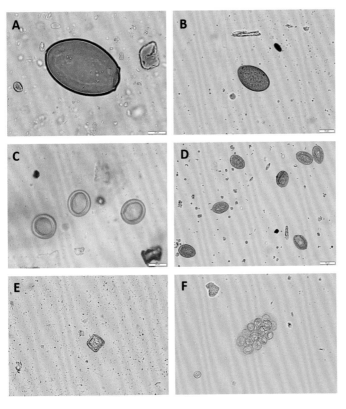

Plate 3 Trematode and cestode eggs: (A) *Paragonimus kellicotti* – dog; (B) *Alaria* sp. – dog; (C) Taeniidae eggs – cat; (D) *Diphyllobothrium* sp. – dog; (E) *Moneizia benedeni* – cattle; (F) *Dipylidium caninum* – dog. Photo: Mani Lejeune.

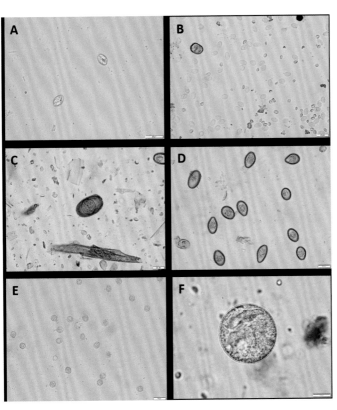

Plate 4 Protozoans: (A) *Sarcocystis* sp. (sporocysts) – dog; (B) *Giardia* sp. (cysts) – white tail deer; (C) *Eimeria intricata* (oocysts) – sheep; (D) *Eimeria* spp. (oocysts) – dog; (E) *Toxoplasma gondii* (oocysts) – cat; (F) *Buxtonella sulcata* (cysts) – horse. Photo: Mani Lejeune.

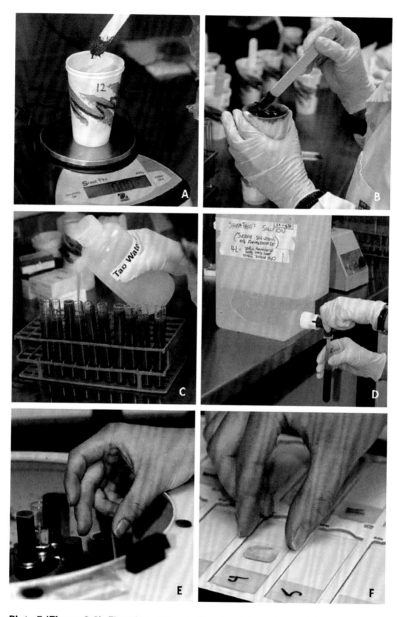

Plate 5 (Figure 3.6) Flotation: (A) weighing 4 g of faecal material; (B) dilution of faecal material in water and filtering through a gauze pad into centrifuge tube; (C) filling of centrifuge tubes to equal level; (D) diluting faecal pellet in Sheath solution after centrifugation; (E) filling tubes enough to form meniscus, cover tubes with cover slips; (F) remove cover slip from tube and transfer to labelled microscope slide. See also Plate 5. Photo: Dr Regula Waeckerlin, University of Calgary, Canada.

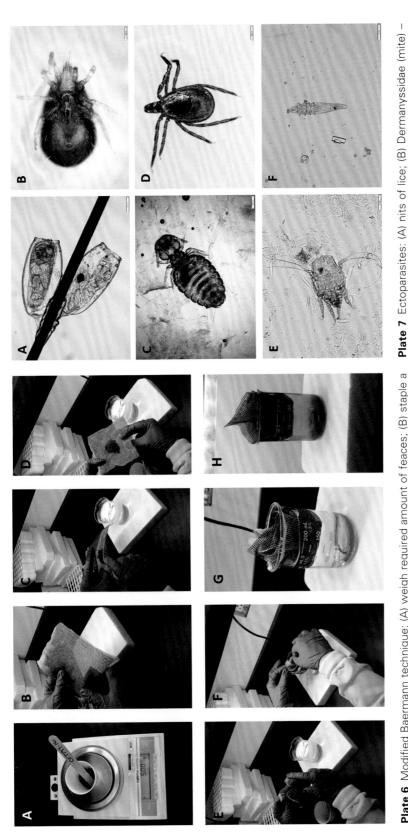

Plate 6 Modified Baermann technique: (A) weigh required amount of faeces; (B) staple a cheese cloth sandwiched between two window screen and make a 8 × 8 cm envelope; (C) place the faeces on the cheese cloth inside the envelope; (D–F) gently squeeze fold the envelope and immerse in 200 ml of water in a beaker; (G) keep it overnight under light; (H) remove the envelope and decant the supernatant without disturbing the sediment and examine the sediment under microscope for larvae. Photos: Mani Lejeune, Holly White and Jaime Hazard.

Plate 7 Ectoparasites: (A) nits of lice; (B) Dermanyssidae (mite) – cat; (C) *Damalinia* sp. (lice) – goat (D); *Ixodes scapularis* (tick) – dog; (E) *Chorioptes bovis* (mite) – cattle; (F) *Demodex canis* (mite) – dog. Photo: Mani Lejeune.

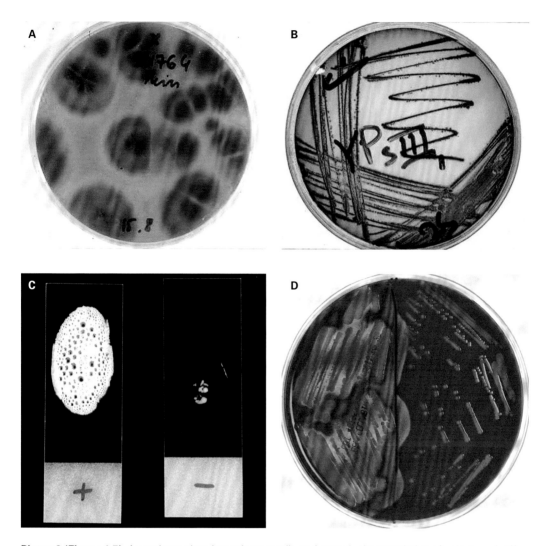

Plates 8 (Figure 4.5) Agar plates showing culture media and growth characteristics of colonies of different microorganisms. (A) Agar plate with a culture of a fungus (*Paecilomyces* sp.) obtained from the skin of a reptile. Fungal identification depends on the colour and type of growth on specialized agar media (for example, sabouraud dextrose agar) and on the morphology of the fruiting bodies or Hyphal structure (see also Figures 4.14a and b). (B) Specialized media (CIN) with a culture of *Yersinia pseudotuberculosis*. This agar is selective for some enteric bacteria, *Yersinia* spp. colonies appear small and pigmented (dark grey). CIN = Cefsulodin-Irgasan-Novobicin agar (Difco). (C) This is an example of a positive (left) and negative (right) catalase test. Catalase is an enzyme which breaks down hydrogen peroxide (H_2O_2) into H_2O and O_2. When a drop of 3% H_2O_2 is added to a colony of catalase positive bacterium, the reaction produces bubbles. Photo: Paul Gadja. (D) This is an example of the appearance of *Staphylococcus* spp. grown on blood agar. On the left side of the plate is *S. aureus* showing characteristic golden colonies surrounded by clear zones of beta haemolysis. On the right side of the plate is *S. xylosus*, which is coagulase negative and does not produce haemolysis. Photo: Karen Liljebjelke.

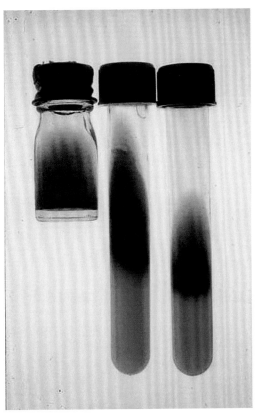

Plate 9 See also Figure 4.11c Simple biochemical screening test (for example, triple sugar iron (TSI) agar slopes) can be used to determine the species of selected bacteria during survey work but in most cases a series of 20–30 tests will be required. Illustrated are the Urease test, lysine iron agar (LIA) and triple sugar iron agar (TSI) slopes used to help identify Enterobacteriaceae.

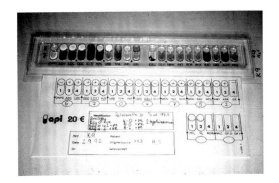

Plate 10 See also Figure 4.11d API strip card used to illustrate the typical biochemical reactions of 'type' cultures representative of bacterial species obtained from the American Type Culture Collection.

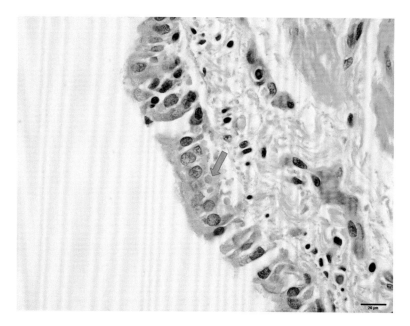

Plate 11 See also Figure 4.18a Canine airway epithelium with intracytoplasmic inclusions of canine distemper virus (the arrow indicates eosinophilic intracytoplasmic inclusion bodies). Photo: Dr Jennifer Davies, University of Calgary, Canada.

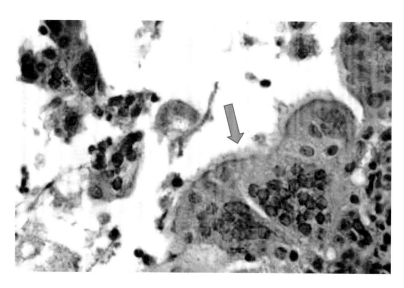

Plate 12 See also Figure 4.18b Trachea of a chicken infected with infectious laryngotracheitis virus with multinucleated syncytial cell formation (the arrow indicates a multinucleated syncytial cell). Photo: Dr M. Faizal Abdul-Careem, University of Calgary, Canada.

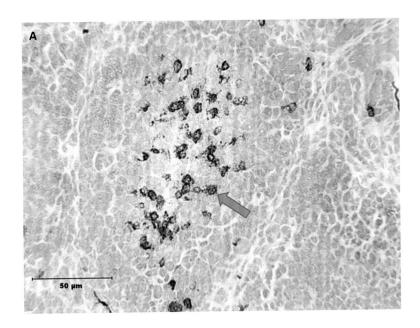

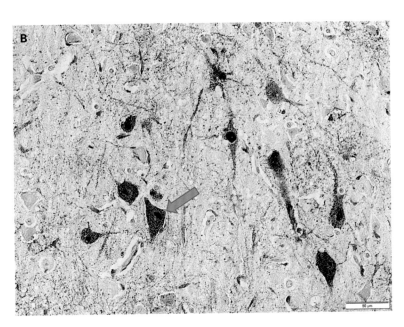

Plate 13 See also Figure 4.19 Immunohistochemistry staining can be performed in frozen sections or formalized sections to visualize viral antigens. (A) A frozen section of Bursa of Fabricius of a chicken infected with Marek's disease virus (the arrow indicates the brown colour stained viral antigens). Photo: Dr Shayan Sharif, University of Guelph, Canada. (B) A formalized brain section of a dog infected with canine distemper virus (the arrow indicates the brown colour stained viral antigens). Photo: Dr Cameron Knight, University of Calgary, Canada.

Plate 14 See also Figure 4.21 Image captured under fluorescent microscope following staining of trachea infected with infectious bronchitis virus demonstrating nuclear antigen of the virus (reddish colour). Arrow points at the epithelial lining facing the tracheal lumen. The section was counterstained with fluorescent dye staining nuclei (blue colour). Photo: Dr M. Faizal Abdul-Careem, University of Calgary, Canada.

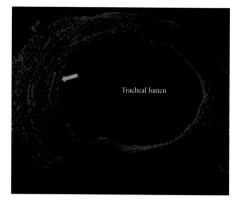

Tracheal lumen

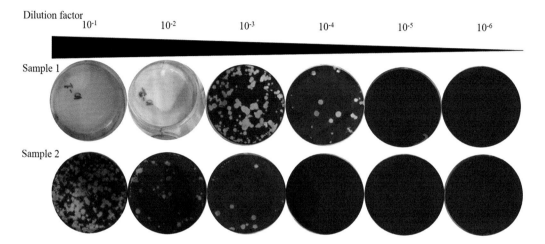

Plate 15 See also Figure 4.23 Avian influenza virus titration has been done in Madin-Darby Canine Kidney (MDCK) cells. After 2 days of inoculation of the MDCK cell monolayer, the cells have been stained with crystal violet to see the extend of cell damage due to the virus replication. The clear areas represent the loss of cells due to viral replication. The number of infectious virus particles in sample 1 is higher than the sample 2. Photo: M. Sarjoon Abdul-Cader, University of Calgary, Canada.

Plate 16 See also Figure 4.27 In SN assay, the unknown serum sample is two-fold serially diluted and titrated against a known quantity of virus. The serum blocks virus infection at the 1 : 2, 1 : 4 and 1 : 8 dilutions, but not at all at 1 : 16. Each serum dilution has been tested in triplicate, which allows for more accuracy. In this sample, the SN titre would be 8, the reciprocal of the last dilution at which infection was completely blocked. Photo: M. Sarjoon Abdul-Cader, University of Calgary, Canada.

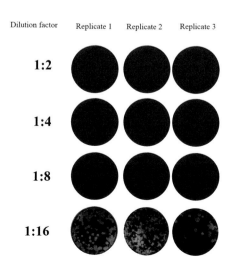

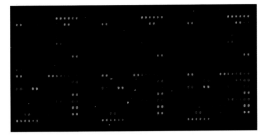

Plate 17 (Figure 4.33b) Scanned array image showing positive (yellow) and negative results (black). Each spot shown in the array corresponds to a specific fluorescent value, determined by the strength of hybridization, in a semi-quantitative manner. Photos: Dr Shayan Sharif and Dr Jennifer Brisbin, University of Guelph, Canada.

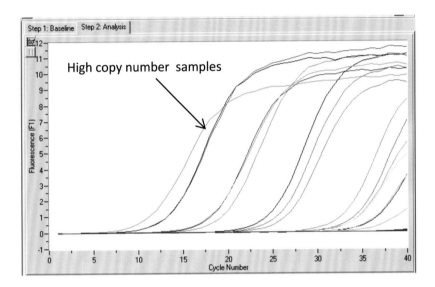

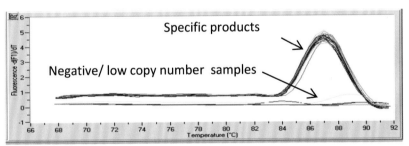

Plate 18 (Figure 4.33b & c) Real-time PCR amplification curves and the melting peaks. (B) Representative amplification curves for the quantification of viral DNA by real-time PCR based on SYBR Green chemistry. The DNA sample was serially 10-fold diluted and each dilution has assayed in triplicate. High DNA copy number samples (lower dilution show amplification curves early during the amplification and the low copy number or negative and no template control samples (higher dilution) show late or no amplification. (C) Melting peak analysis shows the specific amplification. Depending on the sensitivity of the assay low DNA copy number samples yield no peaks as there is no template control. Photos: (b and c) Dr M. Faizal Abdul-Careem, University of Calgary, Canada.

Visual detection of magnesium pyrophosphate

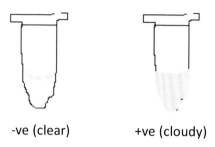

-ve (clear)   +ve (cloudy)

Visual detection using SYBR green

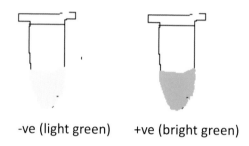

-ve (light green)   +ve (bright green)

Plate 19 (Figure 4.34b) (B) LAMP isothermal amplification products can be visualized using naked eye due to the accumulation of PCR by product magnesium pyrophospahte (cloudy) or colour change using SYBR green.

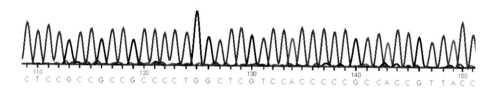

CTCCGCCGCCGCCCCTGGCTCGTCCACCCCCGCCACCGTTACC

Plate 20 (Figure 4.35b) (b) A representative of sequencing output. The blue, red, black and green peaks represent cytosine, thiamine, adenine, and guanine, respectively. Photo: Dr Rikia Dardari, University of Calgary, Canada.

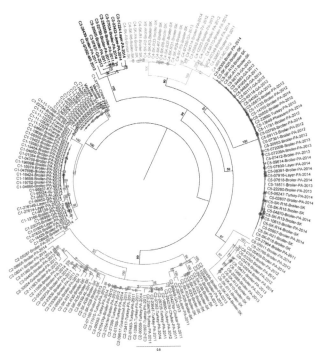

Plate 21 (Figure 4.36) Phylogenetic Tree of a total of over 180 partial σC gene sequences of avian reovirus. The analysis included sequences obtained from the GenBank. Vaccine strains are shown in a bold red colour. The tree shows several clusters of avian reovirus and most of the sequences are far from the sequences of the avian reovirus vaccines. This raises the question whether vaccine induced immunity is protective against the disease induced by these diverse avian reo viral strains. Source: Victor Palomino-Tapia, University of Calgary, Canada.

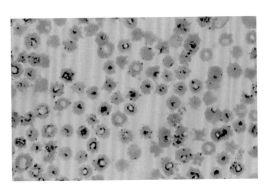

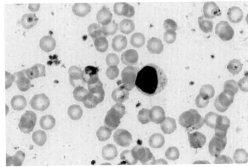

Plate 22a See also Figure 5.6 Blood smear from a cow which later died following fever, haematuria and weight loss over a period of several days. The animal had concurrent *Babesia bovis* and *Theileria* sp. infection following a spring peak in tick populations, Pemagatshel, Eastern Bhutan. Giemsa 20× magnification. The darker pigmented areas in the red blood cells indicate stages of the life cycles of the two protozoal species. See also Chapter 3.

Plate 22b See also Figure 5.7 Poor quality bovine blood smear stained with Giemsa 100× oil immersion. Note the presence of a large mononuclear cell. The red cells are irregular in shape and there is a lot of artefact. The presence of granules suggests that new stain should be prepared.

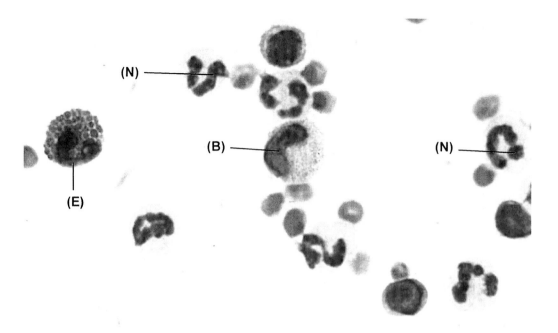

Plate 23 See also Figure 5.8 Equine blood smear viewed under oil immersion (Diff Quick 1000×) illustrating numerous polymorph neutrophils (N), an eosinophil (E) and a basophil (B). Diff Quick™ stain. RBC are also present.

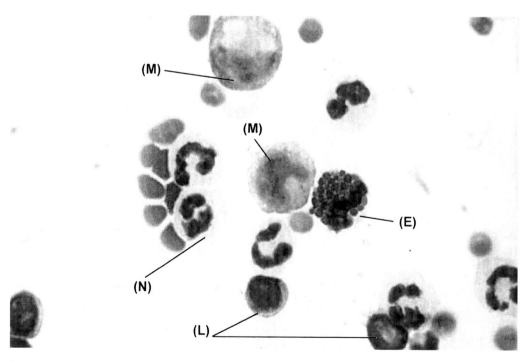

Plate 24 See also Figure 5.9 Equine blood smear viewed under oil immersion 1000× illustrating granulocytes (E an N) and agranulocytes (M and L). (E) – Eosinophil, (N) polymorph neutrophil. It is not always possible to distinguish between monocytes and lymphocytes in blood smears. In this smear, the monocytes (M) are activated and are larger than dormant monocytes or lymphocytes (L). See text. Diff Quick™ stain. RBC are also present.

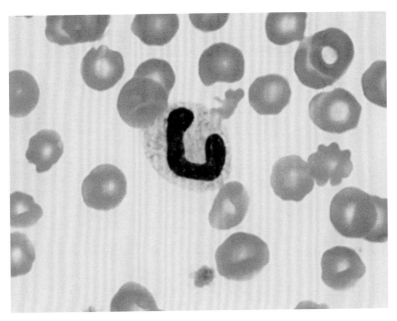

Plate 25 See also Figure 5.10a Canine band neutrophil (immature PMN). Note the band shaped nucleus with smooth sides and lack of constrictions. This cell also shows signs of toxic change with a foamy appearance due to the presence of vacuoles. Modified Wright's stain 100×. Photo: Dr Nicole Fernandez, WCVM.

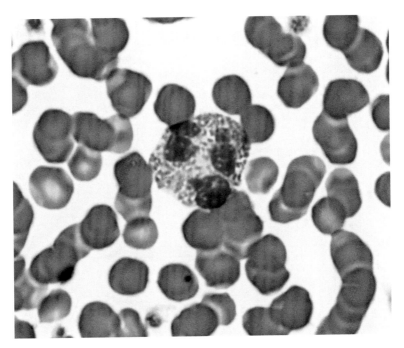

Plate 26 See also Figure 5.10b Feline eosinophil. Note reddish granules in the cytoplasm and segmented nucleus. The nuclear segments are connected by a fine strand. Wright's stain 100×. Photo: Dr Nicole Fernandez, WCVM.

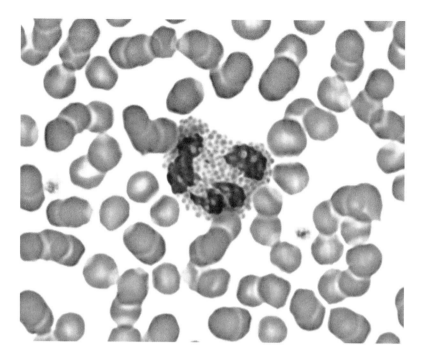

Plate 27 See also Figure 5.10c Feline basophil. Note pale lavender granules in the cytoplasm and segmented nucleus. The colour of the cytoplasmic granules varies from pale grey to dark purple depending on the species. Wright's stain 100×. Photo: Dr Nicole Fernandez, WCVM.

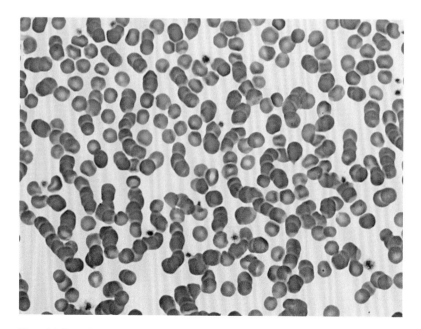

Plate 28 See also Figure 5.10d Feline blood smear showing rouleaux formation. This is commonly seen in equine and feline blood smears. Wright's stain 20×. Photo: Dr Nicole Fernandez, WCVM.

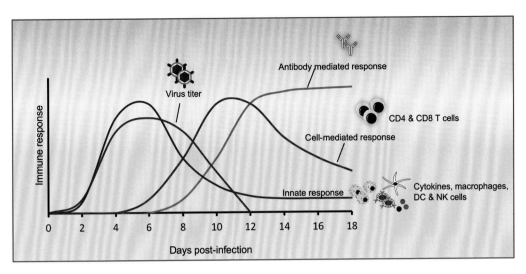

Plate 29 See also Figure 6.1 Immune response generated following exposure to a pathogen can be innate (non-specific) response and adaptive (specific) response. Source: M. Sarjoon Abdul-Cader, University of Calgary, Canada.

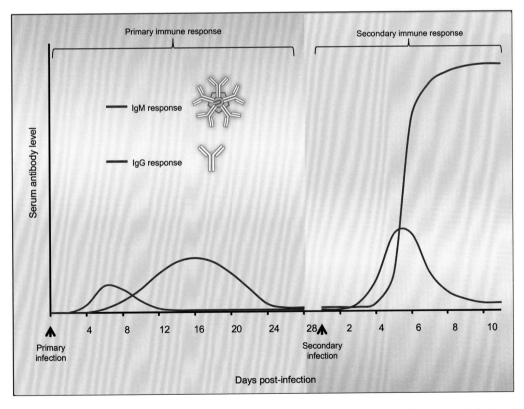

Plate 30 See also Figure 6.5 Serum antibody concentrations following primary and secondary infections. The initial IgM response follows the IgG response. The magnitude of the secondary response is greater and the time taken for the response is shorter. Source: M. Sarjoon Abdul-Cader, University of Calgary, Canada.

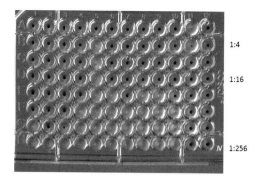

1:4

1:16

1:256

Plate 31 See also Figure 6.9a Photograph of a plate used to perform the HI test (note that this is not the same test plate as that shown in Figure 6.9b). Photo: Dr Davor Ojkic, Animal Health Laboratory, University of Guelph, Canada.

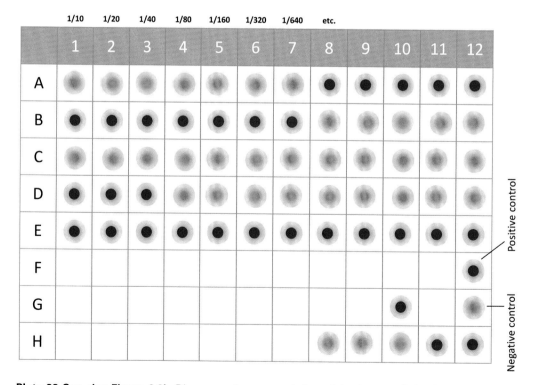

Plate 32 See also Figure 6.9b Diagrammatic representation of the haemagglutination inhibition test. Rows F, G and H contain reagent controls. The button of RBCs indicates no agglutination. Source: M. Sarjoon Abdul-Cader, University of Calgary, Canada.

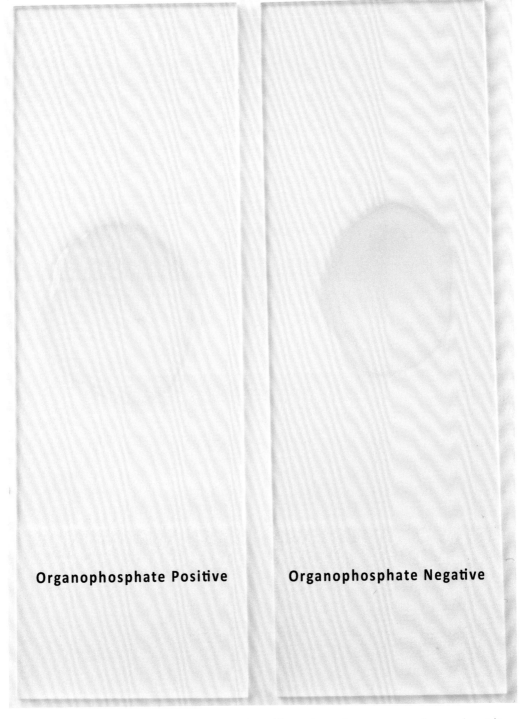

Organophosphate Positive   **Organophosphate Negative**

Plate 33 See also Figure 7.14a Glass slides with positive and negative serum samples are shown in the insert. Photo: Judit Smits/Eugene Janzen, University of Calgary, Canada.

**Plate 34
(Figure 7.15a)**
A formalin fixed bovine brain with polioencephalomalacia in the grey matter, has subtle, sunken yellowish necrotic areas. Photo: Judit Smits/Eugene Janzen, University of Calgary, Canada.

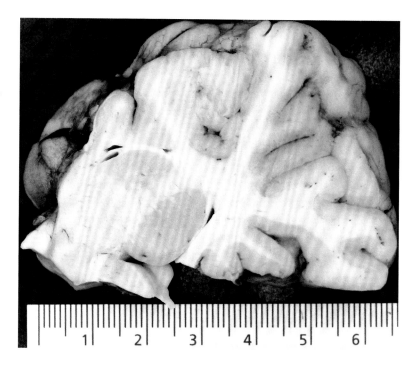

**Plate 35
(Figure 7.15b)**
Bovine polio brain UV light. Photo: Judit Smits/Eugene Janzen, University of Calgary, Canada.

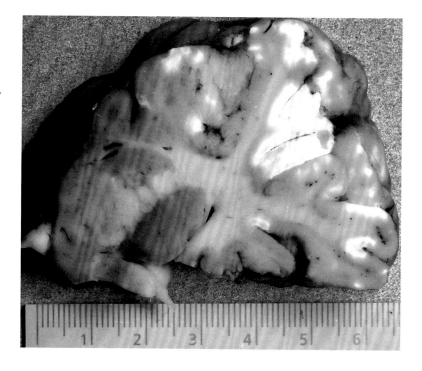

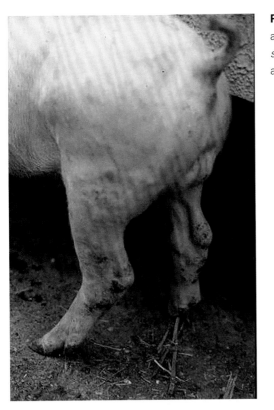

Plate 37 See also Figure 8.19 Skin rash seen in a pig with *Erysipelothrix insidiosa* sp. (now *E. rhusiopathiae*) infection. This pig was treated with antibiotics and subsequently recovered.

Plate 36 See also Figure 8.14 There are often a lot of deaths at lambing time but the presence of the occasional deformed lamb may not necessarily indicate that there is a disease present. Abnormalities like this one-eyed lamb ('cyclops') occur during development, abnormal embryos are usually reabsorbed or aborted early in development. Occasionally some survive to term but die shortly after birth. This lamb was one of two, the other lamb was normal. Some mycotoxins, plants or drugs may result in foetal abnormality (that is, they are teratogenic).

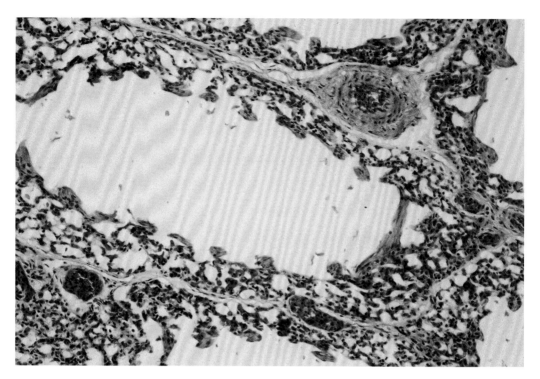

Plate 38 See also Figure 8.21 Histology section H&E 20× of a normal chicken (*Gallus gallus*) lung. H&E = Haematoxylin and Eosin stain (see main text).

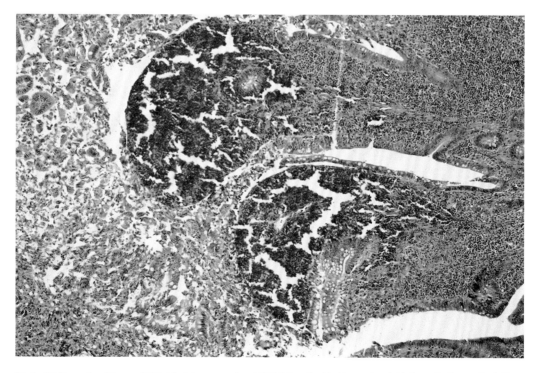

Plate 39 See also Figure 8.22 Histology section H&E 20× of a bird intestine (chicken, Gallus gallus) illustrating haemorrhage secondary to a bacterial infection.

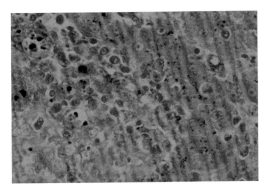

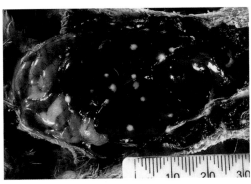

Plate 40 See also Figure 8.23 Histology section of a bird liver 40× stained with Perl's Prussian Blue iron stain to illustrate the presence of iron stored as hemosiderin in hepatocytes and Kupffer (macrophage) cells. The presence of excess hemosiderin can occur as a result of excessive intake of iron or due to excessive iron breakdown as in the case of haemolytic anaemia (that is, can be associated with avian malaria) or subsequent to sever trauma and debilitation resulting in tissue damage and/or muscle breakdown. Histopath 40× Perle›s iron stain of liver.

Plate 41 See also Figure 8.24 Illustration of gross necropsy on a freshly dead aviary bird (parakeet) illustrating an enlarged and discoloured liver with several abscesses. Cultures taken from these abscesses grew a pure culture of *Yersinia pseudotuberculosis* serotype 2. This is not an uncommon cause of death in wild and captive birds especially when predisposing factors such as immune-suppression or concurrent systemic disease are present. Iron storage diseases can also predispose to the development of pseudotuberculosis and other bacterial infections (for more information see Cork (2000).

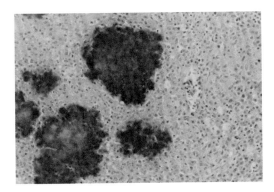

Plate 42 See also Figure 8.25 The use of immune-histochemistry to identify lesions caused by *Yersinia pseudotuberculosis* (serotype 3) in the liver of the case illustrated in Figure 8.24. Immuno-histochemistry is discussed in Chapter 6. Essentially, the cut histology sections are incubated with hyperimmune serum (in this case rabbits anti-Yersinia pseudotuberculosis type 2 antisera) which is bound to an enzyme or conjugate and then rinsed. A substrate is then added to the slides and if the antibody remains bound to the cells (or, in this case, bacterial lesions) on the slide this indicates the presence of antigen (for more information see Cork et al., 1999).

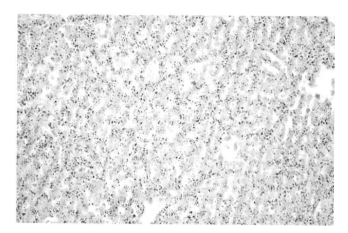

Plate 43 See also Figure 8.26 Histology section of a wild bird liver (New Zealand kokako [Callaeas cinereal]) 10× stained with Perl's Prussian Blue iron stain to illustrate the presence of excess stored iron. Excess iron in the liver and spleen can occur as a result of excessive uptake (usually genetic) or excessive breakdown of blood cells as seen in haemolytic diseases and also in cases of avian malaria where iron from infected and damaged cells is recycled at high rate (Cork, 2000).

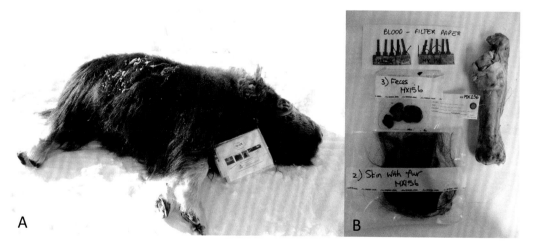

Plate 44 See also Figure 11.4 (A) A hunted muskox cow with a sampling kit that will be used by the hunter to collect a set of biological samples when butchering the carcass. (B) The core set of samples collected using the sampling kit: blood-saturated filter-paper strips, faeces, left metatarsus (or left hind leg), and a piece of skin with hair from the rump. Photos: Matilde Tomaselli.

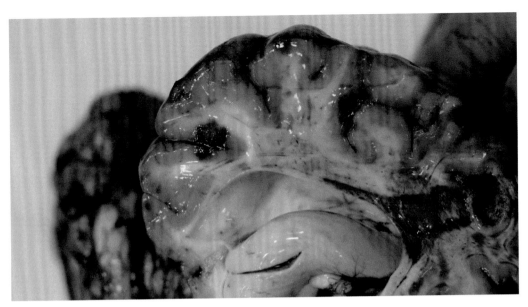

Plate 45 See also Figure A2.2 Bovine cerebral cortex. Note multifocal areas of slightly raised, red areas within meninges and extending into the cortical grey matter corresponding to areas of ischaemic infarction as a result of bacterial vasculitis. Photo: C. MacGowan, University of Calgary, Canada.

Plate 46 See also Figure A2.5 Abdomen, cat. Acute inflammation. The last rib (•) and liver (*) can be seen on the left side of the picture with loops of intestine (†) in the centre. Note abundant stands of tan fibrin (‡) covering the serosal surface of the intestines. Note also the subtle finely granular texture of the intestinal serosa which can in some cases be the only sign of fibrin exudation. Abdomen, cat, acute inflammation. Photo: Dr J. Davies, University of Calgary, Canada.

Plate 47 See also Figure A2.7a Thorax, bovine. Chronic suppurative bronchopneumonia: In this image, the cranioventral parts of the lungs are dark red and multinodular consistent with bronchopneumonia. A chronic process is suggested by the presence of abscesses visible from the pleural surface (‡) and fibrous adhesions between the visceral and parietal pleura (†). Photo: Dr J. Bystrom, University of Calgary, Canada.

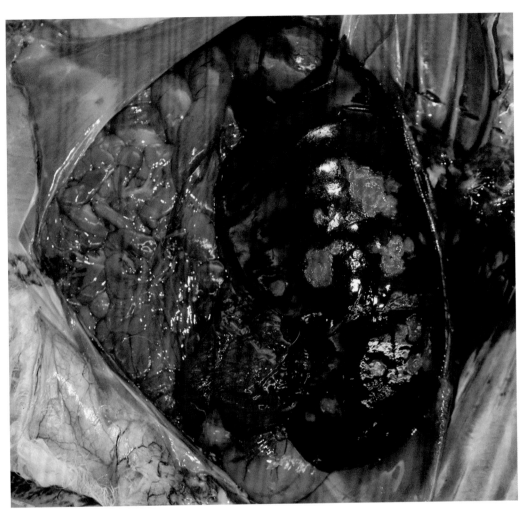

Plate 48 See also Figure A2.8 Liver, sheep. Necrosis: there are multifocal, slightly sunken areas of tan discolouration throughout the liver, visible from the capsular surface. This is a characteristic appearance of hepatocellular necrosis in this case as the result of Fusobacterium necrophorum infection. F. necrophorum usually reaches the liver via the portal circulation from the rumen and is a common sequelae to acute rumenitis. Necrosis can be differentiated from degenerative (glycogenosis or lipidosis) changes which can have a similar homogenous tan appearance by the fact lesions in this case are sunken consistent with tissue loss.

Plate 49 See also Figure A2.10 Liver, sheep. Chronic fascioliasis (Fasciola hepatica). F. hepatica is a liver fluke commonly associated with acute and chronic liver disease in small ruminants. Acute disease manifests as sudden death as a result of overwhelming hepatocellular necrosis caused by large number of larval flukes migrating through the parenchyma. Chronic fascioliasis occurs are a result of chronic cholangiohepatitis as a result of the presence of adult flukes within bile ducts. In this image, a tangle of adult flukes can be seen within the lumen of a large bile duct at the hilus. The left lobe is atrophic, fibrotic bile ducts are visible from the cut section and the right lobe is hypertrophied. This appearance is characteristic but it is unclear why the left lobe is preferentially affected. In addition, the hilar lymph nodes are enlarged as a result of chronic antigenic stimulation. . Photo: Mr R. Irvine, University of Glasgow, UK.

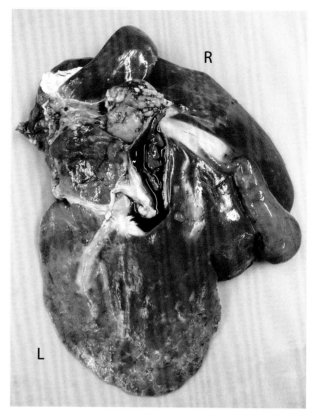

Plate 50 See also Figure A2.11 Kidney, bovine. Autolysis. Autolysis occurs more rapidly in organs containing higher levels of proteolytic enzymes such as the liver and kidney or that contain large amounts of bacteria such as the GI tract. In this case the kidney is pallid and has lost structure appearing diffusely softened. Note also fine emphysematous gas bubbles in the tissue. Photo: Mr R. Irvine, University of Glasgow, UK.

Plate 51 See also Figure A2.12 Pluck, sheep, the trachea is on the left side of the picture (†). Hypostatic congestion. Note the left lung is diffusely dark red and right lung is diffusely pale pink. In this case it appears that the body was placed on the left side after death and blood has pooled by gravity in the dependent lung. This should be differentiated from inflammation or congestion by gently palpating the lungs to detect a change in texture which is a feature of a true lesion. Photo: Mr. R. Irvine, University of Glasgow, UK.

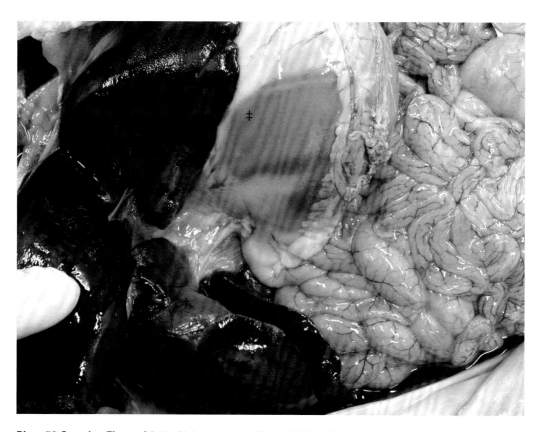

Plate 52 See also Figure A2.14 Abdomen, dog. Bile imbibition: The liver is on the left side of the picture (*) and the left liver lobe has been retracted to reveal the orange green gallbladder (†). There has been extensive imbibition of bile pigment into the serosa of the adjacent stomach (‡). Note the bland appearance of the affected tissue with no signs of an inflammatory response which would have been present had bile leakage occurred pre-mortem. Photo: Mr. R. Irvine, University of Glasgow, UK.

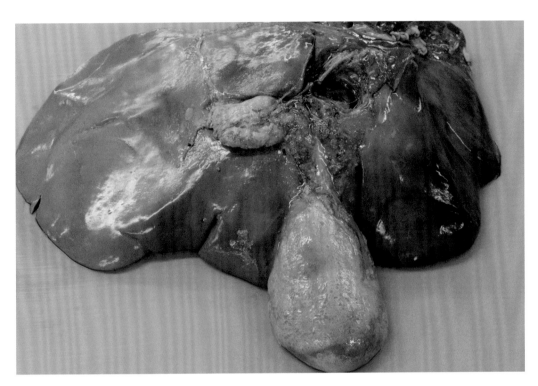

Plate 53 See also Figure A2.15 Liver, sheep. Pseudomelanosis. The liver is viewed from the caudal surface with the ventral aspect at the bottom of the picture. The right and quadrate lobes are noticeably darker in colour than the rest of the liver consistent with build of iron sulphide in the tissues as a result of post-mortem bacterial overgrowth. Note also post-mortem emphysema in the wall of the gallbladder. Photo: Mr. R. Irvine, University of Glasgow, UK.

Plate 54 See also Figure A2.18 Liver, bovine. Post-mortem emphysema. Gas production by bacteria which overgrow post-mortem produce a honeycombed appearance to affected tissues. This should be differentiated from emphysema caused by clostridial infection which can be diagnostically useful. Photo: Mr. R. Irvine, University of Glasgow, UK.

PART III

SPECIAL TOPICS

There are a wide range of special topics that have relevance for setting up a veterinary laboratory and delivering veterinary services. These include the production and testing of veterinary biologicals and vaccines, conducting clinical trials, field based research and disease modelling. These areas of specialist interest are important for the development and delivery of a robust veterinary service but are not routinely part of the mandate for a small diagnostic facility so they will not be discussed in detail here. However, each district and regional diagnostic laboratory will be expected to contribute to the development of epidemiological databases and to submit data to national health and disease surveillance programmes. There may also be a requirement to provide diagnostic support for local clinical research projects, baseline surveys or for local clinical trials of vaccines and medicines. In this chapter, a few selected topics relevant to the development and delivery of an effective veterinary service will be discussed.

In Chapter 9, the practical considerations for designing and performing epidemiological studies are outlined along with definitions of the terminology commonly used. This chapter includes case studies from the field. Additional information on epidemiology and use of disease modelling may be found in the references given at the end of the chapter. In Chapter 10 some common clinical presentations of production limiting diseases in livestock are described. The tables in this chapter outline some of the likely differential diagnoses for these clinical presentations and provide guidance on the selection of samples required to confirm a diagnosis. Additional information about common diseases in livestock, along with recommendations for disease control and prevention, may be found in relevant preceding chapters (3–8) or in bibliographies at the end of the chapters. In Chapter 11, the potential role of wildlife in the spread of infectious disease is considered along with the importance of monitoring the health of wildlife populations. In Chapter 12 we present a One Health approach to tackling antimicrobial resistance (AMR) and in Chapter 13, we cover the important role of the World Organisation for Animal Health (OIE) in supporting the development and sustainability of veterinary diagnostic services. The final chapter (14) outlines some important vector borne diseases and the principles of vector surveillance.

chapter 9

Epidemiology

9.1 Introduction to epidemiology

Susan C. Cork and Sylvia Checkley

Regional and district field programmes play a crucial role in gathering routine data for epidemiological purposes and can provide important information on endemic and emerging diseases. Epidemiological data is used to inform decision making at the regional and national level and may also be required for the purposes of international reporting (see Chapter 10). Unfortunately, field data is often unreliable and/or restricted due to lack of resources at the extension level and a lack of diagnostic services to confirm a diagnosis. However, as a result of concerns surrounding the global spread of highly pathogenic avian influenza (H5N1) in 2003 a concerted effort was made to improve veterinary services, including diagnostic support, at the district level in many countries. This was to ensure early disease detection, reliable reporting and rapid response to safeguard the health of animals and the people who rely on them for their livelihoods. This has resulted in a better network of disease reporting across the globe but there are still many rural and more remote areas where the veterinary infrastructure remains limited or non-existent.

The setting for most epidemiological work is the 'field', that is, where the samples are collected (farm, veterinary clinic and so on) and to a lesser extent, the laboratory where the samples are analysed. Active surveillance and retrospective epidemiological studies rely on good coordination between laboratory and field services if the data gathered is to represent the true 'field' situation. Field based animal health staff and laboratory professionals should be able to appreciate the principles of epidemiological studies and to understand the terms used. Passive surveillance involves the use of data collected for other means, for example, routine laboratory data or notifiable disease reporting. In order to apply this data more broadly, an understanding of the type of reporting, and the rationale for selecting the samples submitted, is necessary.

Epidemiology is the study of the 'patterns of disease', that is, the frequency, distribution and determinants of health and disease in a population of animals. The unit of concern in epidemiological studies is not the individual animal but rather the group (pack, herd, flock), category (age group, sex, breed) or an entire population (cattle in a region, district or country). In some cases, the disease characteristics identified through epidemiology can be used as a diagnostic tool to recognize specific diseases due to the way the disease spreads, the nature and duration of the clinical signs and the outcome. For example, the now eradicated rinderpest (cattle plague) historically killed a large number of animals rapidly (that is, an acute disease with a high mortality) and spread very quickly whereas bovine tuberculosis may spread slowly and causes illness but not necessarily death until the later stages of infection (that is, a chronic disease with morbidity and mortality late in the disease).

Disease determinants are those factors that influence health and disease through complex interactions, varying with the age of the animal(s) and with different stages of management and husbandry. In epidemiology, the term can be used to describe any factor which, when altered, produces a change in the frequency or characteristics of a disease. Few diseases have a single cause and the development and progression of clinical signs can depend on a wide range of intrinsic and extrinsic factors, for example, the age and physiological state of the herd/flock, the management of the animals, changes in weather, recent history of movement to a new region, nutritional status or the presence of concurrent disease. Some of the questions most often asked in an epidemiological study, or during a disease outbreak investigation follow.

1 Which animals (species, breed, age, sex) are affected?
2 How many animals are sick (morbidity) and how many have died (mortality)?
3 When did the disease start and what were the initial and subsequent clinical signs?
4 Have there been any previous health problems and what are the routine preventive programmes (for example, vaccination, worming)?
5 Have there been any recent changes in diet/management?
6 Have any new stock been recently purchased?
7 Are there any signs of the disease on neighbouring farms or in wild animals?
8 Are any people sick with the same disease?
9 Is there a seasonal occurrence (that is, are disease outbreaks associated with changes in climatic conditions and the presence of disease vectors such as ticks)?
10 Did the disease outbreak occur concurrently with wildlife migration?

The subsequent action will depend on the likely cause(s) of the disease outbreak. Once the likely causes of the disease outbreak(s) have been determined it is then important to examine sick and dead animals and their environment and to collect appropriate samples for laboratory testing to confirm the diagnosis (see also Chapter 1.5 Clinical examination, sample selection, submission and clinical diagnosis). The following definitions are provided to assist in building up an epidemiological vocabulary for laboratory and animal health extension staff.

Endemic: The constant presence of a disease or infectious agent within a given geographic area. It may refer to the usual prevalence of a given disease within an area or a relatively new but constant situation. The term enzootic may also be used when discussing disease in animals.

Epidemic: The occurrence in a country or region of cases of an illness (or disease outbreak) clearly in excess of the expected level. The number of cases occurring in an epidemic will vary according to the cause, that is, the infectious agent involved and its method of transmission, the size and type of population exposed, previous exposure to the disease agent and the time and place of occurrence. Even a single case of a disease which has not been recently recognized in an area requires prompt reporting and epidemiological investigation to prevent further cases. The term 'epizootic' can also be used.

Prevalence: This is the number of animals which have a disease (diagnosed on the basis of clinical signs or by laboratory tests) in a stated population at a given time. This is regardless of when the illness began. The prevalence rate is defined as the number of positive cases divided by the number of animals in the population tested at a given point in time.

Mortality rate: This is the proportion of animals in a given population which die during a disease outbreak. Disease specific mortality rate is the proportion of animals whose deaths can be attributed to a specific disease (that is, confirmed by diagnostic tests).

Morbidity rate: This is the proportion of animals in a given population which develop clinical disease. Disease specific morbidity rate is the proportion of animals that develop clinical signs attributable to the disease under investigation and confirmed by diagnostic tests.

Incidence: This is the number of new cases of a disease occurring in a specified period. The incidence rate is the number of new cases of the disease reported in a specific time (for example, per year or month) divided by the number of animals in the population at risk. Incidence does not include cases of disease that were already present in the population at the start of the specified period.

Population at risk: This is the population of animals which are identified as those likely exposed to a disease or 'event'.

Probability: Is the likelihood of an event occurring.

(Case) fatality rate: This is usually expressed as a percentage and is the number of animals diagnosed as having a specific disease that subsequently die as a result of that disease. This term is usually applied in an outbreak of an acute disease in which all cases are monitored for an adequate period of time to include all attributable deaths. The fatality rate gives a measure of the severity of the disease and is not the same as the mortality rate.

It can be understood from the above definitions that the data submitted to an epidemiology unit will only be of value if the correct diagnosis has been made. In many cases more than one disease may be present and the clinical presentation may vary within the population. When attending a disease outbreak, it is important that the veterinary team carefully examine both sick and dead animals and to collect appropriate samples for submission to the laboratory. In cases where laboratory testing, to confirm a diagnosis, is not available the veterinary and animal health extension teams will need to rely on clinical experience and their knowledge of the diseases common to the area.

9.2 Disease surveillance and reporting systems, resources and diagnostic laboratory services

John Woodford

Background

Disease surveillance has become an increasingly important core function of the state veterinary services in most developing countries. The need to establish or strengthen existing active and passive disease surveillance and reporting systems has been driven by several different factors over the recent past. Initially, throughout the implementation of the global programme for the eradication of rinderpest, planning groups relied heavily on analysis of both active and passive disease surveillance data at regional, national and sub-national levels in order to target the limited resources available for prevention and control programmes more efficiently. In the final stages of the eradication programme, passive disease reporting and active disease surveillance, using participatory disease surveillance (PDS) techniques, became key tools used to search for and detect disease outbreaks, especially in the more remote and inaccessible areas of central and east Africa, where the last pockets of wild circulating virus remained. As the disease was progressively brought under control, all countries participating in the joint Food and Agriculture Organization

(FAO)/World Organisation for Animal Health (OIE) global eradication programme were required to provide evidence of freedom from clinical infection and circulating virus using information derived from both active and passive surveillance activities.

In the meantime, the recent highly pathogenic avian influenza (avian influenza type A sub-type H5N1 virus) pandemic, starting in China in 1997, heralded a massive investment on the part of donors to strengthen both active and passive disease surveillance systems, mainly for the purpose of early detection of disease outbreaks, when it was feared that the disease might evolve into a pandemic on a scale similar to that of the 'Spanish flu' epidemic in 1918, which infected one-third of the world's population (c. 500 million) was responsible for the deaths of between 50 and 100 million people. Subsequently, as the epidemiology of highly pathogenic avian influenza became better understood, countries wishing to export live poultry and poultry products have needed to provide evidence derived from both active and passive disease surveillance to prove freedom from the disease and thus regain access to international markets, following new outbreaks.

Indeed, disease surveillance and reporting have become essential core functions of the veterinary services for any countries wishing to trade animals and animal products internationally and to meet the standards set by the World Trade Organization (WTO) and the OIE.

Increasingly, information derived from active and passive disease surveillance data is now being used to inform decision makers utilizing risk-based approaches for the planning and implementation of more cost-effective disease prevention and control programmes.

International standards for trade in animals, animal products and other commodities

The WTO rules ensuring the safe and fair trade in animals, plants and animal and plant materials are set in the Sanitary and Phytosanitary (SPS) Agreement. Under the SPS Agreement, the WTO has identified the OIE as the global body mandated to set the standards for trade in animals, animal products and other biological materials. These standards are provided in the Terrestrial and Aquatic Animal Health Codes of the OIE and cover the (mandatory) requirements to be met by member countries with regard to the notification of their animal health status, the sanitary measures, listed by disease and commodity that can be required to be met by exporting countries and, more generally, all aspects relating to the quality of the veterinary services of member countries.

To be fully compliant with OIE standards, a Member Country must be in a position where it can notify its animal health status accurately at any point in time to the OIE. Disease notification requirements and a 'List of Notifiable Diseases' are provided in Chapters 1.1 and 1.3 of Volume 1 of the Terrestrial Animal Health Code (TAHC), respectively. Chapter 1.6 of Volume 1 of the TAHC also defines the information to be provided and disease surveillance requirements to be met by a Member Country wishing to apply for official (OIE) recognition of freedom from certain animal diseases (AHS, BSE, CSF FMD, CBPP and PPR),[1] that can affect international trade. The sanitary requirements, listed by disease and commodity, for the safe trade of animal commodities are found in Volume 2 of the TAHC.

Role of veterinary paraprofessionals in animal disease surveillance systems in developing countries

Animal disease surveillance involves the use of various tools to determine the presence, absence or distribution of disease or infection within a defined animal population or detecting as early as possible exotic or emerging diseases. The type of surveillance applied depends on the information needed to support decision making. Animal disease surveillance is also used to monitor disease trends, to facilitate the control of disease or infection, to provide data for use in risk analysis, for animal or public health purposes, and to justify investment in disease prevention and control measures.

Active disease surveillance

Active disease surveillance can involve the deployment of veterinarians and/or veterinary paraprofessionals in the field who visit and interact with individual or groups of livestock keepers in order to obtain disease information. The aim of the exercise is to seek information on the presence or absence of one or more specific diseases within a defined population of animals in a defined location at a particular point in time, using methods of enquiry derived from participatory rural appraisal (PRA). Active disease surveillance can also involve the collection of laboratory samples, usually blood/serum, to detect antibodies against specific disease agents, either due to infection or sometimes to confirm sero-conversion following vaccination.

Passive disease surveillance

Passive disease surveillance, in contrast, relies, in the first instance, on the report of the occurrence of a suspected outbreak or case of a disease to a veterinary service provider by a livestock keeper on his farm or in his neighbourhood. The animal health service provider then forwards this report, based on clinical suspicion of a disease, to the nearest veterinary authority. In this respect, the report of the suspected disease occurrence is 'passive' since it was volunteered by the livestock keeper, or his animal health service provider, in the first instance. Normally, if such a report is likely to be a notifiable disease, the suspected case or outbreak would result in the conduct of an outbreak investigation, during which laboratory samples would be collected and submitted for laboratory testing to confirm, or otherwise, the existence of a notifiable disease.

For a country to be in a position to report its animal health status accurately it needs to establish systems to detect, report and confirm, by laboratory diagnosis, all suspected notifiable disease events at the time that, or very soon after, they occur. Such systems involve early detection of suspected disease occurrences, relying largely on passive disease surveillance, an effective communication network, trained personnel to perform outbreak investigations and supporting veterinary diagnostic laboratory services at local through to national levels. Building such disease surveillance networks can be very challenging, especially in countries where resources are limited and where livestock production systems tend to be more extensive. In such situations, effective disease surveillance networks can be built in parallel with the establishment of animal health service delivery systems. An understanding of the socio-economic and geophysical factors that determine the various types of livestock production system is thus important when developing animal health services and disease surveillance systems.

In most developing countries livestock production systems can vary from being, at the one extreme, fully commercial, intensive (usually poultry, pig or dairy) production systems, through the middle range of semi-commercial and smallholder mixed livestock/crop farming systems down to the level of subsistence farming, where any animal products (meat, milk or eggs) would usually be retained for household consumption and only rarely may there be a surplus available to be sold at a local market. Alongside these mainly sedentary farming systems various forms of migratory livestock keeping, where large numbers of cattle, sheep, goats or camels are herded by pastoralists on vast, often remote rangelands that are generally unsuitable for crop cultivation due to irregular, erratic and low rainfall. Migration patterns are determined by a combination of availability of and access to, suitable grazing lands. In between sedentary and migratory livestock production systems, there is transhumance, where part of a family or a group of families migrates with their animals on a seasonal basis, in search of suitable grazing for their animals, while the remainder of the family remain settled at a homestead, where

some crops may be grown and where some family members may become engaged in alternative forms of income generation.

The farming systems described above range between the extremes of very intensive, in terms of animal densities to very extensive. This distinction is important when determining the most appropriate cadre(s) and the numbers of animal health service providers to be deployed, in terms of their level of training, their earnings expectations and the demands of livestock producers. The most intensive commercial livestock production systems are characterized by high input costs and high returns and thus demand a high-quality professional level of veterinary service which can only be provided by a fully qualified veterinarian.

Figures 9.1, 9.2 and 9.3 illustrate examples of very different farming systems – each of which would demand a different level of animal health

Figure 9.1 Intensive commercial dairy farm. Photo: Adobe Stock, ijacky.

service delivery. On the one hand, the highly intensive dairy seen in Figure 9.1 would require a fully trained and experienced veterinarian, while the lower-input/low-output systems illustrated in Figures 9.2 and 9.3 might be satisfied by a less well qualified veterinary paraprofessional service provider working under the supervision of a qualified veterinarian.

Semi-commercial or smallholder mixed livestock/crop producers would sometimes require the presence of a veterinarian, but many simple veterinary interventions can also be provided by a suitably trained veterinary paraprofessional, working under the supervision of a qualified veterinarian. At the other extreme, the nomadic or transhumant management systems, seen in parts of Africa and Asia, which rely almost exclusively on access to low-quality and remote grasslands, are low input and, on a per capita basis, low output. Such livestock production systems are often remote from state veterinarians and livestock keepers are less able to afford a professional level of animal health service. In such extensive production systems, it may be more appropriate to deploy veterinary paraprofessionals trained to perform a defined range of veterinary interventions, who would also be required to be supervised, albeit remotely, by a qualified veterinarian.

In many parts of the world, even where there are smallholder mixed livestock/crop production systems, the state veterinary authorities are unable to afford to deploy sufficient numbers of state veterinarians to reach all farming communities, and owing to the relatively low demand for animal health services, private veterinarians are unable to make a living commensurate with their level of education, social standing or earnings expectations. In all such systems, the deployment of various cadres of veterinary paraprofessionals (see Figures 9.4 and 9.5), whose educational level and earnings expectations are lower, can help to mitigate the lack of access to primary animal health services and at

Figure 9.2 North western Ethiopia – mixed farming system using horses to plough arable land.

Figure 9.3 Northern Kenya, Samburu pastoralists, extensive herding of cattle, sheep and goats.

the same time can contribute towards building a robust animal disease early warning and response network. Whatever the circumstances and whether paraprofessional service providers operate privately or are employed by the state, it is important that the range of interventions to be performed by the various different cadres of veterinary paraprofessional are well defined and that such service providers always work under the direct or indirect, supervision of a qualified veterinarian, in order to ensure correct use of anti-microbials and other medicines and to maintain good standards of practice.

The methods used for the selection and training of veterinary paraprofessionals should be given careful consideration. The most important underlying principles which should be considered when selecting candidates for training as veterinary paraprofessionals is that they should be experienced and well-respected livestock keepers and longstanding members of the community in which they expect to provide animal health services. The training methods used to teach such individuals to provide a basic level of veterinary services and how to manage the treatment of commonly occurring diseases should recognize the innate, indigenous knowledge gained by such individuals over generations of experience of tending their animals from childhood through to adulthood. Such individuals usually have a very good understanding of the common signs associated with livestock illnesses and are often able to identify specific disease entities, without necessarily knowing their scientific names or specific causes. They can easily be trained how to recognize specific types of disease syndrome through observation and thorough clinical assessment of a case and thus provide an appropriate treatment, or if necessary refer the case to his or her supervisor. With additional appropriate training, employing this same approach, paraprofessionals can also become very useful members of a disease surveillance network. Indeed, the recent success of the global eradication of rinderpest was to a

Figure 9.4 Veterinarian working with a group of community-based animal health workers and livestock keepers using PDS tools during the active surveillance for Rinderpest in Kenya in 2001.

large extent achieved through the deployment of community-based animal health workers and other veterinary paraprofessionals who were enlisted by the state veterinary authorities in more remote areas as agents of either 'active' or 'passive' disease surveillance.

In general, active disease surveillance, using participatory methods of enquiry has the advantage that it can be used in a planned programme which has been designed to cover a defined geographical area, over a defined period of time, using a sampling frame that has been calculated to accurately determine whether or not the clinically visible signs of a particular disease have been seen or not by any members of the farming community in the recent past. The method relies on an accurate interpretation of what farmers perceive to have happened in the recent past. The method can thus be used to determine the perceived incidence of one or more disease entities, retrospectively over a defined period of time, or sometimes the prevalence of a particular disease entity at a particular point in time. Data obtained from these enquiries can be subjected to robust statistical analyses and can thus provide reliable information, when the sample size has been correctly determined. As mentioned earlier, 'sero-surveillance' can be used to determine the presence or absence of antibodies against a particular disease agent in a population and thus can detect the (recent) prevalence of infection in the absence of active clinical signs of a disease, or for the purpose of determining the level of immunity to a particular disease either as a result of natural infection or due to recent vaccination. The main disadvantage of (active) PDS is that it is expensive to mobilize sufficient numbers of qualified veterinarians or suitably trained veterinary paraprofessionals to provide an accurate estimate of disease prevalence. Bias can also be introduced due to inaccurate interpretation of farmers' perceptions.

As a means of collecting information on disease incidence and disease prevalence, passive

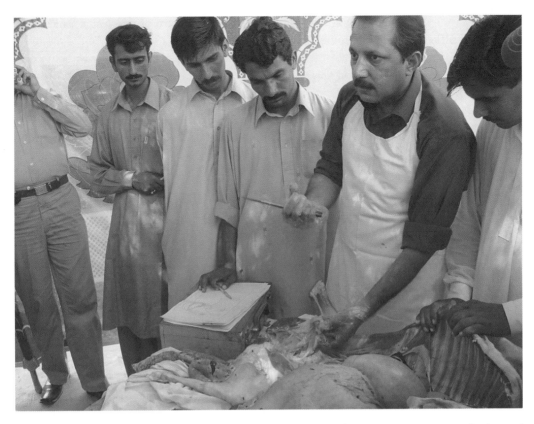

Figure 9.5 Veterinarian training veterinary paraprofessionals to perform a post-mortem examination and how to collect and handle appropriate laboratory specimens, Punjab, Pakistan.

disease surveillance, however, can be much more cost effective because it relies on the fact that a livestock keeper will report an unusual disease event, either unusually high morbidity or high mortality or cases of sudden death or abortion and so on, to their nearest animal health service provider. Such reports may not necessarily be because the farmer knows that there is a disease surveillance system operating, but usually because he wants the service provider to give assistance in the form of curative treatment or advice. Veterinary paraprofessionals can be trained to report suspected notifiable disease events through providing training on the recognition of disease syndromes or clinical signs commonly associated with notifiable disease events rather than specific disease entities.

Extension or awareness campaigns can be used to increase farmers' awareness of the need to report unusual disease events in order to contribute towards the development of risk-based disease prevention and control programmes.

In those countries where reporting of listed notifiable diseases is mandatory, guidelines may be provided under the veterinary legislation to aid field workers to understand when it is necessary to report a disease event and to ensure that all cases of notifiable diseases are reported. Such guidelines may include some or all of the following.

General signs commonly associated with notifiable animal diseases:

1 The disease is acute in onset and causes high mortality.
2 The disease often affects more than one animal in a group of animals and spreads rapidly to affect many other animals in a short period of time (high morbidity).
3 The onset of a notifiable disease on a particular farm or in a discreet animal population can sometimes be associated with the introduction of a new animal or group of animals into a herd or flock, for instance when one or more animals has been purchased from a market or directly from another farmer or trader.
4 In some instances, the introduction of a notifiable animal disease into a particular locality may be associated with the migration of animals through an area when the migratory animals mix with or come into contact with animals that are normally resident in that area.

Specific signs associated with notifiable animal diseases

1 **Sudden death** constitutes an emergency since the cause of death may be **anthrax**, which can be transmitted to humans – an **anthrax investigation must be conducted before the dead animal is moved**.
2 Abortion: all cases of abortion should be reported immediately. Sanitary precautions should be taken since most causes of abortion are **zoonotic** diseases and can cause sickness or death in humans.
3 Acute onset of unusual aggressive or abnormal behaviour or change in normal character or behaviour, especially in dogs but also cattle, horses and sheep and goats.
4 The presence of high fever associated with any of the following additional signs:

i The presence of oral lesions, ulceration or wounds on the lips, in the mouth or around the mouth, with or without excessive salivation and lameness in cattle sheep or goats.
ii Acute respiratory signs including coughing and / or a nasal discharge in cattle, sheep and goats, poultry and horses.
iii Acute diarrhoea either alone or in combination with respiratory signs and / or oral lesions.

Since both active and passive disease surveillance systems described above rely almost entirely on a report based on 'clinical suspicion' provided by a veterinarian or veterinary paraprofessional or the description of a disease provided by a farmer, the data provided has limited value. There is therefore a strong justification for such systems to be supported by diagnostic laboratory services. In most developing countries, it is difficult to reach the level of technology required for the diagnosis of many diseases at the level of district or regional laboratories. However, in such cases, field or district level laboratories can play a useful role in processing samples for diagnostic testing to be carried out at a regional, provincial or national reference laboratory, where a more sophisticated level of technology has been established.

As and when disease reporting is organized on a formal basis the service provider can be trained to capture the salient information relating to such an incident on a standardized reporting form, based on clinical suspicion, which he then forwards to the nearest state veterinary office. Normally, the state veterinarian would then conduct an outbreak investigation and attempt to learn more about the suspected case or outbreak, tracing the possible cause or source of infection and the likely spread and also try to verify the cause by collecting and submitting appropriate laboratory samples to the nearest diagnostic laboratory. In many countries, and especially under circumstances where very few veterinarians are deployed in close proximity to more remote livestock owning communities and where veterinary paraprofessionals have been given appropriate basic training and have acquired good clinical skills such individuals can be trained to conduct a post-mortem examination and undertake a preliminary outbreak investigation in addition to providing a report of the clinical suspicion of a notifiable disease event. This training should include a module on laboratory sample collection and sample handling. At the end of the training the trainees can be supplied with a post-mortem and sampling kit with a plastic laminated copy of Table 9.1 given at the end of this chapter, that gives a list of notifiable disease syndromes, their putative causes and the most appropriate samples to be collected and submitted for a laboratory confirmation to be made.

As outlined in previous chapters in this book, what is important is that any of the veterinary support team, whether for his or her own interest or that of his client, or whether working on behalf of the state as a member of a disease surveillance system, should know which samples would be most appropriate to submit for a laboratory diagnosis to be made.

Both active and passive disease surveillance systems require a robust reporting and recording system in order to allow data to be analysed using statistical software packages which convert the data into information which can be used to inform policymakers or for disease prevention and control strategy formulation exercises.

Examples of standard reporting forms are provided in the appendices at the end of this book. These forms have been carefully designed to provide the necessary information which is required by the OIE disease notification system known as the World Animal Health Information System (WAHIS). Provided that the service providers are literate they can easily learn to fill in these forms with some professional coaching during the first few attempts.

Increasingly, efforts are now underway to develop and test smartphone or tablet applications for disease reporting. Such web-based disease reporting systems, incorporating GPS data and the possibility to send images of animals and pathological lesions to assist the epidemiology unit receiving such a report to make a provisional diagnosis, will gradually replace the paper-based reporting systems currently in use in many countries and will further improve consistency and accuracy of disease reporting as well as achieving real time early warning of notifiable disease events.

Case Study: Afghanistan

Private veterinary paraprofessionals working in partnership with the state veterinary services

In Afghanistan, where there is a well-established network of privately operating veterinarians and veterinary paraprofessionals, working under various levels of professional supervision, covering almost all areas of the country where livestock are being raised, a new system for disease surveillance is being developed. This system involves the award of a 'Sanitary Mandate' contract to veterinary service providers to report, and when requested, to investigate, suspected cases or outbreaks of notifiable diseases. Under this system, the private service provider gives a verbal notification of a suspected notifiable disease occurrence to his nearest provincial veterinary officer using a mobile telephone. This verbal report is then followed up by submission of a written disease report on a standard disease report form (DRF), provided by the veterinary authority. On the basis of the particular circumstances, the provincial veterinary officer decides whether it is necessary to conduct an 'outbreak investigation'. He may decide to do this himself, if the farm is relatively nearby and he has access to a vehicle, or if the disease event involves a large number of animals and appears to be spreading rapidly, or, under other circumstances, he may decide to ask the private service provider to perform the outbreak investigation on behalf of the veterinary authority, under the terms of the sanitary mandate contract. An outbreak investigation involves both forward and backward tracing of the disease event as well as a more detailed on-farm investigation, involving, where appropriate, the conduct of a post-mortem examination and the collection and submission of appropriate laboratory samples according to a provisional or differential diagnosis.

Under such a contract the terms of reference define the exact scope of the services to be performed, the level of remuneration and the conditions of payment. The scheme is now being expanded to include the contracting of other functions such as the involvement of private service providers in providing vaccination services under national disease prevention or control schemes and collecting blood/serum samples for active disease surveillance. Consideration is also being given to contract out other public functions such as providing extension services and undertaking local meat inspection.

A partnership of this type between government veterinary services and private animal health service providers not only helps the state veterinary authority to perform its functions more efficiently, it also enhances the financial sustainability of private service providers, often working in an environment where farmers are either not able or sometimes unwilling to invest in animal health services, to the extent that these can provide a satisfactory level of income for the service provider. Such a system can result in the establishment of a robust surveillance network, where most suspected occurrences of notifiable diseases are reported and investigated to the point where a laboratory diagnosis is made.

This substantially improves the quality of epidemiological information available for accurate disease notification and for planning of national disease prevention and control programmes.

Table 9.1 Examples of common clinical disease syndromes (ruminants) and corresponding possible notifiable diseases or other differential diagnoses and the laboratory samples required to make a definitive laboratory diagnosis (zoonoses in grey). For the most up to date standards on testing for specific diseases in livestock species check the online edition of the OIE Manual of Diagnostic Tests and Vaccines for Terrestrial Animals http://www.oie.int/standard-setting/terrestrial-manual/access-online/.

Clinical disease syndrome	Possible diseases/conditions/causes	Ante-mortem sample	Post-mortem laboratory samples
SUDDEN DEATH	**Anthrax** Clostridial enterotoxaemias Haemorrhagic septicaemia, ovine pasteurellosis Acute poisoning, **Plague** (*Yersinia pestis*) Acute septicaemia, bacterial/viral		**BEFORE OPENING CARCASS: Direct blood smear from ear vein stained with Methylene Blue/Azur Blue** **if negative, then sample:** intestinal contents, heart/jugular vein blood, lung tissue, liver, spleen, kidney, long bone marrow for culture and/or toxin ident
ORAL/PERI-ORAL LESIONS High morbidity	Foot and mouth disease Vesicular stomatitis PPR, sheep/goat pox, Orf, Rinderpest*, etc.	Whole blood – serum Vesicular fluid in PBS, pH 7.2 Epithelium in PBS, pH 7.2 Pharyngeal swab in PBS, pH 7.2 Lesion tissue in PBS, pH 7.2 Ocular/nasal secretions, biopsy of skin papules	Whole blood – serum Vesicular fluid, oral mucosal tissue samples Prescapular or mesenteric lymph nodes Heart muscle tissue, lung, spleen, tonsils skin papules
ABORTION Individual/'abortion storm'	**Brucellosis,** Chlamydiosis **Q fever, Campylobacter** **Toxoplasmosis, Listeriosis** **Leptospirosis,** mycoses, etc.	Serum Milk Vaginal discharge swab	Aborted foetus Foetal stomach contents, foetal liver/lung Cotyledons, placenta, milk/colostrum, vaginal swab; serum samples for culture and serology
ACUTE RESPIRATORY SIGNS Nasal discharge High morbidity	Pasteurella pneumonia Contagious caprine pleuropneumonia (CCPP) PPR, lungworm infections, bacterial/viral bronchitis/tracheitis, calf pneumonia (strep/staph/viral)	Whole blood – serum Nasal discharge swab Faecal sample (lungworm eggs) Cotyledons from aborted placenta	Whole blood – serum Tracheal/bronchial swabs Pleural cavity fluid swabs, lung tissue Placenta, foetal stomach contents

Table 9.1 *continued*

Clinical disease syndrome	Possible diseases/ conditions/causes	Ante-mortem sample	Post-mortem laboratory samples
ACUTE WEIGHT LOSS Diarrhoea Anaemia Jaundice with or without fever, loss of appetite	*E. coli*/perinatal diarrhoea, lamb dysentery PPR, viral diarrhoea (corona/ rotavirus etc.) Fascioliaisis, anaplasmosis, theilleriosis	Whole blood in anticoagulant Thick and thin blood smears Serum, faecal sample	Whole blood, thick and thin blood smears Faecal sample, intestinal tissue, intestinal contents Liver

Note: *Rinderpest has now been successfully eradicated worldwide.

Endnote

1   AHS = African horse sickness, BSE = bovine spongiform encephalopathy, CSF = classical swine fever, FMD = foot and mouth disease, PPR = peste des petits ruminants, CBPP = contagious bovine pleuropneumonia.

OIE (n.d.) Terrestrial Animal Health Code. Volumes 1 and 2. http://www.oie.int/standard-setting/terrestrial-code/access-online/.

Thrusfield, M. (2018) Veterinary Epidemiology, 4th Ed. John Wiley and sons Ltd, Hobroken, NJ

Bibliography

Hassan, A., Rassoul, A.B., Villon, H. Woodford, J., Tufan, M. (2011) An innovative means of establishing a national epidemio-surveillance network in Afghanistan. International Conference on Animal Health Surveillance, International Society for Infectious Diseases (Lyon, May 2011).

chapter 10

Common clinical problems

Susan C. Cork

When animal health and veterinary staff undertake field visits, it is often necessary to provide initial treatment for an animal without confirming the cause of its illness. However, to be sure that the correct medication is given, especially when an entire herd of flock is to be treated, it is good practice to collect suitable samples for submission to the veterinary laboratory. This will improve the chances of identifying the cause of the problem and is especially important where a large number of animals are 'at risk' so that appropriate disease control and prevention measures can be taken. Undertaking a thorough examination and taking a good case history is essential. *Remember that emerging and re-emerging disease can be missed if staff are not observant and thorough.*

The principles of identifying a health problem, that is, taking a case history and performing a clinical examination have already been discussed. Some diseases are associated with a specific presentation (age/class or subgroup) so it is important to collect this type of information in the case history. The principles of good sample collection have also been outlined (see Chapter 1).

In this chapter, the following clinical problems will be discussed in more detail. A summary of the likely differential diagnoses for these clinical presentations are provided in the associated tables:

1 infertility and abortion (Table 10.2 and Figure 10.1)
2 diarrhoea (Tables 10.3, 10.4–10.5)
3 haematuria/haemoglobinuria (Table 10.6)
4 hair loss, itchy skin (Table 10.7, 10.8)
5 ill thrift, weight loss (Table 10.9)
6 neurological signs (Table 10.10).

10.1 Infertility and abortion

The following section will focus on reproductive disorders in cattle but the principles followed are similar for all species.

Infertility in the female

If there is an apparent infertility problem (that is, failure to conceive in the female), it is first necessary to rule out early embryonic death, which may also result in repeated returns to service. To do this there must be reliable records available. Record keeping is usually good if the animals are visiting an artificial insemination (AI) centre. If possible check the dates of returns for AI. If the intervals are longer than the normal oestrus period (19–21 days for cows) there may have been early embryonic loss. If cows continue to come into oestrus after normal service by a bull it is also important to check the fertility of the bull (for example, sperm count, sperm morphology and motility and so on). However, if cows continue to return to service after AI it is also important to check the AI technician's technique and the suitability of the storage facilities, as well as the quality of the semen.

There are many causes of infertility including sub-clinical disease but it is important to initially check for faults in animal husbandry.

Consider the following questions.

1 Are the animals submitted for AI really in heat?
2 Is the technician using the correct technique?
3 Check the quality of the semen, that is, send a sample to the laboratory for expert evaluation of the morphology and motility of the spermatozoa.
4 Examine the animal(s) affected. Is a specific age group affected? (Some diseases are more common in younger animals.) Is the problem restricted to a particular village? What is the general health status of the herd? (Underweight animals will often fail to come into oestrous.)
5 Routine and more specific samples should be taken from affected animals to check for reproductive health and well as general health status, for example, faeces, blood smear, whole blood, serum as well as vaginal swabs/smears and so on.

It may be necessary to call for more specialized veterinary assistance to complete a detailed rectal/vaginal examination of each 'infertile' cow in order to determine the size and shape of the uterus and the presence of abnormalities in the reproductive tract.

Abortion

A number of causes of abortion can affect individual animals (for example, septicaemia) and may not require a specific investigation. However, if there is an outbreak of abortion in an area,

or a larger number of abortions than would be expected in an individual herd (for example, 5%) this can be serious and requires a thorough and detailed investigation. It should be noted that some infectious causes of abortion are zoonotic (that is, brucellosis and leptospirosis) and so precautions should be taken to prevent exposure of field staff and community members to these pathogens. Routine examination of any case of abortion requires taking a full clinical history, examination of the affected animal(s) and the aborted foetus/placenta. Other sick animals in the herd should also be examined. Biosecurity measures (that is, disinfection of hands and footwear, appropriate use of personal protective equipment [PPE], correct disposal of waste products, change of coveralls and so on) should be taken to avoid transmitting a pathogen from one farm or region to another. Although the epidemiological pattern of the disease and the associated clinical presentation can suggest the most likely cause of abortion, a confirmatory diagnosis will depend on the submission of suitable samples to the laboratory.

History

A clinical history should include the following details and answers to the following questions.

1 The date, location, owner and animal details.
2 How many animals are affected and which age group(s) are involved?
3 Record recent changes in management, breeding history (that is, AI or natural service), and nutrition.
4 An outline of recent veterinary procedures, vaccination, worming, treatment with antibiotics and so on.
5 A description of any clinical signs observed in affected animal(s) prior to abortion and estimate at what stage of pregnancy the abortion occurred.

Table 10.1 Gestation periods of common domestic species. These figures are estimates for common breeds but there can be a variation.

Species	Gestation period (days)	Gestation period (approx.)
Cat	58–65	9 weeks
Cow	270–290	9 months
Dog	56–70	9 weeks
Donkey	365	1 year
Goat	145–155	5 months
Horse	330–340	11 months
Pig	112–115	3–4 months
Sheep	144–151	5 months

6 Note what other livestock are kept and if any clinical disease is apparent (now and previously) in these animals?

7 Are there previous offspring from the affected animals?

The normal gestation periods of some common domestic species are given in Table 10.1.

Clinical examination of dam and sire (if natural service)

After performing a full clinical examination of the dam and sire (see Chapter 1) collect a semen sample and preputial wash from the bull to check for the presence of Trichomoniasis (*Trichomonas* sp.) and other microorganisms. Examine vaginal smears from the dam and, if possible, also examine samples collected from the vagina of other cows serviced by the same bull.

Examination of the placenta and foetus

When attending a case of abortion use the appropriate PPE. Examine the placenta and foetus carefully and describe any unusual features, for example, strange colour, smell, texture. Some diseases cause characteristic changes to the placenta (for example, toxoplasmosis in sheep – see Figure 3.33). At what stage of development is the foetus? If the abortion occurred late in the pregnancy, that is, near to full term, it is good to ask the following questions.

1 Did the neonate walk, that is, check hoof tips?

2 Is this an abortion or a perinatal death? (The hoof tips will be worn if the neonate has tried to stand up.)

3 Did the neonate start breathing? (Examination of the lungs will indicate whether or not the neonate took a breath, that is, the lung tissue will float is it contains air.)

4 Did the neonate suckle? (Examination of the stomach will indicate if the neonate took any milk, is it empty?)

Laboratory samples

1 Take as many of the following samples as possible:

- placenta (fresh and fixed)
- foetal stomach contents
- liver, spleen, kidneys, heart blood (fresh and fixed) from the foetus: check if the lung tissue floats in water.

2 Perform a full necropsy on the dead foetus and/or dam – details of any gross findings should be recorded on the sample submission forms (see Appendix 2).
3 From a live dam collect vaginal swabs, a serum sample, milk, blood smear, whole blood/serum.
4 If indicated, send other samples, for example, faecal smear, smear of any discharge material, faecal sample for culture, or feed sample/ suspect poison (some plants and mycotoxins may cause abortion).

Table 10.2 outlines some common causes of abortion in cattle. Common causes of abortion in pigs are outlined in Figure 10.1. Note that, in some cases, especially where a flock or herd has had previous exposure to a disease, the clinical and epidemiological picture may differ from the typical presentation reported in text books.

Table 10.2 Some infectious causes of abortion in cattle.

Disease agent	Clinical features	Time of abortion	Disease in the foetus	Placental lesions	Samples to collect
Brucella abortus	Abortion (up to 90% in susceptible herd)/or no apparent effect, mastitis	6 months	May develop pneumonia	Necrosis of cotyledons, opacity of placenta and oedematous	Foetal stomach contents, placenta (fresh/fixed), milk, semen; paired serum samples
Trichomonas foetus	Infertility, return to service at 4–5 months, abortion and/or pyometra (5–30% of herd affected)	2–4 months	Rotten macerated foetus covered in flocculant material	Serous or flocculant discharge from the uterus and over the placenta	Cervical mucous, foetal stomach contents, uterine exudates (smears must be examined when fresh)
Campylobacter foetus	Infertility, irregular oestrous cycles, abortion (2–20% of herd affected)	5–6 months	Flecks of pus seen over the visceral peritoneum	Semi-opaque surface over the placenta, petechiation and localized oedema	Foetal stomach contents, fresh/fixed tissue and placenta, discharges for culture/smears
Leptospira Pomona, L. hardjo	Abortion at any stage during the acute febrile phase of infection, 25–30% of herd affected, +/– haematuria	> 6 months	Perinatal deaths are common, the dam may develop mastitis with red milk	Avascular placenta, atonic yellow cotyledons, brown fluid between allantois and the amnion	Serum (paired) from dam and other animals in the herd, culture foetal stomach contents, kidney, liver and urine
IBR (virus)	25–50% affected	> 6 months	Autolysed	Autolysed	

Table 10.2 *continued*

Disease agent	Clinical features	Time of abortion	Disease in the foetus	Placental lesions	Samples to collect
Mycotic agents	6–7% may be affected	3–7 months	Often fungal lesions over the foetus	Necrotic maternal cotyledons, soft yellow cushion like lesions, small raised leathery areas across the intercotyledonary area	The fungi (*Aspergillus* sp., *Absidia* sp.) may be seen in direct smears taken from the placenta, foetal tissues and placenta may also be submitted for culture
Listeria monocytogenes	Septicaemia with abortion late in gestation	> 7 months	No obvious abnormality	No obvious abnormality	Submit sera from the dam to rule out other causes and fixed/ fresh foetal and placental tissue
Unknown viruses	30–40% can abort depending on the virus involved and previous exposure	Variable	As above	As above	As above along with whole blood for haematology
Non-specific i.e. *Salmonella* septicaemia	May abort early in gestation, can involve many animals of an individual	Variable, may have early embryonic loss	Variable	Variable	As above

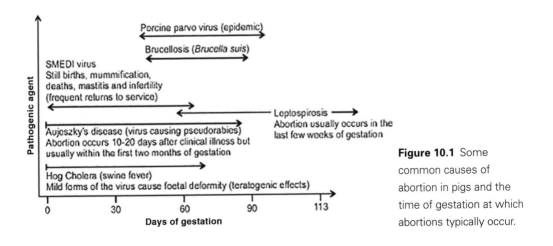

Figure 10.1 Some common causes of abortion in pigs and the time of gestation at which abortions typically occur.

10.2 Diarrhoea

The term diarrhoea is used for soft or partially formed faeces which may be passed more frequently than normal. A wide range of diseases present with diarrhoea as one of the predominant clinical signs. There are various forms of diarrhoea with a range of severity. In severe acute cases animals may pass very watery stools, show signs of abdominal pain and become dehydrated quite rapidly as is seen in many systemic bacterial diseases such as colisepticaemia or salmonellosis. In these cases, there may also be blood and mucus in the stool as well as undigested fibre. In other cases, diarrhoea may be mild, with only a slight indication of abdominal discomfort observed in the animal, possibly a change in intestinal sounds and usually little weight loss. Mild cases of diarrhoea may indicate the presence of parasites or a recent change in diet. Excitement and fear may also cause loose faeces. If an animal frequently passes a large volume of well formed, but incompletely digested faeces, this may indicate an unsuitable diet or it may imply incomplete digestion or malabsorption. If prolonged, even mild diarrhoea may become debilitating and result in significant weight loss and nutrient deficiencies. It is important to take a good clinical history and to perform a full clinical examination on the animal in order to determine whether the diarrhoea is a primary or secondary problem. Determining the cause of the diarrhoea will ensure that treatment options are correctly assessed.

Take a full history

It is very important to take a good clinical history. Determine whether it is an isolated case or part of a wider problem. Check whether or not there have been any recent changes in management, for example, a new diet or a recent introduction of new stock. Has the animal been showing other signs of ill health? Is the problem recent or chronic? Has the animal been treated for parasites recently? Has the farm manager or owner given any other medicine (including traditional remedies) to the animal? Are other livestock or people showing signs of ill health?

Perform a thorough clinical examination

Taking appropriate biosecurity measures, examine the sick animal(s) and others in the herd/village. Perform a full clinical examination and check for any evidence of a rise in temperature, weight loss, dehydration and so on. Examine the faeces and describe the colour, smell and consistency. Is there any blood or mucus present, are there any strips of intestinal lining present? Is the smell of the urine strong? What is the colour of the urine? Urine will be dark if the animal is dehydrated. Check the mucous membranes and skin elasticity – is the animal anaemic? dehydrated? jaundiced? Are there any haemorrhages or blisters in the mouth? Is the animal showing signs of abdominal pain?

Collect laboratory samples

If there are several animals affected or if an animal is very unwell, submit the following samples along with a full clinical history and description of the case (use laboratory submission forms – see Appendix 2 – for this purpose):

1 faecal sample from affected and in-contact animals (5–10 g in labelled bottle or plastic bag)
2 whole blood and blood smear
3 serum from affected and 'in-contact' animals
4 necropsy of dead animals and other samples as indicated.

If the diarrhoea is mild and animals appear well it may not be necessary to take any samples unless faeces are required for parasitology. If there are any dead animals present perform a necropsy (see Appendix 2 and Chapter 8) and submit a full necropsy report along with fresh and fixed tissue samples. Include sections of intestine, especially from the ileocaecal valve (to check for *Mycobacterium avium paratuberculosis* in cattle), and make several smears of intestinal mucus/contents on glass slides (to check for the presence of protozoa). Collect and note the presence of any parasites seen in the intestine and stomach. Make sure that samples of liver, lung, spleen, kidneys and heart are included. Sections for histology should measure approximately 1 × 1 × 1 cm and should be fixed in ten times their own volume of buffered 10% formalin in sealed, labelled jars.

While waiting for laboratory results, if the diarrhoea is severe, it may be necessary to begin supportive treatment for affected animals. If the animal(s) is(are) not showing other clinical signs this may not be necessary but routine worming and a change of diet may help. If affected animals are dehydrated or have a high temperature it may be necessary for the veterinarian to provide fluid replacement therapy and, where indicated, to give an injectable broad-spectrum antibiotic to prevent primary or secondary bacterial infections. It is often advisable to isolate the affected animal(s) from healthy stock while

Table 10.3 Some causes of diarrhoea in horses.

Cause	Clinical signs	Age commonly affected
Salmonella sp.	Acute, profuse, foul smelling watery diarrhoea, pyrexia, many individuals affected	Foals
Actinobacillus equi	Acute diarrhoea	Outbreaks in foals
Corynebacterium equi	Sudden onset, +/- respiratory disease, dehydration, death	Young foals
Clostridial sp.	Acute, profuse, watery diarrhoea	Any age
Aspergillus fumigates	Chronic diarrhoea	Foals
Rotavirus	Acute, profuse watery diarrhoea	Newborn foals
Helminths	Acute, sub-acute or chronic associated with heavy infections (may have hypotroteinuria)	Usually older foals
Any of the above	Acute/chronic diarrhoea	Older debilitated horses
Toxins, antibiotics	Individual animals affected, may develop watery diarrhoea with dehydration and malabsorption	Older horses
Nutritional	Chronic cases may have bulky faeces or profuse diarrhoea. May also have mild cases.	Older horses associated with inappropriate diet
Foal heat diarrhoea	Occurs in mares on heat (when suckling)	Mares 7–120 days after parturition
Enterotoxigenic *E.coli*	Acute, profuse watery diarrhoea, dehydration and acidosis (common in colostrally deprived animals)	Newborn calves < 3 days old
Salmonella spp.	Acute dysentery, fever and deaths occur	All ages, outbreaks can occur precipitated by stress

Table 10.4 Some causes of diarrhoea in cattle.

Cause	Clinical signs	Age affected
Clostridium perfringens type B & C	Severe haemorrhagic enterotoxaemia	Young well-fed calves
Mycobacterium paratuberculosis	Chronic diarrhoea, weight loss	Sporadic in mature cattle
Candida spp.	Acute, profuse, watery diarrhoea	Young calves (following antibiotic use)
Rotavirus and coronavirus	Acute, profuse, watery diarrhoea	Newborn calves, 5–21 days of age
BVD virus	Erosive gastroenteritis	Young cattle, usually sporadic
Rinderpest (largely eradicated as part of a global initiative)	Erosive stomatitis and gastroenteritis, highly contagious	All ages depending on immune status, high mortality
Malignant catarrhal fever virus	Erosive stomatitis and gastroenteritis, enlarged lymph nodes, ocular lesions, may have haematuria and develop terminal encephalitis	Usually adult cattle, sporadic cases or outbreaks can occur
Ostertagiasis	Acute or chronic diarrhoea, hypoproteinaemia, weight loss	Young cattle
Protozoa	Dysentery, pain in abdomen	Calves 3 weeks–12 months, outbreaks occur
Dietary	Varies from chronic malabsorption with a high volume of semi-formed faeces to acute watery diarrhoea	Any age, often in older animals with worn teeth
Toxicity	Varies depending on the cause	Varies

Table 10.5 Some causes of diarrhoea in pigs.

Cause	Clinical signs	Age affected
Enterotoxigenic *Escherichia coli*	Acute profuse diarrhoea, dehydration	Young pigs, weaners
Salmonella spp.	Acute septicaemia or chronic forms	All ages
Clostridium perfringens type C	Acute or per acute enterotoxaemia	Young pigs, newborn
Treponema hyodystenteriae (swine dysentery)	Acute outbreaks of vomiting and diarrhoea	Outbreaks in young pigs
Rotavirus/coronavirus	Acute diarrhoea, dehydration, death	Outbreaks in young pigs
Coccidiosis	Diarrhoea, oocysts in faeces	Common at 5–14 days of age, high morbidity, low mortality
Trichuris suis	Chronic diarrhoea, weight loss	All ages
Ascaris sp.	Mild diarrhoea	More common in young pigs
Iron deficiency	Mild diarrhoea, anaemia	Young pigs, 6–8 weeks of age
Nutritional	Chronic or acute diarrhoea, malabsorption	Any age

waiting for a diagnosis. Note that some causes of gastroenteritis in livestock may infect people so make sure that basic hygiene is maintained, for example, animal handlers and animal health staff should wash their hands after handling sick animals and if indicated, wear the appropriate PPE. Specific diseases which should be considered in horses, cattle and pigs are outlined in Tables 10.3–10.5.

Remember that diarrhoea may be part of a more generalized illness (septicaemia, toxaemia) and not restricted to the intestines. Causes of gastrointestinal upset include:

1 infectious systemic disease (viral, mycotic, bacterial) – there are often many animals affected
2 helminths – often young animals affected
3 protozoa – usually young or debilitated stock affected
4 nutritional – associated with inappropriate diet, food shortage or spoiled food
5 toxic – one or several animals affected
6 physiological – stress/excitement related.

10.3 Haematuria and haemoglobinuria (red urine)

If an animal has red urine this may be due to the presence of red blood cells, haemoglobin or a chemical (medicine or plant). To make a correct diagnosis it is necessary to take a good clinical history and to perform a thorough clinical examination. To confirm the diagnosis, it is usual to collect freshly voided urine and other samples including fresh blood, blood smears and serum. The urine sample may be left to settle overnight or it can be centrifuged in a laboratory. If there are a lot of red blood cells present in the urine the animal has 'haematuria'. A few red cells may be present due to contamination from the reproductive tract in female animals. If there are no red blood cells in the sediment then

urine is probably red/brown due to the presence of haemoglobin. To make sure, however, the sample should be submitted to a veterinary laboratory to allow full sediment examination and urinalysis.

Clinical history

The following are key questions that should be asked.

- What species are involved?
- How many animals are affected?
- What age range of animal(s) are affected?
- Has the problem occurred previously?
- What is the duration of the problem?
- Have any animals died following development of the problem?
- What is the diet? (Do the animals have access to bracken fern or other toxic plants?)
- Has any medication been given recently?
- Have there been any abortions recently?
- Is mastitis present?
- Is there a tick problem in the area?
- What, if any, control measures have been taken?
- Is the red urine persistent or intermittent?

Clinical examination

- Has the animal got a high temperature?
- Are the lymph nodes enlarged?
- Are there any ticks present on the animal?
- Is the animal anaemic/jaundiced?

It is very important to determine whether or not the animal(s) is/are anaemic so check the mucous membranes of the mouth and eyes. If the mucous membranes are yellow the animal is jaundiced, if they are pale it indicates the presence of anaemia. Anaemia and/or jaundice can indicate that the animal has haemolytic anaemia which can be associated with the presence of blood parasites. If there are widespread haemorrhages present in the mucous membranes the animal may have a terminal haemorrhagic disease.

Laboratory samples

Collect the following samples:

1 urine (fresh and fixed in formalin) about 10 ml in a jar (take a mid-flow sample)
2 whole blood and blood smears (fix blood smears in methanol)
3 serum samples
4 milk samples may be required if there is concurrent red milk or mastitis
5 ticks for identification (some ticks transmit blood protozoa and rickettsia which may be responsible for haemolytic anaemia – see section 3.6).

If there are any dead animals it is necessary to conduct a thorough necropsy (see Chapter 8). Pay special attention to the bladder wall and kidneys. Collect a urine sample and submit fresh and fixed tissues (including spleen, lymph node, liver and kidney, ureters and bladder wall) for laboratory examination. If there are aborted foetuses or dead neonates, use appropriate PPE and perform a full necropsy on these as described under the section on abortion. Precautions should be taken to protect both people and other animals from exposure to potential pathogens.

At the laboratory, urine samples are spun down in a centrifuge and the sediment(s) examined. If there are any red blood cells in the sediment this confirms that the animal has haematuria. If there are few or no cells in the sediment then the animal may have haemoglobinuria or is excreting a coloured chemical. The distinction between haematuria and haemoglobinuria is an important one as the causes of haematuria and haemoglobinuria are quite different (see Table 10.6).

Table 10.6 Summary of some diseases to be considered as possible causes of red urine in cattle.

Clinical presentation	Number of animals involved and duration	Diagnosis	Comments
Haematuria			
Haematuria, with or without anaemia	May be several cases in areas where bracken fern is common	Enzootic haematuria	History of long term ingestion of bracken or rock fern
Haematuria, pyrexia, thirst and frequent urination	Individual or clusters of cases	Pyelonephritis/ glomerulonephritis	May have a history of urinary tract problems, may be associated with *Corynebacterium renale*
Haematuria, frequent urination with or without pain	Individual or clusters of cases	Urinary calculi	Calculi can be seen in urinary sediment, can have secondary bacterial infection
Thirst, frequent urination with or without haematuria, anaemia	Individual cases	Renal disease	May have history of urinary tract problems

Clinical presentation	Number of animals involved and duration	Diagnosis	Comments
Haemoglobinuria			
Haemoglobinuria, anaemia, pyrexia	Can have several cases especially where competent tick vectors are common and where the disease is endemic	Babesiosis	Tick borne disease, history of exposure to ticks, can detect organism in blood smears
Acute haemolytic anaemia with, or without, haemoglobinuria after parturition	Individual animals affected	Hypophosphotaemia	Acute onset and suggestive case history
Haemoglobinuria, anaemia, pyrexia	May have an outbreak with concurrent mastitis/abortion	Leptospirosis (especially *L. pomona*)	Acute onset, may occur where pigs are housed close by
Acute haemolytic anaemia, haemoglobinuria	May affect the whole herd/group	Kale and/or rape poisoning	History of dietary changes
Acute haemolytic anaemia with, or without, haemoglobinuria, pyrexia	Usually individual cases	Bacillary haemoglobinuria (*Clostridium novyi*)	Acute onset
Acute haemolytic crisis, haemoglobinuria, with, or without, pyrexia	One of many animals affected	Copper toxicity and other toxins	Check case history and management, usually need to treat systemic signs

If the animal is very anaemic and has a high temperature (pyrexic) it could be a case of acute haemolytic anaemia with secondary haemoglobinuria. In chronic cases of haematuria (for example, in enzootic haematuria) animals may also become severely anaemic but rarely have a high temperature; in these cases, the haematuria is often intermittent and the anaemia has developed slowly. While awaiting laboratory results it may be necessary for the veterinarian to treat the anaemia by using haematinics and vitamin B complex. If there are a lot of ticks present on the animal(s) a tick control programme should be implemented.

10.4 Hair loss and itchy skin

Irritation of the skin may result in mild or severe clinical signs. Itching (pruritus) may result in hair loss, scab formation and secondary bacterial infections as a result of self-mutilation and rubbing which disrupts normal skin resilience. Hair loss may also occur secondary to systemic disease, malnutrition and endocrine disorders, but in these cases, there may be other clinical signs and the skin is not usually pruritic. There are a wide range of factors that may cause skin disease but in most situations the clinical signs are due to a combination of factors.

Itchy skin may be associated with the development of raised red patches, scabs, scaling and evidence of ectoparasites (especially mites). Liver disease and malnutrition (for example, vitamin A deficiency) may increase the susceptibility of livestock to ectoparasites as well as resulting in poor skin and hair condition. Endocrine disorders such as an underactive thyroid gland (hypothyroidism) may be associated with bilateral (and systemic) hair loss, weight gain and susceptibility to cold. In most cases this condition will only occur sporadically. Prolonged systemic diseases and, in sheep, specific conditions such as scrapie may also need to be considered as a cause of poor coat, disrupted wool growth and skin damage. In any case of skin disease, it is essential that a good case history is taken and that the animal(s) is/are given a through clinical examination to look for underlying causes.

1 Take a full clinical history. How many animals are affected? What age group is involved? When did the lesions first appear? Check the diet and other management parameters. What other clinical signs were apparent?

2 Perform a thorough clinical examination. Examine a selection of animals in a group and look for early and developed lesions. The scabbed over lesions may only be evidence of self-trauma and secondary infection. Look for small raised papules and other primary lesions and also look carefully for the presence of ectoparasites. Examine the animals for other signs of disease, for example, malnutrition, liver damage and so on.

3 Collect samples of hair and/or wool. It may be necessary to take several skin biopsies to represent the primary lesions and skin scrapings or hair samples for microbiological examination. Collect ectoparasites for identification. If required, skin biopsy material should be submitted to the laboratory in

Table 10.7 Terms used to identify skin changes.

Description	
Diffuse (widespread) lesions	
Scales	Unbroken skin surface, dry, flaky skin
Excoriations	Traumatic, abrasions and scratches
Fissures	Deep cracks
Hyperkeratosis	Excessive overgrowth of dry horny thickened skin, skin surface unbroken
Eczema	Red itchy skin, weeping dermatitis, scabby disruption of the surface
Discrete (well defined) lesions	
Vesicle (bullae, blister)	Fluid filled blister, 2 mm diameter, superficial
Pustule	Pus filled blister, will rupture, 2–5 mm diameter
Wheals	Swollen and reddened areas, transitory
Papules	Elevated, inflamed areas, tend to point and rupture
Nodules	Elevated, solid 1 cm diameter (variable), in acute and chronic stage, surface unbroken
Scab	Crust of coagulated serum, blood, pus and skin debris raised above skin surface
Pyoderma	May become diffuse, secondary (occasionally primary) infection of the skin and/or follicles

preservative, along with a full case history. Try to send apparently healthy as well as diseased tissues in the submission.

Table 10.7 describes some of the common terms used to identify skin changes. Table 10.8 outlines some of the common causes of skin disease in livestock. For more detailed information consult a specialist veterinary centre or the reference texts listed at the end of the chapter.

Table 10.8 Common causes of hair loss and skin disease.

Clinical sign	Cause	Common host/comments
Pruritis (itchy skin) with or without secondary skin lesions	Sarcoptic and chorioptic mange mites Aujesky's disease (viral) Lice (pediculosis) Ked, itchmite, sheep scab Scrapie (prion)	Cattle and sheep, equine limbs Cattle, other signs in pigs Cattle, sheep, horses Sheep Sheep
Pruritis, hypersensitivity reaction	Sweet itch (midges)	Horse
Pruritis associated with allergy	Various (food and contact)	All species, urticarial wheals
Photosensitivity	Liver disease, toxins	All species
Peri-anal pruritis	Oxyuris equi	Foals
Parakeratosis and hyperkeratosis	Hypovitaminosis A, vitamin B deficiency, zinc deficiency	Pigs and other species
Scurfy skin	Vitamin and other deficiency Ectoparasites	All species
Primary urticaria	Insect bites, stings and contact hypersensitivity, warble flies, drug reaction, etc.	All species
Secondary urticaria	Can occur following respiratory infections in horses Erysipelothrix in pigs	Various
Generalised dermatitis	Widespread fungal (mycotic) and/or bacterial, i.e. *Dermatophilus* sp. infection Ringworm (fungal) Burns and frost bite	Horses, cattle, sheep All species (hair loss patchy) All species, extremities
Photosensitivity	Chemical irritant and sun burn, Systemic toxaemia and drug reactions	All species All species
Hypersensitivity (generalized)	May occur with mange mites (*Sarcoptes* sp., *Demodex* sp.)	Variable lesions, may be severe with secondary infection
Regional dermatitis	Hook worm and larvae of other parasites (larval migrans)	Carnivores and humans
Specific lesions	Various diseases (systemic and localized)	Cattle (ulcerative and/or erosive – foot and mouth disease, BVD, rinderpest, MCF, etc.)

Table 10.8 *continued*

Clinical sign	Cause	Common host/comments
		Sheep (sheep pox, facial pyoderma due to Staphylococcal infection, blow fly myiasis, etc.)
Localised lesions	Wounds and localized infection Abscesses*	All species

Note: *An abscess is a circumscribed infection which occurs under the skin or in an organ secondary to bacterial infection e.g. *Staphylococcus aureus* following damage (i.e. infection/wound).

10.5 Ill thrift and chronic weight loss

There are a number of causes of poor condition and weight loss (see section 10.2, diarrhoea). If only an individual animal is affected the approach to the case will be a little different to that which is required where 'ill thrift' affects a whole herd or flock. In both situations, it is necessary to perform a thorough clinical examination and to take a detailed case history. Ill thrift in individual animals may be due to a wide range of factors such as poor ability to compete, the presence of mouth lesions, poor dentition, neoplastic disease or developmental disorders. If a group of animals are affected it is important to check the husbandry, that is, evaluate the diet, test for the presence of gastrointestinal parasites, examine stocking rates and breeding/replacement policy. Does the animal have a good appetite, and want to eat? The most common cause of weight loss or poor weight gain in livestock is under or malnutrition. Subclinical or clinical disease, for example, gastrointestinal parasitism may make the situation worse but the lack of adequate nutrients may be the main pre-disposing factor. Chronic sub-clinical or clinical gastrointestinal parasitism may be a significant cause of ill thrift in young stock. In older animals, chronic diseases such as tuberculosis (*Mycobacterium bovis*) or Johne's disease (*Mycobacterium avium paratuberculosis*) may cause weight loss along with diarrhoea. Chronic ingestion of toxic substances may also cause poor coat and weight loss in livestock (for example, ragwort poisoning in horses and ruminants). Table 10.9 outlines the common causes of ill thrift, poor weight gain or weight loss in livestock. As with any disease problem, the approach to the problem must be thorough.

1 Take a clinical history. What is the duration of clinical signs? Level of morbidity? Mortality? Which age groups of animal(s) have been affected? Is there any data from previous health screening or necropsy? What is the history of the property, that is, records previous trace element deficiencies, presence of toxic plants and so on?

2 Perform a thorough clinical examination. Always examine the mouth, check faecal samples and urine, check for evidence of liver damage (swollen abdomen, jaundice, oedema, anaemia, behavioural changes). Other clinical signs?

3 Collect clinical samples (see Chapter 1 and the outline for a necropsy in Chapter 8 and Appendix 2).

In disease(s) characterized by chronic wasting, affected animals are often culled and a necropsy can provide valuable information. At necropsy, special attention should be paid to the liver and gastrointestinal system. Look for evidence of irritation, hyperplasia or focal lesions in the intestinal tract, and the rumen and abomasum

Table 10.9 Causes of weight loss and failure to gain weight.

Nutritional causes	Comments
Inadequate food	
Poor quality feed	Diet low in energy or specific nutrients, in monogastric species (pigs, poultry) protein may be the limiting factor
Imbalance of energy intake to energy output due to high production demands	May also see clinical signs of metabolic disease in some animals, i.e. hypoglycaemia
Malnutrition due to mineral, vitamin (B, D, E) and/or trace element (Fe, Mn, Co, Cu, K, Se) deficiency	Shortage of minerals such as magnesium or calcium can cause serious metabolic disorders, other deficiencies (vitamin A, Cu, Zn) can result in chronic disease over a period of time. Diagnosis is based on clinical history and signs and supported by sampling
Inadequate food intake	Check mouth for signs of trauma or lesions, check for any obstructions in the throat, oesophagus or gut
Physiological factors	Stress and cold weather will increase energy demand, intake may decrease with 'stress', e.g. high stocking rates
Excessive loss of protein and carbohydrate or faulty absorption	
Loss of glucose in the urine	Diabetes mellitus or renal disease
Metabolic disease, malabsorption	Digestive upset or liver damage
Protein loss in the urine	Renal disease
Protein loss in the faeces	Helminths (*Ostertagia* sp., *Nematodirus* sp. Paramphistomes) and protozoa
Increased protein and energy utilization	Neoplastic disease
Internal and external parasites	Malabsorption and competition
Chronic bacterial disease	Tuberculosis (*M. bovis*), Johne's disease (*M. paratuberculosis*)
Systemic disease	Fever and inappetence associated with infectious disease may result in ill thrift
Inadequate utilization of nutrients	Secondary to severe diarrhoea, hepatic disease and metabolic disturbance

(in ruminants). Check the consistency of the faeces. If indicated, take smears from the ileocaecal junction and intestinal mucosa for microbiological and parasitological examinations. Neoplasia (tumorous growth) may be evident and may have an infectious cause in some species (for example, bovine leukosis). Examine the lymph nodes of the abdominal viscera carefully.

10.6 Neurological signs

There are a number of causes of neurological disturbances, these can be the result of physical, physiological or pharmacological impacts on the central or peripheral nervous system, for example, cystic lesions in the brain, toxic or viral encephalitis, peripheral nerve damage. The clinical signs observed depend on the location of the lesions or the severity and nature of the physical or chemical disturbance but are typically exhibited by tremors, circling, changes in consciousness (that is, excitability, somnolence or depression), abnormal movement and behaviour (that is, head pressing, paresis, ataxia) and/or recumbency. There are also a range of metabolic disorders and systemic diseases which may directly affect a range of organ systems but are represented by clinical signs suggestive of a neurological disorder, for example, staggering in grass sickness and tremors in milk fever. Damage to the peripheral nervous system, for example, trauma to the radial nerve associated with a fractured humerus (front limb bone), does not usually affect consciousness but will result in temporary or permanent changes to movement. Pain associated with colic and other disorders may also result in abnormal behaviour therefore it is important to take a good clinical history and to perform a thorough clinical examination to identify the cause of the problem. In some diseases, or if the animal is nervous or agitated, it is important to ensure full restraint of the animal, with the assistance of a competent handler, before samples are collected. Special precautions must be taken if rabies is suspected, that is, staff should be vaccinated for rabies if exposure to rabid animals is likely and protective clothing should be worn when collecting samples. Sample collection guidelines and requirements for the handling of suspect cases of rabies will be determined by the local veterinary and public health authorities. If unsure, please contact your regional veterinary officer before handling potentially infectious material.

The epidemiological pattern of the disease under investigation can give some indication of the most likely cause of neurological disturbance in a group of animals but it may be difficult to diagnose the cause of the problem in an individual animal unless a detailed neurological examination is carried out. This may require the involvement of a specialist.

1 Clinical history. How many animals and which species are affected? Which age groups are involved? When were the clinical signs first observed? Are the clinical signs progressive? What is the vaccination history of the animal(s)? What changes in management have occurred recently (that is, diet, grazing, housing, water supply and so on)?

2 Clinical examination. Observe the animal(s) from a distance, watch for changes in gait and posture, check for changes in behaviour and other signs of disease. Are there any ectoparasites present or any evidence of animal bites? Perform a routine health check (temperature, pulse, heart rate and respiration). Check the faeces, urine and mucous membranes. TAKE APPROPRIATE PRECAUTIONS IF RABIES IS SUSPECTED.

3 Sample collection. Take samples to assess the general health status of the animal (faeces, urine, blood), it may be advisable to collect specific samples for diseases common to the area, for example, East coast fever (cerebral complications). In some cases, a thorough necropsy may be required to locate brain lesions and allow confirmation of a diagnosis.

Table 10.10 summarizes some causes of common clinical signs which may be associated with a neurological disorder. The table is not meant to be comprehensive but aims to give some idea of the range of possible causes and the common metabolic and infectious diseases that need to be considered.

Bibliography

Constable, P., Hinchcliff, K W., Done , S., Gruenberg, W. (2016) Veterinary Medicine, 11th edn. A Textbook of the Diseases of Cattle, Horses, Sheep, Pigs and Goats, 2 vols. Saunders Ltd, Philadelphia, PA.

Merck Veterinary Manual (n.d.) https://www.merck vetmanual.com/.

Table 10.10 Summary of some common causes of neurological disturbance in livestock.

Principle sign	Location of lesion and comments	Example
Tremor	Can be moderate and generalized but potentially leading to paralysis (a), or intention tremor (b)	a) Shaker pigs, border disease in lambs b) congenital cerebella hypoplasia (neonates)
Convulsions	Can have independent episodes associated with a lesion in the cerebral cortex (a) or continuous and intermittent leading to paralysis (b)	a) Idiopathic or traumatic epilepsy, b) encephalitis associated with increased intracerebral pressure. May also be seen with hypomagnesaemia in cattle and hypocalcaemia in piglets
Compulsive rolling and sweating	Can be due to disturbance in balance (a) or associated with pain (b)	a) Vestibular disease due to infection or toxic change b) can be seen with colic in horses
Compulsive walking and head pressing	May be associated with the end stage of systemic disease and altered mental state	This can be seen in cases of chronic toxicity, e.g. ragwort poisoning in horses and other causes of liver damage
Circling	Variable presentation, can be due to vestibular disease (may also have nystagmus) (a), may be associated with lesions in the inner ear or vestibular nuclei in the brain (b), can also result in deviation of the head and/or compulsive behavior with lesions in the cerebrum or medulla (c)	a) Brain abscesses due to *Listeria monocytogenes* and other bacteria, b) otitis media due to infection or trauma/ fractures of the skull causing damage to the cranial nerves, c) brain abscesses and cysts in cerebrum or medulla (could be due to a tapeworm cyst, i.e. as in gid or bacterial abscess)
Inability to chew	Can be due to a sensory deficit (a), damage to focal nerves (b), lesions in the mouth (c)	a) Peripheral nerve damage due to trauma to head/jaw or toxin, b) guttural pouch mycoses in horses, toxins c) wooden tongue in cattle due to *Actinobacillus* sp.
Inability to swallow	Variable signs but usually drops food or regurgitates (a), may show distress if unable to swallow i.e. after peripheral nerve damage (b)	a) Could have a foreign body, tumour causing a blockage, b) could be caused by trauma, toxins (e.g. botulinum, tetanus) and also rabies (always be careful when examining the mouth)
Weakness and staggering	Could be associated with primary neurological disease (a) or due to metabolic crises or toxic insult (b)	a) Prion (BSE, scrapie) and viral causes b) milk fever, peracute systemic infection, tick paralysis
Flaccid paralysis	This can involve all four limbs (a) or can be localized to one or two (b)	a) May occur as part of the paralytic signs associated with rabies induced spinal cord meningitis or can occur secondary to fractures of spinal vertebrae b) localized trauma
Tetany	May be generalized due to a toxic insult or a metabolic disease or can occur secondary to acute spinal cord trauma	Tetanus, hypomagnesaemia, trauma

chapter 11

Wildlife health and disease surveillance

Matilde Tomaselli and Patricia Curry

11.1 Introduction

Wild animals are susceptible to the same range of diseases that affect closely related domestic species. Some viral diseases may be species-specific but many, including foot and mouth disease, are readily transmitted between wild and domestic ruminants and other wildlife and livestock. Many infectious pathogens affect a range of species (that is, multi-host pathogens), and wild populations may act as early warning or 'sentinel systems' for the presence of emerging diseases or diseases new to an area (for example, waterfowl sentinel for highly pathogenic avian influenza viruses).

Wild animal populations can be reservoirs for common pathogens and can also be spillover hosts when pathogens are transmitted from domestic animals to immunologically naïve wildlife populations. Spillover hosts can also transmit back, or spill back, the infection to a potential maintenance host. These mechanisms are complex and it is often unclear whether (1) the wild animals were the source of the disease for the domestic livestock, (2) the domestic animals were the source of the disease for the wild population, or (3) both populations were infected by an outside source (for example, through a shared water supply).

Depending on the ecosystem, there might be specific seasons in which resource scarcity can lead to increased interaction between wild and domestic animals, and to enhanced spillover of pathogens in both directions (domestic to wildlife and vice versa). For example, during times of drought, wild bovids and other wild animals frequently move close to watering holes used by domestic cattle, resulting in more direct contact between wild and domestic species. Similarly, pasture loss during prolonged and more intense drought periods due to the warming climate can force cattle to feed in forested areas, allowing more opportunities for pathogen exchange among species. Table 11.1 lists examples of diseases that may be transmitted between wild and domestic bovids, and between other wild and domestic species.

Many zoonotic diseases can also be carried and/or spread by wild species. For this reason, a full study of disease dynamics and risk factors warrants a 'One Health' approach that engages professionals from the human and animal health sectors, as well as disease ecologists, wildlife experts and subject specialists. Disease transmission between humans and wild animals may occur by direct contact, but indirect transmission routes are more common, including sharing of water sources (for example, leptospirosis), contamination of food (for example, Salmonellae) or bedding areas (for example, ringworm, ectoparasites), passage through arthropod vectors (for example, ticks, flies and mosquitoes transmit a range of diseases; see Chapter 13), or contact with infected domestic

Table 11.1 Examples of diseases that can be transmitted between wild and domestic animal populations.

Family/class (species)	Disease name (aetiology)	Comments
Bovidae (cattle, sheep, goats and relatives)	Leptospirosis (*Leptospira* spp.)	Transmission though direct contact (bacteria are passed in urine) and waterborne; various serovars exist, some transmitted via specific hosts (i.e. *L. pomona* in pigs), others carried by various wildlife species (especially rodents). Specific serovar can be zoonotic.
	Tuberculosis (*Mycobacterium bovis*)	Found in wild and domestic animals, wildlife reservoirs (e.g. badgers, brush tailed possums, deer) can make it difficult to control. Zoonotic.
	Salmonellosis (*Salmonella* spp.)	Various serovars exist; rodents and birds can contaminate feed. Specific serovar can be zoonotic.
	Anthrax (*Bacillus anthracis*)	Regional outbreaks, land may remain contaminated with spores for decades; can occur after heavy rainfall. Zoonotic.
	Brucellosis (*Brucella abortus*)	Asymptomatic carriers can occur; high risk of transmission during parturition. Zoonotic.
	Anaplasmosis (*Anaplasma* spp.)	Can be transmitted widely where competent arthropod vectors are present.
	Rabies (Rhabdovirus)	May be endemic in some wildlife species, especially canids; domestic and wild bovids are susceptible. Zoonotic.
	Foot and mouth disease (Picornavirus)	Transmission through direct contact or water/windborne; may have wildlife reservoirs (e.g. wildebeest in Africa).
	Malignant catarrhal fever (Gamma herpesvirus)	May have wildlife reservoirs; clinical severity varies from species to species.
	Parasitic diseases (endo- and ectoparasites)	Exchange of parasites can occur between wild and domestic bovids.
Suidae (pigs, warthogs)	Highly susceptible to foot and mouth disease (amplify virus); Aujeszky's disease; African swine fever and hog cholera; vesicular stomatitis; vesicular exanthema	Peccaries seem resistant to some viruses common in domestic pigs; African wild swine may carry diseases that impact domestic pigs.
Camelidae (camels and relatives)	Susceptible to the same diseases as Bovidae	Domesticated camels are highly susceptible to sarcoptic mange.
Cervidae (deer, caribou, moose, and relatives)	Most species susceptible to viral diseases such as bluetongue, epizootic haemorrhagic disease, malignant catarrhal fever and other diseases carried by bovids	Wild deer are native to Asia, North and South America, and Europe. Some common species are semi-domesticated and classed as livestock (e.g. red deer).

Table 11.1 *continued*

Family/class (species)	Disease name (aetiology)	Comments
Equidae (horses, donkeys, and zebras)	Susceptible to vector-borne diseases, such as equine encephalitis	Affects most wild equids; vector control is important in the control of vector borne encephalitis; mosquitoes and birds often play an important role in the disease ecology.
	African horse sickness (Orbivirus)	Spread by midges.
	Glanders (*Burkholderia mallei*)	Zoonotic.
Aves (wild and domestic birds)	Avian influenza (Orthomyxoviruses)	Low pathogenic (LP) strains are distributed worldwide with a number of migratory wild birds acting as reservoirs; highly pathogenic (HP) strains derive from mutations of LPs. Severe systemic disease with high morbidity and mortality for HPs. Transmission through direct contact and airborne. Some strains are zoonotic and cause mortality in humans.
	Avian pox (Poxvirus)	Often mild and self-limiting disease but some outbreaks have high mortality; most bird species are susceptible; worldwide distribution but higher infection rates in temperate and warm climates; transmitted through direct contact, environmental contamination and vectors (mosquitoes).
	Newcastle disease (Paramyxovirus)	Highly contagious and severe disease with worldwide distribution; some wild birds can carry the virus without becoming ill (e.g. pigeon-like birds shed intermittently the virus for 1 year); transmission through direct contact with infected or carrier birds. Zoonotic.
	Avian cholera (*Pasteurella multocida*)	Transmitted through direct contact or environment contamination; wild birds but also mammals can carry the infection; sudden mortality is observed in the acute form of avian cholera, while localized infections are evident in the chronic form.

animals (for example, rabies). Habitat modification and human encroachment into wildlife habitat can also lead to complex interactions that increase the risk of exposure to zoonotic pathogens for both humans (for example, Ebola virus) and naïve wildlife (for example, human-associated respiratory infections in populations of endangered great apes).

11.2 Practical considerations for designing a disease survey in wildlife

Although the approach to wildlife disease surveillance is similar to that taken with livestock, it is usually more difficult to obtain epidemiological information from free-ranging species. This stems from several practical and logistical challenges. Some of the 'special problems' that limit the detection of diseases in wildlife include, for example, difficulties in defining and achieving an adequate sample size, selection bias in the sampling methods, difficulty in obtaining good quality samples and measurement bias in the diagnostic tests used (that is, lack of validated tests for wild species).

Wildlife surveillance can be conceived as a designed survey (active or targeted surveillance), for example, with the objective of identifying one or more diseases and/or understanding disease prevalence. This relies on the active collection and analyses of data and biological samples from live and/or dead animals (for example, by capturing live animals, by obtaining samples from harvested animals or through field disease investigation activities). In local and traditional knowledge systems, participatory epidemiology (PE) tools can be also used as a form of disease surveillance. In PE, the epidemiological intelligence on wildlife populations, including population health status and the presence of diseases, is gathered in the form of ethnoveterinary knowledge by interviewing key informants (for example, hunters, trappers). This process relies on the use of participatory tools that allow for both qualitative and quantitative assessments (see example in section 11.3).

More commonly, wildlife health and disease surveillance relies on routine reporting, as opposed to the active investigation, of disease. This is called passive surveillance (also known as general or scanning surveillance). For example, an animal is found dead and is submitted to a central laboratory where standardized necropsy, examination of samples and testing are carried out. A diagnosis or suspected diagnosis is determined and a detailed report is written, which is then usually entered into a database. Epidemiologists and others can access the central database to generate information on characteristics, occurrences and patterns of disease at local or larger geographical scales.

Such data are potentially valuable long into the future.

There are, however, issues and biases to consider when relying solely on passive surveillance for disease identification in wild animals. For example, in the wild, carcasses of dead animals tend to be removed or cached by scavengers very quickly (within hours as opposed to days). This means that passive surveillance is likely to detect only a fraction of the total individuals that have died from a disease, if any at all (that is, in the case of small-sized species and carcasses in remote/sparsely populated regions). Additionally, larger or more charismatic animals that are found dead are more likely to be reported than other species. Conversely, there are rare cases in which passive surveillance might lead to overestimation of certain diseases. For example, rabid animals are less likely to avoid danger and have greater risk of being killed by vehicles. The prevalence of rabies in road-killed animals might, therefore, be higher than the prevalence of rabies in the general wildlife population of an area. Selection bias and measurement bias both need to be carefully considered during analyses of wildlife data from passive or active surveillance. For more information, refer to Wobeser (2007) in the bibliography.

The design of a wildlife disease survey will depend on the disease to be studied, the expected prevalence of the suspected disease in the targeted population and the size of the population(s) to be sampled. The principles of survey design are well outlined in various textbooks but it is preferable to discuss any intended survey programme with a statistician, veterinary epidemiologist and other professionals, as well as the laboratories involved in the testing of the samples. When PE tools can be applied, the expected prevalence of the targeted disease(s) can be estimated through participatory techniques and the data used for sample size calculations.

When collecting biological samples and data from live wild animals, it is necessary to work with specialized teams that are experienced in wildlife capture techniques, including the use of tranquillizer darts and humane trapping, as well as specific skills relevant to the data being collected. To minimize stress for the animals and to maximize the sample collection and data quality, it is advisable to clearly define the role of each team member and be ready to manage critical situations (for example, prolonged recovery time from sedation or anaesthesia). Additionally, when there is potential for unexpected lesions, being prepared to collect extra diagnostic samples (for example, swabs, biopsy, smears) can be important for identifying a new disease.

A range of diagnostic samples and data can also be collected from dead wild animals. Examples are when wildlife is hunted for subsistence, recreation (sport hunting/fishing), or as part of population control programmes. Understanding the context and engaging with the appropriate stakeholders is an important aspect to consider for the success of the sampling and, ultimately, the surveillance programme. For example, working with hunters to understand logistics and feasibility for sample collection and sample storage in the field is an essential step in the successful design and implementation of a hunter-based sampling programme.

To maximize the immediate and the long-term benefits of wildlife disease surveillance, it is strongly recommended to store the data and samples appropriately (for example, data stored in digital format, samples identifiable with unique numbers) and to archive sub-samples for future studies. Assessing and understanding the health status of wildlife populations can provide useful information about the integrity of ecosystems and is of particular importance for implementing effective wildlife conservation and management plans. The considerations noted in this section highlight the importance of having ongoing knowledge of the general health of wildlife populations rather than focusing solely on disease. The following section provides an example of participatory wildlife health and disease surveillance implemented in a remote and resource-constrained setting of Canada's North.

11.3 A practical example from Canada's Arctic: the participatory muskox health surveillance in Cambridge Bay, Nunavut

In the Arctic, implementing wildlife surveillance activities is an extremely difficult endeavour due to the harsh environmental conditions and logistical challenges. However, there is urgent need for wildlife health and disease surveillance in Canada's north owing to the declining numbers of key species and the important roles these play in the livelihoods of indigenous peoples. Hunting wild game, or 'country foods', is essential for the livelihood of northern peoples. Commercial food stores in the north have a limited selection of products, and what is available there is expensive. For northern indigenous peoples, wild animals are also profoundly connected with cultural and social values. In this context, wildlife health is closely tied to the physical and mental health of many northerners, and there are growing concerns about zoonotic pathogens (for example, brucellosis) and the sustainability of arctic wildlife.

Wild animals range throughout vast and remote areas of the Canadian Arctic where temperatures remain well below freezing for most of the year and where the dark season during winter is characterized by limited (or no) sunlight. In such a setting, accessing animals for surveillance is logistically difficult (or impossible in some areas/seasons) and the costs are extremely high and prohibitive. Collaborating with multiple stakeholders, including resource users, and integrating quantitative and qualitative approaches and knowledge are essential to overcome these barriers.

In the Arctic, northern peoples have maintained continuity with the traditional use of renewable resources. Although technology has changed over time, people continue to travel on the land and observe and harvest wild game throughout the year. This means that many local residents (for example, elders, hunters) possess holistic experiential and traditional knowledge about the environment and animals they depend on for subsistence. Additionally, they can access biological samples from the animals they harvest at temporal and spatial scales that would not be feasible to achieve otherwise. Collaborating with local people for wildlife sampling and to understand context through their observations is, therefore, crucial for improving wildlife health and disease surveillance. Such an integrative approach has been piloted and implemented since 2014 in the Arctic community of Cambridge Bay (Iqaluktutiaq, Victoria Island, Nunavut, Canada) to understand the health and disease status of muskoxen (Figure 11.1).

Muskoxen (*Ovibos moschatus*) are non-migratory ungulates and key species in the northern ecosystem. In Cambridge Bay, muskoxen are harvested for subsistence and revenue (that is, sport hunting and commercial harvest). These mammals are also a particularly important source of country foods when other wild game, especially caribou, are not available, and for this reason they are extremely valuable with respect to local food security. Several events, including anecdotal reports of local muskox decline, have raised concerns regarding the health status of muskoxen in the area. These led to the initiation of the participatory muskox health surveillance programme in Cambridge Bay.

In this programme, muskox health and disease status are investigated by combining local and traditional knowledge from muskox hunters (key informants) with scientific knowledge derived from analyses of biological samples, targeted scientific studies, or field disease investigations. Hunters actively participate in the programme not only by collecting samples from the harvested muskoxen, but also by providing their knowledge and observations on the status of muskoxen in the area (Figure 11.2).

Social science methods combined with PE techniques provide the framework for robust collection of hunters' knowledge on muskox health and diseases in the form of epidemiological data. Semi-structured interviews (individual interviews followed by group interviews) with the application of PE techniques are used to gather both qualitative and quantitative data. For example, proportional piling tools are used to gather information on the relative prevalence of diseases observed in muskoxen as well as data on population demography (structure and abundance) (Figure 11.3); mapping is used to georeference disease outbreaks; and a combination of seasonal calendar and proportional piling can define epidemic curves of disease/mortality outbreaks. Data are then interpreted by scientists together with the hunters in feedback sessions that are organized as the last phase of the interview process.

The interviews facilitated the co-design with hunters of the hunter-based sampling programme. Specifically, information about the challenges that hunters face in the field has enabled the scientists and hunters to collaborate in finding feasible solutions and approaches. Compact, lightweight sampling kits are given to hunters to take along during the hunt.

The pre-packaged and pre-labelled sampling kits allow for easy collection of a standardized set of samples from the harvested muskoxen, with the core set of samples including the left hind leg (or metatarsus), blood-saturated filter-paper strips, skin with hair from the rump, and faeces (Figure 11.4). Each kit also includes a form to record additional data about the collector and the sampled animal (that is, hunting site, the date, age and sex of the animal, a measure of body condition status, and any abnormal findings noticed when butchering the carcass). The sampling kits are flexible tools because they can be modified over time and other samples can be added. For example, for the purpose of a toxicology study, liver, kidney and muscle tissues have been temporarily added to the list of samples; also the lower jaw has been added to the list after observations of unusual broken teeth in muskoxen. Any changes made to the sampling kits need to take into account hunters' feedback regarding the feasibility of sample collection in the field. The continued communication and co-design of the sampling programme with the

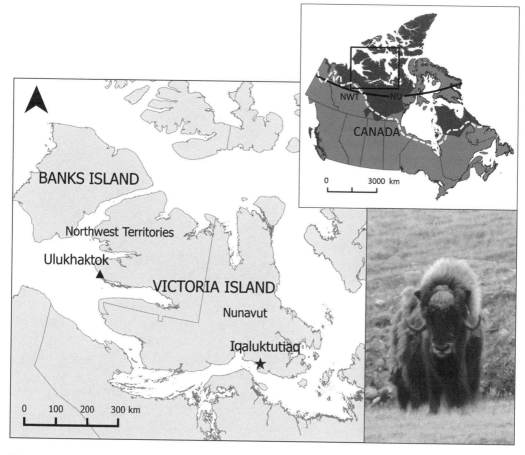

Figure 11.1 Map of Victoria Island in the Canadian Arctic Archipelago showing the only two settlements established on the island, Iqaluktutiaq, or Cambridge Bay, Nunavut (study area) and Ulukhaktok, Northwest Territories (approximately 1700 and 400 people, respectively). Victoria Island has a surface area of approximately 217 km², a similar size to that of the United Kingdom. At top right, Victoria Island is georeferenced within Canada (squared box); the current known distribution of muskoxen in Canada is shown in dark grey (information from Kutz et al., 2017); the Arctic Circle is marked with a black solid line (above which temperatures remain well below 0°C for most of the year and limited, or lack of, sunlight characterizes the lengthy winter season); and the "tree line" (line above which trees do not grow) is indicated with a dashed white line. At bottom right the picture of an adult male muskox from Victoria Island. Map generated in QGIS 2.8.9. Figure reproduced from Tomaselli (2018).

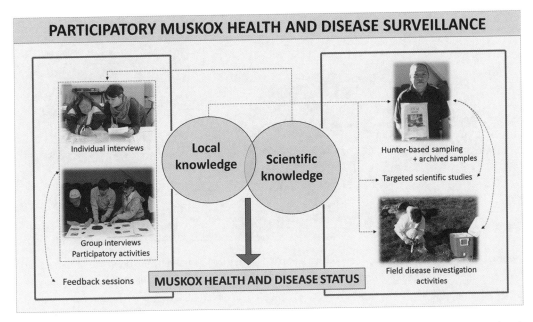

Figure 11.2 Schematic representation of the participatory muskox health surveillance programme developed in the community of Cambridge Bay in the Canadian Arctic. Muskox health and disease status are assessed by combining local knowledge with scientific knowledge. The dotted lines represent how the different components of the system relate to each other. Illustration: courtesy of Matilde Tomaselli.

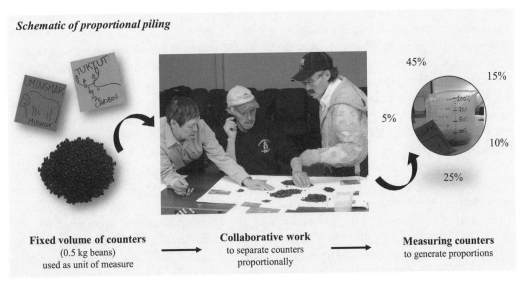

Figure 11.3 Schematic representation of the proportional piling technique used in the group interviews performed with participants of the community of Cambridge Bay (Victoria Island, Nunavut). Proportional piling generally uses 100 counters (beans) as unit of measure which are then counted individually to generate proportions; in this project, we used a larger amount of beans and measurement was done using a measuring cup with a percentage scale. Figure reproduced from Tomaselli et al. (2018a), with kind permission of Elsevier.

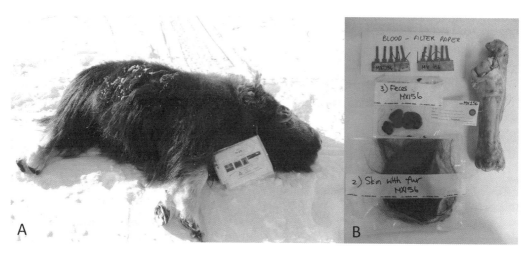

Figure 11.4 (A) A hunted muskox cow with a sampling kit that will be used by the hunter to collect a set of biological samples when butchering the carcass. (B) The core set of samples collected using the sampling kit: blood-saturated filter-paper strips, faeces, left metatarsus (or left hind leg), and a piece of skin with hair from the rump. See also Plate 44. Photos: courtesy of Matilde Tomaselli.

hunters is essential for hunters' compliance with the programme and for the quality and completeness of samples and data collected.

Once the sampling is completed, hunters return the sample kits to a central holding facility (for example, conservation officer's location or Hunters and Trappers Organization's work site) and are provided with a monetary reward for their efforts. The samples are then gathered and sent to southern labs for examination and testing. Once results are compiled, the final and vital step is to report findings back to all stakeholders, including the hunters and the community, so that people are informed about the health of muskoxen in their area and so that management actions can be implemented if needed.

In Cambridge Bay, the interview process has gathered important historic and current information on the status of muskoxen in the study area. This information would have been missed otherwise. Local knowledge has enabled the identification of a decline of muskoxen in the Cambridge Bay area and has characterized the magnitude and time of occurrence of this decline. Changes in demography (that is, sex and

age classes), body condition status, morbidity, and mortality have also been identified through the interview process, along with the relative prevalence and trend over time of endemic and emerging syndromes observed in muskoxen. Finally, for the period 2010–2014, a minimum of 120 mortality events consistent with disease outbreaks have been newly discovered through the interviews. This is stark contrast to only ten dead muskoxen having been identified in 2010 through standard passive surveillance. The hunters' knowledge has proved essential to better understand the status of muskoxen in the area, to formulate hypotheses regarding the mechanisms of the decline, and to prioritize future studies.

Additionally, hunters' knowledge has the potential to increase the timeliness for detecting changes both at individual and population level (that is, provide an early warning system). For example, hunters had already observed scabby lesions on the nose and mouth of muskoxen in 2004, but these were not recorded until 2014 when the participatory muskox health surveillance programme started. Further, the pathogen was not found until hunters who participated in the hunter-based sampling programme observed those same lesions. This triggered a field disease investigation that allowed the pathogen responsible of those reported lesions (orf virus) to be scientifically identified, and that also identified other diseases relevant in the system (that is, brucellosis, lungworm infection) (Figure 11.5).

This last example, among others, illustrates well how complementary use of local and scientific knowledge can greatly enhance the outcomes of a wildlife disease surveillance system. In addition, including resource users in the knowledge-generating process fosters dialogue and trust among parties and promotes shared ownership and shared responsibility. These can improve the sustainability of the surveillance system over time, as well as co-management outcomes for the species.

The integrative and participatory muskox health surveillance programme implemented in Cambridge Bay has been transferred to other northern communities with the aim of increasing the geographic scope of the surveillance system. Although these efforts are in an early stage of implementation, a similar approach is now being used by these communities to assess the health status of caribou in the terrestrial environment and of narwhal in the marine environment.

The surveillance approach described here can also be transferred to other settings and other wildlife species. In such different contexts, the programme will need to be tailored to local realities and take into account the resources available, as well as the challenges and opportunities that exist. The checklist and considerations, in Box 11.1, apply widely and can guide the development of a participatory wildlife surveillance programme to help generate positive outcomes.

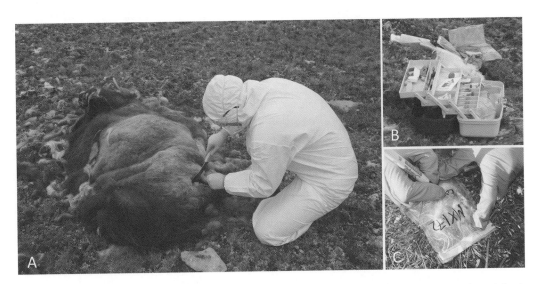

Figure 11.5 Examples of field disease investigations. (A) Field necropsy of a muskox cow found dead. The individual performing the necropsy is wearing personal protective equipment (PPE: goggles, mask, gloves, and disposable coveralls). (B) A toolkit for wildlife disease field sampling: necropsy and sampling tools, extra PPE, disinfectants, and other materials are kept in a modified tackle box that is compact and easy to transport. (C) Every sample collected in the field is stored in a leak-proof bag and identified immediately with the identification number of the animal and a brief description of the tissue sampled. Photos: courtesy of Matilde Tomaselli.

Box 11.1 Checklist for professionals engaged in participatory wildlife surveillance

Knowledge about the local context, culture and traditions

Methods implemented must respect participants' cultural norms and beliefs. For best chance of success, sampling and surveillance/monitoring activities must not violate these and should blend with practices already in use.

Co-identification of relevant stakeholders for the surveillance system

Community engagement is vital beginning at the early stage of the programme. Collaborative efforts are needed to identify all relevant stakeholders.

Co-design the surveillance programme (for example, identification of objectives, resources available, limitations and constraints, data management plan, evaluation plan)

The community/ies involved should actively participate in creating the scope and design of the programme. To ensure long-term sustainability, the purpose of the surveillance activities must be clear and have strong local relevance.

Capacity-building for data gathering, analyses, and communication

The sustainability of the programme is ensured by building local capacity for all components of the surveillance. A detailed plan for capacity-building should be in place for any participatory wildlife surveillance programme.

Implementation of active surveillance activities (for example, local knowledge gathering with participatory epidemiology or PE tools, sample collection by hunters and field disease investigations)

- Personnel involved in PE activities (gathering, analyses or reporting) must be familiar with the ethical principles that apply when working with and gathering knowledge from people;
- Personnel involved in sampling activities must be familiar with safe handling and sampling procedures to minimize health hazards.

For further details refer to next section 'Additional practical considerations for data gathering through local knowledge, hunter-based sampling and field disease investigations'.

Flexibility, adaptability, and optimization

Preparing as much as possible for challenges and unexpected events can help optimize results.

Timeliness in data analyses and communication

A local champion for the programme who is charged with the task of ensuring regular communication can facilitate good and constant communication. Meeting in person at planned times and having all team members report results back to community/ies in person are other vital aspects of communication.

Co-design of appropriate response and mitigation measures

Response is essential. Co-designing the response strategy with all stakeholders is critical for its successful implementation, as well as the success of the overall programme.

Stakeholder and team feedback and programme evaluation

An evaluation and feedback plan should be in place both throughout and at the end of the programme to allow for the continued optimization of the surveillance.

Collaborative programme re-design based on feedback and evaluation

The surveillance programme should not be viewed as fixed and not modifiable; rather, it should be envisioned as a flexible tool that can be adapted according to the modified needs and circumstances. The collaborative re-design of the programme is a process to achieve improved outcomes.

For more information on the application of participatory wildlife health and disease surveillance, refer to Tomaselli (2018) in the bibliography.

11.4 Additional practical considerations for data gathering through local knowledge, hunter-based sampling and field disease investigations

Local knowledge gathering using PE tools

Selection of the correct key informants is of paramount importance in order to collect good quality data. Collaboration with local organizations and leaders is essential in this phase. Methods for gathering local knowledge can vary depending on the objective of the surveillance/monitoring. To collect robust data on health parameters as well as diseases of the targeted species, it is advisable to interview participants until thematic saturation is reached, apply triangulation techniques, and validate the information with participants. The research team must be knowledgeable about the ethical principles that apply when gathering knowledge from people (and possibly from different ethnic groups). In addition, the team must be familiar with social science and PE methods and techniques. Relevant additional resources listed at the end of the section are the website of the Participatory Epidemiology Network for Animal and Public Health (PENAPH) and Tomaselli et al. (2018a). Local knowledge can inform the components of the surveillance that are based on sampling and vice versa. For example, information from local knowledge holders can trigger a targeted disease survey or investigation (for example, the above-described observation and ultimate identification of orf virus); conversely, information derived from sampling can guide the collection of local knowledge (for example, focused questions to understand the ecology of a specific disease and its impact).

Hunter-based sampling

The sample collection process must be practical, straightforward, and doable under the environmental conditions and other constraints that exist in the setting. Sampling kits and samples must be easily transportable in rough conditions and must maintain their integrity during any transport or storage that will be required. If the system involves incentives for local participants, the incentive distribution process must be simple, smooth and immediate upon sample submission. Co-design of the sampling programme, including the training, with hunters and other stakeholders, is essential. Keep in mind that sampling wildlife is inherently difficult and circumstances or conditions beyond the collector's control often compromise samples. Preparing as much as possible for non-ideal samples can help optimize results. For example, if multiple filter-paper blood strips are collected but not all are perfectly saturated, at the time of analysis it may be possible to combine two fully saturated halves to make 'one whole' strip.

Targeted field investigation

In case of reported abnormalities in the sampled/harvested animals or observations of unusual mortalities, a specialized team should act promptly and implement field disease investigation activities whenever possible. Rapid follow-up and action are key to detecting disease outbreaks in wildlife.

Bibliography

Catley, A., Alders, R.G., and Wood, J.L. (2012) Participatory epidemiology: approaches, methods, experiences. The Veterinary Journal 191: 151–160.

Chambers, R. (1994) The origins and practice of participatory rural appraisal. World Development 22: 953–969.

Charron, D.F. (2012) Ecohealth Research in Practice: Innovative Applications of an Ecosystem Approach to Health. International Development Centre, Springer, New York.

Curry, P.S., Elkin, B.T., Campbell, M., Nielsen, K., Hutchins, W., Ribble, C., Kutz, S.J. (2011) Filter-paper blood samples for ELISA detection of Brucella antibodies in caribou. Journal of Wildlife Diseases 47: 12–20.

Kutz, S., Rowell, J., Adamczewski, J., Gunn, A., Cuyler, C., Aleuy, O.A., Austin M., et al. (2017) Muskox Health Ecology Symposium 2016: Gathering to share knowledge on Umingmak in a time of rapid change. Arctic 70: 225–236.

Mariner, J.C., Paskin, R. (2000) Manual on Participatory Epidemiology: Methods for the Collection of Action-Oriented Epidemiological Intelligence. FAO Animal Health Manual. FAO, Geneva.

OIE (2010) Training manual on wildlife diseases and surveillance. Workshop for OIE national focal points for wildlife. Paris, France, http://www.oie.int/fileadmin/Home/eng/Internationa_Standard_Setting/docs/pdf/WGWildlife/A_Training_Manual_Wildlife.pdf.

Participatory Epidemiology Network for Animal and Public Health (PENAPH) (n.d.) https://penaph.net/resources/.

Tomaselli, M. (2018) Improved wildlife health and disease surveillance through the combined use of local knowledge and scientific knowledge. PhD thesis, Faculty of Veterinary Medicine, University of Calgary, Calgary, Alberta, Canada, http://hdl.handle.net/1880/107597.

Tomaselli, M., Dalton, C., Duignan, P.J., Kutz, S., van der Meer, F., Kafle, P., Surujballi, O., Turcotte, C., Checkley, S. (2016) Contagious ecthyma, rangiferine brucellosis, and lungworm infection in a muskox (Ovibos moschatus) from the Canadian Arctic, 2014. Journal of Wildlife Diseases 52: 719–724.

Tomaselli, M., Kutz, S., Gerlach, C., Checkley, S. (2018a) Local knowledge to enhance wildlife population health surveillance: conserving muskoxen and caribou in the Canadian Arctic. Biological Conservation 217: 337–348.

Tomaselli, M., Gerlach, C., Kutz, S., Checkley, S., the Community of Iqaluktutiaq (2018b.) Iqaluktutiaq voices: local perspectives about the importance of muskoxen, contemporary and traditional use and practices. Arctic 71: 1–14.

Wobeser, G.A. (2007) Disease in Wild Animals. Investigation and Management, 2nd edn, Springer, New York.

chapter 12

Antimicrobial resistance: a threat to human and animal health

Niamh Caffrey and Karen Tang

Bacterial, viral, fungal and protozoal infections are common in both human and veterinary medicine, and are often treated with antimicrobial medications. These microorganisms can change or mutate, to develop antimicrobial resistance (AMR), making usual treatments ineffective. In this section, we will focus on a specific type of AMR – antibiotic resistance in bacteria. Bacteria can develop AMR naturally, and resistance genes can also be transferred from one microorganism to another. Antibiotic use has resulted in the rapid rise of AMR, due to selective pressure in which antibiotic-resistant bacteria are more likely to survive compared to non-resistant ones. This allows antibiotic-resistant bacteria to thrive and multiply within the host and in the environment.

Antibiotic-resistant bacterial infections can have serious health consequences in humans and animals. Not only do they result in a delay of effective treatment and an increased risk of treatment failure, they are also associated with increased length of stay in hospital for human patients admitted to hospital, increased risk of requiring admission to the intensive care unit and increased risk of death. Infections with antibiotic-resistant bacteria are also expensive, with medical care, hospitalization, therapy and loss of productivity all contributing to excess cost for public health authorities. By 2050, it is estimated that 10 million human deaths per

year worldwide due to AMR, with a total cost of US$50 trillion (The Review on Antimicrobial Resistance, 2016). AMR is now recognized as one of the biggest threats to human health globally.

12.1 Contributors to AMR

Antibiotic use in humans

Antibiotic use in humans contributes to the development of antibiotic resistance. Individuals have the highest risk of bacteria developing resistance to an antibiotic within 1 month of use of that antibiotic, though the risk is still there for up to 12 months (Costelloe et al., 2010). The association between antibiotic use and resistance is present not just at the individual level, but also at the community, regional and country levels. That is, greater use of antibiotics in a region is associated with greater antibiotic resistance in bacteria isolated from residents of that region, regardless of whether those individuals used the antibiotics themselves. Though antibiotics are an important and effective treatment for bacterial infections, it is estimated that 20% of all outpatient prescriptions for antibiotics are unnecessary, such as when they are given for viral respiratory infections (Belongia

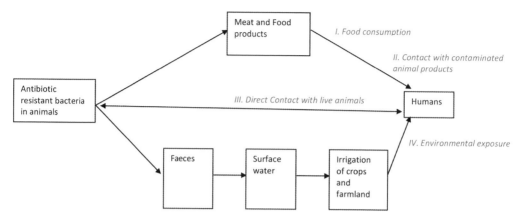

Figure 12.1 Potential mechanisms of transmission of antibiotic resistance from animals to humans.

and Schwartz, 1998). Further, when individuals can access antibiotics without a prescription, as is the case in many countries, this contributes to inappropriate and excess use, leading to increased antibiotic resistance.

Antibiotics used in animals and agriculture

Antibiotics are used commonly for veterinary reasons, to treat pets, farm animals and in aquaculture. They may be used to treat disease (particularly gastrointestinal and pulmonary infections, organ abscesses and mastitis), to prevent disease, and to promote growth and increase feed efficiencies. Many of the antibiotics used in animals are similar, or identical, to those used in humans. About 50% of all antibiotics that are produced worldwide are used for veterinary reasons (Teuber, 2001). Not only does antibiotic use in animals contribute to resistance in animals and agriculture, these resistance genes may also spread to the environment and to humans. It is therefore important to consider veterinary contributions to the antibiotic resistance problem in humans.

Transmission of resistance from animals to humans

Resistant bacteria can infect humans from a number of sources including the environment, other humans and animals. Transmission of resistance from animals to humans occurs through a number of routes including the following (Figure 12.1):

1 direct contact with contaminated animal products
2 cross-contamination between contaminated animal products and other food products in processing plants, kitchens, retail markets, and restaurants
3 direct contact between humans and live animals
4 contamination of the environment, examples include:

  i animal excretions and secretions in the environment
  ii slurry used for fertilization
  iii spills, leakages, and contamination of water sources from slurry.

This can result in contamination of water sources and other food products such as fruits and vegetables.

12.2 What can be done about antimicrobial resistance?

Taking a One Health approach

The concept of One Health recognizes that the health of humans, animals and the environment are interconnected (Figure 12.2). Application of the concept to AMR requires interdisciplinary collaboration and communication between physicians and veterinarians and the engagement of a wide range of expertise including microbiologists, environmental and wildlife experts, farm advisers and social scientists. Owing to the complexity of antimicrobial resistance it is paramount that the issue is addressed by both human health and veterinary authorities in a harmonious and collaborative way.

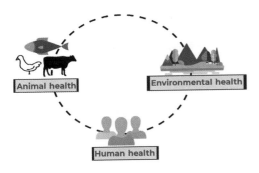

Figure 12.2 The One Health approach considers the interactions in health among humans, animals, and the environment.

Antibiotics and human health

Interventions targeting use of antibiotics in humans

There has been a lot of work done to limit excessive use of antibiotics in human medicine, to slow the rise of AMR. These interventions are often aimed to change prescriber behaviour through education, audit and feedback, promotion of the use of clinical guidelines, restriction of antibiotics (such as requiring an infectious diseases consult, or pre-approval prior to use), and computer assisted decision support tools. Population-level interventions include promotion of vaccinations, to reduce the risk of bacterial infections.

World Health Organization Global Action Plan

The World Health Organization (WHO) has recognized the urgency and importance of addressing the rapid rise of antimicrobial resistance. They have published a Global Action Plan, with five key objectives (World Health Organization, 2014).

1 To improve understanding of AMR, through education and training.
2 To strengthen the knowledge base surrounding AMR through the development and maintenance of surveillance and research. Examples of knowledge gaps include the need to understand how resistance circulates among humans, animals and the environment, and the need to identify non-antimicrobial alternatives for use in agriculture.
3 To reduce infection rates through effective infection prevention strategies.
4 To optimize antimicrobial use in human and animal populations.
5 To make an economic case for sustainable investment, and increasing investment in new medicines, technologies and vaccinations.

In addition to the WHO guidelines on the use of antimicrobials important for human medicine, the World Organisation for Animal Health (OIE) has developed a list of antimicrobial agents of veterinary importance (most recently updated in 2018) (OIE, 2018c), which should be considered when contemplating antimicrobial therapy in animals (Figure 12.3). Antimicrobials of

Figure 12.3 OIE list of antimicrobials of veterinary importance.

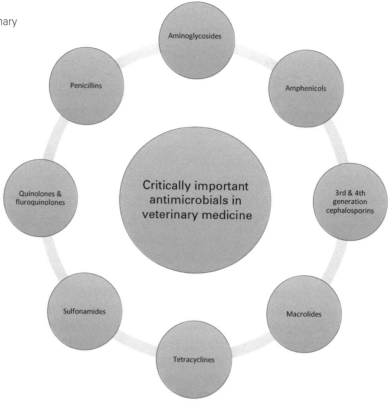

veterinary importance important antimicrobials should not be used as a preventative treatment or in the absence of clinical signs. They should only be used as a first line treatment option if justified with bacteriological testing, and their extra label use should be reserved for instances when no alternatives are available.

Antibiotics and animal health

Antimicrobial use (AMU) in intensive livestock production systems plays a crucial role in controlling infectious diseases. Balancing the need to use antimicrobials to ensure animal health and welfare with the need to reduce AMU due to the risk of developing antimicrobial resistance is a challenge. The evidence that antibiotic use selects for antibiotic resistance is overwhelm-

ing. Bacterial resistance is driven by several selectors, however, it is possible to minimize the transmission of resistant bacteria. Rational and targeted use of antimicrobials will maximize therapeutic effect and minimize the development of AMR, therefore prudent use guidelines, and methods to reduce the use of antimicrobials in food-producing animals are key to reducing dependence on antimicrobials in food animal populations.

Reducing AMU in livestock: how can this be achieved?

Antimicrobial stewardship refers to evidence based prescription of antibiotics to decrease antibiotic overuse or misuse. It involves educating practitioners in veterinary and human disciplines on assessment of the need for antimicrobials and the appropriate selection, dosing, route of administration and duration of antimicrobial therapy.

Prudent use standards outlined by the OIE aim to maintain the efficacy of antimicrobials, ensure their rational use in animals, keep animals in good health, prevent or reduce the transfer of resistant microorganisms within animal populations and from animals to humans and prevent antimicrobial residue in animal products for human consumption (OIE, 2015, Teale and Moulin, 2012). A few guidelines are provided below for the farm and for regional programmes.

ACTIONS ON THE FARM
1 Improving animal health

- Decisions regarding treatment should be made by professional veterinary services or suitably trained individuals. Inaccurate diagnosis of a pathogenic bacteria can lead to overtreatment with antimicrobials, and possibly delay the recognition of serious viral diseases.
- Where feasible, treatment should be at the individual animal level rather than treatment of a population.
- When antimicrobials are to be used to treat infection, accurate calculation of bodyweight is imperative to prevent over- or underdosing. Additionally, considering diseased animals may have a decreased appetite; where medicated feeds containing a therapeutic dose of an antimicrobial are used, intake should be monitored to ensure animals ingest an adequate quantity of the product, and modifications made to the treatment protocol if necessary.
- Clinical experience, the expected susceptibility of the target pathogen, administration route, expected activity at the infection site, and the history of the animals in question will influence antimicrobial selection.

In situations where there is recurrence of disease, or a failure of first line treatment, diagnostic tests should be utilized to determine second-line treatment.

- Narrow spectrum antimicrobial agents should be chosen before broad spectrum products when appropriate.
- Aseptic techniques should be used to minimize the need of perioperative use of antimicrobials.

A note of the use of biocide compounds

Resistance to biocide compounds such as disinfectants, antiseptics, preservatives and sterilants can occur by similar mechanisms to antibiotic resistance. Products previously active against bacterial biofilms are becoming less effective. These include chlorine, quaternary ammonium compounds and aldehydes. The combination of concentration and contact time determines the efficacy of the application. Environmental factors such as temperature can also affect product efficacy. For example, repeated use of low concentrations of a product may allow tolerance to the product to accumulate, ultimately leading to biocide resistance. Usage guidelines for biocide products indicate the dilution of the product, storage, and contact time required for a specific application and these guidelines should be followed carefully for all applications.

2 Improved biosecurity
Biosecurity refers to a set of measures or actions taken to prevent and control the spread of infection and disease. Biosecurity protocols should be in place on all livestock facilities. The following biosecurity guidelines should be implemented where applicable (Figure 12.4).

i Use separate clothes and boots for each unit.

ii Limit access to production areas.

iii Install and use hand washing and disinfection facilities close to the workplace.

iv Remove dead animals in a timely manner.

v Use an all-in-all-out (AIAO) production system (animals of the same age/size enter the production system at the same time (or within a short timeframe) and go through the production system together. Animals all leave the system at the same time, allowing the entire production area to be thoroughly cleaned between production cycles). Using such systems helps to decrease the spread of infectious diseases.

iv Implement regular cleaning and disinfection procedures.

3 Vaccines

The use of an effective vaccination programme is a key component of an animal health programme. Use of vaccines can lead to dramatic improvements in animal health and can lead to significant reductions in the use of antimicrobials. Vaccinations to a wide range of infectious agents are available, although access may depend on region or country regulations.

ACTIONS AT THE REGIONAL OR NATIONAL LEVEL

The OIE recognizes the need for coordinated action between human and animal health and environmental sectors to combat antimicrobial resistance (OIE, 2018a). The OIE Strategy on Antimicrobial Resistance and the Prudent Use of Antimicrobials aligns with the WHO Global

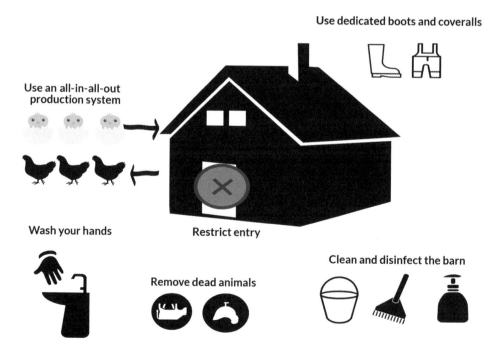

Figure 12.4 Practices to improve biosecurity on farms.

Action Plan and outlines the goals and tactics that the OIE has in place for the fight against AMR. As part of the OIE strategy the onus is on national regulatory authorities to implement regulations relating to the use of veterinary antimicrobials. Regulatory framework to control access, advertising and distribution of antimicrobials should be established. The OIE Terrestrial Animal Health Code (OIE, n.d.) provides recommendations to assist competent authorities with this task. Member Countries of the OIE are provided with tools to help with implementation of the antimicrobial resistance guidelines. These include tools to support evaluation of the performance of veterinary services, and to improve critical competencies for a veterinary service to function effectively.

1   Quality assurance programmes

Many countries have adopted quality assurance programmes to meet increasing consumer preferences for higher standards of animal health, welfare and food safety. Such programmes can incorporate the responsible use of antibiotics on farm, and the maintenance of records of medicine use. Where such assurance programmes are subject to regular audits the measures implemented work towards reducing antimicrobial use (AMU) and encouraging responsible AMU.

2   Veterinary stewardship

There should be no financial incentives between veterinary practitioners, suppliers of antimicrobials and the pharmaceutical industry. This will prevent conflicts of interest that could facilitate unnecessary prescription of antimicrobials. Veterinarians should keep abreast of treatment guidelines provided by national authorities or veterinary professional bodies to assist in making decisions regarding the appropriate antimicrobial, dose and route of administration. Optimal prescribing decisions will take into account regional and local trends in antimicrobial sensitivity.

3   Surveillance

Adequate monitoring of AMU is a necessary component of antimicrobial stewardship. Basic data including the annual weight in kilograms of the active ingredient of the antimicrobial used for each species, the type of use and the route of administration should be recorded. These data will allow evaluation of usage patterns and antimicrobial exposure in animal production systems. OIE standards for harmonization of national antimicrobial resistance surveillance and monitoring programmes can be used to follow trends in AMR, detect the emergence of new AMR mechanisms and provide data needed to conduct risk analyses. The Terrestrial Animal Health Code is a comprehensive document providing guidelines for member countries in how to develop an antimicrobial resistance surveillance and monitoring programme. This includes advice on general aspects such as use of statistically based surveys, sampling strategies, sample size, sample sources, bacteria to target, storage of bacterial isolates, susceptibility testing and recording, and storage and interpretation of data.

Laboratory methodologies for antimicrobial susceptibility testing have been standardized in the OIE Manual for Diagnostic Tests and Vaccines for Terrestrial Animals (OIE, 2018b), allowing for direct comparison of results between different regions or countries. Timely diagnosis of the causative agent when treating infection is the ultimate goal of microbiology. Such point of use diagnostics are becoming more readily available, albeit still cost prohibitive. It is hoped that the cost of such equipment will continue to fall as an increased range of products comes to market, allowing their incorporation into routine diagnostic practices.

12.3 Conclusion

The risks facing the global population from the development of antibiotic-resistant infections are serious. Collaborative efforts to reduce the spread of AMR are necessary if these risks are to be mitigated. Antimicrobial stewardship in both human and veterinary medicine will play a critical role in decreasing these risks. Each country must play their own part in adequately regulating access to and use of antibiotics. Both human and animal health practitioners must make educated and responsible decisions when treatment with antibiotics is necessary. Animal producers should incorporate mitigation strategies to minimize the need for antibiotics when raising livestock, such as incorporating animal health plans and using appropriate biosecurity measures on their premises. The combination of these strategies, if carried out appropriately, should reduce the need for antibiotics in livestock production. Organic or antibiotic-free farming practices are becoming more common, indicating that it is possible to raise livestock successfully without the use of antibiotics. Without a One Health approach to tackling the global problem of antimicrobial resistance the spread of AMR in humans, animals and the environment will become an ever-increasing threat to global health.

Bibliography

Angulo, F.J., Nargund, V.N., Chiller, T.C. (2004) Evidence of an association between use of antimicrobial agents in food animals and anti-microbial resistance among bacteria isolated from humans and the human health consequences of such resistance. Journal of Veterinary Medicine. B, Infectious Diseases and Veterinary Public Health 51(8–9): 374–379.

Bell, B.G., Schellevis, F., Stobberingh, E., Goossens, H., Pringle, M. (2014) A systematic review and meta-analysis of the effects of antibiotic consumption on antibiotic resistance. BMC Infectious Diseases 14(13).

Belongia, E.A., Schwartz, B. (1998) Strategies for promoting judicious use of antibiotics by doctors and patients. BMJ 317(7159): 668–671.

Chang, Q., Wang, W., Regev-Yochay, G., Lipsitch, M., Hanage, W.P. (2015) Antibiotics in agriculture and the risk to human health: how worried should we be? Evolutionary Applications 8(3): 240–247.

Costelloe, C., Metcalfe, C., Lovering, A., Mant, D., Hay, A.D. (2010) Effect of antibiotic prescribing in primary care on antimicrobial resistance in individual patients: systematic review and meta-analysis. BMJ 340: c2096.

Davin-Regli, A., Pagès, J.M. (2012) Cross-resistance between biocides and antimicrobials: an emerging question. Scientific and Technical Review of the Office International des Epizooties 31(1): 89–104.

Kaki, R., Elligsen, M., Walker, S., Simor, A., Palmay, L., Daneman, N. (2011) Impact of antimicrobial stewardship in critical care: a systematic review. ournal of Antimicrobial Chemotherapy 66(6): 1223–1230.

Khachatourians, G.G. (1998) Agricultural use of antibiotics and the evolution and transfer of antibiotic-resistant bacteria. Canadian Medical Association Journal 159(9): 1129–1136.

Lambert, M.L., Suetens, C., Savey, A., Palomar, M., Hiesmayr, M., Morales, I., Agodi, A., Frank, U., Mertens, K., Schumacher, M., Wolkewitz, M. (2011) Clinical outcomes of health-care-associated infections and antimicrobial resistance in patients admitted to European intensive-care units: a cohort study. Lancet Infect Dis, 11(1), 30–38. doi:10.1016/s1473–3099(10)70258–9

Laxminarayan, R., Duse, A., Wattal, C., Zaidi, A. K., Wertheim, H. F., Sumpradit, N., Cars, O. (2013) Antibiotic resistance-the need for global solutions. Lancet Infectious Diseases 13(12): 1057–1098.

Lipsitch, M., Singer, R.S., Levin, B.R. (2002) Antibiotics in agriculture: when is it time to close the barn door? Proceedings of the National Academy of Sciences of the United States of America 99(9): 5752–5754.

Maragakis, L.L., Perencevich, E.N., Cosgrove, S.E. (2008) Clinical and economic burden of antimicrobial resistance. Expert Review of Anti-infective Therapy 6(5): 751–763.

McEwen, S.A. (2012) Quantitative human health risk assessments of antimicrobial use in animals and selection of resistance: a review of publicly available reports. Revue scientifique et technique 31(1): 261–276.

Michael, C.A., Dominey-Howes, D., Labbate, M. (2014) The antimicrobial resistance crisis: causes, consequences, and management. Front Public Health, 2: 145.

Morgan, D.J., Okeke, I.N., Laxminarayan, R., Perencevich, E.N., Weisenberg, S. (2011) Non-prescription antimicrobial use worldwide: a systematic review. Lancet Infect Diseases 11(9): 692–701.

Ocan, M., Obuku, E.A., Bwanga, F., Akena, D., Richard, S., Ogwal-Okeng, J., Obua, C. (2015) Household antimicrobial self-medication: a systematic review and meta-analysis of the burden, risk factors and outcomes in developing countries. BMC Public Health, 15: 742.

Official Journal of the European Union (2015) Commission Notice, Guidelines for the prudent use of antimicrobials in veterinary medicine (2015/ C 299/04). https://ec.europa.eu/health//sites/ health/files/antimicrobial_resistance/docs/2015_ prudent_use_guidelines_en.pdf.

OIE (2015) OIE Standards, Guidelines and Resolution on antimicrobial resistance and the use of antimicrobial agents. http://www.oie.int/fileadmin/Home/eng/ Media_Center/docs/pdf/PortailAMR/EN-book-AMR. PDF.

OIE (2018a) Antimicrobial Resistance, About Antimicrobial Resistance. http://www.oie.int/en/ for-the-media/amr/.

OIE (2018b) Manual of Diagnostic Tests and Vaccines for Terrestrial Animals. http://www.oie.int/en/ international-standard-setting/terrestrial-manual/ access-online/.

OIE (2018c) OIE List of Antimicrobial Agents of Veterinary Importance. http://www.oie.int/fileadmin/Home/ eng/Our_scientific_expertise/docs/pdf/AMR/A _OIE_List_antimicrobials_May2018.pdf

OIE (n.d.) Terrestrial Animal Health Code. http://www. oie.int/international-standard-setting/terrestrial- code/access-online/.

Orand, J.P. (2012) Antimicrobial resistance and the standards of the World Organisation for Animal Health. Scientific and Technical Review of the Office International des Epizooties 31(1): 335–342.

One Health Initiative (n.d.) About the One Health Initiative. http://www.onehealthinitiative.com/ about.php.

One Health Global (n.d.) One Health Global Network Webportal: What is One Health? http://www. onehealthglobal.net/what-is-one-health/.

Page, S.W., Gautier, P. (2012) Use of antimicrobial agents in livestock. Scientific and Technical Review of the Office International des Epizooties 31(1): 145–188.

Ranji, S.R., Steinman, M.A., Shojania, K.G., Gonzales, R. (2008) Interventions to reduce unnecessary antibiotic prescribing: a systematic review and quantitative analysis. Medical Care 46(8): 847–862.

Schwaber, M.J., Carmeli, Y. (2007) Mortality and delay in effective therapy associated with extended-spectrum beta-lactamase production in *Enterobacteriaceae bacteraemia*: a systematic review and meta-analysis. Journal of Antimicrobial Chemotherapy 60(5): 913–920.

Swartz, M.N. (2002) Human diseases caused by foodborne pathogens of animal origin. Clinical Infectious Diseases 34(Suppl 3): S111–S122.

Teale, C.J., Moulin, G. (2012) Prudent use guidelines: a review of existing veterinary guidelines. Scientific and Technical Review of the Office International des Epizooties 31(1): 343–354.

Teuber, M. (2001) Veterinary use and antibiotic resistance. Current Opinion in Microbiology 4(5): 493–499.

The Review on Antimicrobial Resistance (2016) Tackling drug-resistant infections globally – final report and Recommendations. https://amr-review. org/sites/default/files/160518_Final%20paper_ with%20cover.pdf.

Thompson, A.R.C. (2013) Parasite zoonoses and wildlife: One Health, spillover and human activity. International Journal for Parasitology 43(12–13): 1079–1088.

Ungemach, F.R., Müeller-Bahrdt, D., Abrraham, G. (2006) Guidelines for prudent use of antimicrobials and their implications on antibiotic usage in veterinary medicine. International Journal of Medical Microbiology 296(2): 33–38.

van Hecke, O., Wang, K., Lee, J.J., Roberts, N.W., Butler, C.C. (2017) Implications of antibiotic resistance for patients' recovery from common infections in the community: a systematic review and meta-analysis. Clinical Infectious Diseases 65(3): 371–382.

Wilton, P., Smith, R., Coast, J., Millar, M. (2002) Strategies to contain the emergence of antimicrobial resistance: a systematic review of effectiveness and cost-effectiveness. Journal of Health Services Research & Policy 7(2): 111–117.

Wooldridge, M. (2012) Evidence for the circulation of antimicrobial-resistant strains and genes in nature and especially between humans and animals. Revue scientifique et technique 31(1): 231–247.

World Health Organization (2014) Antimicrobial resistance: global report on surveillance. http:// apps.who.int/iris/bitstream/10665/112642/ 1/9789241564748_eng.pdf?ua=1.

World Health Organization (2017) WHO guidelines on use of medically important antimicrobials in food-producing animals. http://www.who.int/ foodsafety/publications/cia_guidelines/en/.

chapter 13

The World Organisation for Animal Health's core missions in relation to laboratories working in the veterinary domain

Jennifer Lasley

The need to fight animal diseases at the global level led to the creation of the Office International des Epizooties, OIE, in 1924. In 2003, the Office became the World Organisation for Animal Health but kept its historical acronym, OIE. The OIE is the intergovernmental organization responsible for improving animal health worldwide and is recognized as a standard-setting organization by the World Trade Organization (WTO). In 2017, the OIE has a total of 181 members and maintains permanent relations with 71 other international and regional organizations and has regional and sub-regional Offices on every continent (Figure 13.1).[1]

13.1 The OIE's core missions

The OIE's mandate is to improve animal health, animal welfare and veterinary public health worldwide. The OIE fulfils this mandate through different core missions: developing international

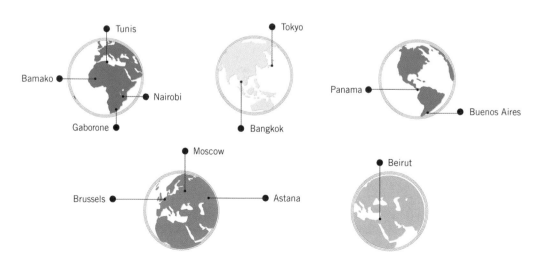

Figure 13.1 OIE's regional and subregional offices around the world, © OIE, 2018.

standards on animal health and welfare; sharing reliable information on the animal disease situation worldwide in real time; collecting, analysing and disseminating veterinary scientific information worldwide; and developing international solidarity to achieve better control of animal diseases in the world.[2]

Developing international standards on animal health and welfare

In order to prevent and control animal diseases, including zoonoses, ensure the sanitary safety of world trade in terrestrial and aquatic animals and animal products, and improve and secure animal welfare by appropriate risk management, the OIE develops international standards on animal health and welfare. These standards are prepared and updated by recognized scientific experts and are adopted at annual general sessions through voting by the World Assembly of Delegates of the OIE.

Sharing reliable information on the animal disease situation worldwide in real time

As a result of increased trade and travel in the era of globalization, infectious diseases can spread more quickly than ever before and up-to-date information on disease occurrence is vitally important. Effective surveillance and enabling early detection of these diseases at their source in animals is therefore crucial so that they can be quickly controlled, thereby protecting animal and human populations. Since its creation, one of the OIE's historic missions has been to ensure transparency and improve knowledge of the global animal disease situation, including zoonoses. This mission is fulfilled on a daily basis thanks to a unique tool, the OIE World Animal Health Information System (WAHIS) that helps

to establish trust through transparency and communication.

Collecting, analysing and disseminating veterinary scientific information worldwide

The OIE collects and analyses the latest scientific information on the prevention and control of animal diseases. This information is then made available to members so that they can apply the most effective methods. The work of the OIE is supported by a worldwide network of expertise in animal health, veterinary public health and animal welfare, including a wide array of national focal points, expanding expertise of the OIE reference centres and diverse scientific expertise in the form of ad hoc and working groups. Permanent exchange of information and the constant strengthening of the scientific and technical competencies of the members of these networks all help to ensure the scientific excellence of the OIE worldwide.

Developing international solidarity to achieve better control of animal diseases in the world

The OIE supports its members and works to ensure the capacity and sustainability of veterinary services in line with the OIE's intergovernmental standards, notably by acting on the quality of the national veterinary services, diagnostic laboratories and veterinary education. To help members deal effectively with health threats, the OIE provides support through a range of programmes, notably within the framework of the PVS (Performance of Veterinary Services) Pathway, aimed at strengthening national animal health systems through systematic assessments during in-country missions.

13.2 The OIE's core missions in relation to laboratories working in the veterinary domain

Each of the OIE's core missions relates directly to laboratories working in the veterinary domain. In the broader context and considering the current trend of globalization, animal health measures have increasing importance to facilitate safe international trade of animals and animal products while avoiding unnecessary impediments to trade.

Laboratories are a cross-cutting function that is critical to the good management of the overall veterinary services. Laboratories play an essential role not only in international trade, but also in every other function carried out by veterinary services, from import to export analysis, from epidemiological investigation to food and environmental residue analysis, from vaccine efficacy to food safety (Figure 13.2).

Improving animal health, animal welfare and veterinary public health worldwide

OIE international standards

The Agreement on the Application of Sanitary and Phytosanitary Measures (SPS Agreement) encourages the members of the WTO to base their sanitary measures on international standards, guidelines and recommendations, where they exist. The OIE is the WTO reference organization for standards relating to animal health and zoonoses.

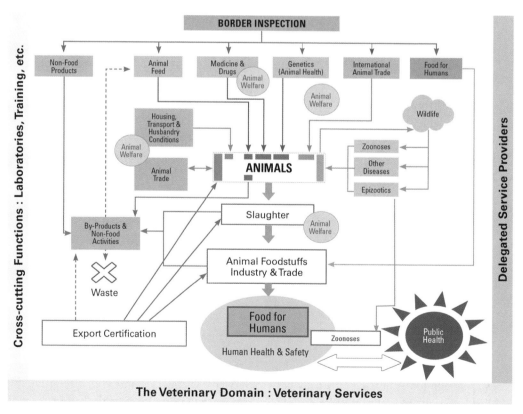

Figure 13.2 The veterinary domain: veterinary services, © OIE, 2018.

Therefore, the OIE international standards relating to laboratories are important to know, understand and implement in order to ensure compliance.

The OIE publishes two *Codes* (*Terrestrial* and *Aquatic*) and two *Manuals* (*Terrestrial* and *Aquatic*) as the principle references for animal health, animal welfare and veterinary public health. The *Terrestrial Animal Health Code* and *Aquatic Animal Health Code* respectively aim to assure the sanitary safety of international trade in terrestrial animals and aquatic animals and their products as well as to provide harmonized standards for surveillance and control of animal diseases. The Manual of Diagnostic Tests and Vaccines for Terrestrial Animals and the Manual of Diagnostic Tests for Aquatic Animals provide a harmonized approach to disease diagnosis by describing internationally agreed laboratory diagnostic techniques (Figure 3.13).[3]

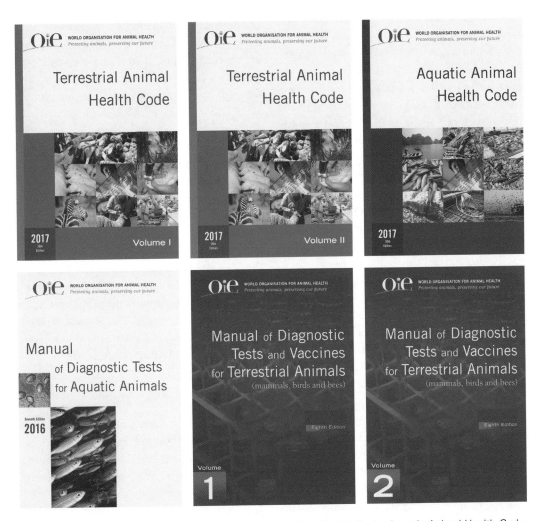

Figure 13.3 OIE's International Standards: Terrestrial Animal Health Code, Aquatic Animal Health Code, Manual of Diagnostic Tests and Vaccines for Terrestrial Animals and Manual of Diagnostic Tests for Aquatic Animals.

Terrestrial Animal Health Code

The glossary of the Terrestrial Code defines a laboratory as 'a properly equipped institution staffed by technically competent personnel under the control of a specialist in veterinary diagnostic methods, who is responsible for the validity of the results. The Veterinary Authority approves and monitors such laboratories with regard to the diagnostic tests required for international trade.'[4]

Laboratories provide the scientific information needed for decision making by the veterinary services. Therefore, the appropriateness of veterinary health actions depends partly on the quality of the laboratories' services. Furthermore, the quality of the analyses depends on the quality of the reagents used, the procedures employed and the competency of the staff. Diagnostic assays provide the basis for decisions with far-reaching consequences and their quality and the legal framework within which they are performed must therefore be clearly defined. Chapter 3.4 of the Terrestrial Code provides standards related to veterinary legislation and specifically Article 3.4.7 states the legislative requirements related to laboratories, facilities and reagents.

The Terrestrial Manual is the reference for harmonized disease diagnosis as it describes the internationally agreed laboratory diagnostic techniques for OIE Listed Terrestrial Animal Diseases.[5] It aims to facilitate international trade in animals and animal products and to contribute to the improvement of animal health services worldwide. The objective is to provide internationally agreed diagnostic laboratory methods and requirements for the production and control of vaccines and other biological products.

The introductory chapters of the Terrestrial Manual (Part 1) set general standards for the management of veterinary diagnostic laboratories and vaccine production facilities. Part 2 comprises specific recommendations and includes eight chapters of recommendations for

validation of diagnostic tests and three chapters of recommendations for the manufacture of vaccines. Part 3 comprises chapters on OIE Listed Terrestrial Diseases and other diseases of importance; and Part 4 is the list of OIE reference centres at the time of publication.[6]

Parts 1 and 2 of the Terrestrial Manual are of interest for anyone working in the laboratory setting. Important information related to the good management of veterinary diagnostic laboratories,[7] including collection, submission, storage[8] and transport[9] of biological materials, biosafety and biosecurity,[10] quality management,[11] validation of diagnostic assays,[12] high throughput sequencing, bioinformatics and computational genomics (HTS-BCG),[13] tests for sterility and freedom from contamination,[14] vaccine production[15] and vaccine banks,[16] is included.

Diagnostic tests are essential tools for confirming the health status of animals and identifying pathogens. They enable the early detection, management and control of animal diseases including zoonosis and facilitate the safe trade in animals and animal products. Part 3 of the Terrestrial Manual provides internationally agreed diagnostic laboratory methods. For all the OIE Listed Animal Diseases as well as several other diseases of importance, the Manual specifies and details the internationally agreed upon diagnostic test methods. The Manual is developed by the Biological Standards Commission and supported by OIE reference centres.

Biological threat reduction

Infectious disease agents and toxins found in animal populations and animal products are a considerable and ongoing threat to animal health, agricultural economies, food security, food safety and public health. By and large most disease outbreaks and food contaminations occur naturally. However, there is also a real risk that disease may be introduced into susceptible human or animal populations following a deliberate or accidental release of an infectious agent or toxin. These 'unnatural' biological threats carry special risks because pathogens may be engineered or released in such a way as to make them more harmful. Although the probability of a deliberate or accidental release may be relatively low, the impact may be catastrophic from the national to the global level.

The response to disease is similar whether it is directed against natural infection, or deliberate or accidental release. Expert investigations carried out by health authorities are needed to establish the cause of a disease outbreak and veterinary laboratories are often the first to discover the source. The most effective and sustainable way to protect against threats from deliberate and accidental releases of animal pathogens is to strengthen existing systems for surveillance, early detection and rapid response, and biosafety and biosecurity in the laboratory and field settings, while fostering scientific networks that work towards altruistic goals. This approach has multiple collateral benefits for animal health, agriculture, public health, poverty alleviation, animal welfare, and economies.[17]

13.3 Sharing, in real time, reliable information on the animal disease situation worldwide

WAHIS

In its leading role to improve animal health and welfare worldwide, one of the OIE's key missions is to ensure transparency of the global animal disease situation. The OIE's WAHIS+ is the global tool to achieve these objectives. WAHIS+ is an open-access internet-based system that processes data on animal diseases in real time and then informs the international community. WAHIS+ enables the collection and the dissemination of data on animal diseases of epidemiological significance in both domestic species and wildlife and is vital for the global dissemination of information on animal disease events, including zoonotic pathogens.

Data reporting in WAHIS+ is reserved for delegates of OIE members and the chief veterinary officers of non-members and their authorized representatives such as national focal points via a secure and restricted-access portal, who use WAHIS+ to notify the OIE of relevant animal disease information. However, the information reported by OIE members in WAHIS+ is accessible and available to the general public. In addition, WAHIS+ consists of an early warning system to inform the international community, by means of alerts, of relevant epidemiological events that occurred in OIE members and non-members and territories, and a monitoring system in order to monitor presence or absence of OIE Listed Diseases over time.[18]

As a part of the monitoring system, six-monthly reports are submitted to inform the presence or absence of OIE Listed Diseases. In addition, annual reports are completed by members to provide information on non-OIE-listed diseases, the impact of zoonoses on humans, animal populations, veterinary services personnel, laboratories and diagnostic tests performed, and vaccine manufacturers and vaccine production. This information is of particular interest to laboratories that seek confirmatory services from other laboratories, that network concerning diagnostic methods or diseases, or that participate in inter-laboratory proficiency testing and, ultimately, adds to the transparency of the animal disease situation in OIE members.

WAHIS+ is an evolving tool which responds to global needs, changes to international standards and has increased pace for innovation and technological change. The OIE continues to look towards the future and evolve WAHIS+ to be easier for OIE members to collect, report and upload data from their own databases. The WAHIS+ interface allows for data to be viewed, analysed and extracted rapidly and in different formats. High-quality and reliable geospatial data, a nimble and flexible platform, interactive and dynamic maps, extended data mining with automatic tools for extraction, and genomic data linked to epidemiological data, are just some of the innovative tools available on the WAHIS+ platform for all to access. Scientists all around the world will have access to these functionalities to undertake comprehensive risk analyses and strengthen disease traceability and contribute to analyses on genetic epidemiology.

13.4 Collecting, analysing and disseminating veterinary scientific information worldwide

National focal point programme for veterinary laboratories

The OIE provides a global capacity building programme for OIE delegates and national focal points. National focal points are designated by the OIE delegate for eight key topics of critical importance to the animal health system: animal disease notification, wildlife, veterinary products and antimicrobial resistance (AMR), animal production food safety, animal welfare, aquatic animals, communication, and veterinary laboratories. The national focal point should assist the delegate in fulfilling his/her responsibilities to the OIE, especially in these areas of specialization, in order to ensure a standard-setting process that is transparent and fully participatory among OIE members.

Regional seminars for national focal points are held in each OIE region and for each topic. These seminars aim to provide participants with knowledge on the standard-setting process, compliance with the OIE international standards and raising awareness about regional and global issues and concerns outside of the national context. The seminars provide specialized and horizontal training across the OIE mandate on

topics relevant to their area of expertise, while also promoting reflection on how national systems can further work towards compliance with the OIE standards.

The OIE works to empower national focal points to share information, provide further training on OIE standards to staff and to better connect veterinary services operating in the veterinary domain but not necessarily under the veterinary authority (for example, aquatics, wildlife, laboratories and food safety).

The global capacity building programme for national focal points for veterinary laboratories has engaged members specifically on the topic of the OIE international standards related to laboratories.

The main goal of the laboratory focal points is to support the careful and responsible management of the national veterinary laboratory network in accordance with OIE international standards as outlined in the terrestrial and aquatic manuals, in close collaboration with the OIE delegate. In practice, the laboratory focal point would ideally be the champion for the implementation of the OIE international standards on biosafety and biosecurity, quality management, validation of diagnostic assays and overall scientific excellence of the laboratory network, as well as the safe collection and transport of biological materials within their country.

An overarching theme of the laboratory focal point seminars is 'Creating and nurturing a culture of safety and quality', which allows a deeper examination of the tenets of good laboratory management as described in the terrestrial manual's introductory chapters: transport of biological materials, biological risk analysis, quality management and laboratory networking. This framework aims to transmit pragmatic approaches to laboratory focal points so that they will be better equipped to translate their knowledge of the OIE international standards into action in their role as the stewards of the national veterinary laboratory network by providing the evidence base for veterinary services' decision making, and, in turn, ensuring good governance of the veterinary services in OIE member countries.

The OIE counts on the full involvement of laboratory focal points in the implementation of the OIE international standards; their continued dedication will inform the OIE's work on additional guidance and tools related to laboratories.

OIE reference centres

An OIE reference centre is designated either as an 'OIE reference laboratory', the principal mandate of which is to function as a world reference centre of expertise on designated pathogens or diseases, or an 'OIE collaborating centre' the principal mandate of which is to function as a world centre of research, expertise, standardization of techniques and dissemination of knowledge on a specialty.[19]

The network of OIE reference centres constitutes the core of OIE scientific expertise and excellence. The ongoing contribution of these institutes to the work of the OIE ensures that the standards, guidelines and recommendations developed by the specialist commissions and published by the OIE are scientifically sound and up-to-date.

OIE reference laboratories should use, promote and disseminate diagnostic methods validated according to OIE standards, recommend tests or vaccines as OIE standards, develop reference material in accordance with OIE requirements, implement and promote the application of OIE standards and store and distribute to national laboratories biological reference products and any other reagents used in the diagnosis and control of the designated pathogens or diseases. These laboratories should also develop, standardize and validate according to OIE standards new procedures for diagnosis and control of the designated pathogens or diseases, provide diagnostic testing facilities, and scientific and technical advice on disease control measures to OIE members.

OIE reference laboratories are expected to maintain a system of quality assurance, biosafety and biosecurity relevant for the pathogen and the disease concerned, establish and maintain a network with other OIE reference laboratories designated for the same pathogen or disease and organize regular inter-laboratory proficiency testing to ensure comparability of results, and organize inter-laboratory proficiency testing with laboratories other than OIE reference laboratories for the same pathogens and diseases to ensure equivalence of results.[20]

In its designated specialty, an OIE collaborating centre must provide its expertise internationally in support of the implementation of OIE standards and seek collaboration with other OIE reference centres. They are also expected to propose or develop methods and procedures that facilitate harmonization of international standards and guidelines applicable to the designated specialty, to carry out and/or coordinate scientific and technical studies in collaboration with other centres, laboratories or organizations, and to collect, process, analyse, publish and disseminate data and information relevant to the designated specialty.[21]

OIE register of diagnostic kits

In order to ensure high-quality results, OIE members and their laboratories need diagnostic kits that are known to be validated according to OIE criteria. In order to improve the quality of kits, to ensure that the diagnostic kits can be used to correctly establish animal disease status and to enhance confidence in diagnostic kits, the OIE developed the OIE register of diagnostic kits. Diagnostic tests validated as fit-for-purpose are included in the register of the OIE. The aims of the OIE procedure for the registration of diagnostic kits are to certify commercial diagnostic kit as validated fit for some specific purpose(s) and to produce a register of recognized diagnostic kits for OIE members. The procedure is open to all diagnostic kits for animal diseases, including zoonosis.

The OIE register is available online.[22] For diagnostic kits certified by the OIE, the OIE logo appears on any device associated with the kit, on the instructions for use and on the sales packaging. The process of producing a register of recognized diagnostic kits has provided greater transparency and clarity of the validation process and provides a means for recognizing those manufacturers that produce validated and certified test methods in kit format.

OIE-approved international standard sera

The OIE Biological Standards Commission coordinates a programme for the preparation, validation and distribution of OIE-approved international reference standards for diagnostic assays for infectious diseases of animals. The standards are prepared by an OIE reference laboratory in accordance with guidelines for antibody and antigen standards drawn-up by the Commission in collaboration with certain reference laboratories. Such standard preparations are designated by the OIE as primary reference standards for use in conjunction with tests described in the OIE terrestrial manual. The aim of the programme is to harmonize diagnostic testing and encourage the mutual recognition of test results for international trade.[23] To further promote the preparation and distribution of standard reagents for diagnostic testing, the OIE launched the creation of an international OIE Virtual Biobank. The OIE Virtual Biobank is a web-based catalogue of the biological resources that are held in biobanks hosted by OIE reference centres around the world. This catalogue represents a source of standardized information to search and locate samples, especially diagnostic reagents and reference reagents, along with associated metadata, and process retrieval requests.

OIE laboratory twinning programme

The OIE applied the concept of twinning between laboratories and implemented the laboratory twinning programme to build capacity and expertise for the most important animal diseases including zoonoses and dangerous pathogens in priority regions, in direct support of the OIE's strategy to improve global capacity for disease prevention, detection and control through better veterinary governance.

The OIE laboratory twinning programme establishes sustainable links between OIE Reference Centres and national laboratories in regions that are currently underrepresented, leading to exchange of knowledge, skills and experiences over a determined project period through staff exchanges, quality assurance (QA) and biosafety training, proficiency testing or other forms of engagement. Each twinning project provides mutual benefits for both reference centres and national laboratories by creating joint research opportunities, and the whole international community will benefit from stronger global disease surveillance networks for the control of animal diseases that cause a serious threat to health and livelihoods.[24]

Through laboratory twinning projects, the OIE provides a more balanced geographical distribution of advanced scientific expertise, allowing more national laboratories to comply with OIE standards and start playing a greater role regionally and globally for early disease detection, confirmation and rapid response and control. The higher level of scientific expertise acquired is also essential to allow countries to formulate science-based animal health control strategies and to maintain national veterinary scientific communities and support the standard-setting process of the OIE and of other WTO and SPS recognized international standard-setting organizations.

13.5 Developing international solidarity to achieve better control of animal diseases in the world

The OIE Performance of Veterinary Services (PVS) Pathway

Veterinary services, including the veterinary laboratories, in developing and in-transition countries may need guidance and support to strengthen the necessary organization, infrastructure, resources and capacities that will enable their countries to benefit from increased, safe trade in animals and animal products via the application of sanitary measures aligned to OIE standards. At the same time, stronger veterinary services provide greater protection for animal health and public health within their own country. The OIE considers veterinary services to be a global public good and their performance and compliance with international standards (structure, organization, resources, capacities, role of paraprofessionals) as a public investment priority.

The PVS Pathway is the OIE's flagship capacity building platform for the sustainable improvement of a country's veterinary services' compliance with OIE international standards. To help ensure the effective performance of the veterinary services of members, the OIE has dedicated chapters,[25,26] of the OIE *Terrestrial Animal Health Code* to the quality of veterinary services.

The OIE international standards and guidelines constitute the basis for independent external country evaluations of the quality of veterinary services and have been democratically adopted by all OIE members. A specific

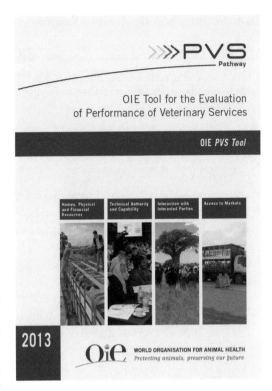

Figure 13.4 OIE Performance of Veterinary Services (PVS) tool.

methodology has been developed and the OIE has published the 'OIE Tool for the Evaluation of Performance of Veterinary Services' (the OIE PVS Tool) as the basis for evaluating performance against the international standards published in the *Terrestrial Animal Health Code* (Figure 13.4).

At the time of writing, the OIE is undertaking a process to evolve the PVS Pathway and provide a wider array of engagement options for its members. These options include: PVS self-evaluation, PVS evaluation with specific content (targeting global priorities such as PPR [peste des petits ruminants] eradication and AMR), formally linking PVS gap analysis with internal national strategic planning cycles, and linkages between the OIE PVS Pathway and the WHO's International Health Regulations Monitoring and Evaluation Framework.

Laboratories and the PVS Pathway

The PVS Evaluation mission uses the OIE PVS Tool,[27] where 47 critical competencies are systematically evaluated via documentation reviews, interviews and physical observations against five graded levels of advancement, each with detailed descriptions or indicators to transparently guide the process.

Evaluation of the laboratory function of veterinary services addresses the technical capability and authority of the veterinary service to address current and new issues including prevention and control of biological disasters based on scientific principles. Four critical competencies are directly relevant to the laboratory function.

- II.1A Access to veterinary laboratory diagnosis. The authority and capability of the VS to have access to laboratory diagnosis in order to identify and record pathogenic agents, including those relevant for public health and that can adversely affect animals and animal products.
- II.1B Suitability of national laboratory infrastructures. The sustainability, effectiveness and efficiency of the national (public and private) laboratory infrastructures to service the needs of the VS.
- II-2 Laboratory quality assurance. The quality of laboratories (that conduct diagnostic testing or analysis for chemical residues, antimicrobial residues, toxins, or tests for, biological efficacy, and so on) as measured by the use of formal QA systems including, but not limited to, participation in relevant proficiency testing programmes.
- II-10 Residue testing. The capability of the VS to undertake residue testing programmes for veterinary medicines (for example, antimicrobials and hormones), chemicals, pesticides, radionuclides, metals, and so on.[28]

PVS Gap Analysis

The PVS Gap Analysis (PVS Costing Tool) is a carefully structured exercise with national veterinary services to determine the priority goals, strategies, activities and investments required to improve veterinary services. During the mission, the country's veterinary services, supported by a team of OIE-certified experts and using the PVS evaluation information as a baseline, develop strategic and costed actions to improve their performance and meet national targets (Figure 13.5).

In terms of the veterinary laboratory, the PVS Gap Analysis provides an analysis of the laboratory's resources based on needs arising from the priorities defined in the official programmes of veterinary services related to trade (for example, specific programmes for zoning or compartmentalization), animal health (for example, disease surveillance programmes, monitoring of vaccination efficacy, and so on), and veterinary public health (for example, residue testing, food safety programmes, and so on). The PVS Gap Analysis also provides an assessment of the number and types of laboratory analysis required for these official programmes and an indicative operational budget for these activities only.

While this information is very important to veterinary services, the PVS Gap Analysis is limited by both the scope of its expertise and the time available to the laboratory function, as part of a much broader whole-of-system planning and costing approach. Therefore, the PVS Gap Analysis report provides a superficial, overarching analysis of national laboratory needs.

The OIE PVS Evaluation and PVS Gap Analysis (Costing Tool) missions allow for the evaluation of veterinary services' official need for laboratory analysis as well as their availability and cost as a first step, but do not allow an in-depth analysis of the pertinence, efficiency, and future and ongoing needs of the national laboratory network. In particular, the substantial cost

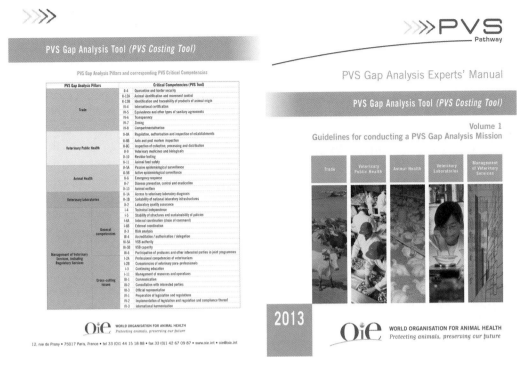

Figure 13.5 PVS Gap Analysis (Costing Tool).

of a national laboratory network is not considered in depth, based on veterinary services' need for laboratory analysis. It is for this reason that the PVS Pathway Laboratory Mission was developed.

PVS Pathway Laboratory Mission

Traditional laboratory evaluations assess technical capacity and suggest improvements, such as implementation of standard operating procedure, equipment maintenance and calibration, additional training in new techniques, purchase of new equipment or proficiency testing exercises, often in the context of short-term donor-funded projects.

As these missions are generally constrained by their context and terms of reference, they often do not allow for an in-depth examination

of the national laboratory network's sustainability, including strategic, management and financial issues. Often this leads to simplified technical recommendations and over-investment in laboratories that are systematically unable to implement the recommendations properly and leaves unaddressed underlying complications, such as insufficient budget allocations, lack of human and physical resources, or procurement difficulties (Figure 13.6).

The PVS Pathway Laboratory Mission was designed therefore to determine the resources needed by national veterinary laboratory networks to meet the needs of VS and to evaluate the pertinence of their structure and viability in the national context in order to present elements needed for strategic decision making. The approach analyses current and prospective demand and the current supply of laboratory services, as well as a range of sustainable management, organizational and

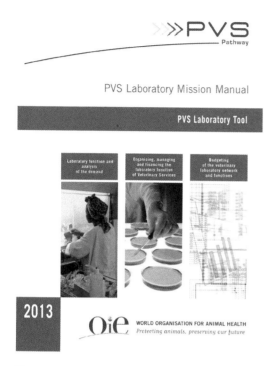

Figure 13.6 PVS Pathway Laboratory Tool.

National veterinary services preserve and develop animal resources and thus contribute to reducing poverty and hunger worldwide through improving rural livelihoods and human nutrition through greater access to animal protein. Their additional impact on global health security further safeguards the planet. For these compelling reasons, supporting the livestock sector through sustainable investments in national animal health systems and their laboratory function, based on international standards, protects and develops all communities, from global to local levels. The OIE is working each day to protect the health of animals, and in doing so, protecting our future.

financial options to facilitate sound decision making on an efficient, fit-for-purpose national laboratory network, including official delegation to private laboratories.

The importance of laboratories working in the veterinary domain

Each of the activities described above is related to the work conducted by veterinary laboratories every day. These activities continue to evolve with the OIE's core missions in mind. In the OIE's view, laboratories play an essential role in international trade and many other functions carried out by veterinary services. Sustainable and functioning veterinary laboratories, acting as a cross-cutting function, strengthen and contribute to the overall good management and quality of veterinary services.

Endnotes

1   http://www.oie.int/about-us/.
2   http://www.oie.int/fileadmin/Home/eng/Media_Center/docs/pdf/Key_Documents/EN_Leaflet OIE_web.pdf.
3   http://www.oie.int/international-standard-setting/overview/.
4   http://www.oie.int/index.php?id=169&L=0&htmfile=glossaire.htm.
5   http://www.oie.int/en/animal-health-in-the-world/oie-listed-diseases-2018/.
6   http://www.oie.int/international-standard-setting/terrestrial-manual/.
7   http://www.oie.int/fileadmin/Home/eng/Health_standards/tahm/1.01.01_MANAGING_VET_LABS.pdf.
8   http://www.oie.int/fileadmin/Home/eng/Health_standards/tahm/1.01.02_COLLECTION_DIAG_SPECIMENS.pdf.
9   http://www.oie.int/fileadmin/Home/eng/Health_standards/tahm/1.01.03_TRANSPORT.pdf.
10 http://www.oie.int/fileadmin/Home/eng/Health_standards/tahm/1.01.04_BIOSAFETY_BIOSECURITY.pdf.
11 http://www.oie.int/fileadmin/Home/eng/Health_standards/tahm/1.01.05_QUALITY_MANAGEMENT.pdf.
12 http://www.oie.int/fileadmin/Home/eng/Health_standards/tahm/1.01.06_VALIDATION.pdf.

13 http://www.oie.int/fileadmin/Home/eng/Health_ standards/tahm/1.01.07_HTS_BGC.pdf.

14 http://www.oie.int/fileadmin/Home/eng/Health_ standards/tahm/1.01.09_TESTS_FOR_STERILITY. pdf.

15 http://www.oie.int/fileadmin/Home/eng/ Health_standards/tahm/1.01.08_VACCINE_ PRODUCTION.pdf.

16 http://www.oie.int/fileadmin/Home/eng/Health_ standards/tahm/1.01.10_VACCINE_BANKS.pdf.

17 http://www.oie.int/fileadmin/Home/eng/ Our_scientific_expertise/docs/pdf/EN_FINAL_ Biothreat_Reduction_Strategy_OCT2015.pdf.

18 http://www.oie.int/animal-health-in-the-world/ the-world-animal-health-information-system/ the-oie-data-system/.

19 http://www.oie.int/our-scientific-expertise/ overview/.

20 http://www.oie.int/our-scientific-expertise/ reference-laboratories/introduction/.

21 http://www.oie.int/our-scientific-expertise/ collaborating-centres/introduction/.

22 http://www.oie.int/en/our-scientific-expertise/ certification-of-diagnostic-tests/the-register-of- diagnostic-tests/.

23 http://www.oie.int/our-scientific-expertise/ veterinary-products/reference-reagents/.

24 http://www.oie.int/en/support-to-oie-members/ laboratory-twinning/.

25 http://www.oie.int/index.php?id=169&L=0& htmfile=chapitre_vet_serv.htm.

26 http://www.oie.int/index.php?id=169&L=0& htmfile=chapitre_eval_vet_serv.htm.

27 http://www.oie.int/en/support-to-oie-members/ pvs-evaluations/oie-pvs-tool/.

28 http://www.oie.int/fileadmin/Home/eng/ Support_to_OIE_Members/pdf/PVS_A_Tool_ Final_Edition_2013.pdf.

chapter 14

Arthropod vectors and arthropod-borne diseases

Regula Waeckerlin and Susan C. Cork

Arthropods (phylum *Arthropoda*) have gained increasing significance in the transmission of viral, bacterial and parasitic (protozoan) diseases to humans and animals. In many cases, pathogens undergo part of their replication cycle in the arthropod. At other times, invertebrates can be merely mechanical vectors. Mosquitoes (class *Insecta*, order *Diptera*, family *Culicidae*) are the most common transmitters of arthropod-borne diseases, followed by ticks (class *Arachnida*, order *Ixodida*), midges (order *Diptera*, family *Nematocera*) and phlebotomid sandflies (order *Diptera*, family *Psychodomorpha*). However, any blood-feeding insect has the potential for disease transmission, for example *Rhodnius prolixus* ('kissing bug') transmitting *Trypanosoma cruzi* (the protozoan causing Chagas disease) or the genus Glossina transmitting *T. conglolensis* and *T. vivax* (the agents causing hagana).

14.1 Monitoring of disease vectors

Arthropod monitoring

Monitoring the presence, abundance and phenology (seasonal variation) of arthropods that can function as disease vectors can provide valuable information on the risk of disease transmission in a certain area, even before the actual disease

is observed. For successful insect monitoring, the following questions have to be answered to decide on the trap type used, the length of the trapping season, the duration of single trapping events, and methods of sample collection and storage.

Ecological questions to consider when setting up arthropod sampling:

- Which species will be sampled?
- What are its life stages (eggs, larvae, pupae, nymphs, adults)?
- Where does the arthropod breed?
- Are there special requirements for each life stage (temperature, humidity, plant coverage)?
- When (life stage/season of the year/time of day) does it bite?
- Does it bite only one or several hosts?
- Which vertebrate species are affected?
- Where is the highest density of affected species?
- What will the collected samples be used for (ecological studies, pathogen isolation)?

Variable components of trapping

- type of trap
- determination of trapping season

- length of trapping season
- frequency and duration of trapping events
- type of collection (dry / in alcohol / in formalin)
- storage of sample.

Mosquitoes and other biting flies

Mosquitoes are the most important arthropod vectors for viral diseases and other pathogens. Table 14.1 gives an overview of the most important mosquito-transmitted viruses, the geographic location, the species that transmit them and important diagnostic techniques. The three stages of a mosquito life cycle are larvae, pupae and adults. Not all mosquitoes are blood feeding, and of the ones that are, only the adult females consume blood before oviposition.

Trapping of mosquitoes, midges and other biting flies is usually performed by the use of baited traps, in which case carbon dioxide and light/black light are most commonly used as

Table 14.1 Important mosquito-borne viruses.

Pathogen	Distribution	Mosquito species	Vertebrate hosts	Diagnostic methods
Alphaviruses				
EEE	North America (east)	*Ochlerotatus sollicitans* *Culex nigripalpus* *Culiseta melanura* *Coquilletidia perturbans*	Passerine birds	Virus isolation (mosquito or mammalian cell lines, suckling mice) Haemagglutination / Haemagglutination Inhibition antigen ELISA RT-PCR, qRT-PCR from blood / CSF during viremia Serology (ELISA, IFT)
	Central & South America	*Melanoconion spp*	Rodents, birds	
WEE	North America (west)	*Culex tarsalis*	Passerine birds	
	California, Utah, Colorado	*Ochlerotatus melanimon* *Ochlerotatus dorsalis*	Jackrabbits	
	Central & South America	*Melanoconion spp*	Rodents, birds	
	Argentina	*Aedes albifasciatus*		
VEE	Central America	*Culex panocossa*	Rodents	
	Central & South America	*Anopheles spp* *Aedes / Ochlerotatus spp* *Culex ocossa*	Rodents	

Table 14.1 *continued*

Pathogen	Distribution	Mosquito species	Vertebrate hosts	Diagnostic methods
	South America (north) Florida	*Culex cedecei*	Rodents?	
	South America (north)	*Culex portesi*	Equid	
	Brazil	*Culex taeniopus*	Rodents	
		Psorophora spp		
		Mansonia spp		
		Deinocerites pseudes	Green heron	
Chikungunya	Asia, India	*Aedes aegypti*		
	Africa	*Aedes africanus*		
	Africa	*Aedes luteocephalus*		
	Europe	*Aedes albopictus*		
O'nyong-nyong	Africa	*Anopheles gambiae*	Human	
		Anopheles funestus		
Ross River	Australia	*Ochlerotatus vigilax* *Ochlerotatus camptorhynchus*	Marsupials?	
	Australia	*Culex annulirostris*		
	Fiji, Cook Islands	*Aedes polynesienses* *Aedes aegypti*		
Sindbis Virus	Europe Africa India Asia Australia	*Culex spp*		
	Scandinavia, Eastern Europe, Russia	*Culiseta ochropters*		
	Egypt	*Culex univittatus*		
Flaviviruses				
Yellow Fever	tropical Africa, South America, Carribbean	*Aedes aegypti*	Human (epidemic, urban cycle)	Virus isolation (mosquito or mammalian cell lines) RT-PCR, qRT-PCR Serology (HI, IFT, ELISA)

Pathogen	Distribution	Mosquito species	Vertebrate hosts	Diagnostic methods
	Africa	*Aedes africanus* *Aedes bromeliae*	Monkeys (enzootic, jungle cycle)	
	Africa (west)	*Aedes vittatus* *Aedes furcifer* *Aedes luteocephalus*	Monkeys (enzootic, jungle cycle)	
	Central & South America	*Haemagogus spp*	Monkeys (jungle sylvatic cycle)	
		Sabethes chloropoterus		
Dengue				
	Southeast Asia Caribbean Central & South America Africa	*Aedes aegypti*		
		Aedes africanus	Monkeys (enzootic, jungle cycle)	
	Southeast Asia	*Aedes albopictus*		
		Aedes albopictus		
	Pacific	*Aedes polynesienses*		
	Malaysia	*Ochlerotatus niveus*	Monkeys (sylvatic cycle?)	
West Nile Virus	North America	*Aedes vexans*	Birds (bridge vector)	
	North America (west)	*Culex tarsalis*	Passerine birds	
	Africa Middle East India Europe Russia North America (east) Central & South America	*Culex pipiens*		
	North America (east)	*Culex quinquefasciatus*	Birds	
	North America (west)	*Culex tarsalis*	Birds	
	North America	*Culex perturbans*	Birds (bridge vector)	

Table 14.1 *continued*

Pathogen	Distribution	Mosquito species	Vertebrate hosts	Diagnostic methods
	Egypt	*Culex univittatus*		
Murray Valley	Australia	*Culex annulirostris*		
	Australia	*Culex annulirostris*	Birds	
Japanese Encephalitis	Asia, Japan, Papua Neuguinea	*Culex tritaeniorhynchus*	Birds, pigs	
		Culex vishnui		
		Culex gelidus		
Saint Louis Encephalitis		*Culex pipiens*		
	North America (east)	*Culex quinquefasciatus*	Birds	
	North Ameica	*Culex tarsalis*	Birds	
	North America	*Culex nigripalpus*		
Bunyaviruses				
Rift Vallet Fever	South & East Africa	*Aedes mcintoshi*	Ungulates	Virus isolation (mosquito or mammalian cell lines, suckling mice) RT-PCR, qRT-PCR Serology (Hemagglutination-Inhibition, Complement Fixation, IFA, ELISA)
	Middle East Africa	*Culex pipiens*	Ungulates	
	Africa	*Culex theileri*	Ungulates (bridge vector)	
Wesselsbron Virus	South Africa	*Aedes mcintoshi* *Aedes circumluteolus*		
California Encephalitis	North America	*Ochlerotatus melanimon* *Ochlerotatus dorsalis*	Rabbits, ground squirrels, rodents	
Jamestown Canyon	North America	*Anopheles spp* *Ochlerotatus stimulans* *Culiseta inornata*	Deer	

Pathogen	Distribution	Mosquito species	Vertebrate hosts	Diagnostic methods
LaCrosse	North America	*Ochlerotatus triseriatus* *Ochlerotatus canadensis*	Chipmunks, squirrels, rodents	
Snowshoe hare	North America	*Ochlerotatus stimulans* *Ochlerotatus canadensis* *Culiseta inornata*	Lagomorphs	
Tahyna	Europe, west Asia	*Ochlerotatus caspius* *Aedes vexans* *Culiseta annulata*	Hares, pigs	
Rocio	Brazil	*Ochlerotatus scapularis*		
	Caribbean	*Psorophora ferox*		

Note: For the most up to date guidelines on testing for specific diseases in livestock species check the online edition of the OIE Manual of Diagnostic Tests and Vaccines for Terrestrial Animals http://www.oie.int/standard-setting/terrestrial-manual/access-online/.

bait. Furthermore, additional attractants such as octenol can be used alongside carbon dioxide. These trapping techniques target mostly blood-seeking females. Alternatives are gravid traps, which selectively collect resting gravid females after a blood meal, and aspiration from the environment or the host, which is less biased for gender of the mosquito, but selectively collects individuals that are flying/feeding at a specific time. *Malaise traps* are larger, tent-like structures that funnel insects and collect them in a vessel at the highest point.

Because mosquitoes need standing water for their larval and pupal development, a trap location in the vicinity of an open, organically rich water source promises the highest catch rates.

Figure 14.1 Carbon dioxide mosquito trap components and trap in the field. (A) lid; (B) fan; (C) CO_2 value; (D) CO_2 input; (E) battery (for fan); (F) collection cup; (G) compressed air CO_2 cylinder.

The principal climatic factors driving mosquito distribution are temperature and precipitation.

Baited trapping can be performed from durations of a 12 h trap night to one week or more. The method of trapping used depends on the questions to be answered and the resources available (Figure 14.1). Collection can be made dry, in ethanol, propylene glycol or formalin depending on the future use. Dry, cool conditions for mosquito storage are preferred for virological work and determination of species composition. However, if cool storage conditions cannot be maintained, moulding of the samples may be a problem. Collection in ethanol and propylene glycol and storage at 4°C will usually still be sufficient for DNA/RNA detection by PCR, depending on the storage duration and the stability of the virus to be detected.

Ticks

Ticks are important vectors of Protozoa, Spirochaeta, viruses and other pathogens. Table 14.2 gives an overview of the most important tick-borne diseases, the geographic location, the species that transmit them and important diagnostic techniques.

Table 14.2 Important tick-borne pathogens.

Pathogen	Tick species	Distribution	Vertebrate hosts	Diagnostic methods
Intracellular bacteria				
Rickettsia				
Rickettsia rickettsii	*Dermacentor variabilis*	North America (mostly east) Central America	Three hosts Small mammals (larvae, nymphs) Dogs, cattle, horses, humans (adults)	Blood / buffy coat / CSF smear (Giemsa, Wright, Dif-Quick stain) IF, IHC from lesion PCR Serology (IFA, ELISA)
Rickettsia rickettsii	*Dermacentor andersoni*	North America (west)	Three hosts Small mammals (larvae, nymph) Large mammals (adults)	
Rickettsia conorii	*Dermacentor reticulatus*	Europe Central Asia (west)	Three hosts Small rodents (larvae, nymph) Deer, dogs (adults)	
Rickettsia conorii Rickettsia slovaca	*Dermacentor marginatus*	Europe	Three hosts Small rodents (larvae, nymph) Large herbivores, carnivores (adults)	
Boutonneuse fever (Rickettsia conorii) Rickettsia rickettsii	*Rhipicephalus sanguineus*	worldwide	Three hosts Dog, wild carnivores, human (adults)	

Pathogen	Tick species	Distribution	Vertebrate hosts	Diagnostic methods
Rickettsia conorii	*Hyalomma truncatum*	Africa	Two hosts Small rodents and other mammals (immature lifestages) Wild and domestic ungulates, equines (adults)	
Rickettsia parkeri	*Amblyomma maculatum*	North America	Three hosts	
Rickettsia africae	*Amblyomma variegatum*	tropics, Africa, Caribbean	Three hosts Small mammals, birds, reptiles (larvae, nymphs) Large ruminants and carnivores (adults)	
Rickettsia africae	*Amblyomma hebraeum*	Africa	Three hosts Small mammals, birds, reptiles (larvae, nymphs) Large ruminants (adults)	
Ehrlichia				
Ehrlichia canis	*R. sanguineus*	worldwide	Three hosts Dog, wild carnivores, human (adults)	Serology (IFT) Buffy coat smear (Monocytes in Giemsa / Wright / Dif-Quick stain) PCR, qPCR (e.g. 16S)
Ehrlichia chaffeensis Ehrichia ewingii	*Amblyomma americanum*	North America	Dog	
Heartwater (*Ehrlichia ruminantium*)	*Amblyomma variegatum*	tropics, Africa, Caribbean	Three hosts Small mammals, birds, reptiles (larvae, nymphs) Large ruminants and carnivores (adults)	Cerebral cortex / cerebellum cell smears (Giemsa stain, presence in endothelial cells) Cell culture from whole blood (endothelial cells) PCR Serology (ELISA, IFT)
Heartwater (*Ehrlichia ruminantium*)	*Amblyomma hebraeum*	Africa	Three hosts Small mammals, birds, reptiles (larvae, nymphs) Large ruminants (adults)	

Table 14.2 *continued*

Pathogen	Tick species	Distribution	Vertebrate hosts	Diagnostic methods
Anaplasma				
Tick borne Fever (Anaplasma phagocytophilum)	*Ixodes scapularis*	North America	Three hosts Lizard, bird, small mammal (larvae, nymphs) Ungulates, humans, other mammals (adults)	Intracytoplasmic inclusion in peripheral blood smear (Giemsa / Wright stain) Culture (mammalian and tick cell lines) PCR, qPCR IF / IHC Serology (IFT, ELISA)
Tick borne Fever (Anaplasma phagocytophilum)	*Ixodes ricinus*	Europe, Western Asia	Three hosts Lizard, bird, small mammal (larvae, nymphs) Ungulates, humans, other mammals (adults)	
Tick borne Fever (Anaplasma phagocytophilum)	*Ixodes pacificus*	North America	Three hosts Lizard, bird, small mammal (larvae) Rodent, human, horse, dog (nymphs) Human, horse, dog, wild ungulates	
Anaplasma marginale A. centrale A. ovis	*Dermacentor spp*			
Anaplasma marginale, A. centrale, A. ovis	*R. annulatus (B. annulatus)*	subtropical / tropical	One host Large ruminants	
	R. microplus (Boophilus microplus)	Southern US South America Africa Asia North Australia	One host Large ruminants	
Anaplasma marginale	*Rhipicephalus spp*			
Anaplasma marginale A. centrale A. ovis	*Hyalomma spp*			

Pathogen	Tick species	Distribution	Vertebrate hosts	Diagnostic methods
Borrelia				
Lyme disease (Borrelia burgdorferi)	Ixodes scapularis	North America	Three hosts Lizard, bird, small mammal (larvae, nymphs) Ungulates, humans, other mammals (adults)	Difficult Serology (ELISA, IFT) PCR / qPCR from peripheral blood, CSF or urine Blood smear (prolonged Giemsa / Wright, followed by crystal violet; acridine orange) In vitro culture
Lyme disease (Borrelia burgdorferi)	Ixodes ricinus	Europe, Western Asia	Three hosts Lizard, bird, small mammal (larvae, nymphs) Ungulates, humans, other mammals (adults)	
Lyme disease (Borrelia burgdorferi)	Ixodes pacificus	North America	Three hosts Lizard, bird, small mammal (larvae) Rodent, human, horse, dog (nymphs) Human, horse, dog, wild ungulates	
Lyme disease (Borrelia burgdorferi)	Ixodes persulcatus	Europe, Northern Asia	Three hosts	
Relapsing Fever (Borrelia duttoni)	Ornithodoros moubata complex	Africa Caucasus Russia	Multi host Pig, human	
Coxiella				
Q-fever (Coxiella burnetii)	Ixodes spp	Africa Europe Middle East Asia Australia / New Zealand	Three hosts Human, sheep, cattle, other ungulates, other mammals	Serology (ELISA, IFT) PCR from blood / milk
Q-fever (Coxiella burnetii)	Dermacentor spp			
Q-fever (Coxiella burnetii)	Rhipicephalus spp			
Q-fever (Coxiella burnetii)	Hyalomma spp			
Q-fever (Coxiella burnetii)	Amblyomma spp			

Table 14.2 *continued*

Pathogen	Tick species	Distribution	Vertebrate hosts	Diagnostic methods
Francisella				
Tularemia (Francisella tularensis)	*Dermacentor andersoni*	North America (west)	Three hosts Small mammals (larvae, nymph) Large mammals (adults)	Culture (chocolate agar, Thayer-Martin media) PCR, qPCR, Sequencing to differentiate from non-pathogenic Francisella spp Serology (IFT)
Tularemia (Francisella tularensis)	*Dermacentor marginatus*	Europe		
Protozoan parasites				
Babesia				
B. microti, B. divergens, B. duncani	*Ixodes scapularis*	North America	Three hosts Lizard, bird, small mammal (larvae, nymphs) Ungulates, humans, other mammals (adults)	Blood smear (erythrocytes in Giemsa / modified Wright / Diff-Quick stain) PCR, PCR-RFLP Serology (IFT)
Babesia divergens	*Ixodes ricinus*	Europe, Western Asia	Three hosts Lizard, bird, small mammal (larvae, nymphs) Ungulates, humans, other mammals (adults)	
Babesia canis	*Dermacentor reticulatus*	Europe Central Asia (west)	Three hosts Small rodents (larvae, nymph) Deer, dogs (adults)	
Babesia canis vogeli	*Rhipicephalus sanguineus*	worldwide	Three hosts Dog, wild carnivores, human (adults)	
Babesiosis (B. bigemina, B. bovis)	*Rhipicephalus annulatus (Boophilus annulatus)*	subtropical / tropical	One host Large ruminants	
	Rhipicephalus microplus (Boophilus microplus)	Southern US South America Africa Asia North Australia	One host Large ruminants	

Pathogen	Tick species	Distribution	Vertebrate hosts	Diagnostic methods
Babesia canis	*D. reticulatus*	Europe Central Asia (west)	Three hosts Small rodents (larvae, nymph) Deer, dogs (adults)	
Babesia canis vogeli	*R. sanguineus*	worldwide	Three hosts Dog, wild carnivores, human (adults)	
Babesiosis (B. bigemina, B. bovis)	*R. annulatus (B. annulatus)*	subtropical / tropical	One host Large ruminants	
	R. microplus (Boophilus microplus)	Southern US South America Africa Asia North Australia	One host Large ruminants	
Theileria				
Cytauxzoon felis	*Dermacentor variabilis*	North America (mostly east) Central America	Three hosts Small mammals (larvae, nymphs) Carnivores, cattle, horses, humans (adults)	Buffy coat / lymph node smear (Giemsa) PCR, qPCR
East Coast Fever (Theileria parva)	*Rhipicephalus appendiculatus*	Africa	Three hosts Large ruminants, carnivores, hares (larvae, nymphs) Large ruminants (adults)	
Bovine Tropical Theileriosis (Theileria annulata)	*Hyalomma marginatum*	Russia Europe subsaharan Africa	Two hosts Small mammals and ground-feeding birds (immature lifestages) Wild and domestic ungulates (livestock, deer, boar) (adults)	
Bovine Tropical Theileriosis (Theileria annulata)	*Hyalomma detrium*	central Asia Mediterranean Middle East	Cattle, horses, sheep, goats, camels (all lifestages)	
Theileria orientalis	*Haemaphysalis longicornis*	Asia, Asia-Pacific (imported)	Cattle (Theileria associated bovine anaemia (TABA)*	
Viruses				
Louping ill	*Ixodes ricinus*	Europe, Western Asia	Three hosts Lizard, bird, small mammal (larvae, nymphs) Ungulates, humans, other mammals (adults)	Virus isolation (chicken embryo, mammalian cell lines, tick cell lines) RT-PCR, qRT-PCR Serology (ELISA)

Table 14.2 *continued*

Pathogen	Tick species	Distribution	Vertebrate hosts	Diagnostic methods
Tick-borne encephalitis	Ixodes persulcatus	Europe, Northern Asia	Three hosts	
Kyasanur Forest Disease	Haemaphysalis spinigera	India	Small mammals, monkeys, birds (larvae) Small mammals, monkeys, birds, humans (nymphs) Ungulates, other large mammals (adults)	
Asfarvirus				
African Swine Fever	Ornithodoros moubata complex	Africa Caucasus Russia	Multi host Pig, human	Virus isolation in cell culture (e.g. pig lymphocytes) Fluorescent antibody test (FAT) from tissue smears or cryostat sections PCR, qPCR, In Situ Hybridization (ISH) Serology (ELISA, IFT)
Bunyavirus				
Crimean-Congo Haemorrhagic Fever (Nairovirus)	Hyalomma marginatum marginatum	Russia Europe subsaharan Africa	Two hosts Small mammals and ground-feeding birds (immature lifestages) Wild and domestic ungulates (livetock, deer, boar) (adults)	Serology (competitive ELISA) RT-PCR
Crimean-Congo Haemorrhagic Fever (Nairovirus)	Hyalomma truncatum	subsaharan Africa	Two hosts Small rodents, hares (immature lifestages) Wild and domestic ungulates (livestock, deer, boar) (adults)	
Crimean-Congo Haemorrhagic Fever (Nairovirus)	Hyalomma detrium	central Asia Mediterranean Middle East	Cattle, horses, sheep, goats, camels (all lifestages)	
Crimean-Congo Haemorrhagic Fever (Nairovirus)	Hyalomma anatolicum	Mediterranean Arabian Peninsula Africa (east)	Two hosts Domestic and wild ungulates, dromedaries (adults)	

Pathogen	Tick species	Distribution	Vertebrate hosts	Diagnostic methods
Crimean-Congo Haemorrhagic Fever (Nairovirus)	*Hyalomma dromedari*		Two or three hosts Hares, small rodents (larvae) Hares, small rodents, large ungulates (nymphs) Camels, cattle, sheep, goat, horse (adults)	
Reovirus				
Colorado Tick Fever (Coltivirus)	*Dermacentor andersoni*	North America (west)	Three hosts Small mammals (larvae, nymph) Large mammals (adults)	Serology (ELISA) PCR IF staining of blood smears

Note: For the most up to date guidelines on testing for specific diseases in livestock species check the online edition of the OIE Manual of Diagnostic Tests and Vaccines for Terrestrial Animals http://www.oie.int/standard-setting/terrestrial-manual/access-online/.
* Emerging pathogen – https://cdn.ymaws.com/www.nzva.org.nz/resource/resmgr/docs/other_resources/Theileria_Handbook1.pdf

Tick life cycles usually include a larval, one (most hard ticks) or several (most soft ticks) nymphal and an adult stadium. Life cycles can be up to several years long. One method for the collection of ticks can be from the environment, where ticks rest (soft ticks) or search for a new host. Environmental sampling is commonly achieved by flagging, a technique where a white piece of cloth is dragged through a specified area, and ticks waiting for a host in the environment attach to the cloth. Ticks are then counted and abundance can be calculated. Traps baited with carbon dioxide (compressed air or dry ice, alone or combined with pheromones) are another method that targets adult ticks in the search of a blood meal. It requires ticks to actively seek a carbon dioxide source rather than just dropping off from the vegetation.

Another possible technique is the collection of ticks *from the vertebrate host*, which targets females that are in the process of blood feeding. This method additionally gives information about the tick burden, the species preference and the variety of tick species on a targeted vertebrate host.

The method of sampling used is determined by tick ecology as well as disease ecology considerations. Figure 14.2 gives an example of a workflow for setting up a tick sampling protocol.

In most cases, arthropod-borne viruses are only detectable from the blood of a vertebrate host for a short time and the peak of viremia coincides with a febrile stage. After this initial viremia, virus isolation can be attempted from cerebrospinal fluid (for viral encephalitises), lymph nodes, or the spleen, but detection rates are much lower. An exception to this is the *Orbivirus* family, which remain detectable in peripheral blood for a long duration after infection, since they adhere to erythrocytes. For example, ungulates for period of up to 60 days (Bluetongue virus) and equines up to 40 days (African Horse Sickness Virus).

The most common diagnostic methods for virus detection are DNA/RNA amplification by PCR or quantitative real-time PCR, as well as virus isolation by cell culture or the inoculation of embryonated eggs.

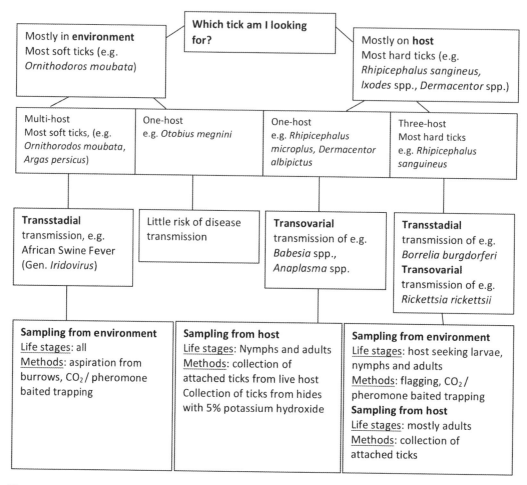

Figure 14.2 Ecological considerations for the sampling of ticks. Special considerations in the diagnostics of arthropod-borne viruses – virology.

Cell culture

In the case of arthropod-borne viruses, virus isolation in insect cells, tick cells or mammalian cells is an option. There are over 40 tick cell lines available, which were all derived from hard ticks (*Ixodidae*). Most commonly used for arbovirus detection are IDE2 cells, derived from *Ixodes scapularis*. Insect cell lines are equally readily available. By far the most popularly used insect cell line for the detection of arboviruses is the C6/36 cell line derived from *Aedes albopictus*.

A wide range of mammalian cells can also be used for arbovirus detection, most commonly kidney cells such as Vero (monkey), BHK-21 (hamster) or LLC-MK2 (pig) cells are used.

While tick-borne viruses rarely replicate in insect cell lines, insect-borne viruses replicate in tick cells quite readily, although sometimes requiring several passages. Both insect- and tick-borne arboviruses usually replicate quite readily in mammalian cells. Tick-borne viruses tend to cause cytopathic effects in mammalian cell lines, but persistent infections in tick cells, so once this is achieved, infected cells can be sub cultured indefinitely.

Molecular diagnostics

Molecular diagnostics such as PCR, RT-PCR and quantitative methods have become a standard diagnostic method for vector borne diseases. Additionally, the multiplexing of assays allows the detection and differentiation of several pathogens in the same assay. In some cases, such as *Francisella tularensis*, pathogenic and non-pathogenic strains of the same agent can be differentiated by sequencing.

A consideration for quality control when attempting PCR from arthropod samples is the use of an internal control. These are usually genomic, vector-specific, and highly conserved DNA or RNA segments. Mitochondrial (such as ND4) or ribosomal gene sequences (such as ITS2) have been used for arthropod internal controls in established assays. There should be a positive signal for the internal control for the assay result to be used for pathogen diagnostics. False negatives can occur due to degradation of the sample or the presence of PCR inhibitors.

Bibliography

Bell-Sakyi, L., Zweygarth, E., Blouin, E.F., Gould, E.A., Jongejan, F. (2007) Tick cell lines: tools for tick and tick-borne disease research. Trends in Parasitology 23(9): 450–457.

Cunha, B.A. (ed.) (2000) Tick-borne Infectious Diseases: Diagnosis and Management. Marcel Dekker, New York.

Hunfeld, K., Hildebrandt, A., Gray, J.S. (2008) Babesiosis: recent insights into an ancient disease. International Journal for Parasitology 38(11): 1219–1237.

Kao, C., King, C., Chao, D., Wu, H., Chang, G.J. (2005) Laboratory diagnosis of dengue virus infection: current and future perspectives in clinical diagnosis and public health. Journal of Microbiology, Immunology and Infection 38(1): 5–16.

Lawrie, C.H., Uzcátegui, N.Y., Armesto, M., Bell-Sakyi, L., Gould, E.A. (2004) Susceptibility of mosquito and tick cell lines to infection with various flaviviruses. Medical and Veterinary Entomology 18(3): 268–274.

Mullen, G.R., Durden, L.A. (eds) (2009) Medical and Veterinary Entomology, 2nd edn. Academic Press (Elsevier), Burlington, MA.

Vernet, G. (2004) Diagnosis of zoonotic viral encephalitis. Archives of Virology 18(Supplement): S231–S244.

PART IV

APPENDICES

appendix 1

Important zoonotic diseases

Susan C. Cork

A1.1 Overview

Zoonotic diseases are diseases caused by pathogens that can be transmitted from animals to humans. Table A1.1 is an abbreviated summary of some important zoonotic diseases. Diseases marked with an asterisk (*) are covered in more detail in the following section. Wild and domestic animals may provide a source of infection to humans either directly or indirectly by the contamination of a shared environment (including food and water supplies). Always seek medical advice from the nearest public health officer or local health clinic if there is any suspicion that a member of staff may have been exposed to one of these diseases. Note that for some infectious diseases, humans may be considered the primary host and are considered to be a source of infection for animals, for example, *Mycobacterium tuberculosis* (a 'reverse' zoonosis).

Many emerging and re-emerging zoonotic diseases are best addressed using a One Health approach with veterinary and public health authorities working together with disease experts. One Health teams will engage other areas of expertise as required, for example, wildlife biologists (for diseases involving wildlife), entomologists (for vector borne diseases), epidemiologists and disease modellers (to design surveillance plans and to assess risk) and social anthropologists (to understand how best to influence human behaviour). Some examples are provided in the texts listed at the end of the section.

Table A1.1 Summary of the source and clinical signs associated with some common zoonotic diseases.

Disease	Agent	Animals involved	Mode of transmission to humans	Clinical signs in humans	Action/ prevention & notes
BACTERIA					
Anthrax*	*Bacillus anthracis*	Warm blooded animals.	Skin wounds, inhalation, ingestion.	Small black scab, if inhaled can cause systemic disease.	Take care when examining cases of sudden death. Do not open carcass, bury in deep pit. Wash hands. Vaccinate livestock.
Brucellosis*	*Brucella* spp.*B. abortus* *B. melintensis* *B. canis* *B. suis*	Livestock species, dogs.	Direct contact with infected body fluids including milk from infected animals.	Undulant fever, aching joints, general malaise, persistent and chronic.	Boil milk before consuming it. Test milking cows for Brucellosis. Take care when examining cases of abortion.
Campylo-bacter	*Campylobacter* spp.	Widespread in mammals and birds.	Faecal contamination of food.	Gastroenteritis, may be chronic and persistent.	General hygiene, wash vegetables in clean water. Cook poultry meat well.
Cat scratch fever	*Bartonella Henselae*	Cats.	Scratches and bites.	Fever, swollen lymph nodes.	Antibiotics, avoid being scratched/ bitten.
Clostridial diseases	*Clostridium spp.* *Cl. Botulinum* *Cl. tetani*	Widespread especially as a soil contaminant.	*Cl. botulinum:* ingestion of botulinum toxins, other Clostridia usually by infection of wounds.	*Cl. botulinum,* neurological signs, may be fatal. *Cl. tetani,* tetanus, other *Clostridia spp.* gangrene.	*Cl. botulinum,* do not eat tinned food out of cans expanded with gas, *Cl. tetani* and others, wash wounds well, for deep wounds seek medical attention. Vaccinate children and adults, (tetanus). Antibiotics and antitoxins. Vaccine available to prevent tetanus.
Erysipelas	Erysipelothrix (*insidiosa*) rhusiopathiae	Poultry (turkeys) and livestock, especially pigs (diamond skin disease).	Wound contaminant.	Small scab usually on hands.	Hygiene, care when handling poultry and meat products.
Glanders	*Burkholderia Mallei* (previously *Pseudomonas mallei*)	Equids, other livestock and wildlife.	Nasal discharges, infected tissues at necropsy. Can live in the soil in tropical areas.	Respiratory and systemic signs.	Hygiene, wear gloves when handling dead animals, avoid aerosol of infectious particles.

Disease	Agent	Animals involved	Mode of transmission to humans	Clinical signs in humans	Action/ prevention & notes
Leptospirosis*	*Leptospira interogans* serovars	Different serovars have specific host species, wildlife and domestic stock.	Direct contact with infected animals, especially urine, abortion fluids. Also indirect contact via soil and water.	Fever, photophobia, aching joints, urinary tract infection. There may be jaundice and haemoglobinuria.	Test stock. Avoid splashing urine in eyes. Boil milk. Care when attending cases of abortion. Test stock and treat. Vaccination for livestock is available. Some strains are endemic in rodents and are present in waterways (Weil's disease)
Listeriosis	*Listeria Monocytogenes*	Widespread.	Food borne.	May cause abortion or birth defects in unborn child.	Wash vegetables in clean water, hygiene. Be careful with chilled produce (salads, soft cheeses soft cheeses, some sea food etc.) and delicatessen meats if pregnant or immunosuppressed.
Meliodosis*	*Burkholderia pseudomallei* (previously *Pseudomonas pseudomallei*)	Widespread in wild and domestic animals, soil contamination especially in rice paddies.	Contamination of wounds.	Systemic disease, fever, chills, death.	Look after wounds and avoid walking bare foot in areas contaminated with animal faeces. Whitmore's disease, more common in the tropics. Vaccines in development.
Plague*	*Yersinia pestis*	Rodent, flea life cycle.	Flea bite, inhalation of fomites.	Bubonic and pneumonic forms. Enlarged lymph nodes, fever and malaise and respiratory disease, death.	Control rodent population and fleas. Avoid overcrowding, contaminated air bedding. General hygiene. Vaccination available, antibiotics.
Psittacosis	*Chlamydophila/ Chlamydia psittaci* (parrot fever).	Birds.	Inhalation, direct contact.	Flu-like symptoms Can be more severe pneumonia asthma-like. Oxytetracycline is the treatment of choice as a lot of other antibiotics will not work.	Take care when handling dead or sick birds. Undertake avian necropsies in a biosafety cabinet.

Table A1.1 *continued*

Disease	Agent	Animals involved	Mode of transmission to humans	Clinical signs in humans	Action/ prevention & notes
Salmonellosis*	*Salmonella* spp.	Widespread, many species.	Generally contamination of food and water supplies. Eggs and meat, especially poultry.	Gastro-enteritis, host adapted human forms (*Salmonella typhi*) may cause severe disease, i.e. Typhoid.	General hygiene. Test poultry (*S. enteritidis*). Cook poultry meat well. Use different chopping boards for raw and cooked food. Vaccination available against Typhoid.
Tuberculosis*	*Mycobacterium tuberculosis* and *M. bovis*	Bovine adapted strains, *M. bovis,* may occur in wild animals. Emerging cases of *M. tuberculosis* in domestic and wild animals.	*M. bovis,* ingestion of contaminated meat or milk (cutaneous forms have been recorded after cuts during necropsy of infected animals). *M .tuberculosis*-aerosol (elephants, primates and other species in Asia are positive for human strains).	*M. tuberculosis* generally respiratory signs and wasting, *M. bovis,* may cause localized disease anywhere in the body, often enlarged lymph nodes.	Test cattle for tuberculosis (Tuberculin test), boil milk, cook meat well. Official inspection of meat before consumption. Vaccination for human tuberculosis available.
Yersiniosis	*Yersinia pseudotuber-culosis* *Y. enterocolitica*	Widespread in birds and mammals.	Contaminated food and water.	*Y. pseudotuberculosis,* causes pseudo appendicitis with localised pain in the abdomen, *Y. enterocolitica* causes gastroenteritis.	General hygiene, wash vegetables in clean water, cook meat well.
RICKETTSIA					
Q fever (query fever)	*Coxiella burnetti*	Various animals.	Milk. Soil. Can be vector borne (ticks).	Fever.	Boil milk. Avoid tick bites. *Coxiella burnetii* is found in cattle, sheep, and goats around the world. Humans typically get infected when they breathe in dust contaminated by infected animals.

Disease	Agent	Animals involved	Mode of transmission to humans	Clinical signs in humans	Action/ prevention & notes
Typhus*	*Rickettsia* spp.	Arthropod borne disease, carried by ticks, lice, etc. Reservoir infections in wildlife.	Tick, flea, mite or louse bite. Tick borne typhus more common in people handling wildlife.	Non-specific, fever, chills, etc.	Control arthropod vectors, avoid tick bites. 'Typhus' is a group of closely related diseases caused by different species of *Rickettsia* sp. transmitted to humans by arthropods. The most important form of typhus is epidemic (transmitted from human to human by lice). Other forms include murine (endemic) typhus, scrub typhus (transmitted by chigger mites) and tick-borne typhus.

FUNGI

Disease	Agent	Animals involved	Mode of transmission to humans	Clinical signs in humans	Action/ prevention & notes
Ringworm*	*Microsporum Trichophyton* spp.	Various animals, contaminated bedding and equipment.	Infected hairs or animal fomites.	Skin lesions, hair loss, red circumscribed areas.	Treat infected animals and wash hands after handling. Use fungicidal washes, in bad cases oral medication may be required. Severe infections are more common in children or immunocompromised individuals.

PROTOZOA

Disease	Agent	Animals involved	Mode of transmission to humans	Clinical signs in humans	Action/ prevention & notes
Leishmaniasis* cutaneous (Tropical sore)/ Visceral form (Kala-Azar)	*Leishmania donovani*, there are human and animal adapted subspecies.	Wild and domestic animals may be a source of infection transmitted by a sandfly vector.	Direct contact or via a sandfly bite.	Cutaneous form skin lesions, unhealing ulcer: visceral form, general malaise, weight loss, enlarged spleen, lymph-adenopathy, anaemia.	Control sandfly vector, avoid being bitten by infected sandflies. Note that some people may not develop clinical signs. Cutanaeous leishmaniasis is more common than the visceral form. Skin sores develop within a few weeks after a sandfly bite. People with visceral leishmaniasis usually

Table A1.1 *continued*

Disease	Agent	Animals involved	Mode of transmission to humans	Clinical signs in humans	Action/ prevention & notes
					become sick within months to years after a sandfly bite.
Sarcocystis	*Sarcocystis* spp.	Protozoal infection of various animals.	Consumption of infected meat.	Myositis, non-specific.	Official meat inspection, cook meat well.
Toxoplasma*	*Toxoplasma gondii*	Many animals can be intermediate hosts, definitive host is the cat.	Ingestion of oocyts in cat faeces, ingestion of cysts in/ handling infected meat.	There may be birth defects in un-born children if the mother has no immunity, general malaise, flu-like signs.	Pregnant women should avoid handling cat faeces/litter tray wastes or infected soil, apply general hygiene when handling meat, cook meat well.
TREMATODES					
Schistosoma*	*Schistosoma japonicum* (also *S. haemolyticum*)	Many animal species.	Contact via contaminated water, percutaneous infection or oral.	Blood and mucous in faeces. Some forms reside in the bladder and cause haematuria.	Wash vegetables in clean water (filter and boil water). Avoid bathing in water where schistosomes are known to be present.
CESTODES					
Beef tapeworm*	*Taenia saginata*	Cattle/human lifecycle.	Consumption of cysts in infected meat.	Often none, tape worm segments in faeces.	Avoid contamination of pasture with human excrement. Use of suitable latrine, cook meat well. The human is the definitive host so control should focus on human hygiene and behaviour.
Hydatids*	*Echinococcu granulosus*	Dog tapeworm, intermediate stage in various wild and domestic animals.	Ingestion of cestode eggs in dog faeces, consumption of uncooked produce (i.e. vegetables and berries etc.) contaminated by canine faeces.	Hydatid cysts anywhere in body especially liver, lungs, brain. Enlarged abdomen. Slow to develop.	General hygiene. Routine worming of dogs. Do not feed dogs raw offal from livestock. Wash fresh produce well. Another similar parasite, *Echinococcus multilocularis* is also an emerging zoonosis, canids are the definitive host but the intermediate hosts are usually rodents.

Disease	Agent	Animals involved	Mode of transmission to humans	Clinical signs in humans	Action/ prevention & notes
Pork tapeworm*	*Taenia solium*	Pig/human life cycle.	Consumption of cysts in infected meat.	Tapeworm segments in faeces, there may be cysts in tissues including the brain (i.e. humans can also act as an accidental secondary host).	As above. Always cook pork well. The human is the definitive host so control should focus on human hygiene and behaviour.
NEMATODES					
Trichinosis*	*Trichinella spiralis*	Pigs and wild boar, horses, bears, other carnivores, rodents.	Ingestion of infected meat (pork, game including some marine mammals).	Myositis, fever, general anaphylactic reaction, death.	Do not feed uncooked wastes to pigs, official meat inspection, cook meat well (note that drying and smoking does not kill *Trichinella*).
Cutaneous LM*	Canine hookworm, *Ancylostoma sp.*	Dogs.	Percutaneous infection of hookworm larvae.	Skin lesions, itching especially of feet.	Hygiene, do not let children play where dogs defecate/wear shoes on beaches, ensure routine worming of puppies and adult dogs.
Visceral LM*	*Toxocara canis*	Dogs especially puppies.	Ingestion of infective stage in dog faeces.	Myositis, non-specific, neurological signs, blindness (rare).	Hygiene, do not let children play where dogs defecate, ensure routine worming of puppies and adult dogs.
ARTHROPODS					
Fleas, Mites, Ticks, Lice	Various arthropod species	Many animal hosts, some human adapted species (e.g Sarcoptic mange).	General nuisance but many also transmit diseases.	Skin lesions, secondary infections, itching/ self trauma.	General hygiene, various powders and sprays to control arthropods on humans, animals and the environment are available. See Chapter 3 (entomology) for specific arthropods and Tables 14.1 and 14.2 for a list of some important vector borne diseases.

Table A1.1 *continued*

Disease	Agent	Animals involved	Mode of transmission to humans	Clinical signs in humans	Action/ prevention & notes
VIRUSES					
Avian Influenza (AI) *	Orthomyxovirus (Highly pathogenic H5N1, H5 and H7 subtypes and H9N2)	Birds and pigs	Aerosol/Oral–faecal.	Fever, flu-like signs. Most cases mild. Some emerging strains have caused human deaths.	Control influenza in poultry, avoid mixing different bird and mammalian species in markets. Vaccines are available but not all cause good cross protection. Humans should be vaccinated against human strains.
Contagious ecthyma	Orf virus, parapox	Sheep/goats	Direct contact with infected animals.	Skin lesion, scabs on hands and face.	Control secondary infections, wash hands.
Cowpox	Pox virus. This virus is in the orthopoxvirus family and is closely related to the vaccinia virus	Cattle.	Direct contact with infected animals i.e. when milking.	Skin lesions, especially on hands.	Control secondary infections, wash hands In humans, cowpox resembles mild smallpox (a serious human pathogen which is now eradicated). This was the basis of the first smallpox vaccines.
Japanese encephalitis (JE)*	Flavivirus	Natural cycle in birds and mosquitoes. Can be amplified in pigs.	Mosquito bites.	Mild malaise, can develop encephalitis.	Control mosquito vectors. Wear mosquito repellent, use mosquito nets. Vaccine available for humans. JE occurs predominantly in Asia in rural or agricultural areas, often associated with rice farming.
Lymphocytic choriomeningtis	Arenavirus	Rodents.	Rodent bite.	Meningitis, coma.	Control rodents.
Nipah virus	Paramyxovirus	Present in some bat population. Can be transmitted to pigs and then to humans.	Contact with infected animals or through contaminated food (e.g. palm juice contaminated with bat faeces).	Subclinical, mild and severe forms (may look like JE).	Avoid contact with bats, reduce contact between wild bats and livestock. Use PPE when handling sick pigs. Prevent bat faeces contaminating food and water supplies. Occurs predominantly in Asia.

Disease	Agent	Animals involved	Mode of transmission to humans	Clinical signs in humans	Action/ prevention & notes
Rabies*	Lyassavirus	Wild carnivores, bats, other animals may also be infected. Urban and sylvatic cycles. For humans, the domestic dog is the main reservoir of infection.	The virus is excreted in saliva and transmitted by bites. Domestic feral dogs are the most common source of infection. Vampire bats in South America. Ruminants and horses can also be infected and bite humans. Inhalation and laboratory accidents are also a risk.	Neurological infection signs, hydrophobia. Dumb and furious stages.	Avoid dog bites/ control feral dogs, vaccinate pet dogs and cats. If bitten, wash wounds well with soap and water, quarantine the animal, follow post exposure prophylaxis procedures. Post exposure treatment is effective if given soon after exposure. Several treatments are required. Preventative vaccines are available for animal handlers and laboratory workers.
Simian herpes virus	Herpes virus	Primates.	Monkey bites.	Possibly fatal encephalitis.	Take care when handling primates, treat all primate bites seriously.
West Nile Virus (WNV)	Flavivirus	Natural cycle in aquatic birds and mosquitoes.	Mosquito bites.	Mild malaise or unapparent, can develop encephalitis.	Control mosquito vectors. Wear mosquito repellent, use mosquito nets Vaccine available for horses.

A1.2 Zoonotic diseases in brief

This section provides more information on some of the diseases that members of the veterinary team may come into contact with when handling animals. Details of the samples which should be submitted to the laboratory to confirm a diagnosis are provided along with some basic recommendations for disease prevention and control in domestic animals. When faced with a clinical case or a disease outbreak, which is caused by a potentially zoonotic pathogen, it is important to take appropriate precautions to ensure the health and safety of all members of the veterinary team and to protect the public. The following information has been compiled from various sources but updated information should always be sought from the veterinary and public health authorities in the locality. Advice can also be sought from the office of the regional veterinary or medical officer.

Anthrax

Cause: *Bacillus anthracis*

Epidemiology: The spores of the organism survive for long periods in the soil. If the soil is later disturbed by ploughing or grazing stock there may be an outbreak of the disease. Such outbreaks usually occur in areas where animals have previously died of anthrax and the vegetative organisms have been exposed to the air and dispersed by scavengers.

Signs in animals: Sudden death. Swollen carcass. Bloody discharges from body orifices (that is, mouth, nostrils, rectum).

Diagnosis: If anthrax is suspected, do not open the carcass or spore formation will occur resulting in long term contamination of the surrounding area. Send a blood smear, taken from an ear vein, to the laboratory. It is important to ask specifically for anthrax to be looked for (microscopy +/– culture) and if possible speak to the nearest veterinary unit about the case. PCR can also be used to confirm a diagnosis of anthrax. Control: Where and when feasible/desirable vaccination is available for livestock. If the diagnosis of anthrax is confirmed the carcass should be buried in a deep pit and the hole filled in with lime.

Avian Influenza

Cause: Orthomyxovirus, influenza A

Epidemiology: Influenza A viruses cycle naturally in wild aquatic birds, especially waterfowl. Many strains are non-pathogenic. The avian influenza viruses are divided into subtypes based on the structure of two proteins on the surface of the virus: haemagglutinin (HA) and neuraminidase (NA). To date, there have been 18 HA and 11 NA proteins identified and these can occur in a wide range of combinations. Most of these can be found in birds but H17 N10 and H18 N 11 have only been isolated from bats. Most mammalian species have common strains of influenza but can occasionally be infected by avian strains. Many infections are asymptomatic. Virulent strains, known as 'highly pathogenic avian influenza' (HPAI) are usually subtypes H5 or H7 and are associated with disease outbreaks in domestic poultry. The virus is transmitted by aerosol but can also be transmitted through the oral-faecal route. Although there are vaccines available against influenza A viruses in humans and animals these are not always fully protective due to changes in the antigenic structure of the virus. This can occur as a result of antigenic drift (common over time) or due to 'antigenic shift' (more rapid but less common) as a result of genetic reassortment between viruses infecting the same host. Poultry and pigs are considered to be potential 'mixing vessels' for the development of strains that can infect humans, this is because strains adapted to pigs (including H9N2) have characteristics that allow attachment to the epithelial lining of the human respiratory tract. In 2003, pan-zoonotic strains of avian influenza H5N1 resulted in a number of human deaths across Asia.

Signs in animals: May be clinically in apparent. In domestic poultry, the virus can act locally causing mild disease (for example, ruffled feathers, off food) but virulent strains are more invasive causing systemic infections with neurological, respiratory and gastrointestinal signs.

Diagnosis: Diagnosis can be confirmed at necropsy using PCR or traditional virological techniques. There are also a number of serological tests including kit tests for rapid diagnosis. Egg inoculation is used to culture strains. Typing is often done using either PCR or HI. (See also Chapter 6.)

Control: Quarantine poultry flocks suspected to have highly virulent avian influenza. Vaccinations are available for poultry and other species. People considered 'at risk', for example, animal handlers should be vaccinated against human strains of influenza to prevent co-infection and subsequent genetic reassortment

Brucellosis

Cause: There are many species of Brucellae. The species of greatest importance for humans are *Brucella abortus* (from cattle) and *Brucella melitensis* (from goats and sheep). *Brucella canis* occurs in dogs and *Brucella suis* in pigs (also reported in other species including deer and occasionally dogs). In some countries *Brucella ovis* is an important cause of infertility in sheep.

Epidemiology: Brucellae are obligate intracellular bacteria with a tendency to infect the reproductive system of animals. Most are fairly host adapted although cross infections can and do occur. The greatest source of environmental contamination is following a *Brucella* induced abortion. Humans may become infected with Br. *abortus* or Br. *melitensis* from abortion fluids or by drinking infected milk as well as from discharges and fluids from apparently normal births.

Signs in animals: There may be no signs. The organism can remain dormant in the tissues of a young cow or nanny goat until she becomes pregnant. Abortion often occurs at about the sixth month of gestation.

Diagnosis: Serum samples (preferably paired, that is, a repeat sample taken 2–4 weeks after the first), milk, placenta, foetal stomach contents, foetal liver/spleen. PCR or culture may also be performed where feasible.

Control: Cattle, especially milking cows (blood and/or milk), should be tested for *Brucella* antibodies. Infected animals should be isolated. Milk should be boiled before consumption. Isolate animals which have aborted and bury any contaminated material after collecting the

appropriate samples for the laboratory. In some countries vaccination for cattle (and goats/sheep) is recommended.

Cutaneous larva migrans

Cause: Hookworm (*Uncinaria* spp., *Ancyclostoma braziliense*, *A. caninum*) Epidemiology: When there is a large population of dogs the environment soon becomes heavily contaminated with dog faeces. Dog to dog transmission of canine parasites is rapid although older dogs usually have a degree of immunity which stops clinical parasitism becoming apparent. Humans and puppies become infected when hookworm larvae penetrate the skin and migrate through the tissues. In humans *A. braziliense* is most commonly implicated.

Signs in animals: There may be few signs although hookworm is more pathogenic than Toxocara. In pups, there may be severe anaemia and sometimes death. In both puppies and dogs there may be signs of dermatitis on the feet and legs due to larval migration.

Diagnosis: Parasite eggs in dog faeces.

Control: Routine worming of dogs as outlined for *Toxocara* sp.

Hydatids

Cause: *Echinoccocus granulosus*

Epidemiology: *E. multilocularis* is fairly widespread in Africa, Asia, South America and the Middle East. It is a small tapeworm of dogs which has the intermediate stages of the life cycle in sheep, cattle, goats, yaks and, occasionally, humans. Humans act as a dead-end host in this cycle but dogs become re-infected by ingestion of cysts in un-cooked infected meat and offal from meat animals (for example, ruminants).

Humans become infected by accidental ingestion of tapeworm eggs in food contaminated with dog faeces and other animals become infected by ingesting contaminated pasture. Humans are not infected by eating infected meat unlike the situation with the human tapeworms (for example, *Taenia saginata*).

Signs in animals: There may be no signs in animals or localized signs due to the location of a cyst. In dogs, there are no clinical signs and diagnosis often requires administration of a purgative to find the eggs in the faeces (for example, arecholine). In humans, the cysts develop slowly but may form huge balloon sized fluid filled cysts in the liver and lungs. If cysts occur in the brain there may be clinical signs of headaches fairly early on but if cysts occur in other organs the disease may not be apparent for several years.

Diagnosis: Cysts are seen in the intermediate host at post-mortem and meat inspection. Hydatid eggs may be seen (under the microscope) in the faeces of purged dogs. Dog faecal samples should be handled with care, note that Taeniad eggs look similar to hydatid eggs and so treatment should be inclusive for all dog tapeworms. CAT (computerized axial tomography) scans, ultrasound and radiography along with exploratory surgery may be required to diagnose the disease in humans.

Control: Routine treatment of dogs with a suitable anthelmintic. Dosing should take place every 6 months of all dogs in the area. Routine meat inspection is important to allow monitoring of the disease in intermediate hosts. Hygiene and careful handling of dog faecal samples. Burial of dog faeces. Wash hands, after handling any dog, and before eating food. Wash fresh produce well if there is a risk of contamination with carnivore faeces. Never feed raw offal to dogs unless it is known to be free of hydatids.

A note on *Echinococcus multilocularis*

The life cycle of *E. multilocularis* involves a definitive host (for example, foxes, coyotes and domestic dogs) and an intermediate host (for example, wild rodents). The life cycle is completed after the definitive host consumes a rodent infected with cysts. Humans can become infected by handling infected animals or by ingesting contaminated food, vegetables, and water. The parasite causes alveolar echinococcosis (AE) in humans and it is hard to diagnose and treat. Lesions can be found in the liver and lungs and can metastasize to other organs. As with *E. granulosis*, humans are a dead-end or incidental host. The incidence of human infestation with *E. multilocularis* is increasing in urban areas, as wild canids migrate to urban and suburban areas and gaining closer contact with human populations. The disease has extended its range in North America and Europe in the last few decades although remains fairly uncommon in humans. Disease prevention includes basic hygiene and preventing domestic dogs catching and ingesting wild rodents. Where possible, food crops should be fenced off to prevent wild and domestic canids contaminating produce.

Japanese B encephalitis

Cause: Flavivirus, related to West Nile Virus

Epidemiology: The virus cycles naturally and silently between marsh birds (for example, night herons) and mosquitoes (for example, *Culex* spp.). More common in marshlands. Accidental mosquito borne infection of humans and horses may occur. Transplacental infection occurs in pigs and newborn piglets may act as amplifier hosts.

Signs in animals: Horses: Sweating, dementia, aimless wandering, photophobia, muscle tremors, ataxia. Mortality rate 5%. Recovery may be complete. Pigs: Infection often unapparent, piglets may show a low-grade encephalitis. Similar signs are seen in cases of West Nile virus which is a related flavivirus.

Diagnosis: Serum samples for serology. Tissues for immunofluorescence or PCR. Also collect whole blood and routine post-mortem samples so that other diseases can be ruled out. Control: Control of mosquito vector, monitor sentinel species, movement control of horses in an outbreak. Vaccinations are available for humans. Horses and pigs are vaccinated for JE in parts of Asia.

Leishmaniasis

Cause: *Leishmania donovani*

Epidemiology: This is a protozoal parasite which is transmitted by a sandfly vector or by direct contact with an infected animal. There are various subspecies. *L. donovani donovani* causes epidemic leishmaniasis in humans and is principally a human disease but *L. donovani infantum* is transmitted by the sandfly to dogs and jackals. Humans may become infected by direct contact. Both of these forms (Visceral leishmaniasis) cause weight loss, malaise, lymphadenopathy and anaemia in humans. Cutaneous leishmaniasis is caused by *Leishmania tropica minor* (human reservoir) and other species (rodent reservoirs) and is transmitted by the sandfly in urban and rural areas respectively. In this form of the disease the main lesion is a small un-healing ulcer.

Signs in animals: Dogs may develop the cutaneous and visceral forms of the disease and can, along with rodents, be a source of infection for humans. The skin lesions may be focal or spread as a crusty area with hair loss. The disease may progress to a wasting form with generalized lymphadenopathy, anaemia, enlarged spleen and terminally, death. Treatment may be difficult.

Diagnosis: Whole blood, serum, skin biopsy. Lymph node biopsy. Fixed and fresh tissues from dead animals including spleen, bone marrow, heart blood, liver and lymph nodes for microscopy and PCR.

Control: Control sandfly vectors (for example, *Phlebotomus* sp. and others).

Leptospirosis

Cause: *Leptospira interogans*. There are many serovars of this organism. In cattle, the most important species are *L. hardjo* and *L. pomona*. In pigs, *L. pomona* is the main serovar. Other serovars are common in wildlife and most serovars will infect a range of species.

Epidemiology: The organism is usually spread in infected urine with consequent contamination of water and pasture. Large numbers of organisms may be passed in abortion fluid. The organism can gain entry through mucous membranes including the conjunctiva. On farms pigs are often an important source of leptospirosis in cattle. Urine from infected rats and other rodents can contaminate human drinking water.

Signs in animals: There may be abortion in pregnant livestock or silent infections in juveniles depending on the serovar involved. Cows which abort may develop red milk. *Leptospira pomona* can cause haemolytic anaemia and haemoglobinuria in calves. Leptospirosis in dogs may result in systemic disease with fever, haemolytic anaemia and jaundice.

Diagnosis: Serum (paired samples preferably taken 2–4 weeks apart), urine (fresh and fixed in formalin), foetal stomach contents and samples of placenta from abortion cases, fresh and fixed tissues from dead animals. Whole blood and milk. Microscopy and culture where possible. Molecular methods can also be used to identify Leptospires.

Control. Isolate sick animals. Streptomycin-penicillin in large doses may be effective in treating cases although some animals become persistent carriers. If *Leptospira* sp. induced abortion is confirmed then all stock (including pigs) should be tested for leptospirosis (using molecular methods or serology) to identify the source of infection. In some countries vaccination is available for livestock.

Meliodosis

Cause: *Burkholeria pseudomallei*

Epidemiology: This organism is common in the environment of rice paddies and wet areas. It is a species of bacteria that is carried by many wildlife species, especially rodents, as well as domestic stock and may be passed in the faeces of healthy animals. Signs in animals: There may be none. In horses, there may be a 'glanders-like disease' with mild respiratory signs or systemic disease with fever and occulonasal discharges. Recovery is often complete. Diagnosis: The Mallein skin test is available to diagnose the disease in horses. Nasal discharge and/or tissues may be submitted for culture/microscopy/PCR screening.

Control: Isolation of clinical cases. Handlers should wash hands after handling sick animals and before handling healthy animals, equipment should not be shared between animals. Note that Glanders in horses is caused by *B. mallei*, this disease is also zoonotic.

Plague

Cause: *Yersinia pestis (Pasteurella pestis)*

Epidemiology: This bacterium can survive in the environment for prolonged periods. The typical urban cycle for bubonic plague involves rodents and the rodent flea with accidental transmission to humans. Humans become infected when bitten by infected fleas, this is more likely when living in crowded rat-infested accommodation. The epidemic plagues (like the black death in the middle ages) are now rare although the disease still occurs sporadically in Asia and the Americas. Wildlife biologists working with ground squirrels have been infected. The pneumonic form of the disease is transmitted from human to human by aerosol.

Signs in animals and humans: Infected rodents develop septicaemia and die; the fleas leave the dead host to infect a new one. Humans and other animals may develop the bubonic form of the disease which is named due to the formation of buboes (enlarged lymph nodes) in the axillae and groin. The bubonic form can be fatal within a few days if not treated with antibiotics. Fever and malaise, enlarged spleen, vomiting and so on also occur as the disease develops. The pneumonic form in humans is initially respiratory and later becomes systemic. In cats and dogs the disease is more frequently localized and presents itself as a skin lesion and enlarged regional lymph node.

Diagnosis: Fixed and fresh tissues for microscopy and culture. PCR can also be used. In the live animal culture of blood may allow isolation of the organism. Whole blood, serum and sputum samples from clinical cases and skin lesion biopsy from affected cats or dogs may be useful.

Control: Control rodents and fleas. Large doses of streptomycin-penicillin combinations may be used to treat cases. Vaccines are available for humans.

Rabies

Cause: Lyssavirus

Epidemiology: Sylvatic (wild) and urban cycles exist. The virus is usually transmitted via the saliva following a bite from a rabid animal. In the sylvatic cycles the main reservoir hosts include small carnivores, such as stoats, weasels, civets and the mongoose, as well as larger carnivores, such as the wolf, wild dog, foxes and jackal. Bats may also act as reservoir hosts and cause a proportion of the rabies cases in humans, infection can occur via a bite or by inhaling infectious material. In urban areas, the most important reservoir host is the domestic dog. Ruminants, horses and humans may become infected if bitten by a rabid wild, or domestic, animal.

Signs in animals: The virus acts on the nervous system and results in unusual behaviour. There may be a furious form and a dumb form in the same animal but often one phase will predominate. In domestic dogs the animal may become aggressive, saliva may form around the mouth and the animal may be unable to eat or drink. Rabid cattle may become more friendly or more aggressive than normal and often bellow.

Diagnosis: Quarantine any animal suspected to have rabies. An infected dog usually dies within 10 days following the onset of signs. Send the entire head (chilled) to the laboratory after notifying the nearest veterinary unit.

Note: diagnosis is possible on a brain specimen even after a degree of necrosis. Examine for rabies by FAT, ELISA[1]. There are also kits available from WHO with swabs to take a brain smear sample without fully opening the skull and a preservative to place the swab in to send to the laboratory. Molecular tools are also now available to confirm a diagnosis.

Control: Vaccinations are available for people considered 'at risk', for example, animal handlers, and for domestic dogs, cats and livestock. In some countries vaccination campaigns are also used to control the virus in wild animal populations (for example, foxes). In urban areas there should be a programme for controlling the number of stray dogs, for example, neutering.

If rabies is suspected it is important to contact the relevant authorities and to wear appropriate PPE. If bitten the wound should be thoroughly washed with soap and water and a series of post exposure prophylaxis treatments given. https://www.who.int/rabies/resources/en/

Ringworm

Cause: *Microsporum canis, Trichophyton* spp. and others

Epidemiology: These fungi live in the keratin layers of the skin and in the hair coat of various domestic and wild animals. Animal to animal and animal to human transmission occurs via direct contact with infected skin and hair or by contact with contaminated bedding and grooming equipment. Signs in animals: There may be few signs or there may be hair loss with scaling of the skin and the formation of plaques. The hair may be brittle and discoloured. The skin lesions are rarely itchy unless there is some secondary bacterial infection. Animals with severe ringworm often have an underlying immune deficiency and therefore other diseases should also be considered.

Diagnosis: Hair and skin scrapings from infected animals. Other samples as indicated for a general health profile. Microscopy and culture.

Control. A number of anti-fungal washes and creams are available. *Dermatophilus* sp. (bacterial skin infection) may cause similar skin lesions in livestock so this should also be considered. Antibiotics for secondary skin infections may be required. Humans should make sure that they wash their hands after handling infected animals. Wash contaminated equipment and bedding and so on using antifungal chemicals.

Salmonellosis

Cause: *Salmonella* spp. There are many species of Salmonellae that are potentially pathogenic to both livestock and humans. In poultry, the host adapted strains are *S. pullorum* and *S. gallinarum*. *S. enteriditis* occurs in the oviducts of poultry and may cause food poisoning in humans (via ingestion of contaminated eggs). In humans, the host adapted strain *S. typhi* is responsible for typhoid fever. *S. typhimurium* is a cause of Salmonellosis in many species and is common in rodents.

Epidemiology: Salmonellae are common in the intestinal tract of many species, a few are non-pathogenic and cause no disease. In some species, a host adapted strain may cause no disease but another strain may result in severe gastro-enteritis and often systemic complications. Generally, salmonellosis is an enteric disease and is contracted through the ingestion of contaminated feed or water. Rodent contaminated fodder may infect cattle and pigs as well as humans. Human to human transmission occurs in typhoid especially where hygiene is poor. Systemic disease with Salmonellae at any time during pregnancy may result in abortion. Uncooked poultry meat and eggs are a special risk to humans.

Signs in animals and humans: There may be no signs or there may be a severe acute gastro-enteritis with vomiting (humans) and diarrhoea. Systemic infections with complications resulting in death may also occur, these infections are characterized by fever and possibly a rash.

Diagnosis: Whole blood, serum, faecal smear and faeces for culture. In dead animals, fresh and fixed tissues for culture and histological examination. Liver, spleen and heart blood should be collected.

Control: Affected animals should be isolated. Hygiene is very important, that is, appropriate food storage to keep rodents out, wash hands after handling animals or using the toilet, good latrine facilities for humans and appropriate facilities for disposing of animal wastes. There are vaccines against some *Salmonella* spp. infections in poultry and other livestock. Meat and eggs should be cooked well prior to consumption.

Schistosomiasis

Cause: *Schistosoma* spp.

Schistosoma mansoni causes intestinal schistosomiasis and is prevalent in Africa and South America. *S. haematobium* causes schistosomiasis of the urinary tract and is prevalent in Africa and the Middle East. *S. japonicum* occurs in Asia and causes intestinal schistosomiasis. Humans are the main hosts for these three species. Other species may cause cutaneous larva migrans or swimmer's itch. Other species of schistosomes predominantly infect animals.

Epidemiology: This trematode organism is often found in waterways especially slow flowing streams and in lakes. Intermediate hosts include snails and other aquatic life. Humans and animals may become infected by swimming in contaminated water or by accidentally ingesting schistosomes on unwashed vegetation. The organisms live in the blood vessels of animals, most commonly the mesenteric blood vessels but *S. haemolyticum* lives in the blood vessels of the bladder.

Signs in animals: There may be few signs in animals but there is sometimes diarrhoea, with or without blood. In humans, the intestinal form of the disease can be more severe leading to damage of the intestinal tract and wasting. In the urinary form of the disease there is haematuria and extensive damage to the urinary system unless the disease is treated early.

Diagnosis: In the human disease (Bilharzia) serum samples, faecal sample, rectal snip, for the urinary form, also urine (fixed and fresh). In animals, faecal samples should be collected (note that schistosomes are often not very pathogenic in animals), there is also a nasal form in animals in which schistosome eggs can be found in nasal secretions.

Control: Wash vegetables well in clean water before eating them. Avoid bathing in contaminated water.

Tapeworm

Cause: *Taenia* spp. (*Taenia saginata, T. solium*). There are also numerous other tapeworms which may infect humans (including those acquired from eating raw fish). *Taenia* species tapeworms also occur in dogs, the cysts of these tapeworms may occur in sheep, cattle, yaks and pigs (occasionally humans) and result in few or no signs, however, the cysts of *T. multiceps* may cause space occupying lesions in the brain resulting in ataxia, abnormal behaviour and recumbency.

Epidemiology: Humans become infected by ingesting infective cysts in beef (*T. saginata*) or pork (*T. solium*). Infection with *T. solium* cysts (*Cysticerus cellulosae*) may also occur by accidentally ingesting the tapeworm eggs in human faeces. (Humans may also become infected by the cysts of the dog tapeworm *T. multiceps* following the accidental ingestion of dog faeces.)

Signs in animals and humans: There are often few signs in animals unless the infective cysts develop in the brain or in a medullary cavity of a bone in which the signs will be localized to that of the space occupying lesion. In humans infected with *T. saginata* there may be few signs except the passing of tapeworm segments in the faeces. *T. solium* infection may result in few signs in humans unless cysts develop causing hepatic, pulmonary or neurological signs.

Diagnosis: Cysts can be seen in meat at the time of meat inspection. In humans, radiology and ultrasound may be used to diagnose the pres-

ence of cysts. Tapeworm segments can be found in human faeces.

Control. General hygiene and appropriate latrine facilities. Do not use human waste as a fertilizer (night manure). Do not contaminate grazing pasture with human excrement. Cook all meat well before consumption. (Dog tapeworm infections should be controlled by routine worming, see also hydatids.)

Toxoplasmosis

Cause: *Toxoplasma gondii*

Epidemiology: This protozoan parasite commonly cycles through rodents and domestic and wild cats. The cat is the end host and frequently shows no outward signs of disease. Rodents, humans and other omnivores become infected by consuming sporulated oocysts in food contaminated by cat faeces or by consuming infective toxoplasma cysts in meat. Herbivores become infected by ingesting oocytes on contaminated pasture.

Signs in animals: In the adult cat, there may be no signs but in kittens there may be a fever and general malaise with myositis and diarrhoea. In dogs and humans there may be a non-specific fever and flu-like signs. In pregnant women, the organism may cross the placenta and cause birth defects. Abortion may occur in infected sheep. Neurological signs may occur in neonates.

Diagnosis: Serum (preferably paired sera taken 2–4 weeks apart) for serology, microscopy, HA,[3] FAT and ELISA, faecal samples from cats. Tissue samples from meat animals.

Control: General hygiene. Pregnant women should not handle contaminated soil or cat faeces. Wash hands after handling meat. Cook all meat products well before eating.

Trichinosis

Cause: *Trichinella spiralis*

Epidemiology: This is a nematode helminth that is common in the muscles of pigs, wild boar and other omnivores including rodents and bears. Humans and other omnivores become infected by eating meat infected with tissue cysts of *T. spiralis*.

Signs in animals: There may be none. There will be cysts in the muscles which may cause a transient myositis. In some animals, including humans, there will be an initial gastroenteritis after consumption of the meat followed by signs of anaphylactic shock up to 2 weeks later when the helminth larvae migrate through the tissues. There will be fever, headache and vomiting.

Diagnosis: The cysts in meat can be identified at official meat inspection but need to be followed up by trichinoscopy at the veterinary laboratory. An ELISA test and molecular tools are also now available to screen meat for this parasite.

Control: Official meat inspection (including pork, horse meat and game). Cook meat well before eating (note that drying and smoking meat does not kill *T. spiralis*). Do not feed pigs uncooked waste food. Control rodents in the piggery.

Tuberculosis

Cause: *Mycobacterium bovis*. Epidemiology: *Mycobacterium bovis* is an obligate intracellular bacterium most commonly identified in cattle but it also infects deer and other ungulate species. Wildlife may also be important for the transmission of the bacterium from one area to another. Once the disease becomes endemic in wildlife it is very hard to eradicate. The organism may be located in focal areas in the liver, lungs and lymph nodes at post-mortem. Humans may be infected by drinking unpasteurized milk from an infected cow or when handling or consuming infected meat.

Signs in animals: Initially there may be few signs. Later in the disease the animal may have enlarged lymph nodes and become thin. In the pulmonary form the animal may show respiratory signs with coughing and laboured breathing.

Diagnosis: The tuberculin test (a skin fold test) may be used to diagnose the infection in live animals. It must be mentioned, however, that infection with non-pathogenic environmental *Mycobacteria* spp., for example, *M. avium* may cause false positive results on this test so advice should be sought from a veterinary officer on performing and interpreting the results of the tuberculin test. Most diagnoses of bovine tuberculosis are made at post-mortem or meat inspection. Send fresh and fixed tissues to the laboratory (especially enlarged lymph nodes). Microscopy and culture should be performed where possible and feasible. Molecular techniques are also commonly used. Some serological screening tests are also under development but the performance of these is highly variable. Control: Tuberculin testing and quarantine procedures. Treatment should not be attempted. Milk should be boiled before consumption by humans and all meat should be officially inspected. Note that in countries where human tuberculosis is endemic livestock may become infected with *Mycobacterium tuberculosis*. In Asia, human tuberculosis is prevalent in domesticated elephants and other species.

Typhus

Cause: *Rickettsia* species

Epidemiology: Most of the agents causing typhus are transmitted by an arthropod vector. Tick typhus (*Rickettsia conori*) and scrub typhus (*Orienta tsutsugamushi*) are transmitted to humans by various tick and mite species (respectively) from wildlife reservoirs (especially rabbits, rodents, also quail). Human typhus (epidemic typhus; *R. prowazekii*) is now less common. It is generally transmitted by the human louse and is not strictly a zoonotic disease. The murine (mouse) form can also be transmitted by lice. Fleas may also act as vectors for rickettsial disease. Signs in animals: There may be no signs or there may be fever and general malaise.

Diagnosis: Evidence of the arthropod vector. Serum and whole blood from human patients for serology. PCR can also be used. Control: Control the arthropod vector and avoid being bitten. Doxycycline has been found effective in treatment.

Visceral larva migrans

Cause: *Toxocara canis* (also *Toxocara cati, Toxascaris leonina*) Epidemiology: This is the common ascarid helminth of the dog (cat) and most puppies are born infected with it. Puppies and to a lesser extent dogs pass parasite eggs out in their faeces. Humans, especially children, become infected accidentally when playing in the environment, when holding infected puppies, or by ingesting contaminated soil or dog faeces containing the infective stage. The worm larvae migrate through the intestinal tract into the liver, lungs and sometimes the brain and eyes.

Signs in animals: There may be few signs unless the infection is large in dogs. Puppies (and kittens) may have diarrhoea and a swollen belly. In heavy infections, there may be sudden death following blockage of the intestine or neurological signs due to aberrant migration of larvae.

Diagnosis: Worm eggs in dog (and cat) faeces.

Control: General hygiene. Routine worming of dogs and puppies. Puppies (and kittens) should be dosed with a suitable anthelmintic every 2 weeks from 2 weeks of age until 4 months old and then 6 monthly. Bitches (and female cats) should be wormed before whelping and then 6 monthly.

Note: Due to the risk of hydatids, dog faeces are considered potentially dangerous. Gloves should be worn when handling dog faeces.

Additional information about any of the diseases listed can be obtained through your local or national public health and veterinary authorities. Some key references are provided below. Owing to the changing nature of some pathogenic agents it is important to seek the most up to date advice with regard to disease prevention and control. For the most up to date guidelines on testing for specific diseases in livestock species check the online edition of the OIE Manual of Diagnostic Tests and Vaccines for Terrestrial Animals.[2]

Endnotes

1 See Chapter 6 for more explanation of the FAT = fluorescent antibody test and the ELISA.

2 http://www.oie.int/standard-setting/terrestrial-manual/access-online/.

3 HA = Haemagglutination tests, FAT and ELISA (see Chapter 6 and index).

Bibliography

Centers for Disease Control and Prevention (2018) https://www.cdc.gov/.

Cork, S.C., Hall, D., Liljebjelke, K. (eds) (2016) One Health Case Studies: Addressing Complex Problems in a Changing World. 5M Publishing. Sheffield.

Krause, D.O., Hendrick, S. (eds) (2011) Zoonotic Pathogens in the Food Chain. CABI, Oxford.

Krauss, H., Weber, A., Appel, M., Enders, B., Isenberg, H.D., Schiefer, H.G, Slenczka, W., von Graevenitz, A., Zahner, H. (2003) Zoonoses. Infectious Diseases Transmissible from Animals to Humans, 3rd edn. ASM Press, Washington DC.

OIE (World Organization for Animal Health) http://www.oie.int/standard-setting/terrestrial-manual/access-online/.

World Health Organization (2018) http://www.who.int/topics/zoonoses/en/.

Zinsstag, J., Schelling, E., Waltner-Toews, D., Whittaker, M., Tanner, M. (eds) (2015) One Health: The Theory and Practice of Integrated Health Approaches. CABI, Oxford.

appendix 2

Necropsy guidelines

Samuel Sharpe

A2.1 Role of anatomical pathology and necropsy in veterinary disease investigation

A diagnostic pathology service should be an integral part of any veterinary service/disease investigation programme. In rural areas, where access to laboratory services is limited or in hot climates where rapid putrefaction of cadavers may make samples unsuitable for ancillary testing gross, post-mortem examination may be the only method of disease investigation. Even in areas where lab services are more readily available, gross necropsy examination should form the basis of any disease investigation.

Thorough necropsy examination allows the following.

- Characterization of disease presentation: Are the presentation and gross lesions typical for the suspected disease process or atypical and warrant further investigation?
- Characterization of subclinical or co-morbidities: The cause of morbidity and mortality is rarely simple and is often multifactorial. Relying on targeted laboratory testing may overlook these important co-morbid factors necessary to understand and remedy an animal health problem.
- Correlation of lesions with results of other testing: ancillary testing is rarely totally sensitive or specific and cannot usually differentiate from true clinical infection, active subclinical infection or post-infectious carrier status. Necropsy examination is required to put ancillary testing results into context.
- Informs a logical direction of further testing: information gathered during a thorough post-mortem examination will allow the refining of initial differential diagnosis list and targeting or staging of often costly ancillary testing.
- Sample collection for ancillary testing including: histopathology, microbiology, toxicology and parasitology.
- Sample collection for tissue/sample banking: Tissues can be collected for histopathology and ancillary testing. In addition, formalin-fixed or fresh-frozen tissue can be archived to provide material for future retrospective assessment.

In this way pathology acts as a bridge between clinical practice and the diagnostic laboratory, providing samples and direction for ancillary testing on one hand and interpreting and contextualizing diagnostic testing results on the other. For these reasons, it is important that veterinarians and paraprofessionals become experienced in necropsy technique and sample collection and can recognize the gross tissue changes or lesions typical for disease conditions common to the area.

A2.2 Reasons for carrying out necropsy-based disease investigation

- Sudden unexplained death of a single or group of animals: This is a very common scenario with which a pathologist may be presented. In these cases, historical information is usually limited and this necessitates an even more thorough and methodical approach to necropsy so important details are not overlooked. The sudden unexpected nature of the death often causes the owner to jump to the conclusion that the animal or animals have been poisoned. While this is a consideration it is relatively rare, and the disease investigation should be approached with an open mind.
- Epidemic disease outbreak: This may include epidemic disease with high mortality or a disease outbreak with high morbidity and low mortality with implications for production efficiency and animal welfare. In either case necropsy protocol may be adapted to focus on the organ or system of interest suggested by clinical signs and history. In these cases, it is very useful to examine several animals who have succumbed to disease along with several morbid animals at various stages of the clinical course euthanized specifically for necropsy examination. Limiting investigation to animals that have died naturally may mean that acute lesions with the most diagnostic utility are overlooked or have been masked by secondary changes.
- Research: Many live animal experiments will be terminal and require necropsy examination to evaluate outcomes and collect samples for further testing. In addition, any unexpected mortalities should be fully investigated both to rule out an unexpected outcome with the experimental model and to investigate the possibility of concurrent disease which may jeopardize the validity of results or negatively impact welfare of experimental animals.

- Surveillance: Surveillance encompasses the aggregation of disease investigation and diagnostic testing data to create a picture of animal disease in a particular area. In many areas, this is based on active surveillance programmes which use molecular tests to look directly for a limited number of economically important diseases. While molecular testing has rapid throughput, high sensitivity and specificity they have a narrow focus and will only detect what is tested for. This approach is good for collating information about known disease syndromes; however, will invariably miss the presence of new or emerging disease. Basing surveillance programmes on necropsy examination and analysis of routine diagnostic specimens is much more likely to provide information on these conditions which may be of significant importance.

A2.3 Equipment and safety

Equipment required:

Complete necropsy examination and sample collection can be performed in the field with very minimal equipment.

Personal protective equipment (PPE):

- coveralls
- rubber gloves
- cut-resistant gloves/arm sleeves
- rubber boots
- protective eyewear
- container for waste sharps.
  Tools:
- knife: with fixed handle, that is, not a folding knife
- scissors
- scalpel blades and handle
- forceps
- rib cutters (long-handled gardening shears are ideal)

- hacksaw, hammer, chisel/flat-headed screw-driver (for brain removal)
- cutting board or tray
- string
- notebook and pen for recording findings
- sampling containers: plastic containers for fixative, clean sample bags or empty plastic containers, vacutainer blood tubes
- 5 ml, 10 ml, 20 ml syringes, 18 gauge and 22 gauge needles
- camera: very useful for documenting lesions for second opinion.

Note that tools and sampling materials should be gathered and prepared prior to beginning the necropsy examination. This includes preparing and labelling appropriate sampling containers, which should be clearly labelled to allow identification of the animal and type of specimen. In addition to improving efficiency of the process, gathering and labelling sampling containers will also reduce the risk that any samples will be overlooked during the necropsy.

Safety considerations:

Performing field necropsy examination involves working with sharp instruments and often very heavy specimens without access to specialized machinery, such as winches or hydraulic tables found in a post-mortem laboratory. In addition, zoonotic risk of potential pathogens encountered should always be a primary consideration.

If anthrax is suspected based on signalment, history or external appearance the carcass should not be disturbed and testing performed before necropsy is performed.

Working with an assistant is always advisable, even when examining small animals, as this will facilitate efficient and accurate dissection and makes handling and manoeuvring larger animals much safer. To reduce the risk of accidental injury and to ensure that all lesions are inspected by the professional in charge of the procedure only essential personnel should take part in the necropsy.

Equipment should be well maintained and kept in good working order. Knives and other cutting tools should be sharp to make work easier and reduce the risk of injury to personnel. Appropriate PPE including designated rubber boots and coveralls should be worn. Thick rubber gauntlets, which can be disinfected and re-used, are recommended over latex examination gloves, which may not be robust enough for necropsy examination. Cut-resistant gloves either made of Kevlar or chainmail can be worn underneath rubber gloves on the non-cutting hand. Eye protection in the form of safety glasses or a plastic face shield is strongly advised to reduce the risk of droplet contamination.

If there is a choice for where the necropsy is to be performed it is strongly advised to select a site that is practical and can be easily disinfected to minimize the risk of disease spread. It is important to know and to adhere to local rules and regulations governing handling and disposal of biohazardous material in your area.

A2.4   Mammalian necropsy technique

Note that the following text will describe a necropsy technique with the animal positioned in left lateral recumbency. This section reflects the demonstration video on the website. It should be noted, however, that a wide variety of techniques have been described and it is up to the prosector how they wish to position the animal for necropsy examination. It is strongly advised to always orientate the animal in the same way and perform the necropsy in the same order. Over time the prosector will become familiar with the normal location and appearance of organs and structures in the chosen orientation and this will aid in the detection of abnormalities.

In addition, the exact necropsy procedure including the order in which organs are removed and sampled will be governed by many factors. In cases where a specific syndrome is suspected examination and sample collection can be directed toward the organs or systems involved. This of importance when investigating diarrheic illness especially if moribund animals have been euthanized specifically for necropsy. In these cases, it is imperative to remove and sample the GI tract as soon as the carcass is opened to preserve mucosal lesions which can be masked very rapidly by autolysis.

1 clinical history

A full history should be acquired from animal husbandry personnel including:

- number of sick/number of dead animals (morbidity and mortality)
- species, age, sex of affected animals
- duration and nature of clinical signs
- location of affected animals
- details of husbandry (feeding, housing, recent movement and so on) particularly if any factors have recently changed.

All details should be included on submission forms if samples are to be sent to a diagnostic laboratory. This information is invaluable in helping lab personnel decide on appropriate testing and to understand your concerns and questions as the submitter and ensure that these are addressed satisfactorily.

2 weigh the carcass

If a scale is available this is a very useful piece of information. Body weight can be compared to published normal values for the age and species of the animal in question. In addition,

comparison of organ weights, particularly heart and liver, to the total body weight can provide useful information about the size of these organs.

3 external examination

The necropsy should begin with a careful examination of the external appearance of the animal. Proceed in a logical order to ensure nothing is missed or not examined. Particular areas to note include the following.

Body condition

- Semi-quantitative body condition scoring (that is, on a point scale out of five or nine) as performed in live animals is not validated for post-mortem specimens, but body condition can still be estimated by palpation of the ribs and bony prominences of the lumbar vertebrae and pelvis (see section on estimating nutritional condition).

Hair coat and skin

- What is the quality of the hair coat? Are there any areas in which pigmentation is reduced? Is the hair easily broken or epilated?
- Are there any areas of hair loss? If yes, is the hair completely absent (true alopecia) or have the hair shafts been broken (possible evidence of self-trauma, which may be indicative of pruritus)?
- Skin should be pale pink: common abnormal colours include red (haemorrhage, congestion, inflammation), yellow (jaundice/icterus), black (hyperpigmentation/melanocytic neoplasia), blue (cyanosis/vascular malformation).
- Are there any areas of surface exudate, ulceration or trauma?
- Are any ectoparasites noted?

In all cases the pattern and area of distribution should be carefully described and noted and ectoparasites collected into 70% ethanol to allow for further identification.

Eyes

- Is there exudate (mucus/pus/blood) on the ocular surface or more likely matted into the periocular hair?
- Is the cornea clear so structures within the anterior chamber can be viewed? (See common post-mortem artefacts.)
- Are there any masses in the anterior chamber?
- Are the eyes sunken? This could indicate severe dehydration or emaciation (see section on estimating nutritional condition).
- If the globe appears abnormal it should be removed and fixed in formalin for histological examination (see the section 'Removing the eye').

Nose and ears

- Is there exudate (mucus/pus/blood) coming from the nose or matted into the hair around the nares or inside the pinnae?

Mouth

- Examine the lip margin/mucocutaneous junction, mucous membranes and teeth.
- Mucous membranes should be pale pink: common colour changes are similar to those in the skin.
- Note condition of the teeth: more complete examination of dentition can be performed after removal of the pluck (see later).

Anus

- If the animal has been recently scouring, faecal material is often matted into the fur around the anus.

Vulva/penis

- Exudate or evidence of urine scalding should be noted.

Mucocutaneous junctions

- It is very important to closely examine all mucocutaneous junctions (palpebral margin, oral cavity, nares, anus, vulva/penis and coronary band of the hooves). Many important infectious diseases (for example, foot and mouth disease) may produce characteristic lesions in these locations. Lesions can be subtle and are rarely pathognomonic requiring thorough examination and sampling for histopathology and ancillary testing.
- Are there any vesicles, plaques, warty growths, areas of ulceration or trauma to any mucocutaneous junctions?

Estimating nutritional condition

Poor nutritional condition can result from simple protein-calorie malnutrition either as a result of simple lack of food or impaired feeding as a result of dental disease or orthopaedic problems which impair foraging behaviour. In addition, increased loss of protein either via damaged gastrointestinal (GI) tract (protein losing enteropathy) or kidneys (protein losing nephropathy) or increased metabolic rate associated with chronic inflammatory processes or neoplasia can also result in cachexia. Nutritional condition can be estimated in several ways.

1 Assessment of body condition by palpation of bony prominences.
2 Examination of the eyes: sunken eyes are associated with dehydration, but also can be seen in emaciated animals as the retro-orbital fat pad is depleted.
3 Subjective assessment of adipose tissue in subcutaneous tissues, pericardium, greater

omentum, falciform ligament and within the retroperitoneal space surrounding the kidneys.

4 Splitting a long bone (usually the femur) and examination and subjective assessment of fat content of the bone marrow. This is the last fat store to be used and if depleted is consistent with severe emaciation. Objective measurement of marrow fat is relatively easy and can be performed by taking a 10–20 g piece of bone marrow which should be weighed and the weight noted. The marrow should be placed in a covered container and allowed to desiccate in room air over a period of days. Weight should be measured every day until several days of no reduction in mass are noted. Marrow fat is then calculated as a percentage of dry weight/wet weight.

4 Reflect limbs and skin the body

Note that when cutting through skin avoid cutting on the skin surface more than necessary. Body hair is very hard and will rapidly blunt the knife. Instead make a stab incision through the skin and cut on the underside of the skin.

- Make a stab incision in the right axilla and continue to cut through the subcutaneous tissues and skeletal muscle of the thoracic girdle until the limb can be reflected over the back of the animal.
- Make a stab incision in the inguinal region. Cut through the subcutaneous tissue and skeletal muscle to locate the coxofemoral (hip) joint. Incise the joint capsule and cut through the ligament of the head of the femur so the limb can be reflected.
- Incise the skin along the ventral midline from the mandibular symphysis (chin) to the pubis.
- Skin back the skin overlying the right half of the animal from the ventral midline to the level of the lateral spinous processes.

Things to note

- Amount and colour of skeletal muscle.
- Amount and colour of subcutaneous fat.
- Is there any fluid within the subcutaneous tissues? What is the colour and consistency of the fluid?
- Is there any evidence of petechial (less than 2 mm diameter), purpural (2–20 mm in diameter) or ecchymotic (greater than 20 mm diameter) in the skin, subcutis or muscles?

Sampling

Formalin:

- Any lesions in skin, subcutis or skeletal muscle.
- A section of skeletal muscle.
- Subcutaneous lymph node (axillary or prescapular nodes are usually encountered during forelimb dissection).
- Sciatic nerve: it is advisable to collect a peripheral nerve and the sciatic nerve is easy to find underneath the caudal thigh muscles running along the caudal border of the femur. The nerve is broad, tan-white and soft to firm. A 2–3 cm long section is sufficient.

5 Open abdomen

- Taking care not to puncture the abdominal viscera cut through abdominal musculature around the curve of the last rib from the xiphoid process of the sternum to the lateral spinous process of the first lumbar vertebra.
- Continue the incision just ventral to the lumbar muscles and around the wing of the ileum to the pubis.
- This will form a triangular flap of abdominal wall which can be reflected ventrally.
- Opening the abdomen in this way will allow for the identification of any fluid which may be present without allowing it to spill out.

Things to note

- Is there any fluid in the abdomen? What is the volume? Any more than a few millilitres of clear, straw coloured fluid is abnormal.
- If fluid is seen: Colour? Clear or opaque? Is there any material (for example, fibrin, clotted blood, feed material) floating in the fluid?
- Are organs in correct anatomical location? If not, this may be a sign of displacement, torsion or volvulus. Note that artefactual displacement can occur when the body is moved post-mortem. True displacements are usually accompanied by evidence of vascular compromise (see section on interpretation of lesions).
- Texture of the serosal (outer) surface of the GI tract. The serosal surface should be smooth and shiny. Note any granularity or material adherent to the serosal surface.
- Amount and colour of fat within the greater omentum and around the kidneys.

Sampling

- Sample any abnormal fluid into a plain (red topped) blood tube.

6 Open thorax

- Lift the rib cage and look behind the liver to inspect the peritoneal surface of the diaphragm.
- The diaphragm should be taut and describe a curve along the inside of the rib cage.
- Make a stab incision through the diaphragm and listen for an in-rush of air:

  - this means that there was negative pressure inside the diaphragm
  - an absence of this finding is consistent with pneumothorax (air within the pleural space) and this finding should be noted.

Causes of pneumothorax

Pneumothorax is a serious clinical problem and is invariably a significant finding at post-mortem. Air can enter the pleural space either via a defect in the chest wall or via a leak in the lung tissue.

Causes include:

- trauma, either as a result of direct injury or as a result of rib fracture resulting in lung laceration
- neoplasia
- abscess or granuloma
- migrating foreign body.

If pneumothorax is detected careful examination of the lungs and chest wall is warranted to find a source of air leakage.

- Cut around the diaphragm along the curve of the last rib.
- Cut through the proximal part of the ribs at the level of the rib head. Rib cutters do not deal well with soft tissue so make a deep cut through the subcutaneous tissues and muscle on the outside of the chest at the level of the rib head.
- Use the rib cutters to cut the ribs one at a time.
- Note the ease with which the ribs can be cut. This is subjective and will become easier to judge with experience. This may be useful in estimating bone density.
- Using the knife (in smaller animals) or the rib cutter, cut the distal part of each rib at the level of the sternebrae and remove the chest wall.

Things to note

- Is there any fluid in the pleural space? If so, what is the volume, colour, consistency of the fluid and is there any material present in the fluid? Any more than a few millilitres of clear, straw coloured fluid is abnormal.

- Are the lungs fully inflated and largely filling the thorax?
- Are any part of the lungs adhered to the chest wall or to the other structures in the chest?

  - If present can the adhesions be easily broken down (fibrin, acute) or are they very tough (fibrous, chronic)?

- Are there any other organs apart from the lungs, heart and mediastinal tissues in the chest? Diaphragmatic hernia can lead to herniation of abdominal organs into the chest, and this should be noted along with the location and size of the diaphragmatic defect.

Sampling

- Sample any abnormal fluid into a plain (red topped) blood tube.

7 Open pericardium

- Using scissors and forceps, make a linear incision in the pericardium from the apex to the base of the heart.
- Record the volume and character of any fluid.
- Note if the pericardium is adhered to any part of the outer surface of the heart.

Sampling

- Sample any abnormal fluid into a plain (red topped) blood tube.

8 Remove pluck

Definition: The pluck is defined as the closely associated organs in the neck and thorax from the tongue to the diaphragm.

- Working on the ventral aspect of the head free the tongue by cutting the muscles in the floor of the mouth.
- With the knife tight against the inside of one of the mandibles make a stab incision until the point of the knife is felt against the roof of the mouth.
- Cut forwards to the mandibular symphysis and caudally to the angle of the jaw.
- Repeat on the other side.
- Using a blunt probe (a knife sharpening steel is most effective) lever the tongue out of the mouth and complete excision by cutting connective tissue at the frenulum.
- In young animals, the mandibular symphysis can be incised using rib cutters allowing more complete examination of the oral mucosa and dentition.

The hard and soft palate is now visible between the mandibles. Carefully examine the gingiva and oral mucosa for lesions. To facilitate examination the mandibular symphysis can be cut with rib cutters.

- Incise the soft palate around the caudal border of the hard palate.
- When the hyoid is encountered, disarticulate one of the joints visible adjacent to the larynx allowing dissection to continue caudally.
- Continue the incision dorsal to the oesophagus and using the tongue as a handle dissect the oesophagus and trachea free from the connective tissues of the neck as far as the thoracic inlet.
- The thoracic contents are attached dorsally and ventrally via reflections of the pleura. These can be incised by cutting just ventral to the vertebral bodies and dorsal to the sternebrae.
- At the diaphragm, locate and ligate the oesophagus with a double ligature. Incise the oesophagus between the ligatures.
- In turn incise the caudal vena cava and aorta and remove the plunk. Set aside for examination.

9 Remove abdominal contents

Monogastric species

- Lift the right lobes of the liver to expose the hilus.

  - In small animals check patency of the bile duct by making a small incision in the wall of the proximal duodenum adjacent to the pylorus and expressing bile by gently applying pressure to the gall bladder.
  - Additionally, in small animals closely inspect vessels entering the liver for aberrant blood vessels linking the portal vein with the systemic circulation. This would be indicative of an extra-hepatic portosystemic shunt.

- Incise the vessels at the hilus of the liver and continue to dissect around the cranial edge of the stomach.
- Locate the cardia and remove the stomach with a small piece of diaphragm associated with the distal oesophagus.

Ruminants

- In ruminants, the very large size and weight of the forestomachs (> 100 kg in large bovines) makes it is advisable to separate the intestines from the forestomachs prior to removing from the abdomen. This makes handling and examination of the GI tract much easier particularly if operating with limited number of helpers.
- With the animal in left lateral recumbency the pylorus of the abomasum will visible in the ventral part of the abdomen.
- Ligate the pylorus with two pieces of string and incise between them to avoid spillage of gut contents.
- Feel around the reticulum at the dorsocranial end of the rumen underneath the rumen and locate the oesophagus. Remove the fore-

stomachs together with a small section of diaphragm associated with the oesophagus and set aside for examination later.

All species

- Free the intestines by cutting through the mesenteric attachments and mesenteric root in the cranial part of the abdomen.
- In horses inspect the abdomen and root of the cranial mesenteric artery at this point. Larvae of *Strongylus vulgaris* larvae mature within the cranial mesenteric artery and can cause arterial occlusion and non-strangulating infarction of the intestines.
- Ligate the rectum with two pieces of string, incise and set the entire GI tract aside.

Spleen

- In all species, the spleen is attached to the right lateral aspect of the stomach and will come out when the GI tract is removed.
- Dissect the spleen free and examine the capsular surface and set aside for examination.

Liver

- The liver is held in place by ligaments on the cranial and dorsal aspect of the organ.
- Dissect around the cranial and dorsal aspect of the liver and remove. Set aside for examination.

10 Locate and remove adrenals

- Adrenals are located at the cranial pole of both right and left kidneys nestled between the aorta and vena cava. In emaciated animals, the glands are prominent and can easily be recognized. In animals with abundant retroperitoneal fat careful palpation in the vicinity of the glands may be required to identify them.

- Dissect out and incise.
- The adrenal cortex around the periphery of the glands will appear tan-orange and slightly waxy while the medulla will be dark red-plum coloured.

Things to note

- The ratio of the thickness of the medulla to the cortex on cut section should be 2 : 1.
- The corticomedullary junction should be uniform.
- Any alterations in cortical : medulla ratio, areas of reddening (haemorrhage), nodules, or masses should be noted.

Sampling

- Place both adrenals in formalin.

11 Remove the urogenital tract

- To remove the urogenital tract intact the pelvis must first be split. This is performed by removing soft tissues from the ventral aspect of the pelvis to expose the pelvic symphysis and obturator foramen. Use rib cutters to cut the ischium and pubis either side of the obturator foramen and dissect the resulting section of bone free.
- Take a urine sample with a needle and syringe inserted through the bladder wall. Take a sample prior to opening the bladder to avoid contamination with blood or other fluid.
- Free the kidneys by carefully dissecting around the lateral and dorsal aspects of the organs.
- Applying gentle traction to both kidneys will allow identification of the ureters as taut bands in the retroperitoneal connective tissue. This is much easier in lean or emaciated animals with minimal retroperitoneal adipose tissue.
- Dissect the ureters to the levels of the bladder.

Intact females

- Locate the ovaries and dissect ovaries and uterine horns and body free of retroperitoneal connective tissue. Incise the uterus at the cervix and set aside for examination.
- Careful dissection along the underside of the sacrum will allow the rectum to be freed along with the intrapelvic parts of the urogenital tract.
- Set aside the urogenital tract for examination.

Bread loaf

All parenchymatous organs and tissue (liver, spleen, lung, kidney, tongue, skeletal muscle and so on) have an external and cut surface both of which must be examined. To examine the cut surface the organ or tissue is 'bread loafed', a term which describes making multiple, transverse cuts through an organ, 1–2 cm apart, to allow the internal structure to be examined.

12 Examining pluck and sample collection

Tongue

- Examine the external surface and bread loaf the tongue transversely along its length into 0.5–1 cm pieces.

THINGS TO NOTE
- Vesicles, ulcers, nodules, masses within the mucosal surface.
- On the cut surface, the substance of the tongue comprises skeletal muscle and fibroadipose tissue, is soft to firm, tan pink, pliable and relatively homogenous. Areas of abnormal colour or texture should be noted.
- Check for the presence of foreign bodies.

SAMPLE COLLECTION

- Take a transverse section including mucosa after bread loafing and place in formalin.

Oesophagus

- Using scissors and forceps open the oesophagus along its length.

THINGS TO NOTE

- The mucosa should be homogenously tan, slightly undulating, soft, with an intact, shiny surface. Mucosal defects, ulcers, vesicles, nodules, areas of congestion or haemorrhage or masses should be noted and sampled.
- The submucosa should be abundant and loose. Masses, strictures or foreign bodies should be noted.
- The muscularis is comprised of soft to firm, pale pink smooth muscle. Masses or foreign bodies should be noted.

SAMPLING

- Place a 2 cm long, opened section of oesophagus in formalin.

Larynx, trachea and mainstem bronchi

- Inspect the laryngeal aditus comprising the epiglottis, arytenoids and vocal folds.
- Using scissors and forceps incise between the arytenoids and bisect the dorsal part of the cricoid cartilage. Continue the incision down the dorsal ligament of the trachea and on into the mainstem bronchi.
- Locate the bronchial lymph nodes adjacent to the mainstem bronchi. Incise and examine the cut surface.

THINGS TO NOTE

- Swelling of the laryngeal mucosa, masses or foreign material which may have resulted in upper airway obstruction.

- Presence of fluid or material including stable foam, purulent or blood tinged exudate, or food material within the tracheal lumen should be noted.
- The mucosa should be shiny, intact, and translucent. Mucosal defects, ulcers, vesicles, nodules, areas of congestion or haemorrhage should be noted.

SAMPLING

- Take a 2 cm long section of opened trachea into formalin.
- Take a cross section of lymph node including the capsular surface, cortex, medulla and hilus.

Thyroids and parathyroids

- Thyroid and parathyroid anatomy varies quite significantly between domestic species; an anatomy text should be consulted if the prosector is unfamiliar.
- Thyroids and parathyroids should be carefully dissected, examined and sampled.

THINGS TO NOTE
Thyroids

- Thyroids are paired glands and left and right thyroids should be of symmetrical and of similar size and shape. Glands are diffusely soft to firm, dark red-brown, with a smooth outer contour.
- Assessment of size and weight is subjective as weight ranges for endocrine organs in veterinary species are not available.
- Bilateral increase in size may indicate goitre. Goitre occurs as a result of accumulation of unconjugated colloid in thyroid follicles which can occur secondary to iodine excess or deficiency.
- Iodine deficiency can occur as a result of absolute dietary deficiency or secondary to ingestion of goitrogenic plants or chemicals

which interfere with iodine absorption from the GI tract or conjugation of colloid.

- Goitre is of particular significance in small ruminants, pigs and cattle. Maternal iodine insufficiency will result in congenital goitre and is a cause of foetal loss, abortion and birth of weak-born neonates.
- Unilateral increase in size may indicate benign or malignant neoplasia.
- Of most significance in cats and dogs.
- Cats frequently suffer from functional thyroid neoplasia resulting in thyrotoxicosis.
- Dogs uncommonly suffer from non-functional malignant neoplasia. Morbidity is associated with locally invasive growth and pulmonary metastasis.

Parathyroids

- Bilateral enlargement can be indicative of hyperplasia resulting in secondary hyperparathyroidism: occurring as a result of dietary deficiency of calcium or vitamin D (nutritional hyperparathyroidism), or as a result of chronic renal failure (renal hyperparathyroidism).
- Identification of bilateral enlarged parathyroids should prompt thorough examination of the kidneys and long bones.
- Unilateral enlargement can be indicative of neoplasia or less commonly a parathyroid cyst.
- Functional neoplasia can result in primary hyperparathyroidism causing derangements in calcium metabolism and chronic depletion of bone mineral.

Lungs

- Lungs should be diffusely soft, air filled, pale pink to red.
- Airways should be empty or contain a very small amount of clear, tan fluid.
- Lungs are particularly well-vascularized with abundant, loose interstitial tissue and as such

are very prone to effects of post-mortem hypostatic congestion (see section of post-mortem artefacts).

- For this reason, colour change can be misleading and gentle palpation of lung tissue is required to detect changes in texture or the presence of masses.
- After this the lungs should be bread loafed via transverse sections perpendicular to the mainstem bronchi.

THINGS TO NOTE

- Excess fluid or material in airways: note character, amount and presence of any material:

  - clear fluid or foam: suggestive of oedema, heart failure or acute alveolar damage
  - opaque, thick, yellow-green: suggestive of purulent exudate, acute of chronic inflammation
  - dark red, thick: suggestive of haemorrhage.

- Changes in texture: this will usually manifest as increased firmness of affected tissue. Note particularly the distribution of lesions as this can particularly useful in identifying the cause of lesions:

  - diffuse: oedema or interstitial pneumonia: consider left sided heart failure, viral infection, toxic lung disease (for example, fog fever)
  - multifocal: embolic process: consider septic emboli (search for primary source, heart valves, caudal vena cava, mammary gland) or neoplastic emboli
  - focal: inflammation (abscess versus granuloma) rather than primary neoplasia
  - cranioventral: consistent with bronchopneumonia (see Figure A2.7a).

SAMPLING

Formalin:

- A 0.5–1 cm thick transverse section from each lung.
- Bronchial lymph node.

Fresh-frozen:

- Larger piece of lung for bacterial culture/ancillary testing.

Heart

- Leave the heart attached to the lungs to facilitate examination of great vessels.
- Make a transverse incision across the ventricles just below the level of the papillary muscle.
- This allows measurement of the thickness of the left and right ventricular (LV and RV) free walls and the interventricular septum. The LV free wall and interventricular septum should be of similar thickness. Ratio of LV : RV thickness should be 2 : 1 to 3 : 1.
- Open the proximal pulmonary artery to check for thromboemboli.
- Complete heart examination by dissecting the heart in the direction of blood flow.
- Begin at the right auricle and open the auricle and atrium parallel to the atrioventricular valves.
- Dissect the right ventricular free wall by carefully incising around the coronary groove up and out of the pulmonary outflow tract.
- Open the left auricle and atrium in the same way as the right.
- Make a single cut in the left ventricular free wall down through the mitral valve.
- Open the aortic outflow tract by cutting up through the septal leaflet of the mitral valve.
- In cattle, it is often useful to make a transverse incision through the papillary muscle of the left ventricle as this is a common site for

Figure A2.1 Bovine left ventricular papillary muscle. Note the multifocal to coalescing areas of reddening on the cut surface of the papillary muscle corresponding to areas of ischaemic infarction as a result of bacterial vasculitis caused by *Histophilus somni* infection. See also Online figure 1. Photo: courtesy of C. MacGowan, University of Calgary, Canada.

necrotizing myocarditis caused by *Histophilus somni* infection (Figure A2.1).

THINGS TO NOTE

Examination of the heart should encompass the myocardium, endocardium, heart valves, and great vessels. In overview, the left ventricle is thick walled and forms a point at the apex of the heart. Note if the silhouette of the heart is rounded as this may indicate myocardial remodelling.

- Myocardium: The heart is a muscle and as such should be diffusely soft to firm, dark red with an intact, shiny epicardial surface. Areas of reddening, pallor, excessive softness or firmness should be noted and sampled.
- Valves: In health both the atrioventricular and semilunar valves have a fibrous core covered by shiny, intact endocardium and are delicate, translucent structures with a sharp free edge. Disruption of the endocardium usually by bacterial infection (vegetative valvular endocarditis, see Figure A2.6) with result in accumulation of fibrin and clotted blood on the valvular surface.

- Endocardium: The inner surface of the heart should be diffusely glistening and translucent with the myocardium visible underneath. Post-mortem clotted blood should be differentiated from significant pre-mortem thrombosis. Post-mortem clots can easily be removed from the heart leaving the endocardium intact. Thrombosis occurs when the endocardium is damaged and so will be adherent and leave a defect when removed.
- Great vessels: The aorta and pulmonic artery are thick walled elastic vessels. Walls should be homogenously tan-white with considerable elasticity.

  - Aberrant vessels: Persistence of the ductus arteriosus, the embryological vessel which allows blood to bypass the foetal lung in utero is the most common aberrant vascular anomaly. In animals, greater than a few hours old (one week on equine neonate) the ductus should be closed and after this gradually fibroses to form the ligamentum arteriosum. In young animals check that no communication exists between the aorta and pulmonic artery just distal to the semilunar valves.

SEPTAL DEFECTS

- Atrial septal defect: arises from failure of the foramen ovale to close post-natally. The foramen ovale allows communication between the atria. Atrial septal defect will be found in the caudal part of the interatrial septum.
- Ventricular septal defect: This lesion occurs as a result of failure of the endocardial cushion to completely fuse. These are embryological structures which form the interventricular septum. Septal defects are most commonly seen high in the interventricular septum adjacent to the septal leaflet of the mitral valve.

WEIGHING THE HEART

- Once the heart has been thoroughly examined it can be excised from the pluck by incising the aorta and pulmonic artery just above the pulmonic and aortic valves and cutting through the atria.
- Gently rinse any clotted blood from the lumen and weigh the heart whole.
- Values for normal heart weights by species are published, but a good rule of thumb is that the heart should be between 0.5–1% of body weight.
- In cats, any heart > 17 g is considered enlarged.
- Increased heart weight is indicative of myocardial hypertrophy and should be noted.

SAMPLING
Formalin:

- A 0.5 cm thick transverse section comprising interventricular septum and the root of the left and right ventricles (T-piece section).
- A 0.5 cm thick section of the papillary muscle of the left ventricle.
- A longitudinal section through the right ventricular free wall encompassing a part of atrioventricular valve.

Fresh-fixed:

- A larger piece of ventricle for bacterial culture/ancillary testing.

13 Examining the liver

- Knowledge of the liver anatomy of the species under examination is imperative to completely examine the liver.
- The liver should be diffusely dark red/orange, soft to firm with sharp caudal borders. Large blood vessels enter at the hilus and arborize into the periphery.

Things to note

- Weigh the liver. Values for normal liver weights by species are published, but a good rule of thumb is that the liver should be about 4% of body weight.
- Lay the liver out with the cranial (surface facing the diaphragm) uppermost and examine the gross anatomy ensuring that all liver lobes are present.
- Note any marked discrepancy in size of liver lobes. Note any diffuse changes in colour or texture: is the tissue softer or firmer than expected? Note if the caudal border of the liver is bulging or rounded.
- Capsular surface: the capsular surface should be homogenous and smooth. Look for areas of red or tan discolouration, material adherent to the capsular surface, depressions, nodules, or mass lesions.
- Cut surface: bread loaf the organ into 1 cm slices and carefully examine the cut section of each piece. The cut section should be homogenously dark red/orange, exude a small amount of red tinged fluid, and contain cross sections of blood vessels and bile ducts. These will have a larger diameter closer to the hilus. Note any areas of discolouration, masses, nodules and if any of the lesions seen on the capsular surface extend onto the cut section.
- Note any thickening of vessel or bile duct walls.

COMMON CHANGES THAT MAY BE NOTED
AFFECTING ALL OR PART OF THE LIVER

- Small, firm, pallid: This may indicate atrophy and fibrosis and is suggestive of a chronic change following inflammation or ischaemia.
- Nutmeg liver: chronic passive venous congestion resulting from impaired venous return to the heart is common. The name is derived from the appearance of the cut surface which takes on a finely mottled appearance reminiscent of a nutmeg. The appearance is imparted

by change within the lobules, the functional liver unit. Dark red points indicative of congestion of central veins and centrilobular sinusoids with hepatocellular loss are surrounded by areas of tan pallor which come from intact but degenerate hepatocytes swollen by abundant intracellular lipid.

- Diffusely enlarged, swollen, dark red, exudative cut surface: this may indicate acute venous congestion.
- Swollen, pallid, rounded caudal borders, soft texture: accumulations of fat or glycogen within hepatocytes as a result of hepatocellular dysfunction or extra-hepatic disease can give the liver a swollen appearance. The tissue will be soft, friable and may have a greasy texture.

14 Examining the Spleen:

- The spleen is a strap-like parenchymatous organ with a fibromuscular outer capsule. The spleen is comprised of the blood filled, sinusoidal spaces which make up the red pulp and lymphoid tissue of the periarteriolar lymphatic sheaths which is termed the white pulp.
- The ability of the splenic capsule to stretch and red pulp sinuses to dilate and fill with blood means the spleen can vary dramatically in size.
- Lymphoid hyperplasia can be grossly apparent as myriad tan nodules (2–3 mm in diameter) visible grossly on cut surface.

Things to note

If the spleen is enlarged it can be useful to characterize gross changes in the spleen into four categories.

1 Diffusely soft and bloody: congestion:

- artefact of barbiturate anaesthesia
- acute septicaemia: anthrax, salmonella

- acute hemolytic disease
- torsion: leading to impaired venous outflow.

2 Diffusely firm and meaty: cellular infiltration, hyperplasia of splenic elements, amyloidosis.

- Cellular infiltration:

    - diffuse, infiltrative neoplasia, for example, lymphoma, histiocytic sarcoma
    - diffuse, infiltrative inflammatory infiltrates, for example, fungal or protozoal infection.

- Hyperplasia of reticuloendothelial cells lining red pulp sinuses:

    - chronic sepsis
    - chronic hemolytic disease.

3 Multiple nodules: soft and bloody-nodular lymphoid hyperplasia, hemangiosarcoma.
4 Multiple nodules: firm and meaty-neoplasia versus granulomatous inflammation.

15 Examining the urogenital tract

- Open the kidneys by incising longitudinally through the greater curvature to the pelvis.
- Peel the capsule off and note if there are any adhesions of the capsule to the underlying kidney. This may indicate acute or chronic inflammation.
- Continue to incise down each ureter to the urinary bladder.
- Open the urinary bladder and continue to incise through the trigone down the urethra into the vagina (female) or down to the end of the penis (male).

Things to note

- Adhesions of the capsule to the cortical surface.
- Increase or decrease in size of the kidney:

    - swollen and 'wet' looking kidneys may indicate acute inflammation or necrosis
    - small, firm kidneys may indicate chronic renal injury which has led to atrophy and fibrosis.

- Areas of reddening or pallor:

    - due to the end-arterial blood supply of the kidneys infarction of renal blood vessels lead to infarction
    - renal infarctions tend to be wedge-shape oriented with the long edge toward the capsular surface
    - acute infarction will be dark red and swollen. Chronic infarction will be shrunken and pallid.

- Mass lesions:

    - multiple small masses: embolic bacterial nephritis versus embolic neoplasia
    - single large mass: primary renal tumour.

- Expansion of the renal pelvis:

    - the renal pelvis should be an almost slit-like space containing at most a small amount of urine
    - hydronephrosis: backpressure due to obstruction of a ureter will result in cystic dilation of the pelvis which can result in partial or complete atrophy of the overlying renal tissue
    - pyelonephritis: ascending bacterial infection.

Sampling

Formalin:

- Take a 0.5 cm transverse section of both kidneys which encompasses the capsular surface, cortex, medulla, papilla and pelvis.

Fresh-frozen:

- A larger piece of kidney to submit for bacterial culture/ancillary testing.
- A piece of kidney and surrounding adipose tissue for toxicology.

16 Examining the GI tract

Stomach

MONOGASTRIC
- Open the stomach along the greater curvature and examine the contents. Note quantity and quality of contents. Take a fresh sample of stomach content for toxicology testing.
- Remove content and gently rinse mucosa under running water.
- Note any areas of reddening, erosions, ulcers, nodules, or mass lesions.
- Take a 2 cm² full thickness section of gastric wall from the fundus and pylorus into formalin.

RUMINANTS
- Open the rumen along the great curvature and examine the contents. Take a sample of rumen content for toxicology testing.
- Remove content and examine the mucosa which should form tall, broad papillae. Note any areas of ulceration, fibrosis and scarring, reddening and the presence of any parasites.
- Open the reticulum and examine the contents which should be very similar to rumen content. This is a common site for foreign bodies to become lodged. Examine the mucosa and note any lesions.

- Open the omasum. The ingesta will be noticeably drier and pressed tightly between tall leaves of mucosa. Note any areas of erosion, ulceration or even complete perforation of omasal leaflets.
- Open the abomasum along its length. Note the amount and quality of ingesta and the presence of any gross apparent parasites.
- Carefully examine the mucosa and note any areas of reddening, erosion, or ulceration, or roughening or nodularity which may be associated with encysted nematode larvae.

Intestine

- Examination of the intestines should include the gut tube and mesenteric lymph nodes.
- Examination of the small intestine can be facilitated by freeing the intestine from the mesentery to linearize the gut tube. Make a stab incision in the mesentery anywhere along the intestine. Grasp the gut and continue to incise the mesentery with the knife blade perpendicular to the long axis of the intestine.
- Lay out the linearized intestine and examine.

Things to note

- The serosa should be translucent and glistening. Note any roughening, areas of discolouration or material (for example, fibrin) attached to the serosal surface.
- In small animals, the entire GI tract can be opened and examined.
- In larger animals, select several 15–20 cm sections of intestine to open. These areas can be chosen at random or you can choose to open areas which have an abnormal appearance on the serosal surface. Using scissors snip the wall and then open longitudinally to expose the gut content and mucosa.
- Note the quality of the ingesta and the appearance of the mucosa. The mucosa should be

glistening with very little material adherent to it.

- Mucosal associated lymphoid tissues (MALT) is most abundant in the distal jejunum and ileum and is also present in the more proximal small intestine, cecum, and colon. They appear as plaque-like or multinodular thickenings on the anti-mesenteric side of the intestine.
- Several important diseases in ruminants including bovine viral diarrhoea virus/mucosal disease, malignant catarrhal fever and salmonellosis will cause prominent lesions in MALT. It is therefore very important that when sampling for both histopathology and ancillary testing that these areas are identified and sampled.
- In ruminants open the cecum and spiral colon and examine contents and mucosa.
- Identify mesenteric lymph nodes: incise and examine the cut surface.

Sampling

Formalin:

- Take 2–3 cm long sections of opened intestine and colon directly into formalin. **Do not touch or brush the mucosal surface** as this will damage the delicate villous structures and hinder histological examination.
- Take a cross section of lymph node including the capsular surface, cortex, medulla and hilus.

Fresh-frozen:

- Tie off 2–3 cm sections of jejunum, ileum, and spiral colon with string and excise.
- Several mesenteric nodes should be saved frozen.

Special techniques

Removing the brain

- Remove the head by disarticulating at the atlanto-occipital joint.
- If available, it is strongly advised to secure the skull in a head vice. If not available have an assistant stabilize the skull while the work is completed.
- Make a full thickness incision in the skin overlying the poll and frontal bone and skin the head to, and remove, the ears.
- De-flesh the occipital and parietal bones by cutting away the masseter muscles bilaterally.
- Using a saw make two slightly divergent cuts in the occipital bone running craniolaterally just inside the occipital condyles (Figure A2.3).
- Using the saw make a transverse cut across the frontal bone just caudal to the zygomatic process of the frontal bone (Figure A2.4).
- Now make two sagittal cuts bilaterally through the occipital, parietal, and frontal bones joining up the cuts in the occipital bone and frontal bones (Figure A2.4).

Figure A2.2 Bovine cerebral cortex. Note multifocal areas of slightly raised, red areas within meninges and extending into the cortical grey matter corresponding to areas of ischaemic infarction as a result of bacterial vasculitis. See also Plate 45. Photo: courtesy of C. MacGowan, University of Calgary, Canada.

Figure A2.3 Removing the brain. Arrows delineate the direction of the first two slightly divergent cuts just inside the occipital condyle. See also Online figure 2.

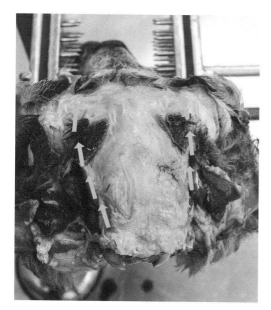

Figure A2.4 Removing the brain. Arrows delineate the direction of the remaining cuts, the first through the frontal bones at the level of the bone just caudal to the zygomatic process of the frontal bone. Following this, two sagittal cuts are made bilaterally through the occipital, parietal, and frontal bones joining up the cuts in the occipital bone and frontal bones. See also Online figure 3.

- Using a flat-headed screwdriver or chisel lever the skull cap up and expose the meninges covering the cerebral hemispheres and cerebellum.
- Incise the dura mater and dissect away from the underlying brain.
- To remove the brain from the skull: tilt the skull backwards as the brain will be freed from front to back. Beginning at the rostral part of the brain cut through the olfactory bulbs. Moving along the ventral surface cut the optic nerves at the chiasm and continue to work ventrally and laterally to sever the other cranial nerves.

SAMPLING
- The brain should NOT be bread loafed prior to formalin fixation.
- If the brain is to be submitted for histopathology it should be immersion fixed in a large volume of formalin, usually in a container separate to other histological samples.
- If fresh brain is required for ancillary testing, make a sagittal cut between the hemispheres of the cortex to bisect the brain along its length. Half of the brain should be fixed and half should be retained frozen. This will ensure all brain areas are represented in both samples.

Opening limb joints

- It is always advisable to open and assess at least three joints during necropsy.
- This is particularly important in neonates in which septic arthritis is a common manifestation of septicaemia.
- Specific technique will vary between joints depending on anatomy but in general the overlying area should be skinned and subcutaneous tissues and muscle should be removed to facilitate identification of the joint space. Using the point of the knife incise the joint capsule and collateral ligaments on

one side of the joint. Applying tension to open the joint space continue to incise peri-articular connective tissue until the joint can be opened.

- Note the quality and quantity of synovial fluid. Normal synovial fluid should be tan to orange and viscous.
- Inspect the synovium and articular cartilage. Note any defects or areas of roughening.

Removing the eye

- To complete necropsy examination both eyes should be removed. This is particularly important if the history suggests visual impairment but should be done in all cases if time allows.
- The eyes will be removed with the extraocular tissues.
- Make a circumferential incision through the eyelids around the edge of the orbit. Using scissors, grasp the sclera, cut the extraocular connective tissue around the outer surface of the globe until the optic nerve can be viewed. Incise the optic nerve and remove the eye.
- Fix the eyes in formalin.

A2.5 Necropsy technique: avian

Follows demonstration video on the website.

External examination

- Weigh the body.
- Plumage: examine the plumage and note any broken or burnt feathers or abnormalities.
- Eyes and infraorbital sinuses: note any ulcers, vesicles, proliferative lesions on the palpebral margins. Note any swelling of the infraorbital sinus.
- Beak: examine the beak and note any proliferative lesions at the mucocutaneous junction.

Open the mouth and examine the tongue and choanae looking for any erosions, ulcers, proliferative lesions or exudate.

- Vent: check the vent and note any soiling of the feathers.
- Feet and legs: closely examine the un-feathered skin of the legs and feet. Note any areas of ulceration or proliferative lesions. Pay particular attention to the soles of the feet as this is a common site for bacterial skin infection ('Bumblefoot') which can be a route of entry for bacteria causing septicaemia.
- Uropygeal gland: the uropygeal or 'preen' gland is located on the dorsum at the base of the tail. It is vestigial or not present in certain orders of birds. Swelling or proliferative lesions should be noted as infection and neoplasia can occur at this location.

Internal examination

1 Before opening the bird it is advisable to wet the feathers with soapy water and then pluck the ventral aspect of the body from the neck to the vent.
2 Skin the ventral aspect of the carcass by making a long incision through the skin from the ventral neck to the vent and completely skin ventral aspect of the bird. Assess hydration status by the tackiness of the subcutaneous tissues.
3 Locate the thymus: in young birds the thymus can be found subcutaneously on both sides of the neck. Thymic tissue is multifocal and is clustered around the jugular vein and vagus nerve. If the thymus cannot be visualized it is advisable to take a section of skin from around the expected location and fix for histology. Serial sections of the area can be taken for histological examination which may reveal thymic tissue.
4 Luxate the hindlimbs by applying pressure to the coxofemoral joints so the bird sits

flatter on the table. At the same time evaluate the coxofemoral joint for degenerative changes and so on.

5 Assess body condition by noting amount of subcutaneous adipose tissue and the prominence of the pectoral muscles. In an emaciated bird the pectoral muscles will be atrophied and the sternum will appear more prominent.

6 Open the body cavity: by incising through the body wall caudal to the sternum. As you lift the sternum inspect the abdominal air sacs.

7 Remove the pectoral muscles bilaterally by filleting away close to the bone.

8 Remove the sternum: cut through the ribs and expose the shoulder joint. Disarticulate the shoulder joint and remove the soft tissues from the scapula. The entire sternum along with coracoid, clavicle and scapula can be removed whole.

Note that the sternum can be removed by cutting directly through the coracoids but this limits examination of the thoracic girdle. This method may be appropriate in domestic poultry but in wild birds where a relatively subtle lesion in the sternal structures can seriously impact the ability to fly it is best to remove the sternum and examine intact.

9 Locate the thyroids which can be found at the root of the common carotid arteries. Dissect, measure and fix in formalin.

10 Open the pericardium: note the volume and character of fluid within and assess for any adhesions between the pericardium and epicardium.

11 Remove the heart: by cutting the great vessels leaving 1 cm of vessel attached to the heart.

12 Examine the heart: The epicardial surface should be diffusely glistening and the heart should have a triangular shape and sharp

apex. Cut the apex off the heart and assess the right and left ventricular wall ratio. Similar to mammals the ratio of left to right should be 2–3 : 1. Cut the heart in half along its long axis to visualize the atrioventricular valves and endocardial surface.

13 Examine the digestive system: remove the lower beak by cutting at the angle of the mandible. Inspect the oral cavity, tongue and choanae.

i   Open the oesophagus and crop: note the quantity and quality of content and inspect the mucosa which should be diffusely glistening. Note any erosions, ulcers or masses. Place a 2 cm long section of opened crop into formalin.

ii   Remove the digestive system: The distal oesophagus, stomachs, intestine, ceca, liver, pancreas and spleen can be removed together. Incise the distal oesophagus just below the tracheal bifurcation and applying gentle traction pull the mass of organs from the cloaca.

iii   Inspect and remove the liver: The liver should be homogenously dark red, soft to firm and both lobes should be relatively similar in size. The caudal border of the liver should not extend past the last rib and if this is present may indicate hepatomegaly. Bread loaf and sample for histopathology, bacteriology and toxicology.

iv   Remove the spleen: The shape of the spleen varies with species of bird but in all should be soft, diffusely dark red with a soft, mildly exudative cut section. Measure the diameter of the spleen and cut in half. Half can be fixed and half retained frozen. Impression smears can be made by pressing the cut section against a clean glass microscope slide. This can be useful in detecting haemoparasites and round cell neoplasia.

v   Linearize the intestine: by gentle traction and sharp dissection to cut the mesentery.

vi Open the proventriculus and ventriculus: note the content and inspect the mucosa. The mucosa should be carefully examined as this is a common site for nematode parasites. Sample a strip of ventriculus and proventriculus including the junction between the two as this is a good place to find lymphoid aggregates.

vii Open the intestine: note the contents and inspect the mucosa. Tie off a section of intestine for bacterial culture. The remainder of the intestine with attached pancreas can be placed in formalin.

viii Open the ceca: inspect the mucosa and sample.

14 Examine the urogenital system: remove the gonad (note the sex), adrenals and kidneys. Bread loaf and fix multiple sections of kidney along with gonads and adrenals. Take a sample of fresh-frozen kidney.

15 Open the cloaca to the vent and examine the contents. In juvenile birds locate the Bursa of Fabricius, examine and place in formalin. The Bursa appears as a soft, tan nodular mass on the dorsal surface of the cloaca. This can be found by pulling the stump of the distal colon ventrally. The bursa is very active in young birds but generally atrophies after 6 months of age in poultry.

16 Examine the respiratory tract.

i   Open the trachea longitudinally and inspect the mucosa. Note the presence of any fluid or material in the lumen. All the way to the bifurcation (common spot for stuff to lodge).

ii Remove the lungs from the coelomic cavity using blunt dissection, palpate, inspect and bread loaf. Note the colour of the lung, diffuse pallor indicates anaemia.

iii Sample a section of trachea and a slice of both lungs into formalin. Take a sample of fresh-frozen lung.

17 Examine the musculoskeletal system.

i   Carefully palpate the wing and leg bones and vertebral column. Vertebral fractures as a result of traumatic injury are a common cause of morbidity in wild birds but can be very subtle. Any suspect lesions can be excised and fixed. Following decalcification subtle fractures can be detected histologically.

ii Attempt to break a long bone. Bones should snap when broken. Any flexing or bending may be consistent with reduced mineralization and be associated with metabolic bone disease.

18 Open the skull and inspect the brain: using rongeurs or scissors cut the top off the skull as a cap. The brain can be removed from the skull and fixed or the head can be fixed whole.

A2.6 Sample collection

Note that in all cases it is strongly advisable to collect a basic set of tissues into formalin and also to be retained frozen even if gross examination was unremarkable (see sampling checklist). Histology can identify significant lesions which were not apparent grossly. This can have serious consequences resulting in missed diagnoses if further testing is warranted but appropriate tissue samples have not been retained.

Histopathology

Histopathology is the study of microscopic morbid anatomy and relies on identification of characteristic microscopic changes associated with various pathological processes. This allows further investigation of the underlying cause of lesions recognized grossly. In some cases, microscopic changes will be pathognomonic and definitive diagnosis can be made from examination of the tissue section. More commonly changes will be suggestive of a pathologic process allowing further refinement of the differential diagnosis list and selection of appropriate ancillary testing.

Histochemical staining

The process of thinly sectioning processed tissue and staining with a variety of dyes and other chemicals for microscopic examination has changed little since its invention several centuries ago.

Routine examination is performed on tissue sections stained with Haematoxylin and Eosin (H+E). H+E differentially stains cellular elements (for example, nuclei stain blue or 'basophilic', cytoplasm stains pink or 'eosinophilic') allowing recognition and interpretation by trained personnel. In addition, cells or tissue elements affected by a specific disease process will display stereotypical colour changes aiding in identification of that process. For instance, an acutely necrotic neuron may appear shrunken, with a more densely staining and darker nucleus and more eosinophilic cytoplasm.

Other histochemical ('special') stains can be used to highlight tissue elements (for example, collagen, GAGs, elastin, reticulin), intra- or extra-cellular accumulations (for example, glycogen, mucin), pigments (for example, hemosiderin, melanin, bile, haemoglobin), minerals (iron, copper, lead) or microorganisms (bacteria, protozoa, fungi).

Immunohistochemistry (IHC) and in situ hybridization (ISH) are both techniques which can be performed on tissue sections which utilize the specificity of antibody binding to detect protein antigen (IHC) or nucleic acid (ISH) in tissues. IHC is commonly used in diagnostic labs to detect antigens in tissue section. In surgical pathology expression patterns of cellular antigens in tumours can be used to give accurate diagnosis and prognostic information. IHC can also be used to detect infectious agents in tissues.

All histochemical and the majority of IHC and ISH techniques can be performed on tissue routinely fixed in 10% NBF. If you have questions regarding sampling and tissue preservation for a specific test please contact your diagnostic laboratory.

Sampling technique and preservation

Tissue samples for histology should be taken into a fixative solution the purpose of which is to stabilize proteins in the sample and inhibit degradation prior to processing for histopathology. In addition, fixation increases tissue stiffness and makes cutting of tissue sections easier.

Commonly used fixatives act by either reducing solubility and disrupting tertiary structure of proteins (so-called denaturing fixatives, usually ethanol or methanol) or by stabilizing proteins by creating covalent bonds within and between protein molecules (so-called cross-linking fixatives, usually formaldehyde).

Ten per cent neutral buffered formalin (10% NBF) is the most widely used fixative in anatomical pathology and can be purchased ready-made in liquid form or as a dry powder ready to be reconstituted.

It is strongly recommended to use ready-made 10% NBF if available, as formalin powder is an irritant to mucous membranes and can be hazardous to work with. Additionally, it is very important to correctly buffer the pH of the solution prior to use as improperly buffered formalin increases the formation of artefactual pigments in the tissue which can complicate examination of tissue sections.

Formalin penetrates tissue by diffusion at a rate of approximately 1 mm/h. Tissue samples placed in formalin should be of an appropriate size to allow formalin penetration and fixation.

Guidelines for sampling tissue into fixative

- Tissue sections should be no more than 1 cm thick.
- Ratio of formalin to tissue volume should be no less than 10 : 1.
- Open all sections of GI tract to expose mucosal surface prior to immersion in fixative.
- If sampling a diffuse lesion or taking routine sections of a grossly normal organ try to take a section which will be representative of tissue architecture or anatomy.
- If sampling a focal or mass lesion within an organ take a sample which includes the border between normal and abnormal tissue as this can be the most diagnostically useful.

Packaging formalin-fixed tissue for shipping

- Samples should be placed in a straight sided, screw-topped container.
- Plastic containers are lighter and much less likely to break compared with glass and are therefore more suitable.
- Remember that fixation will cause tissues to become much stiffer and if a narrow-necked container is used samples will become trapped inside after fixation.
- Containers should be placed with a quantity of absorbable material (for example, paper towel) inside a sealable plastic bag.
- Alternatively, the sample can be fixed for 24–48 h and then shipped in a sealable bag

with a small amount of formalin or formalin-soaked paper towel to keep the sample moist. This negates the need to send large volumes of hazardous formalin through the mail and is especially useful if the samples are very large.

- In cold weather, there is a risk that the formalin may freeze during transport to the laboratory. This will produce catastrophic freeze–thaw artefact in affected tissues making examination of tissues very difficult or impossible. In these cases, it is recommended to fix the samples in formalin for 24 h and then transfer to 70% ethanol for transportation. Ethanol has a lower freezing point and this will reduce the risk of freeze–thaw artefact developing.
- All containers should be labelled appropriately:

  - ID of case and tissues submitted and date
  - warning labels to notify that sample contains 10% neutral buffered formalin
  - labels should meet requirements on labelling of hazardous chemicals in your area.

Microbiology

Sampling

Sections of organs containing suspect lesions should be sampled and submitted for ancillary testing. Samples should be taken into clean or ideally sterile bags or containers. Ideally samples should be refrigerated during shipping to the laboratory as freezing can reduce the sensitivity of culture-based techniques. In addition, lesions can be swabbed with appropriate microbiological swabs which can then be submitted for testing. Ideally a section of lesional tissue along with a swab taken from the lesion should be submitted for testing.

Sampling of GI tract involves using string to tie off a short (5–10 cm long) section of bowel with contents. This can then be excised and submitted for testing. Warning labels which meet requirements in your area should be affixed to sample containers to notify that sample contains potentially bio-hazardous materials.

Cross contamination

Infectious agents will be detected by either growing the agent *in vitro* (that is, bacterial or fungal culture or virus isolation in cell culture) or by a molecular test (for example, PCR, ELISA and so on). In either case false positives can occur if contamination of the sample occurs.

To minimize the chance of contamination, ensure the following.

- Use fresh instruments (that is, blades, forceps, scissors and so on) when collecting samples.
- **Package each sample in a separate container**. Do not submit all samples together in the same container as contamination will occur.
- Take samples of other organs prior to opening the GI tract.

Parasitology

Faecal egg count: Estimation of parasite burden can be provided by submitting a fresh faecal sample for a faecal egg count. Fresh faeces (5 g) should be collected into a clean plastic or glass container and refrigerated prior to submission to the laboratory. **Do not freeze**: as this may damage eggs within the sample and produce an artefactually low count.

If more detailed investigation of the number and types of parasites present is required, all gut contents should be collected according to location in the GI tract and submitted for total worm count.

Toxicology

Toxicological testing will be performed on fresh tissue. Samples taken for this purpose should *not* be fixed in formalin. Samples can be taken into leak proof containers and shipped to the laboratory on ice packs. If a delay in testing is expected, samples can be frozen at $-20°C$. Appropriate tissues will depend on the toxin of interest and on the specific requirements of the toxicology laboratory. In general, it should be noted that investigation of possible intoxication should try to establish ingestion, metabolism, and excretion of the toxin in question.

Note that toxicological assessment requires a large volume of sample therefore at least 10 g of tissue or another sample should be retained if possible.

Ingestion

Careful gross examination of stomach contents may identify the presence of toxic material (for example, bait containing a poison, metallic object, or suspect plant material). If not, a sample of stomach contents should be taken and kept frozen prior to submission for toxicological analysis.

Metabolism

Many drugs and toxins are metabolized in the liver and a liver sample should always be retained in these cases.

Fat soluble agents will be concentrated in tissues with high lipid content: for this reason, adipose tissue and brain should be retained (see section on sampling the brain). Testing for organophosphates (OPs) is performed by measuring the activity of acetylcholinesterase in the brain. OPs inhibit acetylcholinesterases and OP intoxication is indicated by reduced acetylcholinesterase activity.

Aqueous humour in the anterior chamber of the eye is a dialysate of blood plasma and remains stable for up to 48 h post-mortem. This can be particularly useful if investigating suspect fatal calcium or magnesium deficiency in cattle. The aqueous humour should be collected via needle and syringe and stored in a plain blood tube prior to laboratory analysis.

Excretion

Many toxins are excreted in the urine. Fresh kidney and urine samples should always be retained in these cases.

A2.7 Description of findings

It is beyond the scope of this chapter to cover in detail how to describe and interpret lesions. However, it is useful to have some idea as to how to describe lesions and begin to broadly categorize them according to the most likely underlying pathogenesis. It is also useful to be able to systematically and logically describe the gross appearance of a lesion for several reasons. First, you may simply be looking for the input and advice of a more experienced pathologist and need to describe the lesion in question accurately and concisely to them. Second, by going through the process of forming an accurate description it can aid the prosector in interpreting and understanding a lesion.

Descriptions of each lesion should be in the follow terms.

- Size: if a lesion is focal or focally extensive (for example, a mass within an organ) it should be measured as accurately as possible using metric units. If a lesion is diffuse it can be useful to weigh the affected organ. If the lesion comprises a change in colour or texture of the surface of an organ or a mucosal surface, giving an estimate of percentage of the total area affected can be very useful.

- Shape: keep it simple: round, oval, irregular, smooth, papilliferous, nodular, fusiform, reniform, flat, miliary and so on.
- Colour: again, keep it simple: red, brown, green, yellow. Reserve using 'black' for obviously melanocytic lesions and 'white' for bone or mineralized tissue. If something is very pale but not mineralized use 'tan' or 'off-white'.
- Consistency: simple. Again. Soft, firm, hard. As an example, forehead is hard, tip of nose is firm, and lips are soft.
- Location: be as accurate anatomically as you can to describe where abouts in an organ or tissue the lesion is located.
- Number: if lesions are multiple count them up to ten. After that use 'tens', 'hundreds', 'thousands' to give a more accurate picture rather than using 'few' or 'many'.
- Inflammation: in life inflammation is characterized by an increase in blood flow to an affected tissue (reddening), increase in vascular permeability (oedema), leading to outflow of plasma proteins (fibrin) and inflammatory cells. All of these changes persist after death and can be used to detect an inflammatory process in an organ.
- Acute: the prime hallmark of acute inflammation post-mortem is the presence of fibrin. Fibrin is a polymerized plasma protein produced during blood clotting and the acute inflammatory response. It often forms strands of soft, tan, friable material but in less abundant quantities can impart a subtle rough texture to the surface of affected organs (Figures A2.5 and A2.6). Affected tissues may also seem reddened or swollen with an excessive amount of fluid exuding from the cut section.
- Chronic: Fibrosis is the hallmark of chronic inflammation and should be considered if an organ or tissue is shrunken and firm. In addition, inflammatory mass lesions such as abscesses (a fibrous capsule filled with

Figure A2.5 Abdomen, cat. Acute inflammation. The last rib (•) and liver (*) can be seen on the left side of the picture with loops of intestine (†) in the centre. Note abundant stands of tan fibrin (‡) covering the serosal surface of the intestines. Note also the subtle finely granular texture of the intestinal serosa which can in some cases be the only sign of fibrin exudation. See also Plate 46. Photo: courtesy of Dr J. Davies, University of Calgary, Canada.

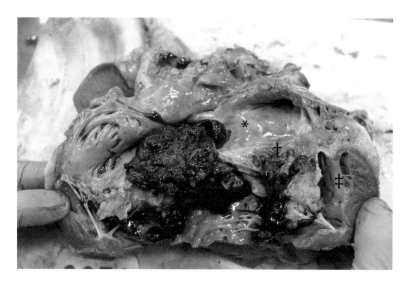

Figure A2.6 Heart, bovine. Acute inflammation: the right ventricle (‡) has been opened exposing the tricuspid valve (†) which is largely effaced by a mass of inflammatory tissue, fibrin, and clotted haemorrhage which extends into the right atrium (*). This is a characteristic appearance of vegetative valvular endocarditis. Bacteria localize to the heart valve following a period of bacteraemia disrupting the endocardium and inciting an inflammatory process and blood clotting which produces the gross appearance. This should be differentiated from post-mortem blood clots which often form in the heart but can be easily removed leaving an intact, glistening endocardial surface. See also Online figure 4. Photo: courtesy of Dr C. Knight, University of Calgary, Canada.

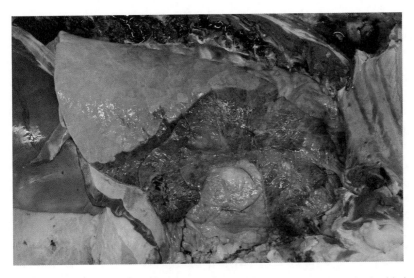

Figure A2.7a Thorax, bovine. Chronic suppurative bronchopneumonia: In this image, the cranioventral parts of the lungs are dark red and multinodular consistent with bronchopneumonia. A chronic process is suggested by the presence of abscesses visible from the pleural surface (‡) and fibrous adhesions between the visceral and parietal pleura (†). See also Plate 47. Photo: courtesy of Dr J. Bystrom, University of Calgary, Canada.

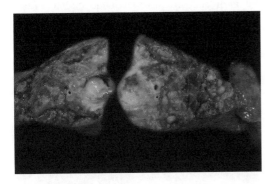

Figure A2.7b Cut surface, lung from Figure A2.7a. In this image, the extent of abscess formation which could be seen from the pleural surface is evident on the cut surface. See also Online figure 5.

purulent material) or granulomas (firm nodular mass lesions with homogenously firm or caseous centres) are characteristic of chronic inflammation (see Figure A2.7a and b).

- Degeneration/necrosis: degenerative changes include intracellular accumulations of fluid, glycogen or lipid, and impart a pallor to affected tissue. Necrotic lesions usually have a similar pallid appearance but will usually be sunken or depressed consistent with tissue loss which is characteristic (Figure A2.8).
- Vascular: acute vascular lesions will usually appear dark red and swollen due to pooling of blood in affected tissues (see Figures A2.1 and A2.2). In chronic lesions tissue will appear shrunken, firm and pallid as a result of atrophy of devitalized tissue with replacement fibrosis.

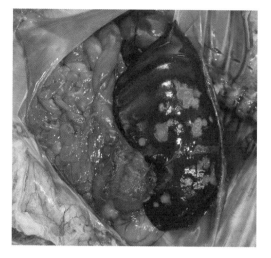

Figure A2.8 Liver, sheep. Necrosis: there are multifocal, slightly sunken areas of tan discolouration throughout the liver, visible from the capsular surface. This is a characteristic appearance of hepatocellular necrosis in this case as the result of *Fusobacterium necrophorum* infection. *F. necrophorum* usually reaches the liver via the portal circulation from the rumen and is a common sequelae to acute rumenitis. Necrosis can be differentiated from degenerative (glycogenosis or lipidosis) changes which can have a similar homogenous tan appearance by the fact lesions in this case are sunken consistent with tissue loss. See also Plate 48.

A2.8 Disorders of growth

Hyperplasia/hypertrophy

Hypertrophy is defined as an increase in size of an organ or tissue caused by an increase in *size* of component cells. In contrast hyperplasia is defined as an increase in size of an organ caused by an increase in *number* of component cells. In both cases the change is reversible and the organ will revert to normal size if the stimulus is removed. In either case the lesions noted grossly will be an increase in size of a tissue or organ in part or whole. Hypertrophy is usually physiological in response to increased demands

on a tissue or organ, for instance hypertrophy of nephrons in one kidney if the other kidney has been removed or damaged or myocardial hypertrophy in response to exercise.

Not all hypertrophic change is beneficial however. Dilated cardiomyopathy in small animals is an idiopathic condition which results in progressive cardiac myocyte dysfunction and loss. In the initial stages of the disease cardiac function is maintained by remaining cardiac myocytes. After time, however, the increased energy demands of hypertrophic myocytes and loss of efficiency results in impaired myocardial function and congestive heart failure.

Hyperplasia can be further subdivided into physiological or pathological change. Physiological hyperplasia can occur as a result of hormonal stimulation in response to changing demands such as mammary gland hyperplasia during lactation, or be compensatory such as in the liver if part of the organ is removed or irrevocably damaged the remaining hepatocytes will divide to restore functional mass. Examples of pathological hyperplasia include benign prostatic hyperplasia in entire

Figure A2.9 Urinary tract, dog. Benign prostatic hyperplasia. Note the enlarged prostate (closed arrow) surrounding the proximal urethra at the neck of the urinary bladder (open arrow). The smooth outer contour and symmetrical appearance of the right and left lobes of the gland are consistent with hyperplasia. See also Online figure 6. Photo: courtesy of Dr J. Davies, University of Calgary, Canada.

male dogs in which chronic androgen stimulation results in hyperplasia of the gland which gradually narrows and occludes the urethra resulting in dysuria (see Figure A2.9). While hyperplasia is reversible in some cases it can be viewed as a pre-neoplastic process increasing the chance of neoplastic transformation in a tissue.

Hyperplasia must be differentiated from neoplasia. This may require putting the lesion into context with other features of the case or require histopathology.

Neoplasia

Neoplasia results from irreversible alteration in mechanisms controlling the cell cycle and cell division. Neoplasia can be benign characterized by localized, expansile growth or malignant associated with invasive growth, metastasis, and other paraneoplastic effects. Benign neoplasia is usually associated with the formation of a mass within an organ or tissue and will be circumscribed or encapsulated. Dependent on location mass effect of a benign growth can be associated with clinical signs. Peripheral nerve sheath tumour for example will often be associated with neuropathy. Malignant neoplasia will grossly appear infiltrative and is often associated with areas of central necrosis as a result of rapid growth leading to ischaemia.

Atrophy

Atrophy is the reduction in size of an organ or tissue as a result of reduction in size or number of component cells. This occurs as a result of withdrawal of trophic factors which may include blood supply, innervation, or hormonal stimulation. As with hyperplasia this may be physiological, as in atrophy of hyperplastic mammary tissue at the end of lactation or be pathological as may occur in skeletal muscle

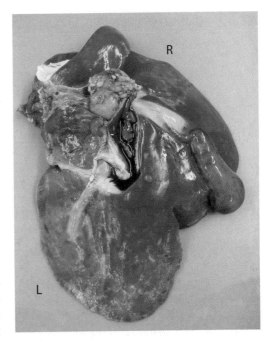

Figure A2.10 Liver, sheep. Chronic fascioliasis (*Fasciola hepatica*). *F. hepatica* is a liver fluke commonly associated with acute and chronic liver disease in small ruminants. Acute disease manifests as sudden death as a result of overwhelming hepatocellular necrosis caused by large number of larval flukes migrating through the parenchyma. Chronic fascioliasis occurs are a result of chronic cholangiohepatitis as a result of the presence of adult flukes within bile ducts. In this image, a tangle of adult flukes can be seen within the lumen of a large bile duct at the hilus. The left lobe is atrophic, fibrotic bile ducts are visible from the cut section and the right lobe is hypertrophied. This appearance is characteristic but it is unclear why the left lobe is preferentially affected. In addition, the hilar lymph nodes are enlarged as a result of chronic antigenic stimulation. See also Plate 49. Photo courtesy of Mr R. Irvine, University of Glasgow, UK.

after chronic disuse or denervation. Pathological atrophy is often seen as a sequelae to chronic inflammation (Figure A2.10).

A2.9 Post-mortem artefact

Post-mortem artefacts that occur as a result of tissue degradation and decomposition can mimic lesions and confuse the unsuspecting prosector. It is not uncommon to receive samples at the lab from veterinarians in the field along with necropsy reports of their gross findings which consist entirely of misinterpreted artefactual lesions. The next section will describe the appearance of several of the most common post-mortem artefacts so that if they are encountered during a post-mortem they can be identified and discounted.

Autolysis

After death, proteolytic enzymes within tissues along with bacterial action cause the breakdown of tissues (Figure A2.11). Tissues with high levels of proteolytic enzymes (kidney and liver)

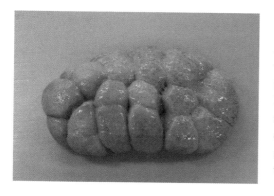

Figure A2.11 Kidney, bovine. Autolysis. Autolysis occurs more rapidly in organs containing higher levels of proteolytic enzymes such as the liver and kidney or that contain large amounts of bacteria such as the GI tract. In this case the kidney is pallid and has lost structure appearing diffusely softened. Note also fine emphysematous gas bubbles in the tissue. See also Plate 50. Photo: courtesy of Mr R. Irvine, University of Glasgow, UK.

or containing large amounts of bacteria (GI tract) will autolyse more rapidly. In addition, autolysis will proceed more rapidly in warmer ambient temperatures and in larger animals or animals with large amounts of adipose tissue, heavy fleece or hair coat which will cool more slowly than smaller carcasses. Affected tissues will lose structure and appear soft and in later stages become entirely liquefied. Even at this stage it is not an entirely lost cause to collect tissues for histopathology as it may be possible to discern some important features even if most tissue detail is lost. Furthermore, molecular testing (for example, PCR) can still be attempted on autolysed tissue although sensitivity will be reduced. Autolysed tissue is not appropriate for bacterial culture as the tissue will be heavily contaminated with bacteria which have invaded post-mortem making interpretation of the significance results impossible.

Cloudy cornea

Transparency of corneas is maintained by the constant removal of water by active transport. After death, this process ceases and fluid enters the cornea by osmosis resulting in gradual opacity. This process will be hastened if the carcass has been frozen and thawed prior to examination. It should be differentiated from true corneal stroma oedema or scarring. Samples should be submitted for histopathology if it is unclear whether this represents a true lesion.

Hypostatic congestion

After death, blood will pool by gravity onto the dependent side of the carcass. This can occur in any organ including the skin and skeletal muscle but is most pronounced in well-vascularized tissues with abundant, loose connective tissue such as the lung (Figure A2.12). It must be differen-

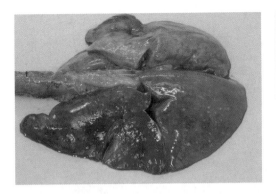

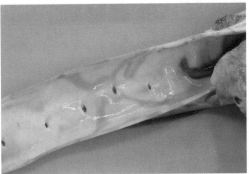

Figure A2.12 Pluck, sheep, the trachea is on the left side of the picture (†). Hypostatic congestion. Note the left lung is diffusely dark red and right lung is diffusely pale pink. In this case it appears that the body was placed on the left side after death and blood has pooled by gravity in the dependent lung. This should be differentiated from inflammation or congestion by gently palpating the lungs to detect a change in texture which is a feature of a true lesion. See also Plate 51. Photo: courtesy of Mr R. Irvine, University of Glasgow, UK.

Figure A2.13 Aorta, sheep. Haemoglobin imbibition. The aorta has been opened longitudinally to expose the endothelial surface which is discoloured by feathered areas of pale red discolouration. This represents post-mortem haemoglobin imbibition in which pigment has leached from vessels into adjacent tissue. In this case the pattern of imbibition follows the vasa vasorum in the wall of the aorta from which haemoglobin pigment has originated. This should be differentiated from true haemorrhage. Note in this case the pale red colour, feathered appearance following course of blood vessels, and the bland appearance of affected tissue which is unremarkable apart from the colour change. See also Online figure 7. Photo courtesy of Mr R. Irvine, University of Glasgow, UK.

tiated from true congestion or inflammation. Striking lateralization with one lung affected and the other unaffected, particularly if it is known on which side the animal has been placed after death is consistent with post-mortem change. Hypostatic congestion will also not affect the texture of the lung whereas inflammation or haemorrhage in the lung will result in increased firmness of the affected tissue. When occurring in the skin or subcutis reddening often has a very sharp, straight line of demarcation between affected and unaffected tissue which is highly unusual in genuine cutaneous congestion.

Haemoglobin imbibition

Haemoglobin pigment will slowly diffuse from blood vessels after death giving a pale red, feathered appearance to tissues. This must be differentiated from true petechial or ecchymotic

haemorrhage. Haemoglobin imbibition occurs around blood vessels and so will often have a dendritic or ramifying pattern or be homogenous and bland with no other alterations in appearance of the tissue (Figure A2.13). True haemorrhage appears as multiple pinpoint areas of reddening (petechiae), larger splashing areas of haemorrhage (ecchymoses) or as paint-brush areas of reddening. Affected tissues may also be swollen or distorted by accumulated blood.

Bile imbibition

Bile pigment can leach from the gall bladder or proximal duodenum resulting in orange or

Figure A2.14 Abdomen, dog. Bile imbibition: The liver is on the left side of the picture (*) and the left liver lobe has been retracted to reveal the orange green gallbladder (†). There has been extensive imbibition of bile pigment into the serosa of the adjacent stomach (‡). Note the bland appearance of the affected tissue with no signs of an inflammatory response which would have been present had bile leakage occurred pre-mortem. See also Plate 52. Photo: courtesy of Mr R. Irvine, University of Glasgow, UK.

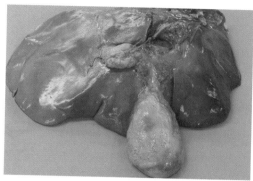

Figure A2.15 Liver, sheep. Pseudomelanosis. The liver is viewed from the caudal surface with the ventral aspect at the bottom of the picture. The right and quadrate lobes are noticeably darker in colour than the rest of the liver consistent with build up of iron sulphide in the tissues as a result of post-mortem bacterial overgrowth. Note also post-mortem emphysema in the wall of the gallbladder. See also Plate 53. Photo courtesy of Mr R. Irvine, University of Glasgow, UK.

green discolouration of surrounding tissues. This must be differentiated from pre-mortem bile leakage which is a serious clinical issue. In post-mortem bile imbibition, the affected tissues appear bland and homogenously stained (Figure A2.14). Bile is very irritant and so when leakage occurs pre-mortem it is associated with significant inflammation producing a localized or diffuse peritonitis.

Pseudomelanosis

Activity of bacteria involved in autolysis can produce iron sulphide which will impart a very dark grey-black colour to tissues and should not be confused with increased pigmentation (Figure A2.15). Melanosis is sometimes seen as sharply demarcated areas of black discolouration in otherwise unremarkable organ particularly in lungs and meninges.

Barbiturate salt precipitate

If euthanasia is performed by the intravenous administration of barbiturates characteristic lesions are associated. Pentobarbital is solubilized in glycerol and ethanol which can impart a strong alcoholic smell to tissues of euthanized animals. In addition, the solution has a low pH and will coagulate blood into a gritty, pale brown paste particularly in the right ventricle. Barbiturates will also come out of solution and crystallize on the endocardial surface or on the pleura or other parts of the lungs if direct intracardiac injection is attempted (Figures A2.16 and A2.17).

Figure A2.16 Thoracic wall, cat. Barbiturate salt crystals. Note the fine, grey-white crystals formed on the parietal pleural surface of the thorax. Crystals can be found on any serosal surface exposed to barbiturate euthanasia solution and are most commonly seen on the pleura, pericardium or large airways following intracardiac injection. Hypercalcaemia. See also Online figure 8. Photo: courtesy of Dr J Davies, University of Calgary, Canada.

Figure A2.17 Pluck, cat. Barbiturate salt crystals. In this case barbiturate salt crystals have precipitated onto the tracheal mucosa. See also Online figure 9. Photo: courtesy of Dr J Davies, University of Calgary, Canada.

Emphysema

Bacterial putrefaction particularly by clostridial organisms will produce a large amount of gas which will give a honeycomb appearance to affected tissue (Figure A2.18). This should be differentiated from true emphysema which is often a feature of clostridial infection. True lesions are always associated with other signs of acute inflammation, vascular comprise and necrosis, such as swelling, haemorrhage and fibrin accumulation (Figure A2.19). In addition, gas will collect inside the GI tract resulting in post-mortem bloat which can result in post-mortem prolapse of intestines or reproductive tract. This must be differentiated from true bloat or prolapse. In true bloat the rumen becomes distended with gas and can result in death as a result of impaired venous return to the heart and compression of lungs and will be attended by obvious signs of pulmonary atelectasis, vascular congestion in the intestines and abundant oedema in the inguinal region.

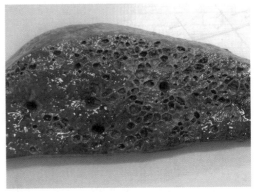

Figure A2.18 Liver, bovine. Post-mortem emphysema. Gas production by bacteria which overgrow post-mortem produce a honeycombed appearance to affected tissues. This should be differentiated from emphysema caused by clostridial infection which can be diagnostically useful. See also Plate 54. Photo courtesy of Mr R. Irvine, University of Glasgow, UK.

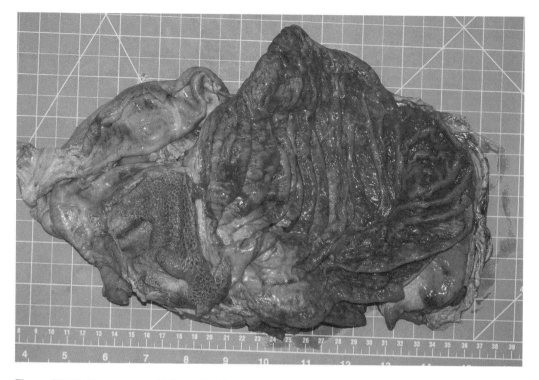

Figure A2.19 Abomasum, calf. Clostridial abomasitis. Acute necrotizing abomasitis associated with clostridial infection is an important cause of disease in calves and small ruminants. Predisposing factors include rapid change in diet, feeding high energy feeds, frozen feed or contaminated milk replacer. Clostridial infection is often associated with significant gas production resulting in emphysema in affected tissues. In this image, submucosal emphysema is evident along with other lesions consistent with necrosis and acute inflammation including ulceration, haemorrhage, and fibrin exudation characteristic of clostridial infection. The presence of these additional lesions can be used to differentiate significant and post-mortem emphysema. See also Online figure 10.

A2.10 Necropsy & sample submission forms

Examples of sampling checklist, sample submission form and post-mortem form for use in the laboratory and for field staff are on the following pages. Each laboratory should design specific forms to meet the needs of the area and the laboratory facilities available.

Sampling Checklist

Histology (formalin-fixed):

- All grossly apparent lesions
- Heart (T-piece)
- Right and Left Lung, cranial and caudal lobes
- Liver
- Right and Left Kidney
- Spleen
- Lymph Nodes: subcutaneous and internal
- Peripheral nerve

- Half of brain: sagittally sectioned between hemispheres
- Thyroids
- Adrenals
- Skin
- Eye
- Tongue
- Trachea
- Skeletal Muscle
- Bone Marrow

GI tract:

Monogastric species:

- Stomach-fundus
- Stomach-pylorus

Ruminants:

- Reticulum
- Rumen
- Omasum
- Abomasum

All Species:

- Duodenum
- Jejunum
- Ileum
- Colon

Hindgut Fermenters:

- Dorsal colon
- Ventral colon

Fresh-Frozen

Mammals:

- Lung
- Spleen
- Liver
- Kidney
- Half of brain: sagittally sectioned between hemispheres
- GI tract

Birds:

- Trachea
- Lung
- Spleen (if big enough after other samples)
- Liver
- Kidney
- Half of brain: sagittally sectioned between hemispheres
- Skin if suspicious (e.g. pox)
- GI tract

Reptiles:

- Lung
- Spleen
- Liver
- Kidney
- Brain (if big enough)
- Skin if suspicious (e.g. Ophidiomyces)
- GI tract

Amphibians:

- Lung/gill
- Spleen
- Liver
- Kidney
- Brain
- Skin (swab for chytrid PCR swab as well)
- GI tract

Sample submission form

DATE: CODE:

Submitter reference:

Owner name & address:

Laboratory reference:

Telephone number:

E-mail:

Fax:

Species: Sex:

Breed: Identification:

Age:

Number of animals affected: Number of animals dead: Population at risk:

Case History

Previous test results and/or relevant necropsy findings (if any available)

Samples submitted/tests requested

Circle the tests you require and indicate how many samples submitted:

DISCIPLINE	HAEMATOLOGY	BIOCHEMISTRY	PARASITOLOGY	SPECIALIST
SAMPLE(S): TESTS	Total blood count	Health screen profile	(faeces / blood / skin scrapings) Helminth egg count (general)	
REQUESTED HISTOPATHOLOGY	Differential white cell	Liver function profile	Fluke egg count	Special stains
MICROBIOLOGY (swab / tissue / fluid / skin scrapings)	Red cell indices Cell morphology	Downer cow profile Electrolytes	Larval culture Faecal smears for protozoa	Immuno-fluorescence Rabies
Bacterial culture and identification	Haemoparasite exam. (see parasitology)	Renal function profile	Examination for trypanosomes	
Antibiotic sensitivity	ESR	Plasma pepsinogen	Specific tests for haemoparasites	
SEROLOGY (contact laboratory for screening tests) Anaerobic culture		Urinalysis	Ectoparasite identification	
Special media CYTOLOGY Fungal culture and identification		Tissue fluids	Skin scrapings Total worm count (PM)	
Viral isolation (please contact laboratory to check media requirements)		Please note specific requests		Please contact laboratory for special requests

Comments:

Sample necropsy form

Submitter:

Owner name & address:

Laboratory reference:

Species:

Breed:

Age:

Sex:

Identification:

Case History

Number of animals affected: Number of animals dead: Population at risk:

Appearance of the carcass (gross findings)
Gastrointestinal system
Respiratory system
Circulatory/lymphatic system
Urogenital system
Musculoskeletal system
Neurological system
Other organs

Samples submitted

Histopathology:

OTHER SAMPLES SUBMITTED

Comments:

Signature of consulting veterinarian:

appendix 3

Some micro-organisms commonly isolated from animals

Susan C. Cork

The following table lists some of the micro-organisms* which are commonly isolated from animals with clinical disease. These lists are not comprehensive but have been prepared to give some guidance for preliminary microbiological assessment of clinical cases (see also specific health problems in Chapter 10).

* Bacterial agents and *Mycoplasma* spp., *Chlamydophila* (*Chlamydia* spp.)/fungi, yeasts and viruses. See Chapter 4 for more details on the laboratory diagnosis of bacterial, fungal and viral diseases.

** Many of the viruses listed cause a range of clinical signs but it may not be possible to successfully isolate them from sick animals or from observed lesions.

*** See Chapter 3 for details of parasitic diseases.

**** Note that the taxonomy of microorganisms is constantly changing and so it is always advisable to contact an expert to be sure that you are using the most up to date nomenclature when identifying specific organisms.

Table A3.1

Respiratory tract	Bacteria	Viruses**
Ruminants See also lungworm*** Cattle/other ruminants Sheep	*Pasteurella multocida* *Pasteurella haemolytica* *Trueperella (Actinomyces) pyogenes* *Mycoplasma* sp. *Histophilus somni* *Haemophilus somnus* *Bordetella bronchiseptica* *Mycobacterium sp.* *Chlamydophila psittici*	Infectious bovine rhinotracheitis, respiratory syncitial virus, Parainfluenza virus, adenovirus, reovirus, rinderpest Maedi Visna, Jaagsiekte
Pigs	*Pasteurella* sp. *Trueperella (Actinomyces) pyogenes* *Mycoplasma* sp. *Actinobacillus (Haemophilus) pleuropneumonia* *Bordetella bronchiseptica*	Atrophic rhinitis, African swine fever, inclusion body rhinitis, porcine reovirus, swine fever
Horses: Always check for evidence of infection in the Guttural pouch (A space on each side of throat connected to the pharyngeal cavity by a narrow opening)	*Streptococcus equi* (i.e. Strangles and others, *S. zooepidemicus*) *Rhodococcus equi* (especially in foals) *Pasteurella* sp. *Actinobacillus equi* *Bordetella* sp. *Cryptococcus neoformans* *Aspergillus* sp. *Pseudomonas* sp. *(glanders)* *Klebsiela* sp.	African horse sickness, equine influenza, equine herpes virus
Poultry	*Haemophilus paragallinarum* (infectious coryza) *Pasteurella multocida* (fowl cholera) *Mycoplasma* spp. *Aspergillus* spp.	Newcastle Disease (NCD), virus, infectious Bronchitis, infectious Laryngotrachietis, fowl pox, influenzavirus, avian paramyxovirus etc.
Dogs and/or cats	*Bordetella* spp. *Pasteurella* spp. *Klebsiella* spp. *Nocardia* spp. *Cryptococcus neoformans*	Distemper virus, Parainfluenza virus
Cats	*Chlamydophila* spp.	Cat flu (calicivirus/ virus), herpes virus

Table A3.1 *continued*

Gastrointestinal tract	Bacteria	Viruses**
Ruminants (also *Coccidia* sp.***)	*Salmonella* sp. *Escherichia coli* (various types) *Clostridium* sp. especially *Cl. perfringens* types B and C *Mycobacterium avium* sub-species *paratuberculosis* Others including bovine tuberculosis (*M. bovis*)	Cattle: bovine viral diarrhoea/mucosal disease complex Foot and mouth disease, Blue tongue, malignant Catarrhal fever, infectious bovine rhinotracheitis, rotavirus, coronavirus
Pigs	*Salmonella* sp. (especially *S. choleraesuis*) *Clostridium perfringens* type C *Campylobacter* sp. *Treponema hyodysenteriae* *Bacillus anthracis* *Yersinia enterocolitica*	Transmissible gastroenteritis (TGE), rotavirus, Aujeszky's disease, porcine reovirus, porcine rotavirus, porcine epidemic diarrhoea, porcine reovirus (may also be associated with vomiting in pigs, with or without diarrhoea)
Horses	*Salmonella* sp. *Rhodococcus equi* (foals) *Actinobacillus* sp.	Rotavirus, some systemic viral infections
Poultry ***Check for coccidia (see Chapter 3)	*Salmonella* spp. *Campylobacter* spp. Clostridial species *Escherichia coli*, bacterial septicaemia associated with Gram positive cocci Candidiasis (yeast/thrush)	Newcastle Disease (NCD) Coronavirus, rotavirus, viral hepatitis
Cats and/or Dogs	Salmonellae *Campylobacter* spp. *Yersinia* spp. *Candida* spp. Yeasts (neonates) Other	Parvovirus, Distemper, Infectious Hepatitis, coronavirus, etc. (dogs) Feline infectious enteritis (panleucopaenia in cats)

Skin flora and causes of infection (abscesses etc.)	Bacteria	Viruses**
The most common causes of skin infections/abscesses in all mammalian species are members of the Streptococci and the Staphylococci. See also, the causes of ringworm (Chapter 4) and ectoparasitic diseases (Chapter 3). Many skin problems are secondary to nutritional deficiency (e.g. vitamin A and vitamin B complex), endocrine disorders (e.g. thyroid) or general debilitation (e.g. old age, physiological stress, liver damage etc.)		
Cattle	*Trueperella (Actinomyces) pyogenes* *Actinomyces bovis* (lumpy jaw) *Actinobacillus lignieresii* (wooden tongue)	All species. Warts and other epidermal viruses (e.g. papilloma virus, pox virus) etc. Bovine leucosis Wooden tongue

Skin flora and causes of infection (abscesses etc.)	Bacteria	Viruses**
Sheep	*Corynebacterium pseudotuberculosis* (caseous lymphadenitis)	Scrapie, sheep pox, Orf virus
Pigs	*Trueperella (Actinomyces) pyogenes* Streptococci (types E,P, U,V)	Swine pox (other viruses may cause skin lesions, e.g. swine fever)
Horses	*Corynebacterium* spp. *Histoplasma* spp. Others	Papilloma virus, sarcoids (mixed causes)
Dogs and Cats	*Nocardia* spp. *Blastomyces* spp. Others	

Genito-urinary tract: Urine is often contaminated by bacteria from the gastrointestinal and/ or reproductive tracts. See Chapter 10 for causes of infertility and infections of the reproductive tract.

Cattle	*Corynebacterium* sp. (see Chapter 10 for causes of abortion) Others	IBR/infectious pustular vulvo-vaginitis
Sheep See also Toxoplasmosis (Chapter 3)	Various, urinary tract infection uncommon. Causes of abortion/infertility include *Brucella* ovis., *Listeria* sp. *Chlamydophila* abortus, Salmonellae, etc.	Various (e.g. Border disease)
Pigs	See Chapter 10 for causes of abortion.	
Horses	Various, urinary tract infection uncommon. Causes of abortion/ infertility include Salmonellae, *E. coli*, *Klebsiella* sp., *Streptococcus zooepidemicus*	Equine herpes virus and Equine viral arteritis (may be isolated from body fluids following an abortion
Dogs (urinary infection relatively rare in cats, may be secondary to crystal formation)	*Escherichia coli* *Proteus* spp. *Pseudomonas* spp. *Enterobacter* spp. Faecal Streptococci	Transmissible venereal tumour (lesions on the vulva, or the penis and preputial mucous membranes in dogs)

Eyes: Some bacteria which have been isolated from eyes. Viral diseases associated with eye discharge and conjunctivitis

Cattle	*Moraxella bovis (infectious bovine keratoconjunctivitis (pink eye)* *Branhamella (Neisseria) ovis* *Chlamydophila psittici*	Malignant catarrhal fever, mucosal disease, infectious bovine rhinotracheitis

Table A3.1 *continued*

Eyes: Some bacteria which have been isolated from eyes. Viral diseases associated with eye discharge and conjunctivitis

Sheep	*Branhamella ovis* *Moraxella* spp. *Chlamydophila psittaci.* *Mycoplasma* spp.	
Pigs		Atrophic rhinitis, inclusion body rhinitis (may cause eye discharge), equine viral arteritis
Horses	*Streptococcus equi* *Streptococcus equisimilis* *Staphylococcus aureus*	Equine viral arteritis
Cats and dogs	*Staphylococcus* spp. *Candida* spp. Yeasts *Chlamydophila psittici*	Cat flu viruses (herpes and caliciviruses)

appendix 4

Examples of laboratory equipment and reagent suppliers

Susan C. Cork

Sigma-Aldrich – general and specialised reagents and consumables
http://www.sigmaaldrich.com/customer-service/worldwide-offices.html

BioMerieux – microbiology and immunodiagnostics
http://www.biomerieux.com/servlet/srt/bio/portail/home

BDH Analytical VWR – general chemicals and supplies
https://us.vwr.com/store/content/external ContentPage.jsp?path=/en_US/vwr_bdh_chemicals_catalog.jsp

BDSL – diagnostic kits and supplies
http://www.bdsl2000.com/

Beckman Coulter – specialized equipment
https://www.beckmancoulter.com/wsrportal/wsr/company/about-us/corporate-overview/index.htm

BD DIFCO – reagents, media and stains
http://www.bd.com/ds/productCenter/index.asp

Thermo Fisher (Life Sciences) – culture media, molecular and immunodiagnostics
https://www.thermofisher.com/us/en/home/life-science.html

Thermo Fisher Scientific
http://www.thermofisher.com/ca/en/home.html

Thermo Scientific – oxoid-microbiology reagents
http://www.oxoid.com/uk/blue/index.asp

Most of these companies can source reagents from a range of suppliers and have regional representatives to facilitate easy ordering and service delivery globally. Company names change as result of mergers and acquisitions but the web links often remain functional.

Colibacillosis in poultry: example with diagnostic flowchart

Karen Liljebjelke

Escherichia coli is a ubiquitous bacteria present in many environments. It is an opportunistic bacterium in poultry, usually secondary to other diseases, such as infectious bronchitis, mycoplasma or management problems which predispose the birds to infection. Colibacillosis is a major cause of morbidity and mortality in poultry, and causes significant economic losses worldwide. Avian pathogenic *E. coli* (APEC) infection causes a variety of syndromes in poultry, including airsacculitis, colisepticaemia, cellulitis, arthritis, salpingitis and peritonitis in poults and laying hens, and omphalitis and yolk sac infection in chicks.

E. coli infection leads to increased mortality in laying hens with obvious lesions in the pericardium, liver, spleen and air sacs. A drop in egg production may or may not occur in the flock. Free range, colony cage or cage-free housing systems are challenging to manage and may increase spread of infection with APEC. There are limited options for control of *E. coli* infections during the laying period because there are few antibiotics approved for use during lay that have zero withdrawal time for eggs. Extra-label use of antibiotics will result in residues in the eggs.

Bacteriological testing (see Figure A5.1) will confirm the presence of APEC. There are six strains of *E. coli* that are most commonly associated with avian disease: O1, O2, O5, O8, O18 and O78. The serotypes O1, O2 and O78 account for up to 70% of APEC infections in poultry. Sero-grouping allows for identification of a limited number of APEC strains, but will identify the six most common serogroups isolated from lesions. Sero-grouping is not necessary to make a diagnosis of colibacillosis. If the blood sample or swab sample is carefully collected from a freshly dead or recently euthanized carcass with lesions consistent with *E. coli* infection, the isolation of pure *E. coli* is supportive of a diagnosis of colibacillosis.

Autogenous vaccines can be used to immunize a laying flock. The vaccine will reduce mortality, but may not reduce morbidity or the drop in egg production. It can take weeks to match the field strain and produce an autogenous vaccine. In addition, handling and vaccination of each individual bird may be prohibitively time and labour expensive.

Diagnostic scheme for *E. coli*

1 Use sterile technique to collect a swab sample from fresh post-mortem tissue lesions. Alternately, a fresh blood sample from an affected bird may be used.
2 Streak the swab sample onto MacConkey agar and incubate at 37°C for 18–24 h. If

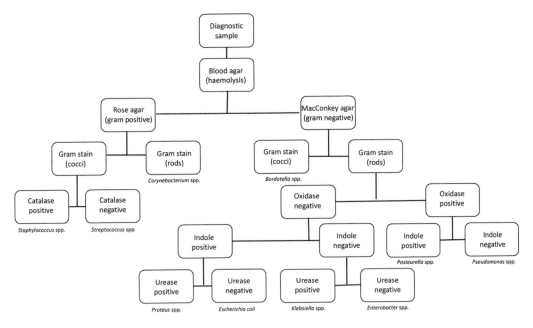

Figure A5.1 Diagnostic scheme for common veterinary bacterial pathogens.

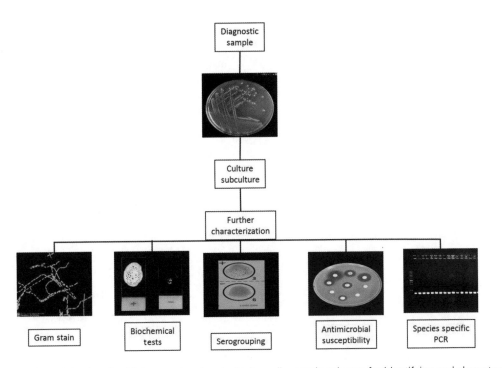

Figure A5.2 A basic microbiology and molecular biology diagnostic scheme for identifying and characterizing bacteria. Photos: Dr Karen Liljebjelke and Paul Gadja, Faculty of Veterinary Medicine, University of Calgary, Canada.

using blood, dilute the sample 1 : 10 in brain-heart infusion broth (BHI), and then streak onto MacConkey agar.

3 Samples may also be plated onto blood agar plates. Dome shaped yellow colonies displaying alpha or beta haemolysis should be gram stained and tested for identification of *Staphylococcus aureus*.

4 Examine MacConkey agar plate for white to pink colonies, some of which may have a hazy zone surrounding them. If there are pink colonies use these for the rest of the diagnostic testing.

5 Streak a pink colony onto blood agar and incubate overnight at 37°C. Colonies from the blood plate should be used for the gram stain, oxidase and indole tests.

6 Gram stain of colonies should show gram negative rods.

7 The gram negative rods: *E. coli*, *Enterobacter* spp. or *Klebsiella* spp., are all oxidase test negative.

8 *E. coli* is indole reaction positive, while *Enterobacter* spp. and *Klebsiella* spp. are indole negative.

9 Sero-grouping of O-group antigens may be performed if the information is required for autogenous vaccine production, or if the information is needed for other purposes.

10 Antimicrobial sensitivity testing is an important ancillary test, as *E. coli* are inherently resistant to the macrolide class of antibiotics, and in general tend to be multi-drug resistant, with a high proportion resistant to the beta-lactams. Disc diffusion is the least expensive and least technical method for antimicrobial sensitivity testing. Methods for creating antibiotic impregnated filter discs in-house are available as a cost-sensitive alternative to purchasing ready-made discs for use in testing.

Bibliography

Dho-Moulin, M., Fairbrother, J.M. (1999) Avian pathogenic *Escherichia coli* (APEC). Veterinary Research 30(2–3): 299–316.

Gross, W.B. (1991) Colibacillosis. In: Calnek, B.W., Barnes, H.J., Beard, C.W., Reid, W.M., Yoder, H.W. (eds) Diseases of Poultry, 9th edn, pages 138–144. Iowa State University Press, Ames, IA.

Arthropod – Invertebrate animals of the phylum Arthropoda, characterized by an exoskeleton made of chitin and a segmented body with pairs of jointed appendages. Arthropods include the insects (that is, lice, flies, mosquitoes) and arachnids (that is, ticks and mites), and are the largest phylum in the animal kingdom. Some species of arthropods have the potential to transmit pathogenic microorganisms from one animal host to another, the term 'vector' is used for these arthropods and the diseases that they transmit are known as 'vector borne diseases', examples include Blue tongue virus (transmitted by *Culicoides* sp. midges), West Nile virus (transmitted by *Culex* spp., *Culiseta* spp. and other mosquitoes) and Lyme disease (transmitted by *Ixodes* spp. and other ticks).

Biosafety – The application of knowledge, equipment and techniques to prevent personal, laboratory and environmental exposure to potentially infectious agents or biohazards. Biosafety requirements determine the containment conditions under which infectious agents can be safely manipulated. The objective of containment is to confine biohazards, to ensure biosecurity, and to reduce the potential exposure of laboratory workers, the public, and the environment to potentially infectious agents.

Based on the Centre for Disease Control (CDC) guidelines, there are four levels of Bio-containment (BSL1–4) to deal with pathogens in the four main risk groups (see below). Most regional diagnostic laboratories are designated as BSL 1 and BSL 2.

- **Biosafety Risk Group 1**: This includes agents that are not associated with disease in healthy adult humans. The work practices in BSL 1 facilities are based on standard microbiological practices. A representative microorganism manipulated in this type of laboratory is non-pathogenic *Escherichia coli*.
- **Biosafety Risk Group 2**: This includes agents that are associated with human disease but which are rarely serious and preventive or therapeutic interventions are available. The work practices in BSL 2 facilities are as above but include limited access, posting of biohazard warning signs, use of sharps precautions and requirement for a biosafety manual and standard operating procedures (SOPs) defining waste decontamination and health monitoring policies. Additional safety equipment would include class I or II biological safety cabinets or other physical containment devices to be used for all manipulations of infectious agents that cause splashes or aerosols of infectious materials. Personal protective equipment (laboratory coat, gloves, face and eye protection) is also required. Representative microorganisms include most of the *Salmonellae* and *Toxoplasma*.
- **Biosafety Risk Group 3**: This includes agents that are associated with serious human disease for which there are therapeutic inter-

ventions or preventative measures available, that is, there may be a high individual risk but a low community risk. Work practices are as for BSL-2 but also include controlled access and thorough decontamination of all waste. Additional safety equipment includes physical containment devices for all manipulations of agents, and use of additional personal protective equipment as needed. Facilities are usually as for BSL-2 with added physical separation from access corridors, self-closing double door access, exhausted air not recirculated and negative airflow into the laboratory. Representative microorganisms include *Mycobacterium tuberculosis*, West Nile virus and *Coxiella burnetii*.

- **Biosafety Risk Group 4**: This includes agents that are likely to cause serious or lethal human disease and for which preventive or therapeutic interventions are not available. This group poses both a high individual risk and high community risk. Work practices and safety equipment include complete isolation of the laboratory worker from aerosolized infectious materials through working in a Class III biosafety cabinet or in a full-body, air-supplied positive-pressure personnel suit. BSL-4 facilities are generally a separate building or completely isolated zone with complex, specialized ventilation requirements and waste management systems to prevent release of viable agents to the environment. Representative microorganisms include Ebola, Marburg and Congo-Crimean haemorrhagic fever viruses.

Note that organisms placed in each risk group can vary from country to country and from one organization to another. There are a number of reasons for the discrepancy but in most cases it relates either to different assessments of Biosecurity risk (that is, whether or not the agent is already present in the country) and also due to different activities conducted in the laboratory, that is, in facilities where antigen or vaccine are being prepared there will be a higher risk of human exposure as compared with that likely in a routine diagnostic facility. A useful summary of the definitions used by different agencies can be found at https://my.absa.org/Riskgroups and biosafety protocols are outlined in http://www.who.int/csr/resources/publications/biosafety/Biosafety7.pdf

Biosecurity – The application of knowledge, equipment and procedures to prevent the transmission of infectious diseases, parasites and pests from one area of containment, or from a geographical region, to another. The procedures developed to mitigate the risk of hazard release, exposure and subsequent consequences, should be based on a scientific risk assessment. Trade standards developed under the mandate of the OIE are designed to maintain biosecurity when animals and animal products are traded between countries.

Climatic zones – There are various definitions used to define climatic regions but typically, the main climatic zones are roughly demarcated by lines of latitude, into which the earth can be divided on the basis of climate. The latter has an impact on the species (arthropod vectors, vertebrate hosts, parasites and so on) that can successfully inhabit these geographical zones. The eight key zoogeographical zones include the following Nearctic (North America), Neotropical (South America), Western Palearctic (Europe), Afrotropical (Africa), Eastern Palearctic (North East Asia), Oriental (Southern Asia), Middle Eastern (Middle East), Australasian (Asia – Pacific). For additional information see http://climate-zone.com/.

Community engagement – An approach or process by which community organizations and individuals work alongside experts and expert groups, government and non-governmental organizations (NGOs) for the purpose of applying a collective vision for the benefit of a community. This is an approach that is now favoured by international development agencies to ensure that appropriate solutions are identified and implemented to resolve community human, animal and environmental health problems.

Disease – A term used for any condition (internal or external) that impairs the normal functioning of an organism or body. The cause may be infectious (that is, where a pathogen is transmitted from one individual to another) or non-infectious (that is, vector borne diseases, metabolic diseases, toxins and so on).

ELISA – Enzyme-linked immunosorbent assay is now a well-accepted and popular immunoassay format. Many ELISA type assays are now available in kit form and/or pen side tests so this technology is much more readily attainable for both the regional veterinary laboratory and field staff.

Endemic – The constant presence of a disease or infectious agent within a given geographic area. It may refer to the usual prevalence of a given disease within an area or a relatively new but constant situation. The term enzootic can also be used.

Epidemic – The occurrence in a country or region of cases of an illness (or disease outbreak) clearly in excess of the expected level. The number of cases occurring in an epidemic will vary according to the cause, that is, the infectious agent involved and its method of transmission, the size and type of population exposed, previous exposure to the disease agent and the time and place of occurrence. The term 'epizootic' can also be used.

Emerging disease – A disease that has appeared in a population for the first time, or that may have existed previously but is rapidly increasing in incidence or geographic range.

Fatality rate – Usually expressed as a percentage and is the number of animals diagnosed as having a specific disease that subsequently die as a result of that disease. This term is usually applied in an outbreak of an acute disease in which all cases are monitored for an adequate period of time to include all attributable deaths. The fatality rate is not the same as the mortality rate.

FAO – Food and Agriculture Organization of the United Nations – see http://www.fao.org/.

Incidence – The number of new cases of a disease occurring in a specified period. The incidence rate is the number of new cases of the disease reported in a specific time (for example, per year or month) divided by the number of animals in the population at risk.

Morbidity rate – The proportion of animals in a given population that develop clinical disease. Disease specific morbidity rate is the proportion of animals which develop clinical signs attributable to the disease under investigation and confirmed by diagnostic tests.

Mortality rate – The proportion of animals in a given population that die during a disease outbreak. Disease specific mortality rate is the proportion of animals which die following a specific disease (that is, confirmed by diagnostic tests).

OIE – The Office International des Epizooties, OIE, was created in 1924 in an effort to address emerging animal health issues at the global level. In 2003, the Office became the World Organisation for Animal Health but kept its historical acronym, OIE. The OIE is the intergovernmental organisation responsible for improving animal health worldwide and is recognised as a standard setting organisation by the World Trade Organization (WTO). In 2019, the OIE had a total of 182 Members and maintains permanent relations with 71 other international and regional organisations and has Regional and sub-regional Offices on every continent (http://www.oie.int/).

One Health – Over the past decades numerous organizations have recognized the need to integrate human and animal health, and the health of the environment, and have taken steps to develop new programs and form new partnerships to support that integration. The concept of 'One Health' is not new and essentially refers to an integrated multidisciplinary approach to solving complex health problems that impact, or are impacted by, animals, humans and the environment.

Pandemic – An epidemic (a sudden outbreak) that becomes very widespread and affects a whole region, a continent, or the world.

Pathogen – An agent that causes disease, especially a living microorganism, such as a virus, bacterium or fungus or an endo/ectoparasite. Some pathogens are obligate (that is, nearly always cause disease) and others are opportunistic (that is, cause disease only when the host's immunity is compromised). Pathogens can be transmitted via various routes from one host to another including the faecal-oral route, by aerosol, by direct and indirect contact (that is, via the environment) and by arthropod vectors.

PCR – Polymerase chain reaction is a technique in molecular genetics that permits the analysis of a short sequence of DNA (or RNA) even in samples containing only minute quantities of genetic material. PCR is used to reproduce (amplify) selected sections of DNA or RNA for analysis. PCR technology is now quite varied and widely available but although PCR is a sensitive tool for diagnostics the results are only reliable if the correct conditions are available to undertake testing. It should be noted that the presence of genetic material in a sample from a case doesn't indicate a diagnosis unless the findings are consistent with the clinical findings, that is, a PCR positive reaction doesn't necessarily indicate that a live pathogen is present.

Population at risk – The population of animals which are identified as those likely to develop a disease.

Prevalence – The number of animals which have developed a disease (diagnosed on the basis of clinical signs or by laboratory tests) in a stated population at a given time. This is regardless of when the illness began. The prevalence rate is defined as the number of positive cases divided by the number of animals in the population tested at a given point in time.

Probability – The likelihood of an event occurring.

Risk – The potential that a chosen activity or action will result in an undesirable consequence. Almost any human endeavour carries some risk, but some are much riskier than others. Essentially, risk = likelihood × consequence.

Risk assessment – The identification, evaluation and estimation of the likely risk associated with an action, event or procedure. Typically, in animal health, this consists of hazard identification and characterization, a release assessment, exposure assessment and then risk characterization. **Risk Analysis** includes risk assessment, risk communication and risk management. Risk assessments are an important part of developing risk mitigation recommendations to ensure that biosafety and biosecurity requirements are met both in the field and in the laboratory.

Sanitary and Phytosanitary (SPS) Agreement – see http://www.wto.org/english/tratop_e/sps_e/sps_e.htm.

WHO – World Health Organization of the United Nations see http://www.who.int/en/.

Zoonotic disease – An infectious disease that can be transmitted from animals to humans. A significant number of infectious diseases, including those caused by viruses, bacteria and parasites, can be transmitted from animals to people through a variety of infection routes. Common modes of transmission include animal bites, aerosol, faecal–oral, arthropod vectors and direct or indirect (that is, via the environment) animal-to-human contact. Examples of common zoonotic diseases include West Nile virus, rabies, Salmonellosis, ringworm, Brucellosis, Lyme disease and Leishmaniasis. Some diseases can also be transmitted back from humans to animals, for example, *Mycobacterium tuberculosis* and influenza A viruses (these can be considered 'reverse zoonoses').

Index